IET ENERGY ENGINEERING 132

Power Line Communication Systems for Smart Grids

Other volumes in this series:

Power Line Communication Systems for Smart Grids

Edited by
Ivan R.S. Casella and Alagan Anpalagan

The Institution of Engineering and Technology

Published by The Institution of Engineering and Technology, London, United Kingdom

The Institution of Engineering and Technology is registered as a Charity in England & Wales (no. 211014) and Scotland (no. SC038698).

The Institution of Engineering and Technology
Michael Faraday House
Six Hills Way, Stevenage
Herts, SG1 2AY, United Kingdom

www.theiet.org

British Library Cataloguing in Publication Data
A catalogue record for this product is available from the British Library

ISBN 978-1-78561-550-4 (hardback)
ISBN 978-1-78561-551-1 (PDF)

Typeset in India by MPS Ltd
Printed in the UK by CPI Group (UK) Ltd, Croydon

Contents

13 Performance evaluation of PRIME PLC modems over distribution transformers in Indian context   349
Konark Sharma and Lalit Mohan Saini

14 Analysis of hybrid communication for smart grids   363
Fabiano Salvadori, Camila S. Gehrke, Fabrício B.S. de Carvalho, and Alexandre C. Oliveira

Chapter 1
Introduction
Ivan R. S. Casella[1] and Alagan Anpalagan[2]

The constant growth of the consumption and demand of electric energy in the world has represented a challenge for the energy sector. In Europe, electricity demand in residences has risen by 1.0 percent per year over the past 30 years and, according to the International Energy Agency, European demand for electricity is expected to increase by 1.4 percent per year by 2030, unless countermeasures are taken [1,2]. In the United States of America (USA), the situation is a bit different. Although demand for electricity has always shown significant growth rates, since 1990, it has been surprisingly declining until it reached negative levels in 2017. The current projection is that demand will return to grow again and reach rates of about 0.9 percent per year by 2030. Nevertheless, it still causes concern and requires investments [3].

In addition, power utilities have to use their resources more efficiently. For example, 20 percent of US generation capacity is used only 5 percent of a year to comply with peak demands. Conventional power grids suffer lack of automated analysis, slow response time, limited control and poor coordination between generation and consumption of energy, resulting in several major blackouts in the past decades. Disruption of electricity negatively affects consumers with different levels of impact based on the nature of the activity [4].

Another important issue is the current global trend of reducing the pollution caused by the use of fossil fuel and its consequences. Global climate changes and greenhouse gas emissions (GHGEs) generated by the electricity and transport industries put more pressure on the conventional power grid model, as most power plants of several countries, such as the USA, are based on fossil fuel sources.

According to the Kyoto Protocol, the European Union (EU) 2020 strategy established a reduction of 20 percent of the 1990s GHGEs levels by 2020. Actually, EU members stated to meet the "20-20-20 Targets" defined in the EU package "Energy-Climate Change", representing 20 percent reduction in GHGE, 20 percent improvement in energy efficiency and 20 percent share for renewable sources in the EU energy mix [5,6]. However, most of today's conventional power grids spread around the world was built more than a century ago. Despite their growth in size

[1]Center for Engineering, Modeling and Applied Social Sciences (CECS), Federal University of ABC (UFABC), Brazil
[2]Department of Electrical and Computer Engineering (ECE), Ryerson University, Canada

and capacity, since then, they are not able to overcome easily and efficiently these important energy challenges.

In this way, a new concept of power system is urgently needed to address these challenges, which has motivated the development of the smart grids (SGs). SGs can be considered as an evolution of the current energy model to optimally manage the relationship between power supply and demand in order to overcome the real problem of energy contingency of the modern world. They are based on new interactive (bidirectional) information and communication technologies (ICTs) with real-time monitoring, control and automatic intervention capability to obtain a significant increase in efficiency, reliability and security of the infrastructure of generation, transmission, distribution and consumption of energy.

Thus, for the success of the SGs deployment, it is necessary to develop a complex ICT infrastructure with a strong interaction to ensure effective monitoring, control and protection of the whole grid. The core of the ICT infrastructure can be based on wireless or wired communication technologies, in a complementary or competitive manner.

Wireless communication technologies appear as an interesting solution as they can provide many benefits to the SGs, such as low deployment cost, ease of expansion, ability to use the technologies currently applied in mobile phone systems, flexibility of use and distributed management. However, they also present some limitations, such as the degrading characteristics of wireless propagation channels (e.g., attenuation, noise, multipath fading, Doppler propagation effect) that hamper quality communication with smart devices, especially those located inside buildings and tunnels (e.g., difficulty in lighting control and loads monitoring and control) or below street level (e.g., difficulty in supervising underground cables), difficult coverage in remote areas (e.g., rural areas, wind farms), and the cost, reliability and safety constraints of using the wireless infrastructure of mobile service providers (if this strategy of wireless communication is adopted). Wired communication technologies, such as, synchronous digital hierarchy (SDH), digital subscriber line (DSL), gigabit ethernet (GbE), wavelength division multiplexing (WDM), also appear viable because they offer high data rates, high reliability and high security, but most of them have several disadvantages for SGs deployment, for example, high deployment costs, high maintenance costs, low flexibility and difficulty in expanding and accessing remote areas.

On the other hand, power line communications (PLCs) are a wired technology that appear as a strong candidate to integrate the ICT infrastructure of the SGs for economic and technical reasons. PLCs are well-established technologies that allow the transmission of data through electrical lines and provide some advantages that make them both a useful complement and a strong competitor to wireless solutions. For example, they can exploit the existing electric grid infrastructure to reduce deployment costs, provide a low-cost alternative to complement existing technologies in the search of ubiquitous coverage, establish high data rate communication through obstacles that typically degrade wireless communications and, also, use technologies currently applied in other sectors (e.g., PLC Internet access).

1.1   Motivation for this book

The emergence of the SG concept has triggered a revolution in the energy sector. This revolution is only possible with the use of a sophisticated bidirectional ICT infrastructure, with real-time monitoring, control and protection capabilities. PLC technologies appear as a very promising and affordable solution to efficiently interconnect, with different levels of granularity, all layers of the ICT infrastructure of the SGs and, as a result, have recently received significant attention and expressive investments. Among the advantages presented, their low deployment cost when electrical wiring infrastructure already exists and their ability to reach remote (e.g., rural areas) or difficult access (e.g., tunnels, underground facilities) electrified areas stand out over other technologies mainly for SG applications. For example, it would be too complex, expensive or even impossible to provide connectivity for power monitoring and control to some of these areas through wireless technologies.

PLC technologies are quite old, but only with the recent use of advanced digital communication techniques, they have overcome the difficulty of transmitting information through a medium (power lines) that was not ideally designed for this purpose with the necessary reliability and security to be used widely in data networks and SG applications. With these technological advances, they can transmit high-quality data in indoor (e.g., inside industrial or residential facilities, vehicles) and outdoor (e.g., electric power transmission and distribution network) scenarios for both narrowband and broadband applications. In addition, various standards have been proposed to achieve performance improvements and interoperability.

Despite all the advances, the use of PLC technologies for applications in SGs is still quite challenging and requires new research and the development of new state-of-the-art techniques to improve their performance and adequacy to this new paradigm.

In this way, this book aims to present a comprehensive introduction to the basic principles involved in the use of PLCs in the ICT infrastructure of the SGs and show how they can benefit from these technologies to improve energy monitoring, control, security and management, especially when renewable energies sources are employed.

The development of the ICT infrastructure of the SGs using PLC technologies is extremely complex since it involves the joint knowledge of the areas of communication and energy, which are areas that are not normally treated together. Thus, this book has been organized to cover from basic concepts of modern digital communications, important coding techniques, specific features of PLC channels and the fundamentals of SGs, to the differences between narrowband PLC (NB-PLC) and broadband PLC (BB-PLC) technologies for SG applications, major PLC standards and some state-of-the-art works. It was conceived with a more didactic approach to some fundamental topics in order to be useful not only to researchers but also to students, academics and engineers.

1.2   Chapters overview

This book was divided into three main parts. In addition to this introduction, the first part presents the fundamental concepts of digital communications needed to understand most of the current PLC technologies; the second part describes the major characteristics and standards of PLC; and the last part presents some basic concepts of the SGs and some interesting applications of PLCs in SGs, as described next.

Part I

1. **Introduction**
    This chapter will present an overview of what will be covered throughout the book.

2. **Fundamentals of digital communications**
    The basic concepts involved in the transmission of information through the communication channel will be presented in this chapter. It will discuss the differences between lowpass and bandpass modulation, transmission by bandlimited channels, communication channels and error probability in AWGN (additive white Gaussian noise). Some simulation results will be presented to illustrate the performance of these systems in AWGN.

3. **Basis of error correction coding**
    This chapter will present the main forward error correction schemes found in the literature, including some modern techniques such as turbo and low density parity check, focusing on the coding techniques adopted by the main modern PLC standards.

4. **Principles of orthogonal frequency division multiplexing and single carrier frequency domain equalization**
    In this chapter, the authors will describe orthogonal frequency division multiplexing (OFDM) and single-carrier frequency domain equalization, which can be employed in modern PLC systems and some of their specific characteristics such as cyclic prefix, frequency domain equalization and peak-to-average ratio. Some simulation results will illustrate the performance of these systems in AWGN and in multipath fading channels.

Part II

5. **Modern power line communication technologies**
    In this chapter, PLC will be introduced and then the operating principle of PLC technologies, how they work, how they are classified, what are the main frequency bands, what are the current modern PLC technologies, what are their main characteristics, what are their advantages and disadvantages and what are their applications.

6. **Power line communication channel models**
   The main NB-PLC and BB-PLC channel models discussed in the scientific literature will be presented in this chapter, taking into account the effects of AWGN, impulsive noise and multipath fading.

7. **Narrowband power line communication systems**
   In this chapter, the authors will describe the main characteristics of the most important NB-PLC standards [e.g., powerline intelligent metering evolution (PRIME), G3-PLC, IEEE 1901.2]. Some simulation results will be shown in order to compare the performance between some of these standards in multipath fading channels with AWGN and impulsive noise.

8. **Broadband power line communication systems**
   This chapter will describe the main characteristics of some important BB-PLC standards (e.g., high definition PLC, HomePlug and IEEE 1901-2010). Some simulation results will be presented in order to compare the performance between some of these standards in multipath fading channels with AWGN and impulsive noise.

Part III

9. **Power line communications for smart grids applications**
   In this chapter, the main parts of a power grid and how they are connected to each other will be presented. Then, the main characteristics and problems of the conventional networks and how they can evolve towards the SGs will be discussed. Within the SG concept, ICTs have played an important role in providing real-time monitoring, control and self-recovery capability. Among existing potential telecommunications technologies, it will be shown that PLCs appear as a strong candidate to integrate the ICT infrastructure of SGs by having several interesting features such as to exploit the existing electric grid infrastructure to reduce deployment costs, provide a low-cost alternative to complement existing technologies in the search of ubiquitous coverage and establish high data rate communication through obstacles that typically degrade wireless communications.

10. **An overview of quad-generation system for smart grid using PLC**
    Electricity produced through conventional power systems is both expensive and inefficient. A portion of useful fuel is wasted as heat and GHGEs. The concept of SGs opens many corridors in terms of energy transfer, and decentralized energy systems have facilitated the use of combined cooling, heating and power (CCHP) systems. CCHP systems are not only more energy efficient but also help to reduce GHGEs. They can additionally be integrated with carbon dioxide extractor to extract carbon dioxide from exhaust gases. This new power generation systems are called quad-generation system. Also, in combined heating and power (CHP) systems, different energy requirements are satisfied from a generation system coupled with a heat recovery system. In this chapter, the operation and planning of CHP, CCHP and quad-generation systems are reviewed. Mathematical

formulations of commonly used objective functions namely cost minimization, efficiency maximization and GHGEs minimization are introduced. Also, various optimization algorithms and the simulation tools being used to solve the optimization formulations are presented. The chapter can serve as a foundation stone for the beginners in this research area. It can also serve as a guide for the practitioners to optimally design, deploy and operate the multi-generation power systems.

11.   **Demand side management through PLC: concepts and challenges**
In this chapter, the authors will present an overview of demand side management strategies with a focus on demand response. Concepts, strategies and implementation requirements will be discussed. Moreover, the role of power electronics in SG will be highlighted and challenges related to PLCs will be underlined.

12.   **PLC for monitoring and control of distributed generators in smart grids**
A concept for remote monitoring and control of distributed generation in modern SGs based on signaling through electricity distribution network with PLC will be presented in this chapter. Safety-related issues such as loss-of-mains (LoM) protection and, more specifically, network islanding that is one of the LoM scenarios will be discussed. State-of-the-art grid LoM detection methods will be characterized as their benefits and disadvantages will be analyzed. Following the presentation of conventional anti-LoM methods, a detailed description of PLC-based LoM detection methods will be given and their feasibility will be justified. A software-defined radio platform will be presented to test and verify the PLC-based LoM concept in the frequency band below 1 MHz. Benefits and disadvantages of the platform will be depicted. Concept design and implementation options for physical layer will be discussed. Modern channel access techniques, such as direct sequence spread spectrum and OFDM, will be analyzed. Field test results for the PLC-based solution in an electricity distribution grid consisting MV (medium voltage) grid segments and LV (low voltage) networks conclude the chapter.

13.   **Performance evaluation of PRIME PLC modems over distribution transformers in Indian context**
The past few years have witnessed a tremendous development in PRIME technology for high speed data communication across MV and LV transmission/distribution networks based SG applications. PRIME technology also elucidates the importance of employing robust modulation schemes across distribution-transformers and motivates research in this direction. Indeed, the aim of this chapter will be to present an investigative study on PRIME channel measurements through MV/LV distribution transformers by implementing experimental tests to analyze the signal-to-noise ratio and the system performance over multipath PLC channels in Indian context.

14.   **Analysis of a hybrid communication for smart grids**
This chapter will deal with hybrid communication systems to support SGs. In SGs, intelligent electronic devices play an important role. They have to efficiently process data locally and communicate to each other, employing different wired

or wireless technologies, to provide the required infrastructure to offer the capability of real-time monitoring and control in important operations of the system. To avoid possible disruptions in the SGs due to unexpected failures, a highly reliable, scalable, secure, robust and cost-effective hybrid communication network (PLCs and wireless) will be considered. This high-performance hybrid communication network should also guarantee very strict quality-of-service requirements to prevent possible power disturbances and outages.

15. **Direct torque control for DFIG based wind turbines employing power line communication technology in smart grid environments**
    In this chapter, a control technique for a wind doubly fed induction generator (DFIG), based on direct torque control (DTC) with power references sent remotely via a PLC technology, will be presented. DTC achieves high dynamic performance, allowing independent control of DFIG electromagnetic torque and rotor flux magnitude. In this way, active and reactive powers can be controlled by the voltage applied to the rotor independently. In order to operate in an SG environment, the proposed system employs PLC technology for transmitting the power references to the controller of the wind generator through power cables. The complete control system (controller and PLC system), implemented in an experimental bench, will be presented together with the practical results obtained that validated the adopted control strategy.

16. **MIMO systems design for narrowband power line communication in smart distribution grids**
    Multiple-input–multiple-output (MIMO) technology is an efficient way to increase system data rate communication reliability without the need to increase bandwidth. Combining NB-PLC and MIMO can be an interesting way to meet the higher data rates demanded by the new applications promised by the SGs. In this chapter, data transmission in distribution grids will be investigated by means of MIMO NB-PLC system utilizing OFDM technology. Multiconductor transmission line theory will be used for accurate MV NB-PLC channel characterization and measurement-based black-box transformer models will be developed to characterize the MV to LV PLC channel. The achievable data rates will be systematically calculated for different scenarios, revealing also the possibility for an extensive and reliable application of MIMO NB-PLC communications through distribution transformers in SGs.

References

[1]   European Commission. European Technology Platform SmartGrids – Vision and Strategy for Europe's Electricity Networks of the Future. Office for Official Publications of the European Communities; 2006.

[2]   Rosselló-Busquet A. Ghnem for AMI and DR, 2012 International Conference on Computing, Networking and Communications (ICNC), Maui, HI,

2012, pp. 111–115. Available from: http://ieeexplore.ieee.org/stamp/stamp.jsp?tp=&arnumber=6167382&isnumber=6167355

[3]   US Energy Information Administration. Annual Energy Outlook 2018 with projections to 2050. Office of Energy Analysis U.S. Department of Energy; 2018.

[4]   Simoes MG, Roche R, Kyriakides E, *et al.* A Comparison of Smart Grid Technologies and Progresses in Europe and the U.S. IEEE Transactions on Industry Applications. 2012 Jul;48(4):1154–1162.

[5]   Barzola J. A Hypothetical Migration Analysis of the PLC Based on IEEE 1901.2 Standard. In: World Multi-Conference on Systemics, Cybernetics and Informatics; 2017. p. 86–91.

[6]   Zhou S, Brown MA. Smart Meter Deployment in Europe: A Comparative Case Study on the Impacts of National Policy Schemes. Journal of Cleaner Production. 2017 Feb;144:22–32.

Chapter 2
Fundamentals of digital communications
Ivan R. S. Casella[1]

Digital communications are the foundations of modern telecommunications. They have provided an efficient way to reach high data rate transmissions and multimedia features with high reliability against the degrading effects of the communication channel such as noise, interference and multipath fading.

In this chapter, a brief review of the key fundamentals of digital communications is presented to assist in understanding the power line communication (PLC) technologies that will be explored in the next chapters.

2.1 Introduction

Digital communications have caused a real revolution in the way information is transmitted over short and long distances and are the basis of modern telecommunications. Without them, for example, it would be virtually impossible to receive and send reliable, quality messages on space missions. They are an evolution of analog communications, offering numerous advantages by exploiting the fact that information is in digital format. Unlike analog communication systems, which need to identify in the receiver, an infinite number of possible transmitted waveforms through the received signal, digital systems have to identify only a finite number of possible waveforms to be transmitted and, consequently, they are much more powerful than any analog system [1].

The main benefits of digital versus analog communication systems are as follows:

- Higher immunity to noise and channel distortions (performing several techniques such as error correction coding, equalization, diversity techniques);
- More reliable long-distance communication in low quality channels;
- Simpler encryption of information;
- Easier storage of information;
- More effective compression of information;
- Higher dynamic range of information signal;
- Multimedia capability;
- Modern digital signal processing techniques (e.g., software-defined radio).

[1]Center for Engineering, Modeling and Applied Social Sciences (CECS), Federal University of ABC (UFABC), Brazil

A basic digital communication system can be represented by the processes of converting the digital information symbols into signal waveforms best suited to travel through the communication channel (at the transmitter), transmission of those waveforms through the channel and, lastly, retrieval of the transmitted digital information symbols (at the receiver).

In general, for short-distance (e.g., universal serial bus) and some long-distance (e.g., optical communications) wired transmissions, baseband digital communication systems are employed. In this type of communication, each digital information symbol causes variations in the amplitude, width or position of a pulse waveform, so that each symbol is represented in a unique way for later, in the receiving process, the symbols can be recovered from the received waveform without ambiguity. This conversion process is called baseband digital modulation or, sometimes, digital coding and will be presented more deeply in Section 2.4.

On the other hand, for both long-distance wireless (e.g., cellular systems) and some wired transmissions (e.g., digital cable TV and PLCs), bandpass digital communication systems are commonly used. In this case, each digital symbol modifies the amplitude, phase or frequency (or a combination thereof) of a sinusoidal carrier waveform in a distinct manner so that the symbols can be retrieved later in the receiver correctly. This process, called bandpass digital modulation, will be covered in detail in Section 2.5.

2.1.1 Communication system model

As mentioned in the introduction, a basic digital communication system can be represented by the simple model presented in Figure 2.1, composed by the following main blocks:

- The transmitter, which is responsible for converting the digital information symbols into signal waveforms suitable for transmission through the communication channels;
- The communication channel, which is the medium through which information propagates;
- The receiver, which is responsible for recovering the transmitted digital symbols from the received signal waveforms.

For these systems, the information signal can be represented by a time sequence of symbols of T_s seconds. In binary digital systems, each symbol corresponds to only one bit ($n_b = 1$) and T_s is equal to T_b, the bit duration.

Differently, in M-ary digital systems, each symbol is part of a finite alphabet of M different symbols (i.e., the signal set) and corresponds to more than one bit

Figure 2.1   Simple digital communication system model

$(n_b = \log_2(M))$, each one with duration of $T_b \cdot \log_2(M)$ [2,3]. In this case, the symbol rate, also known as baud rate, can be defined as

$$R_s = \frac{1}{T_s},$$ (2.1)

and the bit rate can be easily obtained by

$$R_b = R_s \cdot \log_2(M).$$ (2.2)

For baseband systems, the transmitter encompasses the process of baseband modulation (i.e., digital encoding) and the receiver includes the process of baseband demodulation (i.e., digital decoding). On the other hand, for bandpass systems, the transmitter encompasses the process of bandpass modulation and the receiver of bandpass demodulation. An example of representation of baseband and bandpass modulated waveforms is shown in Figure 2.2.

2.1.2 Communication channels

As stated in Section 2.1.1, the communication channel corresponds to the medium through which the information signals (waveforms) propagate. Examples of communication channels are the air in Wi-Fi (wireless fidelity) and LTE (long-term evolution) systems, the space in satellite transmissions, optical fibers in Gigabit Ethernet, telecommunication wires in digital subscriber line (DSL) and electric lines in PLC systems. Although very different in nature, all communication channels exhibit, at least, attenuation, distortions and some kind of noise, thus modifying the transmitted signal waveform.

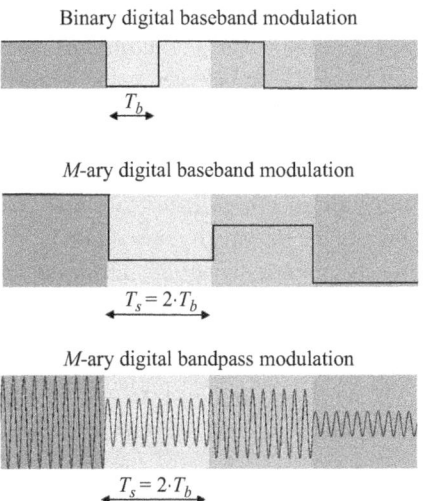

Binary digital baseband modulation

T_b

M-ary digital baseband modulation

$T_s = 2 \cdot T_b$

M-ary digital bandpass modulation

$T_s = 2 \cdot T_b$

Figure 2.2   Example of baseband and bandpass digital modulation waveforms

The attenuation corresponds to the energy losses that a signal suffers when propagating through a communication channel. Hence, the channel attenuation is a function of the distance between transmitter and receiver, since the greater this distance, the higher the attenuation. Consequently, the energy of the signal at the receiver side is always lower than that at the transmitter side. Attenuation is also a function of frequency. For instance, copper wires used for telephony and Internet access through DSL systems present attenuations that decay approximately exponentially with frequency [4–6].

Noise is the term generally used to indicate any random signal that corrupts the transmitted waveforms. The most common kind of noise is *thermal noise*, present in any communication channel. This noise is produced by the random motion of electrons in a medium and its intensity increases with increasing temperature, being zero only at the absolute zero [7]. Because thermal noise is the result of the random motion of many independent electrons in the medium, it may be modeled as a Gaussian random process by the central limit theorem. Another characteristic of this Gaussian noise is that its power spectral density (PSD) is approximately flat up to frequencies on the order of 10^{13} Hz [7]. Thus, thermal noise may also be considered a white noise in this frequency range. As mathematical modeling of communication channels considers that thermal noise has an additive effect on the transmitted signals, this noise is most often known as additive white Gaussian noise (AWGN). In addition to the AWGN, different channels may possess other specific noises, such as the impulsive noise present in wireless [8,9] and PLC [10,11] channels, and the short noise encountered in optical fiber channels [12,13].

Communication channels may also distort the transmitted waveforms as they propagate toward the receiver. Since the channels cannot pass infinite frequencies, any sharp corners of the waves are rounded [7,14]. In fact, all communication channels have a cutoff frequency beyond which the transmitted signals are almost entirely attenuated. There are also many channels that exhibit a low-frequency cutoff. Thus, channels possessing just high-frequency cutoffs are usually modeled as lowpass filters, while bandlimited channels are modeled as bandpass filters [15].

The filtering effect may cause the transmitted waveforms to widen, possibly resulting in an overlap between pulses sent in different time instants if these pulses are not sufficiently apart. This overlapping, known as intersymbol interference (ISI), is one of the main factors limiting the performance of a digital communication system.

The ISI can also be caused by multipath propagation [2,10,13,14,16,17]. In wireless communications, for instance, a signal may reach the receive antenna through many different paths due to atmospheric scattering and refraction, or reflections from buildings and other objects. The signals arriving along different paths will present different attenuations and delays and might add at the receiver either constructively or destructively [2,15–17], thus generating ISI. In PLC systems, multipath propagation is due to reflections produced by impedance mismatch between elements of the lines [10].

In general, ISI channels are modeled by linear transverse filters [2,10,17]. Hence, the received signal, except from the AWGN, can be modeled by the convolution of the transmitted signal with the impulse response of the channel. Some communication channels, such as optical fibers, however, may present severe nonlinear effects [12]. In these cases, linear transverse filters are not well suited and nonlinear channel models must be employed.

It is important to highlight that communication channels may also be either time-invariant or time-varying. In time-invariant channels, such as the wire cables used for telephony and DSL systems, the impulse response or, equivalently, the frequency response remains practically unchanged with time. On the other hand, time-varying channels have their impulse and frequency responses changing with time. This can occur, for example, in wireless communications, where transmitters and receivers might be mobile. The relative motion between transmitter and receiver leads to Doppler effect, which reflects on a time variation of the channel [2,16–18].

2.2 Review of fundamentals

This section will briefly introduce some fundamental concepts for the design and analysis of digital communication systems. At first, it is worth providing a definition for continuous- and discrete-time signals.

A continuous-time signal $s(t)$ is a signal defined on the continuum of time values, i.e., $s(t)$ is a function of a continuous independent variable.[1] On the other hand, a discrete-time signal $s(n)$ is defined only for specific time values, i.e., $s(n)$ is a function of a discrete independent variable [19,20].

Besides this classification, signals may also be classified according to the nature of their amplitudes. Signals whose amplitudes may assume any values in a continuous range are called analog, while signals whose amplitudes may assume just a finite number of values are called digital. Hence, the terms continuous-time and discrete-time refer to the nature of the independent variable, while analog and digital to the nature of the dependent variable.

The bridge from analog-to-digital world can be established by the so-called Nyquist sampling theorem.

2.2.1 Nyquist sampling theorem

Baseband and bandpass digital communication systems discussed in Section 2.1.1 consider the transmission of discrete-time digital information symbols. Although digital information naturally arises in computer-to-computer communication, many other important signals, such as voice, music, pictures, and video, are inherently

---

[1] In fact, a "continuous-time" signal could be used to represent a function of any continuous independent variable and not just a function of time (e.g., a "continuous-time" signal could be used to represent an image, as a function of two spatial variables).

continuous in time (or space) and amplitude. Therefore, to benefit from the advantages of digital communications, analog continuous-time signals must be accurately represented by digital discrete-time signals. This can be performed by the analog-to-digital conversion (ADC) process, composed of the sampling, quantizing and coding operations.

Consider a bandlimited analog continuous-time signal $s(t)$, whose maximum frequency component is f_m. ADC first generates an analog discrete-time signal by the sampling operation, which takes samples of $s(t)$ at regular time intervals, so that

$$s(n) = s(nT_{sa}), \qquad (2.3)$$

where n is the time index, T_{sa} is the sampling period and $f_{sa} = 1/T_{sa}$ is the sampling frequency.

The resulting discrete-time signal can then be written as [13,19]

$$\hat{s}(t) = \sum_{n=-\infty}^{\infty} s(n)\delta(t - nT_{sa}) = s(t) \sum_{n=-\infty}^{\infty} \delta(t - nT_{sa}), \qquad (2.4)$$

where $\delta(t)$ is the Dirac delta function.

The sampled signal will correctly represent the continuous-time signal provided that they have the same spectrum between $-f_m$ and f_m. This is only possible if

$$f_{sa} - f_m > f_m \Rightarrow f_{sa} > 2f_m. \qquad (2.5)$$

If the sampling frequency f_{sa} is lower than $2f_m$, which is known as the *Nyquist frequency*, portions of the spectrum will overlap causing the phenomenon of *aliasing*, and it will not be possible to recover the original signal from the sampled one [7,19].

The inequality in (2.5), referred to as the *Nyquist sampling theorem*, establishes a lower bound on the sampling frequency in order that the sampled discrete-time signal correctly represents the original continuous-time signal.

Theorem 4.1. *To accurately represent a bandlimited analog continuous-time signal by its samples, the sampling frequency (f_{sa}) must be greater than two times the maximum frequency component (f_m) of the signal, i.e., $f_{sa} > 2f_m$.*

It is worth noting that the sampling theorem assumes the existence of a bandlimited signal at the input of the sampling operation. However, physical signals, such as voice and music, are limited in time and thus unlimited in frequency. To ensure that an analog continuous-time signal is properly bandlimited prior to sampling, a lowpass filter called *anti-aliasing filter* is normally used [19,20].

After the sampling operation, quantization and coding operations are started. In the quantization operation, the analog samples are transformed into digital by rounding off the amplitudes of the samples to several possible finite levels. Lastly, in the coding operation, each level is then labeled, i.e., coded, with binary numbers resulting in a binary digital discrete-time signal [7,14,21].

2.2.2 Bandwidth

Considering the Nyquist sampling theorem presented in Section 2.2.1 and the use of ideal pulse shaping (i.e., sync pulse), which will be discussed in Section 2.6.1, the bandwidth occupied by a digital baseband signal transmitting symbols at R_s can be defined as

$$B = \frac{R_s}{2}. \tag{2.6}$$

Moreover, a digital bandpass signal can be obtained by translating, in the frequency domain, a digital baseband signal to a given frequency f_o. This process, usually denoted as heterodyning, is illustrated in Figure 2.3.

Consequently, if the bandpass system is linear, the bandwidth occupied by a digital bandpass signal can be defined as

$$W = R_s. \tag{2.7}$$

In a later section, a more realistic pulse shaping will be introduced and a new evaluation of the signal bandwidth will be presented.

2.2.3 Power and energy

Energy and power are important resources of communication systems. Energy is the capacity to produce work, which basically means in the communication scenario, to

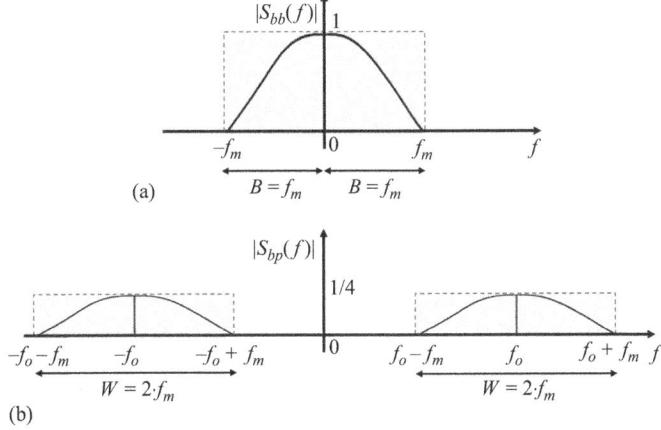

Figure 2.3 Spectrum of (a) baseband and (b) bandpass signals

send information from a transmitter to a receiver. The performance of digital communication systems depends on the received signal energy, which can be determined as [21]

$$E_r = \int_{-T_r/2}^{T_r/2} r^2(t)dt, \tag{2.8}$$

where T_r represents the received signal time duration.

In addition, power can be defined as the rate of transmission of energy and is given by

$$P_r = \frac{1}{T_r} \int_{-T_r/2}^{T_r/2} r^2(t)dt. \tag{2.9}$$

In the analysis of digital communications systems, it is common to deal with energy signals, that is, signals with finite energy for all time such as a pulse waveform (e.g., used to represent a bit or symbol) [21]. In this case, the energy is obtained by

$$E_r = \int_{-\infty}^{\infty} r^2(t)dt. \tag{2.10}$$

However, in order to deal with signals which have intrinsically infinite energy and are very important for communications, such as periodic signals and random signals, it is convenient to define a new class of signals called power signals, i.e., signals with finite power. In this case, power can be defined as [14,21]

$$P_r = \lim_{T_r \to \infty} \frac{1}{T_r} \int_{-T_r/2}^{T_r/2} r^2(t)dt. \tag{2.11}$$

One important power-based parameter in the design and analysis of communication systems is the signal-to-noise ratio (SNR), which expresses the ratio of average received signal power (P_r) to average noise power (P_n) at the receiver, as shown next [14,21]:

$$\text{SNR} = \frac{P_r}{P_n}. \tag{2.12}$$

Nevertheless, in digital communications, a normalized energy-based version of SNR, denoted by E_b/N_0, is usually preferred to analyze and compare different digital systems. E_b/N_0 is defined as the ratio of the bit energy over the noise PSD and can be represented by [14,21]

$$\frac{E_b}{N_0} = SNR \cdot \frac{B_w}{R_b}, \tag{2.13}$$

where R_b is the bit rate and $B_w = B$ or $B_w = W$, respectively, for baseband or bandpass communication systems.

E_b/N_0 is usually employed in digital communications because the detection and decision processes are based on symbols rather than signals. It means that E_s and E_b are more relevant to the detection process than signal strength. The usual choice for E_b is because it makes possible a fairer comparison between different M-ary digital communication systems (how to compare systems with different quantities of bits using a bit-based comparison?).

2.2.4 Measuring efficiency of communication systems

A desirable digital communication system should present the following features [17]:

- Lowest error probability;
- Narrowest bandwidth possible;
- Easy and cost-effective implementation.

In general, existing digital communication systems do not simultaneously meet all of these requirements, making the search for solutions to these issues very important and challenging.

In order to analyze the compliance with the presented requirements, the performance of digital communication systems can be measured in terms of power efficiency (η_P) or energy efficiency (η_E), and bandwidth efficiency (η_{Bw}).

The η_P and η_E measure the ability of a communication system to preserve digital information at low power or energy levels, respectively [2,21]. Specifically, η_E is usually preferred in the analysis of digital communication systems and can be expressed as

$$\eta_E = \frac{E_b}{N_0}, \text{ for a given bit error probability } (P_b). \tag{2.14}$$

In practice, P_b can be estimated by the bit error rate (BER), which is the ratio of the number of information bits received in error to the number of transmitted information bits.

On the other hand, η_B, also known as spectral efficiency, measures the ability of a digital communication system to transmit digital information within a specific bandwidth. It is usually defined as

$$\eta_{Bw} = \frac{R_b}{B_w}. \tag{2.15}$$

The η_{Bw} of a digital communication system can be associated with its capacity [17]. The fundamental Shannon's channel coding theorem states that for an arbitrarily small BER, the η_{Bw} is limited by the noise power of the communication channel and an upper limit can be obtained by the channel capacity formula [22]:

$$\eta_{Bw} < \frac{C}{B_w} = \log_2(1 + SNR) = \log_2(1 + \eta_E \cdot \eta_{Bw}), \tag{2.16}$$

where C is the channel capacity.

In the design of a digital communication system, very often there is a tradeoff between η_{Bw} and η_E. For instance, by adding error correction coding to the transmitted information, the bandwidth occupancy is increased, thus reducing the η_{Bw}. At the same time, it reduces the required received power for a particular BER. On the other hand, the use of M-ary schemes decreases bandwidth occupancy but increases the required received power and, hence, trade η_{Bw} for η_E [21].

2.3   Vector signal space

Digital communication encompasses the transmission of waveforms $s_m(t)$, belonging to a finite signal set S, through the communication channel and the recovery of the transmitted bits by choosing the most likely waveforms from the received signal. Each waveform $s_m(t)$ in S has a duration T_s and represents a group of n_b information bits.

Thereby, in binary digital communication systems, the signal set S is composed by only two signal waveforms and each one represents just one bit of information ($n_b = 1$). On the other hand, in M-ary digital communication systems, the signal set S is composed by M signal waveforms and each one represents $n_b = \log_2(M)$ information bits.

A fundamental approach proposed in [23,24] represents the elements of S as points in a vector space. This widely used representation is particularly general and can simplify the performance analysis of different digital communication systems by converting a continuous-time detection problem into a discrete-time finite-dimensional detection problem.

2.3.1   *Definition of the signal space*

Based on a Euclidean geometry point of view, any finite set of signal waveforms can be represented by a linear combination of N orthonormal waveforms, which form a basis to an N-dimensional vector space. Thus, when the signal waveforms are replaced by vectors of appropriate dimension N, a signal communication system can be fully described by an equivalent vector communication system [24].

Definition 2.1. *Let Φ be the set composed by N different waveforms in the time interval $t_0 \leq t \leq t_0 + T_s$, given by*

$$\Phi = \{\phi_1(t), \ldots, \phi_N(t)\}. \tag{2.17}$$

Definition 2.2. *Let SPAN$\{\Phi\}$ be the set of all signal waveforms $s_m(t)$ formed by the linear combination of the elements of Φ in the time interval $t_0 \leq t \leq t_0 + T_s$, such that*

$$s_m(t) \in \text{SPAN}\{\Phi\} \Leftrightarrow s_m(t) = \{s_{m,1} \cdot \phi_1(t), \ldots, s_{m,N} \cdot \phi_N(t)\}, \tag{2.18}$$

where $s_{m,1}, \ldots, s_{m,N}$ are weighting coefficients.

Thus, the set Φ is linearly independent **if and only if** just the trivial combination can result in

$$s_{m,1} \cdot \phi_1(t) + \cdots + s_{m,N} \cdot \phi_N(t) = 0 \tag{2.19}$$

If Φ is *linearly independent*, then it forms a *basis* for $S = \text{SPAN}\{\Phi\}$ and the dimension of S is the number of elements of Φ.

Definition 2.3. *Let the inner product between two elements of Φ be given by*

$$\langle \phi_j(t), \phi_k(t) \rangle = \int_{t_0}^{t_0+T_s} \phi_j(t) \cdot \phi_k^*(t) \, dt, \tag{2.20}$$

where $\phi_k^*(t)$ represents the complex conjugate of $\phi_k(t)$.

Definition 2.4. *If Φ is an orthogonal basis, all pairs of elements of Φ satisfy*

$$\langle \phi_j(t), \phi_k(t) \rangle = 0, \, j \neq k. \tag{2.21}$$

Definition 2.5. *If Φ is an orthonormal basis, each of its elements $\phi_n(t)$ is normalized to have unit energy, i.e.,*

$$E_n = \langle \phi_n(t), \phi_n(t) \rangle = \int_{t_0}^{t_0+T_s} |\phi_n(t)|^2 dt = 1. \tag{2.22}$$

In a similar way, any two signal waveforms $s_j(t)$ and $s_k(t)$ of the SPAN$\{\Phi\}$ are orthogonal if they satisfy

$$\langle s_j(t), s_k(t) \rangle = 0, \, j \neq k. \tag{2.23}$$

The projection of a given signal waveform $s_m(t)$ of the SPAN$\{\Phi\}$ in one of the components $\phi_n(t)$ of the set Φ is given by

$$s_{m,n} = \langle s_m(t), \phi_n(t) \rangle = \int_{t_0}^{t_0+T_s} s_m(t) \cdot \phi_n^*(t) dt. \tag{2.24}$$

Thus, if a set of M signal waveforms $s_m(t)$, $m = 1, \ldots, M$ can be decomposed into an N-dimensional orthonormal basis (i.e., a *signal space*), where $N \leq M$, then $s_m(t)$ can be represented by the following vector notation:

$$\mathbf{s}_m = \begin{bmatrix} s_{m,1} & \cdots & s_{m,N} \end{bmatrix}^T, \tag{2.25}$$

where each element of $\mathbf{s}_m$ corresponds to the projection of $s_m(t)$ on each of the components $\phi_n(t)$ of the signal space.

In this way, the energy of $s_m(t)$ can be computed as

$$E_{s_m} = |\mathbf{s}_m|^2 = \sum_{n=1}^{N} |s_{m,n}|^2. \tag{2.26}$$

Also, the correlation between two different signals $\mathbf{s}_j$ and $\mathbf{s}_k$ can be obtained by

$$E_{s_j, s_k} = \langle \mathbf{s}_j, \mathbf{s}_k \rangle = \sum_{n=1}^{N} s_{j,n} \cdot s_{k,n}^*, \tag{2.27}$$

while the distance between $\mathbf{s}_j$ and $\mathbf{s}_k$ is given by

$$d_{s_j,s_k} = |\mathbf{s}_j - \mathbf{s}_k| = \sqrt{\sum_{n=1}^{N} |s_{j,n} - s_{k,n}|^2}. \tag{2.28}$$

Hence, the waveforms of a basis can be considered as a coordinate system for the vector space. The Gram–Schmidt (GS) procedure, described in the sequel, can provide a systematic way of obtaining the vector space for a given set of signal waveforms.

2.3.2   Gram–Schmidt and the geometric representation of signals

Given a signal set S composed by M signal waveforms $s_m(t)$ with energies E_{s_m}, respectively, GS orthogonalization technique can be employed to define an N-dimensional *orthonormal signal space*, composed by N waveforms $\phi_n(t)$, $N \leq M$, to represent each element of S in a vector space, according to the following procedure [2,3,14].

- To determine $\phi_1(t)$, consider without any loss of generality, that $\phi_1(t)$ is equal to $s_1(t)$ normalized to unit energy:

$$\phi_1(t) = \frac{s_1(t)}{\sqrt{E_{s_1}}}. \tag{2.29}$$

- To determine $\phi_2(t)$, consider that

$$g_2(t) = s_2(t) - s_{2,1} \cdot \phi_1(t), \tag{2.30}$$

$$\phi_2(t) = \frac{g_2(t)}{\sqrt{E_{g_2}}}, \tag{2.31}$$

  where

$$s_{2,1} = \langle s_2(t), \phi_1(t) \rangle. \tag{2.32}$$

- To determine a generic $\phi_n(t)$, consider that

$$g_n(t) = s_n(t) - \sum_{k=1}^{n-1} s_{n,k} \cdot \phi_k(t), \tag{2.33}$$

$$\phi_n(t) = \frac{g_n(t)}{\sqrt{E_{g_n}}}, \tag{2.34}$$

  where all coefficients $s_{n,k}$ can be determined by

$$s_{n,k} = \langle s_n(t), \phi_k(t) \rangle. \tag{2.35}$$

After defining the set of the N orthonormal waveforms $\phi_n(t)$, each of the M signal waveforms $s_m(t)$ can be represented by the corresponding signal vector $\mathbf{s}_m$, whose elements are the coefficients $s_{m,n}$ described in the GS procedure:

$$\mathbf{s}_m = \begin{bmatrix} s_{m,1} & \cdots & s_{m,N} \end{bmatrix}^T, \quad m = 1, \ldots, M \tag{2.36}$$

2.3.3 *Karhunen–Loève and the geometric representation of noise*

Unfortunately, the GS procedure cannot be directly applied to random signals. In this case, the Karhunen–Loève (KL) expansion offers a powerful tool to obtain an orthonormal basis for a random process $w(t)$ through the solution of the following integral equation [14,25]:

$$\lambda_n \cdot \phi_n(t) = \int_{t_0}^{t_0+T_s} R_w(t,\tau) \cdot \phi_n^*(\tau)d\tau, \qquad (2.37)$$

where $R_w(t,\tau)$ is the autocorrelation function of $w(t)$, and λ_n and $\phi_n(t)$ are the eigenvalues and eigenfunctions of (2.37), respectively.

One of the most important random processes employed in the study of communication systems is the AWGN [2,14]. The AWGN is an uncorrelated, zero mean Gaussian random process with variance $\sigma_w^2 = N_0/2$. All its Zth-order probability density functions (PDFs) are Z-variate Gaussian random variables given by [25]

$$f_{\mathbf{W}_z}(\mathbf{w}_z) = \frac{1}{(2\pi)^{Z/2} \cdot |\det(\boldsymbol{\Sigma}_w)|^{1/2}} \cdot e^{-(1/2)[\mathbf{w}_z - \overline{\mathbf{w}}_z]^T \cdot \boldsymbol{\Sigma}_w^{-1} \cdot [\mathbf{w}_z - \overline{\mathbf{w}}_z^*]}, \qquad (2.38)$$

where $\mathbf{W}_z = [w(t_1), \ldots, w(t_Z)]$, $\mathbf{w}_z = [w_1, \ldots, w_Z]$ and $\overline{\mathbf{w}}_z = 0$ is the mean of $\mathbf{w}_z$. Due to the properties of the AWGN process, the autocovariance matrix $\boldsymbol{\Sigma}_w$ converges to the following diagonal autocorrelation matrix:

$$\boldsymbol{\Sigma}_w = \begin{bmatrix} \sigma_w^2 & & 0 \\ & \ddots & \\ 0 & & \sigma_w^2 \end{bmatrix}. \qquad (2.39)$$

Since the AWGN random process is considered white, its PSD can be represented by [14,25]

$$S_w(f) = \frac{N_0}{2}, \quad -\infty \le f \le \infty. \qquad (2.40)$$

Also, in accordance with the Wiener–Khinchin theorem, the autocorrelation function can be obtained by the inverse Fourier transform of the PSD, resulting in [2,14,25]

$$R_w(t,\tau) = \frac{N_0}{2} \cdot \delta(t-\tau). \qquad (2.41)$$

In this way, (2.37) can be reduced to

$$\lambda_n \cdot \phi_n = \frac{N_0}{2} \cdot \phi_n. \qquad (2.42)$$

This result implies that any orthonormal basis can be used to represent an AWGN random process as, for example, the one obtained by the GS procedure. However, in accordance with the KL expansion, to perfectly represent an AWGN process, it is necessary to employ an infinite-dimensional orthonormal basis, i.e.,

$$w(t) = \sum_{n=1}^{\infty} w_n \cdot \phi_n(t), \quad t_0 \le t \le t_0 + T_s, \tag{2.43}$$

where $w_n = \langle w(t), \phi_n(t) \rangle$ is the projection of $w(t)$ over the component of the basis $\phi_n(t)$.

One very important and useful result in the analysis of random signals is established by the *Theorem of the Irrelevancy* [24]:

Theorem 4.2. *Only the components of a white Gaussian random process that are projected on the signal space affect the decision process.*

Proof. The proof can be viewed in [24]. □

Therefore, from the signal detection point of view, the noise representation presented in (2.43) can be reduced to a more tractable finite N-dimensional signal space by using

$$w(t) = \sum_{n=1}^{N} w_n \cdot \phi_n(t), \quad t_0 \le t \le t_0 + T_s. \tag{2.44}$$

Note that the joint PDF of the components w_n of the AWGN can be obtained by [25]

$$f_{\mathbf{W}}(\mathbf{w}) = \frac{1}{(\pi \cdot N_0)^{N/2}} \cdot e^{|\mathbf{w}|^2 / N_0}, \tag{2.45}$$

where $\mathbf{W} = [w_1, \ldots, w_N]$ and $\mathbf{w} = [w_1, \ldots, w_N]$. The elements of $\mathbf{W}$ are independent and identically distributed (i.i.d.) zero mean Gaussian random variables with $\sigma_w^2 = N_0/2$ and the elements of $\mathbf{w}$ are the values assumed by the corresponding random variables.

In the next section, an optimum receiver for M-ary systems in AWGN channels will be introduced, considering the geometric representation of signals and noise discussed previously.

2.3.4 Optimum receiver structure (MAP/ML criteria)

An M-ary communication system transmits digital information from a transmitter to a receiver through a communication channel. Usually, the channel causes some different kinds of undesirable degradations in the transmitted information. Due to the random nature of the degradations, they are usually described by their statistical distributions.

One of the most common sources of degradation presented in a communication system is the thermal noise. This kind of undesirable signal, as already mentioned in Section 2.1.2, is generated by the random motion of electrons in a conductive material (e.g., electronic devices) at temperatures higher than zero Kelvin and is

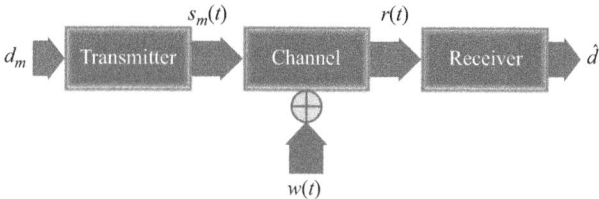

Figure 2.4 Digital communication system model in AWGN channels

normally added to the information signal at the receiver. Often, the thermal noise is modeled as an AWGN random process [2,14].

A generic communication system model is depicted in Figure 2.4. In this model, d_m denotes one of the M possible symbols that compose the symbol alphabet and that is transmitted to the receiver by means of the corresponding signal waveform $s_m(t)$. Each symbol carries n_b different information bits.

Considering, without any loss of generality, the symbol d_i is transmitted through the corresponding signal waveform $s_i(t)$ and assuming that the system is memoryless, as the corresponding transmitted signal is just corrupted by an AWGN $w(t)$, the received signal $r(t)$ can be represented by

$$r(t) = s_i(t) + w(t). \tag{2.46}$$

The geometric approach can be used to simplify the design of the optimum receiver for AWGN channels. As presented previously, the received signal can be represented in an N-dimensional vector signal space by

$$\mathbf{r} = \mathbf{s}_i + \mathbf{w}, \tag{2.47}$$

where

$$\mathbf{r} = \begin{bmatrix} r_1 & \cdots & r_N \end{bmatrix}^T, \tag{2.48}$$

$$\mathbf{s}_i = \begin{bmatrix} s_{i,1} & \cdots & s_{i,N} \end{bmatrix}^T, \tag{2.49}$$

$$\mathbf{w} = \begin{bmatrix} w_1 & \cdots & w_N \end{bmatrix}^T. \tag{2.50}$$

The optimum receiver has to decide which is the most likely transmitted symbol, given that $\mathbf{r}$ was received. Considering that $\mathbf{d}_i$ was transmitted, the conditional probability of making the correct decision $\mathbf{d} = \mathbf{d}_i$, given that $\mathbf{r}$ was received, is commonly denoted as the *a posteriori* probability (APP) of $\mathbf{d}_i$ and can be represented by [14]

$$P(\text{Correct decision} = \mathfrak{C}\,|\mathbf{r}) = P(\mathbf{d}_i \text{ was transmitted}\,|\mathbf{r} \text{ was received}), \tag{2.51}$$

while the unconditional probability of making a correct decision, independently on the received signal $\mathbf{r}$, is given by

$$P(\mathfrak{C}) = \int_{\mathbf{r}} P(\mathfrak{C}\,|\mathbf{r}) \cdot f_R(\mathbf{r}) d\mathbf{r}. \tag{2.52}$$

As $f_R(\mathbf{r}) \geq 0$, the integral is maximum when $P(\mathfrak{C}\,|\mathbf{r})$ is maximum. Therefore, given that the decision is to $\mathbf{d}_i$, the decision error probability can be minimized if $P(\mathbf{d}_i\,|\mathbf{r})$ is maximized. This maximization procedure, known as the maximum *a posteriori* probability (MAP) criterion, is described as follows:

- Receive $\mathbf{r}$;
- Evaluate all M APP of $\mathbf{d}_m$;
- Decide to the symbol $\mathbf{d}_i$ that presents the largest APP, according to

$$P(\mathbf{d}_i\,|\mathbf{r}) > P(\mathbf{d}_m\,|\mathbf{r}), \text{ for all } m \neq i. \tag{2.53}$$

Thus, the optimum receiver, in the sense of minimizing the decision error probability, is the MAP receiver.

Using Bayes' rule, the MAP criterion can be formally represented by

$$\underset{\mathbf{d}_m}{\operatorname{argmax}} \left[\frac{P(\mathbf{d}_m) \cdot f_R(\mathbf{r}\,|\mathbf{d}_m)}{f_R(\mathbf{r})} \right]. \tag{2.54}$$

The decision will be in favor of symbol $\mathbf{d}_i$ if (2.54) is maximum for $m = i$. As $f_R(\mathbf{r})$ is common to all $\mathbf{d}_m$, the MAP criterion simplifies to

$$\underset{\mathbf{d}_m}{\operatorname{argmax}} \left[P(\mathbf{d}_m) \cdot f_R(\mathbf{r}\,|\mathbf{d}_m) \right], \tag{2.55}$$

where $P(\mathbf{d}_m)$ is the *a priori* probability of sending a specific symbol $\mathbf{d}_m$ and $f_R(\mathbf{r}\,|\mathbf{d}_m)$ is the PDF of $\mathbf{r}$ be received conditioned to the transmission of $\mathbf{d}_m$.

Under these conditions, the corresponding transmit vector $\mathbf{s}_m$ is considered constant during the symbol interval T_s and the conditional PDF is given by

$$f_R(\mathbf{r}\,|\mathbf{d}_m) = \frac{1}{(\pi \cdot N_0)^{N/2}} \cdot e^{(|\mathbf{r}-\mathbf{s}_m|^2)/N_0}. \tag{2.56}$$

Therefore, the decision function is given by [2,14]

$$L_m = P(\mathbf{d}_m) \cdot f_R(\mathbf{r}\,|\mathbf{d}_m) = \frac{P(\mathbf{d}_m)}{(\pi \cdot N_0)^{N/2}} \cdot e^{(|\mathbf{r}-\mathbf{s}_m|^2)/N_0}. \tag{2.57}$$

As all the terms of the decision function are positive, it can be expressed by means of the natural logarithm. Making some simplifications, the resulting logarithmic decision function can be represented by

$$\ell_m = \ln[P(\mathbf{d}_m)] - \frac{|\mathbf{r} - \mathbf{s}_m|^2}{N_0}. \tag{2.58}$$

Note that $|\mathbf{r} - \mathbf{s}_m|^2$ is the square of the distance between $\mathbf{r}$ and $\mathbf{s}_m$.

Expanding the terms of (2.58) and making some additional simplifications, the decision function can finally be expressed as [2,14]

$$\ell_m = \xi_m + \langle \mathbf{r}, \mathbf{s}_m \rangle, \tag{2.59}$$

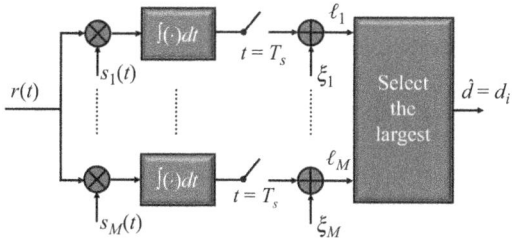

Figure 2.5   Optimum MAP receiver based on correlators

where $\langle \mathbf{r}, \mathbf{s}_m \rangle$ is the correlation between $\mathbf{r}$ and $\mathbf{s}_m$, and ξ_m is given by

$$\xi_m = \frac{N_0}{2} \cdot \ln[P(\mathbf{d}_m)] - E_{s_m}, \tag{2.60}$$

with E_{s_m} representing the energy of $\mathbf{s}_m$.

Thus, the optimum receiver based on the MAP criterion, illustrated in Figure 2.5, consists in calculating ℓ_m, $m = 1, \ldots, M$, and deciding to $\hat{\mathbf{d}} = \mathbf{d}_i$ if the decision function is maximum to $m = i$.

One interesting result obtained in this analysis is that the optimum MAP receiver for AWGN channels, that minimizes the error probability, is a linear system.

The optimum receiver can also be implemented through a different approach based on a special filter designed to perform equivalently to the operation of correlation $\langle \mathbf{r}, \mathbf{s}_m \rangle$.

If $r(t)$ is applied to a filter with impulse response $h_m(t)$, such that

$$h_m(t) = s_m(T_s - t), \tag{2.61}$$

the output of the filter at instant T_s will be

$$y_m(T_s) = \int_0^{T_s} r(\tau) \cdot s_m(\tau) d\tau. \tag{2.62}$$

This output is exactly the same obtained by a correlator at time instant T_s and this filter is usually called matched filter (MF).

Thus, the bank of correlators of the optimum MAP receiver in Figure 2.5 can be replaced by a bank of MF, as shown in Figure 2.6, without any performance degradation.

Another possible implementation of the optimum MAP receiver is shown in Figure 2.7. In this case, the terms $\langle \mathbf{r}, \mathbf{s}_m \rangle$ can be obtained by first projecting the received signal $\mathbf{r}$ in each one of the N components of the signal space ϕ_n and then calculating the sum of the product of each projection r_n with all the components $s_{m,n}$ of the signal vector $\mathbf{s}_m$, as shown by

$$\langle \mathbf{r}, \mathbf{s}_m \rangle = \sum_{n=1}^{N} r_n \cdot s_{m,n}. \tag{2.63}$$

In the same way, the bank of correlators of the implementation of the MAP receiver based on orthogonal basis can also be replaced by a bank of MF, as shown in Figure 2.8.

If $N < M$ and the components of the basis are easily generated, then the optimum receiver implementation based on orthonormal basis should be chosen.

If the symbols $\mathbf{d}_m$ are equiprobable, all *a priori* probabilities are equal and the MAP criterion presented in (2.55) simplifies to the *maximum likelihood* (ML) criterion given by

$$\underset{\mathbf{d}_m}{\operatorname{argmax}} \left[f_R(\mathbf{r} \mid \mathbf{d}_m) \right]. \tag{2.64}$$

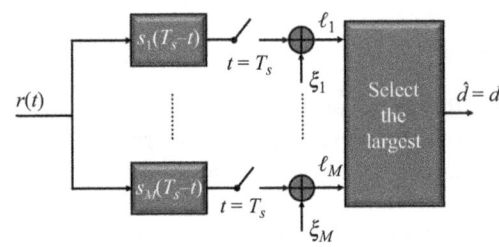

Figure 2.6   Optimum MAP receiver based on MF

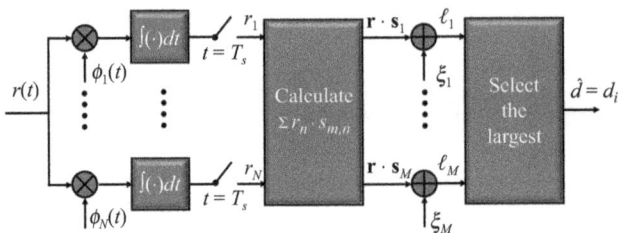

Figure 2.7   Optimum MAP receiver based on orthonormal basis with correlators

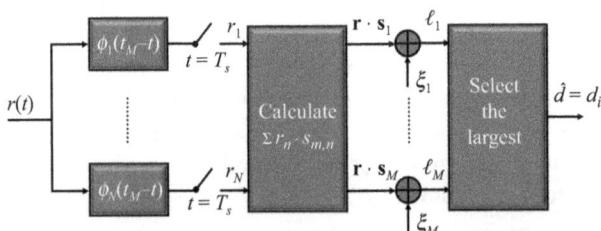

Figure 2.8   Optimum MAP receiver based on orthonormal basis with MF

In this case, (2.58) can be simplified to

$$\ell_m = -|\mathbf{r} - \mathbf{s}_m|^2, \tag{2.65}$$

and the resulting decision function for the ML criterion is given by

$$\ell_m = \langle \mathbf{r}, \mathbf{s}_m \rangle - E_{s_m}. \tag{2.66}$$

This means that the optimum receiver in the ML sense is simply a bank of correlators or MF, followed by a bank of subtractors with the symbol energy E_{s_m}. The ML criterion is summarized below:

- Receive $\mathbf{r}$
- Evaluate all M correlations $\langle \mathbf{r}, \mathbf{s}_m \rangle$
- Subtract E_{s_m} from each correlation output
- Decide to the symbol $\mathbf{d}_i$ that presents the largest value

2.3.5 Decision region and error probability

A very important figure of performance in digital communications is the symbol error probability (P_e), usually measured in practice by the symbol error rate (SER). To determine P_e for the MAP receiver, the N-dimensional signal space is split in M disjoint regions $\Omega_1, \Omega_2, \ldots, \Omega_M$, $N \leq M$, which represent all possible transmit symbols, as illustrated in Figure 2.9 for a hypothetical system with $N = 2$ and $M = 4$.

If a signal waveform $\mathbf{s}_i$ is transmitted and the received signal $\mathbf{r}$ falls in the region Ω_j, $i \neq j$, then the decision will be wrongly made in favor of $\mathbf{d}_j$ and a decision error will occur.

The decision regions should be chosen to minimize P_e in accordance with the MAP decision function presented in (2.58).

Considering for simplicity that all possible transmit symbols are equiprobable, the MAP receiver simplifies to the ML receiver, meaning that the optimum receiver

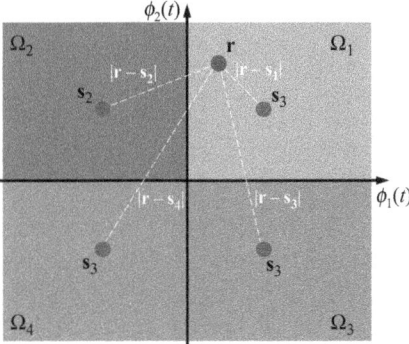

Figure 2.9   Decision regions for N = 2 and M = 4

has just to choose the symbol $\mathbf{d}_m$ that maximize $-|\mathbf{r} - \mathbf{s}_m|^2$. In fact, it is equivalent to choose the symbol $\mathbf{d}_m$ that minimize the distance between $\mathbf{r}$ and $\mathbf{s}_m$, given by $|\mathbf{r} - \mathbf{s}_m|$.

The decision process based on the ML criterion has a very nice interpretation in the vector signal space: the decision is made in favor of symbol $\mathbf{d}_i$, if the signal vector closest to the received vector $\mathbf{r}$ is the signal vector $\mathbf{s}_i$.

Therefore, the probability of making the right decision, given that $\mathbf{s}_i$ was sent, is obtained by

$$P(\mathfrak{C} \,|\, \mathbf{d}_i) = P(\mathbf{r} \in \Omega_i \,|\, \mathbf{d}_i). \tag{2.67}$$

On the other hand, the probability of making the right decision is given by

$$P(\mathfrak{C}) = \sum_{m=1}^{M} P(\mathbf{d}_m) \cdot P(\mathfrak{C} \,|\, \mathbf{d}_m) = \frac{1}{M} \sum_{m=1}^{M} P(\mathfrak{C} \,|\, \mathbf{d}_m), \tag{2.68}$$

and the error probability is easily obtained by

$$P_e = 1 - \frac{1}{M} \sum_{m=1}^{M} P(\mathfrak{C} \,|\, \mathbf{d}_m). \tag{2.69}$$

In the vector space, this probability can also be obtained geometrically by

$$P_e = 1 - \frac{1}{M} \sum_{m=1}^{M} P(\mathbf{r} \in \Omega_m \,|\, \mathbf{d}_m). \tag{2.70}$$

P_b, usually estimated in practice by BER, is another very important figure of merit of digital communication systems and can be easily derived from P_e. For linear systems, such as M-ASK (M-ary amplitude shift keying), M-ary PSK (M-ary phase shift keying) and M-ary QAM (M-quadrature amplitude modulation), employing gray encoding [3,14], P_b can be obtained by

$$P_b = \frac{P_e}{\log_2(M)}. \tag{2.71}$$

For nonlinear systems, such as M-FSK (M-ary frequency shift keying)[2], P_b can be computed as

$$P_b = \frac{M}{2 \cdot (M-1)} \cdot P_e. \tag{2.72}$$

2.3.6   Error probability bounds

Depending on the geometric representation of a given digital communication system, the determination of P_e can be very difficult. In this case, the use of bounds may be very useful.

---

[2]Linear and nonlinear bandpass systems will be discussed in more detail in Section 2.5.1.

One of the most well-known bounds for obtaining an approximation of P_e is the union bound (UB). The UB is an upper bound based on the following probability identity [3,16]:

$$P\left(\bigcup_{m=1}^{M} A_m\right) \leq \sum_{m=1}^{M} P(A_m), \qquad (2.73)$$

where A_m is the mth event of the sample space and $\bigcup$ represents the union operator.

Considering that $A_{m|i}$ corresponds to the error event $|\mathbf{r} - \mathbf{s}_m| < |\mathbf{r} - \mathbf{s}_i|$, given that $\mathbf{s}_i$, $i \neq m$, was sent, the UB for equally likely transmit symbols can be expressed as [2,14]

$$P_e \leq \frac{1}{M} \sum_{i=1}^{M} \sum_{\substack{m=1 \\ m \neq i}}^{M} Q\left(\frac{D_{m,i}}{\sqrt{2 \cdot N_0}}\right), \qquad (2.74)$$

where $Q(\cdot)$ is the Q-function[3] [2,14] and $D_{m,i}$ is the distance between $\mathbf{s}_m$ and $\mathbf{s}_i$.

Another very useful bound is the nearest neighbors bound (NBB). The NBB determines an approximation of P_e by considering just the neighboring signals that are at the minimum distance D_{min} of a given signal $\mathbf{s}_m$. Assuming equally likely transmit symbols, the NBB is given by [16]

$$P_e \approx N_{min} \cdot Q\left(\frac{D_{min}}{\sqrt{2 \cdot N_0}}\right), \qquad (2.75)$$

where $D_{min} = \min_{m \neq i}[D_{m,i}]$ is the minimum distance between all possible pairs of signals $\mathbf{s}_m$ and $\mathbf{s}_i$ and N_{min} is given by

$$N_{min} = \frac{1}{M} \cdot \sum_{m=1}^{M} N_m, \qquad (2.76)$$

with N_m indicating the number of neighbors at a distance D_{min} from $\mathbf{s}_m$.

2.4 Baseband digital communication systems

A baseband digital communication system, represented in Figure 2.10, converts digital information symbols into baseband pulse waveforms that are suitable to be transmitted directly over lowpass channels. As mentioned before, this process is denoted as baseband digital modulation or simply digital coding.

A variety of baseband digital communication systems have been proposed to reach some desirable properties such as good bandwidth and power efficiencies, adequate timing information and error detection capability.

---

[3]The Q-function is defined as

$$Q(x) = \frac{1}{\sqrt{2\pi}} \int_{x}^{\infty} e^{-t^2/2} dt, \quad x \geq 0.$$

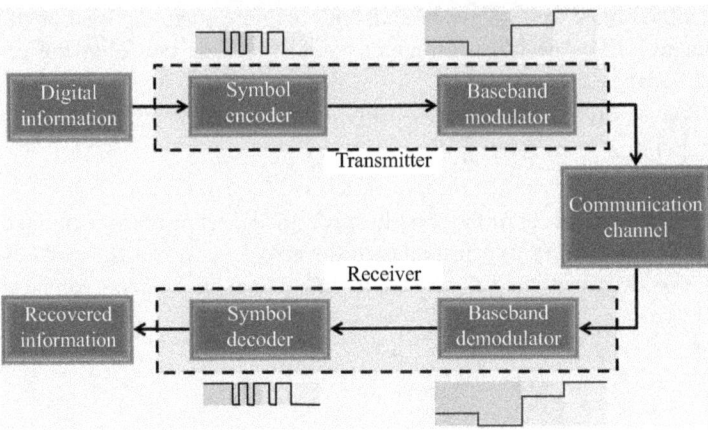

Figure 2.10   Baseband digital communication system diagram

2.4.1   Line coding

Line coding was developed in the past for digital transmission over telephone cables and digital recording on magnetic medias. Recent developments are primarily concentrated on applications in local area networks, including transmissions over unshielded twisted pairs and fiber optic cables.

In general, line coding is a baseband scheme employed to represent digital information by means of different pulse waveforms of an M-dimensional signal set, according to some specific rules [14].

Any line coding waveform can be represented by the following M-ary pulse amplitude modulation (PAM) signal:

$$s_{\text{PAM}}(t) = \sum_i d_i' \cdot h_p(t - i \cdot T_s), \tag{2.77}$$

where d_i' is ith real-valued information symbol obtained by a specific encoding rule and $h_p(t)$ is the pulse shape.

2.4.2   Complex-valued M-ary PAM

A more general and powerful representation of M-ary PAM can be obtained by considering that any information symbol d_i may have complex values [2]:

$$s_{lp}(t) = \sum_i d_i \cdot h_p(t - i \cdot T_s), \tag{2.78}$$

where d_i is the ith complex-valued information symbol.

This representation, also called lowpass equivalent (LPE), is a very useful tool to simplify simulations in computational environments and theoretical analysis of bandpass digital communication signals and systems.

2.5   Bandpass digital communication systems

Digital communications also encompass the transmission of digital information through wireless and wired channels that essentially act as analog passband systems. In this case, digital communication systems are generally referred to as bandpass digital communication systems or digital modulation systems.

As shown in the previous section, a baseband digital communication system converts digital information symbols into baseband pulse waveforms that are suitable to be transmitted directly over a lowpass channel. On the other hand, a bandpass digital communication system, represented in Figure 2.11, converts digital information symbols into bandpass sinusoidal waveforms that are suitable to be transmitted through bandpass channels. Thus, bandpass communication systems can be viewed as systems that map baseband signals into bandpass signals. This operation encompasses the shift of the spectrum of a baseband signal to a higher frequency [2,13,14,21].

Bandpass digital communication systems or simply digital modulation systems are obtained by switching (keying) amplitude, frequency, phase or a combination of amplitude and phase of a high frequency sinusoidal carrier in accordance with the digital information. Thus, the digital modulation techniques could be grouped into the following four major categories:

*   ASK, achieved by amplitude variations of the carrier waveform according to information symbols;
*   PSK, obtained by phase variations of the carrier in accordance with the information symbols;
*   FSK, generated by frequency variations of the carrier in accordance with the information symbols;
*   QAM, originated by amplitude and phase variations of the carrier in accordance with the information symbols.

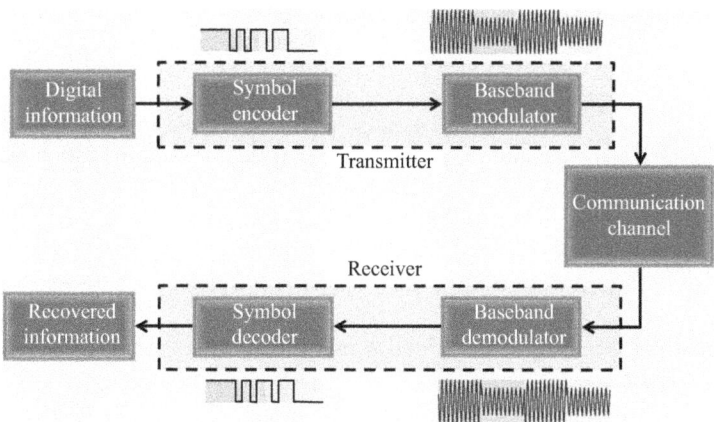

Figure 2.11   Bandpass digital communication system diagram

Other modulation categories can be obtained by combining different parameters of the carrier, but they are not commonly found in practice and are out of the scope of this analysis.

Digital bandpass systems can be classified as linear and nonlinear. Linear communication systems follow the superposition theorem and encompass ASK, PSK and QAM schemes. Typically, they show a good bandwidth efficiency. Nonlinear communication systems do not satisfy the superposition theorem and are represented basically by different FSK schemes. Usually, they show a good energy efficiency.

Any bandpass communication system can be represented by

$$s(t) = \Re \left\{ s_{lp}(t) \cdot e^{j2\pi f_o t} \right\}, \tag{2.79}$$

where f_o is the carrier frequency and $s_{lp}(t)$ is the LPE representation given by (2.78). In the receiver, the MAP detector can be employed to recover the transmitted information.

There are several factors that influence the selection of a digital modulation technique including bandwidth efficiency, energy efficiency required for detection, BER and SER at reception and circuitry complexity. Many of these factors are correlated, and an improvement in one of them generally causes a degradation in another [3,17]. Thus, an appropriate choice depends on the communication system requirements.

2.5.1   Some important bandpass digital schemes

In this section, some important bandpass digital communications systems will be presented and their representation in the signal space will be introduced.

2.5.1.1   Binary amplitude shift keying

As discussed before, in binary ASK (BASK), the amplitude of the carrier is modified in accordance with the digital information symbols. One of the most important and simple BASK schemes is on–off keying (OOK). The OOK is widely used in fiber optic applications mainly because of its simple optical implementation (i.e., light/nonlight).

The OOK signal waveform can be represented by

$$s_{OOK}(t) = \sum_i s^i_{OOK}(t), \tag{2.80}$$

where the ith OOK transmitted waveform, corresponding to information symbol d_i of duration T_s, is given by

$$s^i_{OOK}(t) = \begin{cases} 0 & \text{for } d_i = 0 \\ \sqrt{\frac{2 \cdot E_s}{T_s}} \cdot \cos(2\pi f_o t), \ i \cdot T_s \leq t \leq (i+1) \cdot T_s & \text{for } d_i = 1 \end{cases} . \tag{2.81}$$

Employing GS procedure, the following orthonormal basis can be defined for an OOK scheme:

$$\phi^i_1(t) = \sqrt{\frac{2}{T_s}} \cdot \cos(2\pi f_o t), \ i \cdot T_s \leq t \leq (i+1) \cdot T_s. \tag{2.82}$$

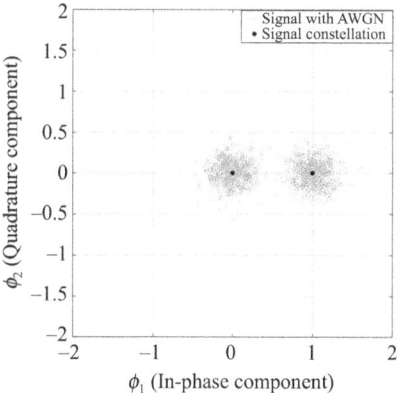

Figure 2.12   OOK constellation diagram

Thus, the LPE representation of an OOK signal can be written as

$$s_{lp}(t) = \sum_i d_i \cdot \Pi\left(\frac{t - i \cdot T_s}{T_s}\right), \tag{2.83}$$

where $\Pi(\cdot)$ is the unit gate pulse and $d_i \in \{0, \sqrt{E_s}\}$.

Therefore, the OOK signal waveform could also be expressed by

$$s_{OOK}(t) = \Re\left\{\sum_i d_i \cdot \Pi\left(\frac{t - i \cdot T_s}{T_s}\right) \cdot e^{j2\pi f_o t}\right\}. \tag{2.84}$$

Figure 2.12 presents an example of the constellation diagram for an OOK scheme and the corresponding received signal corrupted by AWGN for an $E_b/N_0 = 10$ dB.

One interesting observation is that ASK is also widely used in some wireless applications (e.g., gate control), even requiring low efficiency linear amplifiers and presenting low immunity to interferences. The main reason for this is its extremely low cost and low complexity, besides its ability to use simple noncoherent detection.

2.5.1.2   Binary phase shift keying

As mentioned before, in binary PSK (BPSK), the phase of the carrier is modified in accordance with the digital information symbols. The BPSK scheme is widely used in wireless applications that require low P_e at low E_b/N_0. However, BPSK presents low bandwidth efficiency, as any binary scheme.

The ith BPSK transmitted waveform, corresponding to information symbol d_i, can be represented by

$$s_{BPSK}^i(t) = \begin{cases} -\sqrt{\frac{2 \cdot E_s}{T_s}} \cdot \cos(2\pi f_o t), \ i \cdot T_s \leq t \leq (i+1) \cdot T_s \ \text{for } d_i = 0 \\[2ex] \sqrt{\frac{2 \cdot E_s}{T_s}} \cdot \cos(2\pi f_o t), \ i \cdot T_s \leq t \leq (i+1) \cdot T_s \ \text{for } d_i = 1 \end{cases}. \tag{2.85}$$

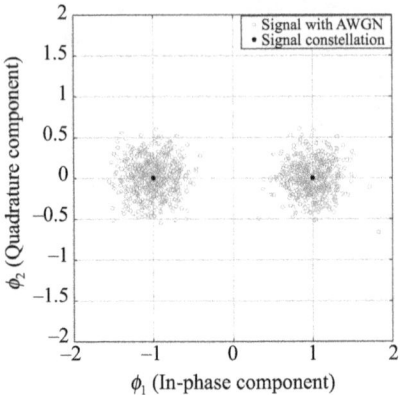

Figure 2.13   BPSK constellation diagram

Using GS procedure, an orthonormal basis for a BPSK scheme can be defined as

$$\phi_1^i(t) = \sqrt{\frac{2}{T_s}} \cdot \cos(2\pi f_o t), \ i \cdot T_s \leq t \leq (i+1) \cdot T_s. \tag{2.86}$$

Thus, the LPE representation of a BPSK signal can be written as

$$s_{lp}(t) = \sum_i d_i \cdot \Pi\left(\frac{t - i \cdot T_s}{T_s}\right), \tag{2.87}$$

where $d_i \in \{-\sqrt{E_s}, \sqrt{E_s}\}$.

In Figure 2.13, an example of the constellation diagram for a BPSK scheme is presented, as well as the corresponding received signal corrupted by AWGN for an $E_b/N_0 = 10\,\mathrm{dB}$.

2.5.1.3   Quaternary phase shift keying

The ith quaternary PSK (QPSK) transmitted waveform, corresponding to information symbol d_i of duration T_s, can be written as

$$s_{QPSK}^i(t) = \sqrt{\frac{E_s}{T_s}} \cdot \cos(2\pi f_o t - \theta_i), \ i \cdot T_s \leq t \leq (i+1) \cdot T_s, \tag{2.88}$$

where

$$\theta_i \in \left\{\frac{(2 \cdot m - 1) \cdot \pi}{4}\right\}, m = 1, \ldots, 4.$$

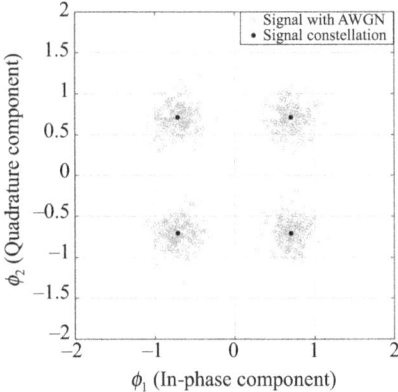

Figure 2.14   QPSK constellation diagram

Using GS procedure, an orthonormal basis for a QPSK scheme can be computed as

$$\phi_1^i(t) = \sqrt{\tfrac{2}{T_s}} \cdot \cos(2\pi f_o t), \ i \cdot T_s \le t \le (i+1) \cdot T_s$$

$$\phi_2^i(t) = \sqrt{\tfrac{2}{T_s}} \cdot \sin(2\pi f_o t), \ i \cdot T_s \le t \le (i+1) \cdot T_s$$

(2.89)

Thus, the LPE representation of a QPSK signal can be obtained by

$$s_{lp}(t) = \sum_i d_i \cdot \Pi\left(\frac{t - i \cdot T_s}{T_s}\right),$$

(2.90)

with

$$d_i \in \left\{ \sqrt{\frac{E_s}{2}} \cdot (1+j), \sqrt{\frac{E_s}{2}} \cdot (1-j), \sqrt{\frac{E_s}{2}} \cdot (-1+j), \sqrt{\frac{E_s}{2}} \cdot (-1+j) \right\}.$$

Figure 2.14 shows an example of the constellation diagram for a QPSK scheme and the corresponding received signal corrupted by AWGN for an $E_b/N_0 = 10$ dB.

2.5.1.4   *M*-ary phase shift keying

The ith M-PSK transmitted waveform, corresponding to information symbol d_i, can be expressed as

$$s_{MPSK}^i(t) = \sqrt{\frac{E_s}{T_s}} \cdot \cos(2\pi f_o t - \theta_i), \ i \cdot T_s \le t \le (i+1) \cdot T_s,$$

(2.91)

with

$$\theta_i \in \left\{ \frac{2 \cdot \pi \cdot (m-1)}{M} \right\}, m = 1, \ldots, M.$$

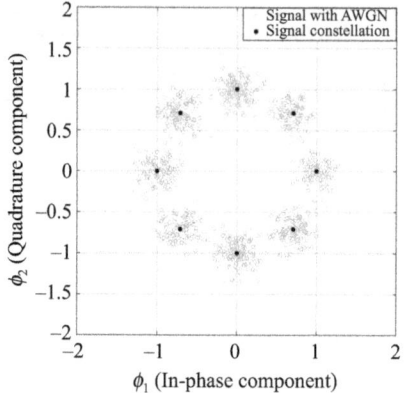

Figure 2.15   8-PSK constellation diagram

An orthonormal basis for an M-PSK signal can be defined as

$$\phi_1^i(t) = \sqrt{\tfrac{2}{T_s}} \cdot \cos(2\pi f_o t), \ i \cdot T_s \le t \le (i+1) \cdot T_s$$

$$\phi_2^i(t) = \sqrt{\tfrac{2}{T_s}} \cdot \sin(2\pi f_o t), \ i \cdot T_s \le t \le (i+1) \cdot T_s$$

(2.92)

Thus, the LPE representation of an M-PSK signal is given by

$$s_{lp}(t) = \sum_i d_i \cdot \Pi\left(\frac{t - i \cdot T_s}{T_s}\right),$$

(2.93)

where $d_i \in \{a_m \cdot \sqrt{E_s} + jb_m \cdot \sqrt{E_s}\}$, $\sqrt{a_m^2 + b_m^2} = 1$ and the pair $(a_m, b_m) = (\cos(2\pi \cdot (m-1)/M), \sin(2\pi \cdot (m-1)/M))$, $m = 1, \ldots, M$. It is worth noting that the ith symbol d_i could also be represented in a two-dimensional vector space by the signal vector $\mathbf{d}_i \in \{[a_m \cdot \sqrt{E_s}, b_m \cdot \sqrt{E_s}]\}$, $m = 1, \ldots, M$.

In Figure 2.15, an example of the constellation diagram for an 8-PSK scheme and the corresponding received signal corrupted by AWGN for an $E_b/N_0 = 10$ dB is shown.

2.5.1.5   *M*-ary quadrature amplitude modulation

The ith transmitted waveform, corresponding to information symbol d_i of duration T_s, of a generic M-QAM scheme, can be represented by

$$s_{M-QAM}^i(t) = \sqrt{\frac{2 \cdot E_i}{T_s}} \cdot \cos(2\pi f_o t + \theta_i), \ i \le t \le i \cdot T_s,$$

(2.94)

with $E_i \in \{E_m\}$ and $\theta_i \in \{\theta_m\}$, $m = 1, \ldots, M$.

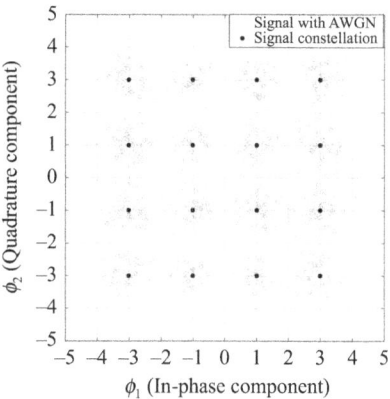

Figure 2.16   16-QAM constellation diagram

An orthonormal basis for an M-QAM signal can be written as

$$\phi_1^i(t) = \sqrt{\tfrac{2}{T_s}} \cdot \cos(2\pi f_o t), \; i \cdot T_s \le t \le (i+1) \cdot T_s$$

$$\phi_2^i(t) = \sqrt{\tfrac{2}{T_s}} \cdot \sin(2\pi f_o t), \; i \cdot T_s \le t \le (i+1) \cdot T_s$$

(2.95)

Specifically for a square M-QAM signal, the LPE representation is given by

$$s_{lp}(t) = \sum_i d_i \cdot \Pi\left(\frac{t - i \cdot T_s}{T_s}\right),$$

(2.96)

where $d_i \in \left\{a_m \cdot \sqrt{E_{min}} + jb_m \cdot \sqrt{E_{min}}\right\}$, a_m and $b_m \in \left\{-\sqrt{M} + 1, -\sqrt{M} + 3, \ldots,\right.$
$\left.\sqrt{M} - 1\right\}$, $m = 1, \ldots, \sqrt{M}$, and E_{min} is the energy of the symbol with minimum energy.

The ith symbol d_i can also be represented in a two-dimensional space by the signal vector $\mathbf{d}_i \in \left\{\left[a_m \cdot \sqrt{E_{min}}, b_m \cdot \sqrt{E_{min}}\right]\right\}$, $m = 1, \ldots, M$.

Figure 2.16 presents an example of the constellation diagram for a 16-QAM scheme and the corresponding received signal corrupted by AWGN for an $E_b/N_0 = 10$ dB.

2.5.1.6   *M*-ary frequency shift keying

The ith orthogonal M-FSK transmitted waveform, corresponding to information symbol d_i, can be described by

$$s_{M-FSK}^i(t) = \sqrt{\frac{2 \cdot E_s}{T_s}} \cdot \cos(2\pi f_i t), \; i \le t \le i \cdot T_s,$$

(2.97)

where $f_i \in \{f_o + m/(2 \cdot T_s)\}$, $f_o = N_{fo}/(2 \cdot T_s)$, $m = 1, \ldots, M$ and $N_{fo} \in \mathbb{N}$ to keep the orthogonality of the symbol waveforms.

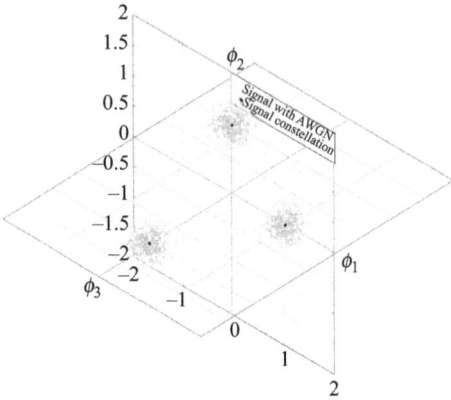

Figure 2.17   3-FSK constellation diagram

Since the set of possible M-FSK transmitted waveforms is orthogonal, an orthonormal basis can be simply obtained by normalizing the own set of M-FSK signal waveforms, i.e.,

$$\phi_m^i(t) = \sqrt{\frac{2}{T_s}} \cdot \cos(2\pi f_i t), \ 0 \le t \le T_s. \tag{2.98}$$

Thus, the LPE representation of an M-FSK signal can be written as

$$s_{lp}(t) = \sum_i d_i \cdot \Pi\left(\frac{t - i \cdot T_s}{T_s}\right), \tag{2.99}$$

with

$$d_i \in \left\{\sqrt{E_s} \cdot \cos\left(\frac{\pi \cdot m \cdot t}{2 \cdot T_s}\right) + j\sqrt{E_s} \cdot \sin\left(\frac{\pi \cdot m \cdot t}{2 \cdot T_s}\right)\right\}, m = 1, \ldots, M.$$

The ith symbol d_i can also be represented in an M-dimensional vector space by the signal vector $\mathbf{d}_i \in \{[0, \ldots, d_{i,m}, \ldots, 0]\}$ and $d_{i,m} = \sqrt{E_s}$ is the mth vector element, with $m = 1, \ldots, M$. Therefore, an M-FSK scheme requires an M-dimensional signal space for its correct representation.

As it is not possible to graphically represent M-FSK schemes for $M > 3$, Figure 2.17 shows a simple example for a 3-FSK scheme ($M = 3$) corrupted by AWGN.

2.5.2   *Performance of bandpass digital schemes in AWGN*

As discussed in the beginning of this chapter, the performance of a digital communication system can be measured in terms of its η_{Bw} and η_E.

In the design of bandpass digital communication systems, there is a tradeoff between η_{Bw} and η_E that must be taken into account in the choice of the best scheme

Bandpass modulation	Null-to-null bandwidth
M-ASK, M-PSK, M-QAM (rectangular pulse)	$W = \dfrac{2R_b}{\log_2(M)}$
Coherent M-FSK (rectangular pulse)	$W = \dfrac{(M+3)\cdot R_b}{2\cdot\log_2(M)}$

Figure 2.18   Bandpass modulation bandwidth occupancies

Bandpass modulation	Symbol error probability (P_e)	Bit error probability (P_b)
BPSK	$P_e = Q\left(\sqrt{\dfrac{2\cdot E_s}{N_0}}\right)$	$P_b = Q\left(\sqrt{\dfrac{2\cdot E_b}{N_0}}\right)$
QPSK	$P_e \approx 2\cdot Q\left(\sqrt{\dfrac{E_s}{N_0}}\right)$	$P_b \approx Q\left(\sqrt{\dfrac{2\cdot E_b}{N_0}}\right)$
M-PSK	$P_e \approx 2\cdot Q\left(\sqrt{\dfrac{2\cdot E_s}{N_0}}\cdot\sin\left(\dfrac{\pi}{M}\right)\right)$	$P_b \approx \dfrac{2}{n_b}\cdot Q\left(\sqrt{\dfrac{2\cdot n_b\cdot E_b}{N_0}}\cdot\sin\left(\dfrac{\pi}{M}\right)\right)$
M-QAM	$P_e \approx \dfrac{4\cdot\left(\sqrt{M}-1\right)}{\sqrt{M}}\cdot Q\left(\sqrt{\dfrac{3}{(M-1)}\cdot\dfrac{E_s}{N_0}}\right)$	$P_b \approx \dfrac{4\cdot\left(\sqrt{M}-1\right)}{n_b\cdot\sqrt{M}}\cdot Q\left(\sqrt{\dfrac{3\cdot n_b}{(M-1)}\cdot\dfrac{E_b}{N_0}}\right)$
Coherent M-FSK	$P_e \approx (M-1)\cdot Q\left(\sqrt{\dfrac{E_s}{N_0}}\right)$	$P_b \approx \dfrac{M}{2}\cdot Q\left(\sqrt{\dfrac{n_b\cdot E_b}{N_0}}\right)$

Figure 2.19   Bandpass modulation error probabilities

for a given application, since hardly a bandpass digital communication system will present good bandwidth and power efficiencies at the same time.

In Figure 2.18, the null-to-null bandwidth expressions of some bandpass modulation schemes [3,14] employing, for simplicity, rectangular pulse shaping instead of Nyquist pulse shaping are presented. In a complementary manner, the error probability expressions of some bandpass modulation schemes [14,26] are shown in Figure 2.19.

Hence, with the presented specifications of bandwidth occupancy and BER for a given E_b/N_0, the choice of the most suitable scheme for a given application can be made based on η_E, represented by

$$\eta_E = \frac{E_b}{N_0}, \text{for a given BER} \tag{2.100}$$

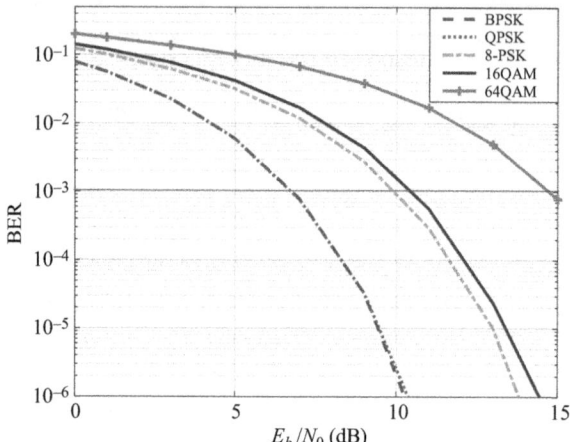

Figure 2.20   BER of some important bandpass modulation schemes

and on η_{Bw},

$$\eta_b = \frac{R_b}{W}. \tag{2.101}$$

Defined in (2.14) and in (2.15), and repeated here for convenience.

Systems that present lower η_E are considered more energy efficient, since they require lower E_b/N_0 to achieve a given BER. On the other hand, systems that present larger η_{Bw} are more bandwidth efficient, since they transmit higher bit rates per unit bandwidth. Ideally, a good system is the one that offers the highest η_{Bw} at a given η_E or the one that requires the lowest η_E at a given η_{Bw} [2].

In practice, as mentioned earlier, the P_b can be estimated by the BER measurement. In order to verify the performance of some important bandpass modulations currently used in wired and wireless systems (e.g., Wi-Fi, LTE and PLC systems), for different noise conditions (AWGN), some Monte-Carlo simulations were performed to the BPSK, QPSK, 8-PSK, 16-QAM and 64-QAM schemes to obtain the BER curves as a function of E_b/N_0. The obtained results are shown in Figure 2.20.

If a BER of 10^{-3} is used as performance reference, one can verify that BPSK and QPSK schemes will need an E_b/N_0 of 7 dB, 8-PSK will require an E_b/N_0 of 10 dB, 16-QAM will require an E_b/N_0 of 11 dB, and 64-QAM will necessitate an E_b/N_0 of 14.5 dB.

2.6   Bandlimited transmission

In the discussion so far, time-limited unit gate pulses $\Pi(t)$ have been used as the pulse shaping filter $h_p(t)$ in the LPE representation of baseband and bandpass digital systems. As the gate pulses are not limited in frequency, part of their spectra will be suppressed by a bandlimited channel, thus resulting in pulse distortion and, possibly, ISI at the receiver, as already discussed in Section 2.1.2. To solve this problem, one can try to replace $\Pi(t)$ with bandlimited pulses. However, bandlimited pulses are not time-limited. Hence, they will overlap in time, also causing ISI.

One solution adopted by most digital communication systems consists in employing pulse shaping and equalization techniques. While pulse shaping refers to the design of bandlimited LPE waveforms in such a way that despite pulse spreading, there is no ISI at the decision-making instants [14], equalization deals with the design of filters to compensate for the distortions imposed by the channel within the frequency band of the transmitted signals [13,14]. These two techniques will be briefly described in the sequel.

2.6.1 Nyquist criterion for zero ISI

In Section 2.5, it was shown that the digital modulation schemes can be represented by the LPE signals given by (2.78), repeated here for convenience:

$$s_{lp}(t) = \sum_i d_i \cdot h_p(t - iT_s), \tag{2.102}$$

where d_i is the ith complex-valued digital information symbol, $h_p(t)$ is the pulse shape and $1/T_s$ is the symbol rate.

Consider for a moment the transmission of the LPE signals $s_{lp}(t)$ over an ideal (distortion-free and noiseless) channel. Even in this case, ISI might occur if the pulse shape is not chosen appropriately. To see this, assume the receiver will try to recover the symbols d_i by sampling $s_{lp}(t)$ at multiples of the symbol period. Then, the nth sample is given by [13]

$$s_{lp}(nT_s) = \sum_i d_i \cdot h_p(nT_s - iT_s) = d_n * h_p(nT_s), \tag{2.103}$$

where the $*$ operator represents the discrete-time convolution of the symbol sequence with a sampled version of the pulse shape.

This convolution sum can be further decomposed into two parts as [13]

$$s_{lp}(nT_s) = d_n \cdot h_p(0) + \sum_{i \neq n} d_i \cdot h_p(nT_s - iT_s), \tag{2.104}$$

where the first term on the right-hand side contains the desired symbol and the second term represents the ISI, i.e., the interference from neighboring symbols.

From (2.104), it is clear that there will be no ISI only if the second term on the right-hand side is zero. This happens if the sampled pulse shape reduces to a Kronecker delta function [19,27] in the sampling instants, i.e.,

$$h_p(nT_s) = \delta_n. \tag{2.105}$$

Remembering that $h_p(nT_s)$ is the sampled version of $h_p(t)$, it is possible to show that the equivalent frequency domain representation of (2.105) can be expressed by [2,13,14]

$$\frac{1}{T_s} \sum_i H_p\left(f - \frac{i}{T_s}\right) = 1, \tag{2.106}$$

where $H_p(f)$ is the Fourier transform of $h_p(t)$.

Equation (2.106) represents the so-called Nyquist criterion for zero ISI and a pulse $h_p(t)$ that satisfies this equation is called the *Nyquist pulse* [2,13].

Thus, the digital symbols transmitted by the use of Nyquist pulses can be correctly retrieved by appropriate sampling of the received waveform at regular intervals of T_s, even with the overlapping of the neighboring pulses, since they are designed to be zero at adjacent symbol instants.

There are an infinite number of pulses that satisfy the Nyquist criterion. The sync pulse is the one with the smallest bandwidth. This bandwidth, known as the Nyquist bandwidth, is half of R_s, i.e., $1/2T_s$ [2,13,14].

Although a minimum-bandwidth pulse is desirable, it is not realizable. A practical pulse will have a bandwidth larger than the minimum value by a factor $1 + \alpha$, where α is known as *excess-bandwidth parameter* or *roll-off factor*. For instance, a pulse with $\alpha = 1$ has 100% excess bandwidth, that is, it occupies twice the band of a sinc pulse. The most commonly used Nyquist pulses are, perhaps, the *raised-cosine pulses*, defined as [2,13,14]

$$h_p(t) = \left(\frac{\sin(\pi t/T_s)}{\pi t/T_s} \right) \left(\frac{\cos(\alpha \pi t/T_s)}{1 - (2\alpha t/T_s)^2} \right), \tag{2.107}$$

which have Fourier transform

$$H_p(f) = \begin{cases} T_s, & |f| \le \frac{1-\alpha}{2T_s} \\ T_s \cos^2 \left[\frac{\pi T_s}{2\alpha} \left(|f| - \frac{1-\alpha}{2T_s} \right) \right], & \frac{1-\alpha}{2T_s} < |f| \le \frac{1+\alpha}{2T_s} \\ 0, & |f| > \frac{1+\alpha}{2T_s} \end{cases} \tag{2.108}$$

Consequently, the resulting baseband signal bandwidth can be estimated by

$$B = (1 + \alpha) \cdot \frac{R_s}{2}, \tag{2.109}$$

and the corresponding bandpass signal bandwidth, for linear bandpass systems, is given by

$$W = (1 + \alpha) \cdot R_s. \tag{2.110}$$

Figure 2.21 shows the time and frequency representations of raised-cosine pulses for some values of α.

In practice, it is common to implement the raised cosine pulse into two stages, one on the transmit side and the other on the receive side. The frequency response of each pulse is designed to be equal to the square root of the response of the raised cosine pulse, reason why they are called square root raised cosine (SRRC) pulses, so that the overall response of the two pulses is equivalent to the response of the raised cosine pulse and also meets the Nyquist criterion for zero ISI:

$$H_{p_{tx}}(f) \cdot H_{p_{rx}}(f) = H_p(f), \tag{2.111}$$

where $H_{p_{tx}}(f)$ and $H_{p_{rx}}(f)$ are, respectively, the frequency responses of the transmit and receive pulses, and $H_{p_{tx}}(f) = H_{p_{rx}}(f) = \sqrt{H_p(f)}$.

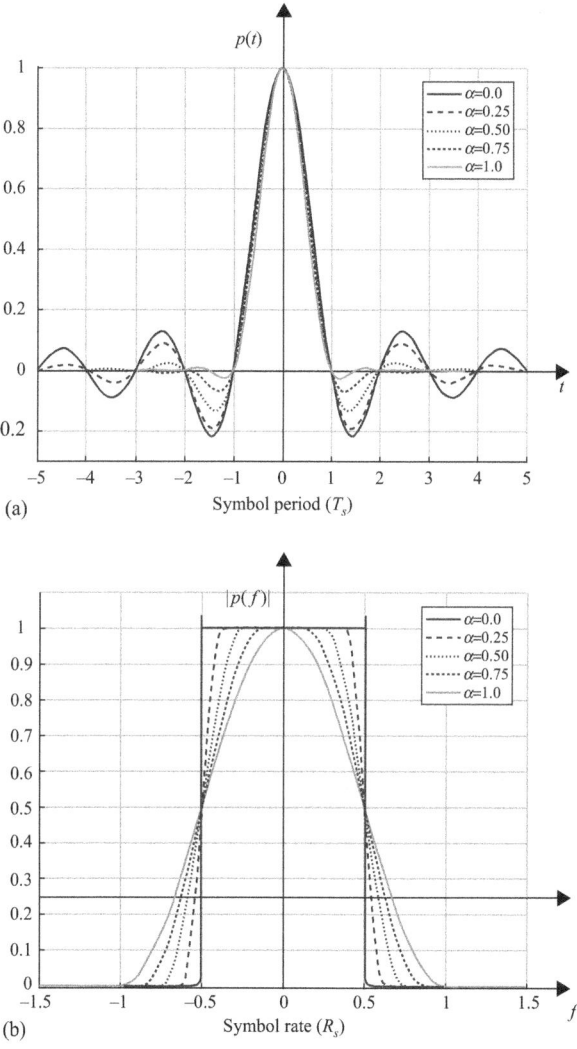

Figure 2.21   Raised cosine pulse (a) time domain and (b) frequency domain

2.6.2   Multipath fading channels

Communication channels can distort transmitted bandlimited signals, even if special pulses such as raised cosine pulses are used. As discussed in Section 2.1.2, in multipath propagation environments, multiple replicas of the transmitted bandlimited signal can reach the receiver through distinct propagation paths, causing each replica to have a different amplitude and delay (and hence phase) values.

It means that, depending on the characteristics of the propagation channel in relation to the parameters of the symbols carried by the transmitted signal, the symbols may arrive at the receiver distorted (multipath fading). For instance, the transmitted symbols may be widened, generating ISI [2,14,16,17].

To better understand the effects of multipath propagation over the transmitted symbols, the following parameters can be defined: delay spread (τ_{spr}), coherence bandwidth (B_{coh}), Doppler spread (D_{spr}) and coherent time (T_{coh}) [16–18]. Since they are random quantities, they are usually represented by statistical values or stochastic functions (e.g., mean value, root mean square, correlation function).

The τ_{spr} expresses the temporal widening of the received symbols, resulting from the time delays between the multiple replicas received from the transmitted signal. The τ_{spr} is a time domain parameter generally extracted from the channel power delay profile, and it has a dual in the frequency domain called of B_{coh}, which represents the maximum frequency separation in which the channel acts on the transmitted signal in a correlated manner (i.e., the bandwidth where the channel frequency response is constant).

These parameters are used to describe the temporal dispersion nature of the channel in a local area. However, they provide no information on how the channel is varying over time due to the relative movement between transmitter and receiver. To quantify the time varying nature of the channel, D_{spr} and T_{coh} are used. The D_{spr} represents the spectral widening of the received signal. It is a frequency domain parameter and its dual in the time domain is the T_{coh}, which represents the maximum time interval in which the channel affects the transmitted signal in a correlated way (i.e., the time that the channel impulse response remain invariant).

Depending on the relationship between symbol parameters (e.g., T_s, R_s) and channel parameters (e.g., τ_{spr} and D_{spr}), different signals may suffer distinct types of multipath fading in a given transmission. While the τ_{spr} causes temporal dispersion and frequency selective fading, D_{spr} causes spectral dispersion and time selective fading. Since the τ_{spr} and D_{spr} are independent, time and frequency dispersions can occur in four different ways, as shown in Figure 2.22, according to the nature of the transmitted signal, channel and relative movement between transmitter and receiver.

If the channel presents constant magnitude and linear phase over the whole signal band at a given time instant, then the received signal will suffer flat frequency fading. In this case, $B_w \ll B_{coh}$ and, consequently, $\tau_{spr} \ll T_s$. Although the received

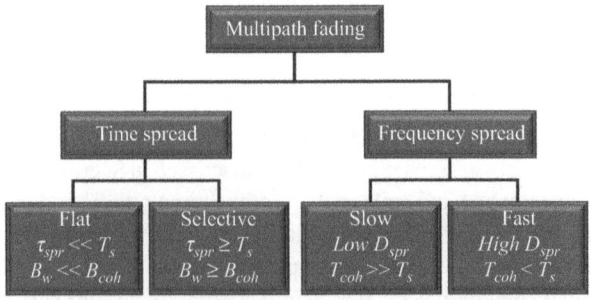

Figure 2.22   Multipath channel classification

signal does not suffer distortion, its intensity and phase can vary over time and space (can reach values near zero during fading instants). The statistical distribution of the instantaneous gain of flat frequency fading channels is important for radio link design and is usually modeled by Rayleigh and Rice distributions. On the other hand, the phase is usually modeled by a uniform distribution [16–18].

Considering that the channel presents constant magnitude and linear phase only on part of the signal band, the received signal will undergo frequency selective fading. In this case, the spectral characteristics of the transmitted signal are modified when the signal arrives to the receiver. This implies that $B_w \geq B_{coh}$, causing different frequency components of the signal to undergo uncorrelated fading (some components are attenuated and others may be reinforced). Consequently, $\tau_{spr} \geq T_s$, that is, the received signal is composed by numerous versions of the signal attenuated and delayed, causing signal distortion. Since frequency selective fading is caused by the temporal dispersion of the transmitted symbols, it results in ISI.

Whether the impulse response of the channel remains practically constant during one or more T_s, varying slowly over time, the received signal will suffer slow fading.[4] In this case, the channel $T_{coh} \gg T_s$ of the transmitted signal. In the frequency domain, this means that D_{spr} is negligible.

Suppose, the channel impulse response changes during T_s, then the received signal will suffer fast fading. In this case, $T_{coh} < T_s$, and it causes frequency dispersion due to D_{spr}, causing signal distortion. In the frequency domain, signal distortion enlarges with increasing D_{spr} in relation to B_w.

In general, multipath channels can be modeled by linear filters with time-varying impulse responses. So, disregarding noise effects for simplicity, the transmitted signal $s(t)$ arriving to the receiver, after passing through the channel with impulse response $h_c(t, \tau)$, can be represented as[5]

$$r(t) = \int_{-\infty}^{\infty} s(\tau) \cdot h_c(t, \tau) d\tau. \tag{2.112}$$

In this case, each path may be associated with attenuation, delay, phase and Doppler frequency factors which vary over time. Thus, the time-varying impulse response of the channel at time t due to an impulse applied at time $t - \tau$ can be defined as [16]

$$h_c(t, \tau) = \sum_i \alpha_i(t) \cdot e^{-j\varphi_i(t)} \cdot \delta[t - \tau_i(t)], \tag{2.113}$$

where $\alpha_i(t)$ is the attenuation in the ith path at time t, $\tau_i(t)$ is the delay in the ith path at time t, $\varphi_i(t)$ is the phase rotation in the ith path at time t, given by

$$\varphi_i(t) = 2\pi f_o \cdot \tau_i(t) - \phi_i^D(t), \tag{2.114}$$

---

[4]It should be noted that the relative movement between transmitter and receiver and the bandwidth of the transmitted signal determine if the fading is slow or fast.
[5]Note that this integral becomes the convolution integral if the channel impulse response is time-invariant. For this case $h_c(t, \tau)$ simplifies to $h_c(t - \tau)$.

and $\phi_i^D(t)$ is the Doppler phase shift in the ith path at time t, obtained by

$$\phi_i^D(t) = \int_{t_0}^{t_0+t} f_i^D(x)dx. \tag{2.115}$$

Suppose, a bandpass digital signal represented by

$$s(t) = \Re\left[s_{lp}(t) \cdot e^{-j2\pi f_o t}\right], \tag{2.116}$$

is applied to the multipath channel, the signal at the channel output will be

$$r(t) = \Re\left[\sum_i \alpha_i(t) \cdot e^{-j\varphi_i(t)} \cdot s_{lp}[t - \tau_i(t)] \cdot e^{-j2\pi f_o t}\right], \tag{2.117}$$

whose LPE representation is given by

$$r(t) = \Re\left[\sum_i \alpha_i(t) \cdot e^{-j\varphi_i(t)} \cdot s_{lp}[t - \tau_i(t)]\right]. \tag{2.118}$$

Considering the channel is time invariant (LTI), the LPE representation of its impulse response can be expressed by

$$h_c(t) = \sum_i \alpha_i \cdot e^{-j\varphi_i} \cdot \delta[t - \tau_i], \tag{2.119}$$

where, due to the invariance of the channel, $\varphi_i = 2\pi f_o \cdot \tau_i$.

In addition, by considering the LPE representation, if a bandlimited signal with LPE bandwidth of B [2] is transmitted through an LTI channel, the signal replicas from different paths that arrive at the receiver within time slots smaller than $\Delta_{T_s} = 1/2B$ will be merged into the receive filter, behaving as they were a single replica of a single path. Each of the resulting paths, for distinct time intervals, is called the resolvable path. Thus, for this case, the LPE of the impulse response of the LTI channel simplifies to

$$h_c(t) = \sum_{i=1}^{L_{path}} \beta_i \cdot \delta\left[t - \frac{i}{2B}\right], \tag{2.120}$$

where $\beta_i = \alpha_i \cdot e^{-j\varphi_i}$ and L_{path} is the number of resolvable path [16].

This representation, illustrated in Figure 2.23, corresponds to a linear tapped delay line (TDL) transverse filter [10,15]. Hence, the received signal, except from the AWGN, can be modeled by the convolution of the transmitted signal with the impulse response of the channel. Some communication channels, such as optical fibers, however, may present severe nonlinear effects [12]. In these cases, linear transverse filters may not be well suited and nonlinear channel models must be employed.

2.6.3   Equalization

The growing demand for new data services (e.g., multimedia, high-speed Internet) has led to the development of digital communication systems with ever-increasing

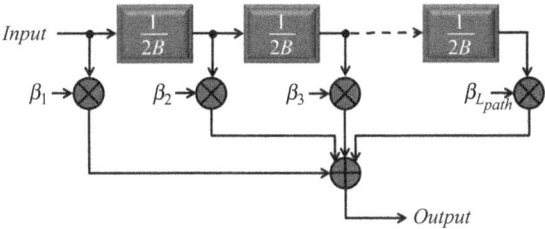

Figure 2.23 Representation of the LPE of the impulse response of the LTI channel

data rates. However, as seen in the previous section, transmitting at a rate $R_s \geq B_{coh}$ (or equivalently, $\tau_{spr} \geq T_s$), the channel becomes very time dispersive and frequency selective, causing ISI.

In this situation, the channel response within the signal band will not be flat and the signal transmitted will be distorted even if special Nyquist pulses as raised cosine were employed. This signal distortion has a more critical impact on the BER increase than the AWGN, since an increase in the power of the transmitted signal will increase the SNR but will not reduce the ISI.

Hence, to solve this problem, digital communication systems often employ a filter called equalizer before the detection process to compensate signal distortions caused by the channel and mitigate ISI. By carefully designing the equalizer, a pulse shape can be obtained at the detector input that has no ISI at the sampling instants, regardless of the channel conditions.

As the channel can vary randomly over time, equalizers should be implemented adaptively to track channel changes. An adaptive equalizer may include a training step to adjust its parameters (i.e., coefficients) periodically to compensate for the effects of the channel on the receiver (also called channel inversion).

Equalizers can be implemented in time domain and in frequency domain. Time domain equalizers (TDEs) are usually indicated to digital communication systems operating in relative low selective fading channels, which results in a simpler filter equalizer (e.g., few coefficients). On the other hand, frequency domain equalizers (FDEs) are commonly used for systems operating in high frequency selective fading channels.

2.6.3.1 Time domain equalization

TDEs can either be linear or nonlinear. Linear TDEs are simpler structures, usually implemented by transverse filters, whose coefficients can be adjusted to generate zero ISI at the sampling instants. Nonlinear TEDs generally have a more complex structure and are able to mitigate ISI in multipath channels with more severe fading conditions, where linear TDEs are generally not effective (e.g., presence of spectral nulls).

The two most common linear TDEs are zero forcing (ZF) and minimum mean square error (MMSE). The former is able to adequately mitigate the ISI but may considerably increase noise during the occurrence of deep fading caused by spectral nulls.

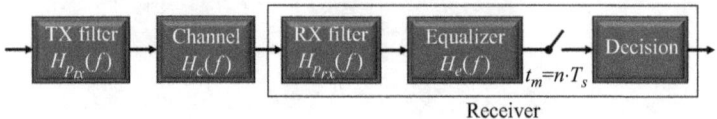

Figure 2.24   ZF TDE system diagram

The latter minimizes the expected mean square error (MSE) between the transmitted and detected symbols at the equalizer output, thus providing a better compromise between ISI mitigation and noise enhancement. Because of this, MMSE TDEs tend to have a lower BER than ZF TDEs [16].

Considering the LPE system diagram presented in Figure 2.24, the overall frequency response $H_o(f)$ is given by

$$H_o(f) = H_{p_{tx}}(f) \cdot H_c(f) \cdot H_{p_{rx}}(f) \cdot H_e(f) \tag{2.121}$$

where $H_e(f)$ is the frequency response of the equalizer and $H_{p_{tx}}(f)$ and $H_{p_{rx}}(f)$ are, respectively, the frequency responses of the transmit and receive filters.

By the Nyquist criterion, presented in (2.106), $H_o(f)$ should meets

$$\sum_i H_o\left(f - \frac{i}{T_s}\right) = T_s, \tag{2.122}$$

If transmit and receive processes use SRRC filters to meet zero ISI criterion, it is only necessary that

$$H_e(f) = \frac{1}{H_c(f)}. \tag{2.123}$$

Thus, to mitigate the ISI, the TDE should act as a simple inverter filter. The frequency response of the equalizer should be the inverse of the frequency response of the channel, repeated at a rate of $1/T_s$.

This result can be investigated more deeply by analyzing the operation of a linear TDE in the time domain.

Let the linear structure of a TDE based on a TDL transverse filter with $2N + 1$ taps spaced of T_s, shown in Figure 2.25, whose impulse response is given by

$$h_e(t) = \sum_{n=-N}^{N} c_n \cdot \delta[t - n \cdot T_s], \tag{2.124}$$

where c_n is the nth coefficient of the filter.

Thus, by applying a distorted pulse $h_u(t)$ to the TDE and sampling its output at $t_k = (k + N) \cdot T_s$ so that the peak of $h_u(t)$ is at the central tap of the TDE, the resulting discrete-time output is given by [2,21]

$$h_e(t_k) = \sum_{n=-N}^{N} c_n \cdot h_u[k - n], \tag{2.125}$$

where $h_u[k - n]$ represents $h_u(kT_s - nT_s)$.

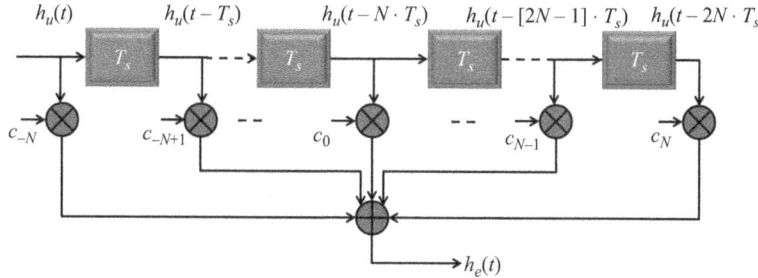

Figure 2.25   Linear TDE model based on transverse filter

Considering the Nyquist criteria for zero ISI, the output of the TDE should be

$$h_e(t_k) = \begin{cases} 1 & \text{for } k = 0 \\ 0 & \text{for } k \neq 0 \end{cases}. \tag{2.126}$$

However to meet this condition, it is necessary a TDE with infinity number of coefficients. Since there are only $2N + 1$ coefficients, the zero ISI condition can be approximated by making [2,21],

$$h_e(t_k) = \begin{cases} 1 & \text{for } k = 0 \\ 0 & \text{for } k = \pm 1, \dots, \pm N \end{cases}. \tag{2.127}$$

Since the function of this TDE is to set the central peak to 1 and force the N values on each side of the central peak to zero (forcing ISI to zero), it is called ZF-TDE.

The ZF-TDE coefficients can be obtained by

$$h_e(t_k) = \sum_{n=-N}^{N} c_n \cdot h_u [k - n] = \begin{cases} 1 & \text{for } k = 0 \\ 0 & \text{for } k = \pm 1, \dots, \pm N \end{cases}. \tag{2.128}$$

The solution of this set of equations in vector notation is given by

$$\mathbf{c} = \mathbf{H}_u^{-1} \cdot \mathbf{h}_e, \tag{2.129}$$

where $\mathbf{h}_e = [0, \dots, 1, \dots, 0]^T$, $\mathbf{c} = [c_{-N}, \dots, c_0, \dots, c_N]^T$ and

$$\mathbf{H}_u = \begin{bmatrix} h_u[0] & \cdots & h_u[-2N] \\ \vdots & & \vdots \\ h_u[N] & \ddots & h_u[-N] \\ \vdots & & \vdots \\ h_u[2N] & \cdots & h_u[0] \end{bmatrix}. \tag{2.130}$$

One of the main disadvantages of ZF-TDE is the performance degradation due to increased noise level during spectral nulls [2,28]. In general, to reduce this effect, the zero ISI constraint must be relaxed, allowing the TDEs to have some residual ISI at their outputs. A more robust TDE based on this approach can be obtained by using the MMSE criterion instead of the ZF [28].

In the MMSE-TDE, the coefficients are adjusted to minimize the MSE between the transmitted symbol d_k (training symbol) and the detected symbol $h_u[k]$ [2,21,28]. In this way, the cost function can be defined as

$$J_e[n] = \mathrm{E}\{|e[k]|^2\} = \mathrm{E}\{|d_k - h_e[k]|^2\}, \tag{2.131}$$

where $\mathrm{E}\{.\}$ is the expectation operator.

Thus, $J_e[n]$ is given by

$$J_e[n] = \mathrm{E}\{|d_k - \mathbf{c}_k \cdot \mathbf{h_u}[k]| \cdot |d_k - \mathbf{c}_k \cdot \mathbf{h_u}[k]|^H\}, \tag{2.132}$$

where $\mathbf{h_u}[k] = [h_u[k], \ldots, h_u[k - 2N + 1]]^T$ and $\mathbf{c}_u[k] = [c_0[k], \ldots, c_{2N+1}[k]]^T$.

The solution of this problem is given by the Wiener's equation [21,28]:

$$\mathrm{E}\{\mathbf{H}_{u_{ext}}^H \cdot \mathbf{d}_k\} = \mathrm{E}\{\mathbf{H}_{u_{ext}}^H \cdot \mathbf{H}_{u_{ext}}\} \cdot \mathbf{c}. \tag{2.133}$$

where $\mathbf{c}$ is the coefficients vector of the MMSE-TDE, $\mathbf{d} = [d_{-N}, \ldots, d_N, \ldots, d_{2N+1}]^T$, and $\mathbf{H}_{u_{ext}}$ is an extended version of matrix $\mathbf{H}_u$ with dimension $(4N + 1) \times (2N + 1)$ given by

$$\mathbf{H}_{u_{ext}} = \begin{bmatrix} h_u[-N] & \cdots & 0 \\ \vdots & & \vdots \\ h_u[N] & \ddots & h_u[-N] \\ \vdots & & \vdots \\ 0 & \cdots & h_u[N] \end{bmatrix} . \tag{2.134}$$

Furthermore, defining the autocorrelation matrix of the received symbols as

$$\mathbf{R}_{uu} = \mathrm{E}\{\mathbf{H}_{u_{ext}}^H \cdot \mathbf{H}_{u_{ext}}\}, \tag{2.135}$$

and the crosscorrelation vector between the transmitted and received symbols as

$$\mathbf{h}_{ud} = \mathrm{E}\{\mathbf{H}_{u_{ext}}^H \cdot \mathbf{d}_k\}, \tag{2.136}$$

it follows that

$$\mathbf{h}_{ud} = \mathbf{R}_{uu} \cdot \mathbf{c}. \tag{2.137}$$

Thus, the TDE coefficients can be obtained by

$$\mathbf{c} = \mathbf{R}_{uu}^{-1} \cdot \mathbf{h}_{ud}. \tag{2.138}$$

As mentioned at the beginning of this section, in general, linear TDEs do not perform as well as nonlinear TEDs when the channel has severe fading conditions. Among nonlinear TDEs, one of the most common is decision-feedback equalizers (DFEs), since it is simple to implement and does not undergo noise enhancement. However, DFE suffers from error propagation when channels have a low SNR, which leads to poor performance [16]. On the other hand, ML sequence equalizer (MLSE) is optimal in the sense of minimizing BER. Unfortunately, the complexity of this equalizer grows exponentially with the length of the channel [16].

2.6.3.2   Frequency domain equalization

As discussed earlier, the development of digital communications systems with high data rates can result in an $R_s \gg B_{coh}$ and hence a rather severe selective fading with ISI in tens or hundreds of symbols.

Linear and nonlinear TDEs can be used on the receive side to compensate, in the time domain, for distortions caused by frequency selective channels. However, depending on the type of TDE, computational processing may increase linearly or exponentially with channel length (e.g., due to the required increase in the number of coefficients of a ZF-TDE or the trellis of an MLSE), not being a good choice for severe frequency selective fading channels.

A promising alternative to TDEs are the FDEs. They are an effective way to improve the system performance under severe frequency selective fading conditions without significantly increasing the complexity of the system. This is possible by associating the operations of fast Fourier transform (FFT) and inverse FFT (IFFT) with a one-tap equalizer (OTE) to compensate for distortions caused by the channel in the frequency domain. In this way, convolution in time is replaced by a simple multiplication in frequency, offering a lower increase of complexity as a function of the length of the impulse response of the channel in comparison with the TDEs. Furthermore, adaptive techniques generally converge more rapidly and are more stable in the frequency domain, making adaptive TDEs an interesting option for time-varying channels [16,29].

A FFT operation on the receive side can map the received symbols to the frequency domain and divide the frequency spectrum of the signal into subcarriers, also called subchannels. In this way, a frequency selective fading channel can be converted into several flat frequency fading subchannels. The OTE can then compensate for channel distortions simply by multiplying the FFT result of each subchannel by the inverse of the frequency response value at the corresponding subchannel. This process requires an estimate of the frequency response of the channel, which can be performed by transmitting training symbols [30].

Considering that the impulse response of the channel is given by (2.120), the channel frequency response can be obtained by

$$\mathbf{f}_c = FFT\{h_c(t)\}, \tag{2.139}$$

where $\mathbf{f}_c = [f_c(1), \ldots, f_c(N_{FFT})]^T, f_c(m)$ is the frequency response complex coefficient at the mth subchannel and N_{FFT} is the FFT block size.

If the received signal waveform is $h_u(t)$ (that can correspond to more than one data symbol), its frequency representation is expressed by

$$\mathbf{f}_u = FFT\{h_u(t)\}, \tag{2.140}$$

where $\mathbf{f}_u = [f_u(1), \ldots, f_u(N_{FFT})]^T$ and $f_u(m)$ is the frequency domain information at the mth subchannel.

Thus, the information corresponding to the mth subchannel free of the distortions caused by the channel can then be restored through the following ZF operation performed in the OTE [30,31]:

$$d_m = \frac{f_u(m) \cdot f_c^*(m)}{|f_c(m)|^2},$$

(2.141)

or by the following MMSE operation [31]:

$$d_m = \frac{f_u(m) \cdot f_c^*(m)}{|f_c(m)|^2 + (1/SNR)}.$$

(2.142)

The FDEs can be used in single carrier (SC) systems and multicarrier systems, such as orthogonal frequency division multiplexing (OFDM). It is important to note that if the presented procedures (i.e., FFT and OTE) are executed in SC systems, an additional IFFT operation on the d_m corresponding to all subchannels will be required after OTE to retrieve the original data symbols. On the other hand, if these procedures are performed in OFDM systems, all d_m already correspond to the transmitted data symbols. The effectiveness of this procedure can be ensured if a cyclic prefix (CP) is employed in the transmission process, as will be discussed further in Chapter 4 [29,30].

2.7 Synchronization

The demodulation process, performed at the receivers of digital bandpass systems, can be classified as coherent and noncoherent. In coherent demodulation, frequency and phase information of the carrier signal is required to the correct recovery of the transmitted information. On the other hand, in noncoherent demodulation, the information of the carrier signal is not required in the demodulation process.

Coherent demodulation usually has the advantage of presenting a lower P_e compared with noncoherent demodulation for the same received signal power, although at the cost of more receiver complexity [32].

Also, both coherent and noncoherent digital demodulation schemes require timing information from the received signal to perfectly recover the digital transmitted symbols.

In real bandpass digital communication systems, the output of the demodulator must be sampled periodically at a rate of at least once per symbol period, in order to recover the transmitted information. Since the propagation delay from the transmitter to the receiver is generally unknown at the receiver, symbol timing must be derived from the received signal in order to synchronously sample the output of the demodulator. The propagation delays of the communication channel also result in offsets in both frequency and phase of the carrier, which must be estimated at coherent receivers [33].

Therefore, to reach optimal carrier and timing estimation, synchronization schemes are employed at the receiver. ML estimation is one of the most employed approaches since it can provide an unified framework for developing optimal synchronization algorithms for digital communication receivers [33].

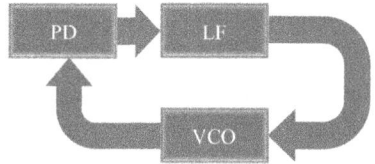

Figure 2.26   Generic phase-locked loop

In the following, carrier and timing synchronization are discussed for coherent receivers.

2.7.1   Carrier synchronization

In general, coherent digital demodulation requires knowledge of the sinusoidal basis functions at the receiver. It means that the receiver has to estimate the frequency and the phase of the carrier from the corresponding received signal. This estimation process is called carrier synchronization.

There are many possibilities for implementing carrier synchronization in a band-pass digital communication system, but the core of most of them is a phase-locked loop (PLL).

A generic PLL, represented by the block diagram of Figure 2.26, is a dynamic control system, the controlled parameter of which is the phase of a locally generated replica of the incoming carrier signal.

The main components of a generic PLL are as follows:

- The phase detector (PD), which is responsible for measuring the phase difference between the local carrier generated by the voltage controlled oscillator (VCO) and the PLL input signal;
- The VCO, which is a variable-frequency oscillator whose output signal frequency is controlled by an input voltage signal (voltage-frequency converter). In normal operation, the free running frequency of the VCO, corresponding to an input voltage of zero, should be equal or close to the frequency of the PLL input signal;
- The loop filter (LF), which is basically a lowpass filter. Usually, first or second order filters are employed.

The PD generates an error signal resulting from the phase difference between the local carrier and the PLL input signal. This error signal is then fed to an LF, which is designed to track the changes in the error signal and to reduce the effects of noise. The filtered error signal controls the frequency of the VCO so that the local generated carrier signal reaches the same phase of the PLL input signal. As the frequency is the derivative of the phase, keeping the phases of the PLL input signal and of the VCO output signal locked, their frequencies will be also locked. A detailed description of the PLL operation can be found in [34].

2.7.2   Timing synchronization

Coherent and noncoherent digital demodulations require knowledge of the timing information of the received signal to perfectly recover the transmitted information. This estimation process is called timing synchronization. The main purpose of the timing recovery is to obtain symbol synchronization.

Symbol decisions are based on the MF output at the end of each symbol period T_s. The detector samples the MF output and uses this information to decide which symbol is most likely to have been sent. In order to make these decisions, the detector must know when the symbols begin and end.

Basically, there are three main techniques to determine the optimum sampling point [32]:

- The first method finds the point where the slope of the MF output is zero by considering ML criterion. If the current timing estimate is too early, then the slope of the MF output is positive indicating that the timing phase should be advanced. If the current timing estimate is too late, then the slope of the MF output is negative indicating that the timing phase should be retarded.
- The second method, commonly denoted as Gardner loop, uses the zero crossings in the oversampled MF output, with oversampling rate of two samples per symbol, to estimate the times in between the optimum sampling points. Zero crossings are found by searching for sign changes between the previous y_{i-1} and following MF y_{i+1} outputs with respect to current analyzed output y_i. A sign change means that the current MF output resides on a zero crossing trajectory, whose error can be estimated by $e_i = [y_{i+1} - y_{i-1}] \cdot y_i$. This method has the advantage of being insensitive to carrier offsets and timing recovery can be locked first, simplifying the task of carrier recovery.
- The third method, usually called dither loop, estimates the optimum sampling point by finding the position that minimizes the variance of the MF output. Typically, the search is performed by estimating the variance at the next interpolation point. If the variance increases, then the timing estimate is not advanced. If the variance decreases, then the timing estimate is advanced.

2.8   Conclusion remarks and trends in digital communications

New communication systems promise to offer a wide range of multimedia applications, which demand high data rates and low error probabilities. These requirements, generally characterized by a demanded quality of service, combined with the ever-increasing demand for high speed Internet access, motivate the development and the use of new technologies.

In wireless communication and PLC systems, for instance, a number of new structures and techniques have been proposed in recent years to mitigate the degrading effects of multipath channels and different sources of interference, thus allowing for an increase in system capacity and reliability and, consequently, in the data transmission rates.

Among these techniques, OFDM has risen as a very efficient, flexible and cost-effective modulation technique [13,18]. In OFDM, digital information can be first mapped to a high-order M-ary scheme to improve system throughput. Then, IFFT and FFT operations are performed in the transmission and reception processes, respectively. By combining IFFT/FFT and CP, OFDM can mitigate ISI, converting a frequency selective fading channel into multiple orthogonal frequency flat fading channels. Consequently, the fading can be compensated by a simple one-tap FDE at the receiver [14,30].

Another interesting technique is multiple-input and multiple-output (MIMO) systems. The theoretical studies presented in [35–37] demonstrated that MIMO systems can offer a significant increase in system capacity without sacrificing precious bandwidth and power resources. The main idea of MIMO systems is to exploit not only the time dimension of the information signals but also the space dimension provided by the use of simultaneously multiple transmissions by using multiple receivers. By jointly processing the signals from the multiple transmissions, the interference can be effectively suppressed, thus increasing the system capacity.

One of the most effective and practical ways to improve the quality of the received signals in MIMO systems is to mitigate the fading effects by using space-time coding (STC) [18,38,39]. As the name suggests, the encoding process is done in both temporal and spatial domains, thus introducing correlation between signals transmitted by different spatially separated sources (e.g., antennas) in different time instants. Among the existing STC schemes, Orthogonal space-time block coding is of particular interest because their ML receiver consists of a simple linear combiner followed by a symbol-by-symbol decoder [38–41]. MIMO systems can also be used to increase the data rate of a digital communication system by the use of spatial multiplexing techniques [18,38,42].

It is important to highlight that although nearly associated with wireless communications, recently MIMO systems have been employed in other systems, such as fiber optic [43,44], DSL [45–48] and PLC [49–52].

In order to increase the robustness of transmitted signals to noise and interferences, forward error correction (FEC) coding, also called channel coding, is normally used. FEC coding techniques add controlled redundancy to the transmitted message to allow the receiver to detect and/or correct errors caused by the communication channel during transmissions.

Although FEC techniques have been used for a long time, a variety of powerful coding schemes have been designed in the last decades. Among them, turbo coding [53–55] and low density parity check (LDPC) coding [54–58] stand out because they can offer a capacity close to the Shannon limit with a feasible complexity [2,13,54,55]. Turbo and LDPC techniques have been adopted by various wireless communication (e.g., IEEE 802.11n/ac) and PLC (e.g., IEEE 1901.2, IEEE 1901-2010) standards.

References

[1]   Eisencraft M, Attux R, Suyama R. Chaotic Signals in Digital Communications. 1st ed. Boca Raton, FL: CRC Press; 2014.

[2]   Proakis JG. Digital Communications. 5th ed. New York, NY: McGraw-Hill; 2008.

[3]   Ziemer RE, Peterson RL. Introduction to Digital Communication. 2nd ed. Upper Saddle River, NJ: Prentice Hall; 2001.

[4]   Karipidis E, Sidiropoulos N, Leshem A, *et al.* Crosstalk Models for Short VDSL2 Lines from Measured 30 MHz Data. EURASIP Journal on Applied Signal Processing. 2006; 2006(85859):785–800.

[5]   van den Brink RFM. Cable Reference Models for Simulating Metallic Access Networks. ETSI; 1998. STC TM6(97)02.

[6]   Werner JJ. The HDSL Environment. IEEE Journal on Selected Areas in Communications. 1991;9(6):785–800.

[7]   Roden MS. Analog and Digital Communication Systems. 4th ed. Upper Saddle River, NJ: Prentice-Hall, Inc.; 1996.

[8]   Batur OZ, Koca M, Dundar G. Measurements of impulsive noise in broad-band wireless communication channels. In: Research in Microelectronics and Electronics, 2008. PRIME 2008. Ph.D.; 2008. p. 233–236.

[9]   Blackard KL, Rappaport TS, Bostian CW. Measurements and models of radio frequency impulsive noise for indoor wireless communications. IEEE Journal on Selected Areas in Communications. 1993;11(7):991–1001.

[10] Dostert K. Powerline Communications. Englewood Cliffs, NJ: Prentice Hall; 2001.

[11] Nassar M, Gulati K, Mortazavi Y, *et al.* Statistical modeling of asynchronous impulsive noise in powerline communication networks. In: Global Telecommunications Conference (GLOBECOM 2011), 2011 IEEE; 2011. p. 1–6.

[12] Agrawal GP. Fiber-Optic Communication Systems. 3rd ed. New York, NY: Wiley Interscience; 2002.

[13] Barry JR, Lee EA, Messerschmitt DG. Digital Communications. 3rd ed. Norwood, MA: Kluwer Academic Publishers; 2004.

[14] Lathi BP, Ding Z. Modern Analog and Digital Communication Systems. 4th ed. New York, NY: Oxford University Press; 2010.

[15] Jeruchim MC, Balaban P, Shanmugan KS. Simulation of Communication Systems. 2nd ed. New York: Kluwer Academic Press; 2000.

[16] Goldsmith A. Wireless Communications. New York, NY: Cambridge University Press; 2005.

[17] Rappaport TS. Wireless Communications Principles and Practice. 2nd ed. Upper Saddle River, NJ: Prentice Hall; 2002.

[18] Tse D, Viswanath P. Fundamentals of Wireless Communication. New York, NY: Cambridge University Press; 2005.

[19] Lathi BP. Linear Systems and Signals. 2nd ed. New York, NY: Oxford University Press; 2009.

[20] Oppenheim AV, Willsky AS, Hamid S. Signals and Systems. 2nd ed. Upper Saddle River, NJ: Prentice Hall; 1996.

[21] Sklar B. Digital Communications Fundamentals and Applications. 2nd ed. Upper Saddle River, NJ: Pearson Prentice Hall; 2001.

[22] Shannon EC. A Mathematical Theory of Communication. Bell System Technical Journal. 1948;27:379–423.

[23] Kotelnikov VA. The Theory of Optimum Noise Immunity. Moscow: Molotov Energy Institute; 1947.

[24] Wonzencraft JM, Jacobs IM. Principles of Communication Engineering. Prospect Heights, IL: Waveland; 1990.

[25] Papoulis A, Pillai SU. Probability, Random Variables and Stochastic Processes. 4th ed. New York, NY: McGraw-Hill; 2002.

[26] Letaief KB, Chuang JCI, Liou ML. M-PSK and M-QAM BER Computation using Signal-Space Concepts. IEEE Transactions on Communications. 1999;47:181–184.

[27] Oppenheim AV, Schafer RW. Discrete-Time Signal Processing. 3rd ed. Upper Saddle River, NJ: Pearson Prentice Hall; 2009.

[28] Haykin S. Adaptive Filter Theory. 4th ed. Upper Saddle River, NJ: Pearson Prentice-Hall; 2002.

[29] Falconer D, Ariyavisitakul SL, Benyamin-Seeyar A, *et al.* Frequency Domain Equalization for Wireless Systems. IEEE Communications Magazine. 2002;1(4):152–156.

[30] Casella IRS. Analysis of turbo coded OFDM systems employing space-frequency block code in double selective fading channels. In: IEEE International Microwave and Optoelectronics Conference. 2007; p. 516–520.

[31] Sari H, Karam G, Jeanclaude I. Frequency-domain equalization of mobile radio and terrestrial broadcast channels. In: Global Telecommunications Conference (GLOBECOM 1994), IEEE; 1994. p. 1–5.

[32] Dick C, Harris F, Rice M. Synchronization in software radios – carrier and timing recovery using FPGAs. In: IEEE Symposium on Field-Programmable Custom Computing Machines. 2000; p. 195–204.

[33] Franks LE. Carrier and Bit Synchronization in Data Communication – A Tutorial Review. IEEE Transactions on Communications. 2000;28:1107–1121.

[34] Rice M. Digital Communications – A Discrete-Time Approach. 1st ed. Upper Saddle River, NJ: Pearson Prentice Hall; 2009.

[35] Foshini GJ. Layered Space-Time Architecture for Wireless Communication in a Fading Environment when using Multi-element Antennas. Bell Labs Technical Journal. 1996;1:41–59.

[36] Foshini GJ, Gans MJ. On Limits of Wireless Communications in a Fading Environment when using Multi-element Antennas. Wireless Personal Communications. 1998;6:311–335.

[37] Telatar IE. Capacity of Multi-Antenna Gaussian Channels. European Transactions on Telecommunications. 1999;10:585–595.

[38] Duman TM, Ghrayeb A. Coding for MIMO Communication Systems. 1st ed. Hoboken, NJ: John Wiley & Sons; 2007.

[39] Larsson EG, Stoica P. Space-Time Block Coding for Wireless Communications. 1st ed. New York, NY: Cambridge University Press; 2003.

[40] Alamouti SM. A Simple Transmit Diversity Technique for Wireless Communications. IEEE Transactions on Information Theory. 1998;16:1451–1458.

[41] Tarokh V, Seshadri N, Calderbank AR. Space-Time Block Codes from Orthogonal Designs. IEEE Transactions on Information Theory. 1999;45:1456–1467.

[42]   Bölcskei H, Gesbert D, Papadias CB, *et al.* Space-Time Wireless Systems: From Array Processing to MIMO Communications. 1st ed. New York, NY: Cambridge University Press; 2006.

[43]   Shah AR, Hsu RCJ, Tarighat A, *et al.* Coherent Optical MIMO (COMIMO). Journal of Lightwave Technology. 2005;23(8):2410–2419.

[44]   Greenberg M, Nazarathy M, Orenstein M. Multimode Fiber as Random Code Generator – Application to Massively Parallel MIMO Transmission. Journal of Lightwave Technology. 2008;26(8):882–890.

[45]   Ginis G, Cioffi JM. Vectored Transmission for Digital Subscriber Line Systems. 2002;20(5):1085–1104.

[46]   Cendrillon R. Multi-User Signal and Spectra Co-ordination for Digital Subscriber Lines [PhD. thesis]. Katholieke Universiteit Leuven; 2004.

[47]   Tauböck G, Henkel W. MIMO systems in the subscriber-line network. In: Proc. 5th Int. OFDM-Workshop; 2000. p. 18.1–18.3.

[48]   Filho DZ, Lopes RR, Ferrari R, *et al.* Achievable Rates of DSL with Crosstalk Cancellation. European Transactions on Telecommunications. 2009;20:81–86.

[49]   Schneider D, Schwager A, Speidel J, *et al.* Implementation and results of a MIMO PLC feasibility study. In: Power Line Communications and Its Applications (ISPLC), 2011 IEEE International Symposium on; 2011. p. 54–59.

[50]   Schwager A, Schneider D, Baschlin W, *et al.* MIMO PLC: theory, measurements and system setup. In: Power Line Communications and Its Applications (ISPLC), 2011 IEEE International Symposium on; 2011. p. 48–53.

[51]   Chrvsochos AI, Papadopoulos TA, Papagiannis GK, *et al.* MIMO-OFDM narrowband-PLC transmission through distribution transformers: modeling and achievable data rates. In: 2015 IEEE International Conference on Smart Grid Communications (SmartGridComm); 2015. p. 109–114.

[52]   Berger LT, Schwager A, Pagani P, *et al.* MIMO Power Line Communications. IEEE Communications Surveys & Tutorials. 2015;17(1):106–124.

[53]   Berrou C, Glavieux A, Thitimajshima P. Near Shannon limit error-correcting coding and decoding: turbo-codes. In: IEEE International Communications Conference. 1993; p. 1064–1070.

[54]   Lin S, Jr DJC. Error Control Coding. 2nd ed. Upper Saddle River, NJ: Pearson Prentice Hall; 2004.

[55]   Ryan W, Lin S. Channel Codes: Classical and Modern. 1st ed. New York, NY: Cambridge University Press; 2009.

[56]   MacKay DJC, Neal RM. Near Shannon Limit Performance of Low-Density Parity-Check Codes. IET Electronics Letters. 1996;32:1645–1646.

[57]   MacKay DJC. Good Error Correcting Codes Based on Very Sparse Matrices. IEEE Transactions on Information Theory. 1999;45:399–431.

[58]   Richardson T, Shokrollahi A, Urbanke R. Design of Capacity-Approaching Low-Density Parity-Check Codes. IEEE Transactions on Information Theory. 2001;47:619–637.

Chapter 3
Basis of error correction coding
Marco A. Cazarotto Gomes[1] and Murilo B. Loiola[1]

The main goal of any communication system is to reliably transmit information from an emitter to a receiver. To accomplish such a task, a communication system must have certain elements that make possible the transmission of information. Figure 3.1 shows a basic block diagram of a general communication system.

In this figure, the information source, or emitter, could be a person or a machine, such as a computer or a smartphone. The output of the information source, which constitutes the message to be sent to the receiver, could be a continuous signal (like our voice) or a sequence of symbols (like letters in a text). The goal of the source encoder is to transform that message in a sequence of bits, called information bits. This transformation must be realized in a way that the number of bits needed to represent the original message is minimized. The channel encoder, on the other hand, is designed to insert a controlled redundancy in the bits' sequences generated by the source encoder to protect the information from errors that may occur during its transmission.

The modulator is responsible to transform the binary sequence at the output of the channel encoder into waveforms suitable for propagation through the channel. Typical channels include telephone lines, optical fibers, air and space. Besides distorting the modulated waveforms, the channel may corrupt the transmitted signals by natural and/or man-made noises.

At the receiver side, the demodulator plays a role opposed to that of the modulator, generating a discrete-time signal from the distorted waveforms arriving from the channel. The equalizer, in its turn, tries to mitigate those distortions, providing a sequence only corrupted by noise to the channel decoder, which uses the redundancy added by the channel encoder to detect and, if possible, correct some bit errors. Finally, the source decoder generates an estimate of the original message.

Once this chapter is intended to provide an overview of channel coding techniques, it is sufficient to consider an equivalent discrete-time baseband model of the system presented in Figure 3.1. In this new model, modulator, channel, demodulator and equalizer are combined into a single block, forming an equivalent channel, whose input is a binary sequence and whose output is composed of a noise-corrupted bit sequence. Source and source encoder are also combined into one block, generating

[1]Center for Engineering, Modeling and Applied Social Sciences (CECS), Federal University of ABC (UFABC), Brazil

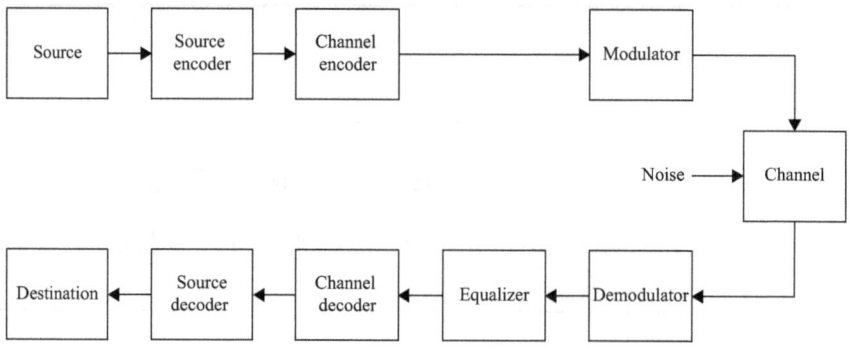

Figure 3.1   Block diagram of a basic communication system

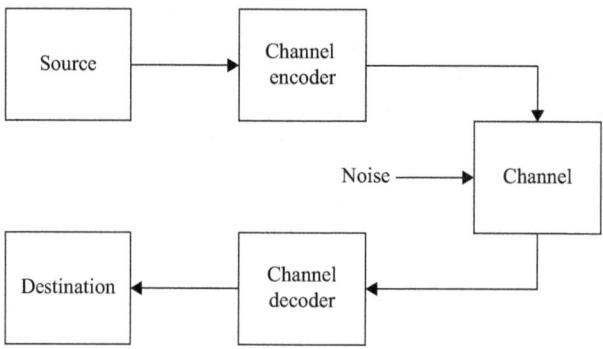

Figure 3.2   Equivalent communication system model

a binary source. This simplified communication system is represented by the block diagram shown in Figure 3.2.

Based on a channel model similar to that in Figure 3.2, Shannon stated, in his channel coding theorem [1], that if we transmit data through a noisy channel at a rate less than the channel capacity, then there exist channel codes that allow an arbitrary small probability of error at the receiver. Unfortunately, the coding scheme proposed by Shannon in his paper is not practical. For that reason, practical channel codes that attain the Shannon limit have been searched since then. Therefore, this chapter presents an overview of such codes, with emphasis in the foundations of block and convolutional codes, which constitute the two main classes of error-correcting codes [2–4].

Block codes, as their names suggest, are codes that encode blocks of information bits. They are memoryless in the sense that the coded outputs, named *codewords*, do not depend on previous input message blocks. In other words, the codewords are computed based solely on the bits composing the current message block, i.e., the current input to the encoder. In contrast, coded output sequences in convolutional codes depend not only on the current encoder input, but also on a number of previous inputs, i.e., the codes have "memory." Both classes of codes are widely used in powerline communications (PLC), as can be seen in [5–14], which constitute a small sample of works on channel-coding technique applied to PLC systems.

Hence, Section 3.1 presents the main concepts behind linear block codes, a family of block codes widely used in communication systems. The low-density parity-check (LDPC) codes, which are some of the most powerful error-correcting codes in use nowadays, and Reed–Solomon codes belong to that family. Convolutional codes are then presented in Section 3.2, which includes some discussion on turbo codes, another powerful class of channel codes used in communication systems.

3.1 Linear block code

Block codes were the first error-correcting codes developed to protect information from the noise present in the communication channel. Among the families of block codes, one of the most important is that of the linear block codes. As it is clear from its name, in a linear block code, the outputs are computed as linear combination of its inputs.

To accomplish that task, the message to be transmitted is firstly divided into blocks of k information bits. A message block is represented by a binary sequence $\mathbf{u} = (u_0, u_1, \ldots, u_{k-1})$ [2,15]. In total, there are 2^k messages. The encoder then transforms each message $\mathbf{u}$ in a sequence $\mathbf{v} = (v_0, v_1, \ldots, v_{n-1})$ known as *codeword*. Therefore, a codeword has n symbols, where $n \geq k$. The difference $m = n - k$ indicates the number of redundant (or parity) bits introduced by the encoder to protect the information bits. The set of 2^k codewords[1] of length n is known as an (n, k) block code. The code rate is defined as $R = k/n$.

By considering a sequence of n binary symbols as a vector, it is easy to see that the set of all possible binary sequences of length n forms a vector space. Thus, a linear block code, i.e., the set of 2^k codewords of length n, can be thought of as a subspace of the n-dimensional binary vector space.

In linear algebra, a vector space is usually described by its basis vectors, which are linearly independent vectors that span the space. In the same way, a linear block code can be represented by its "basis" vectors. Therefore, each codeword can be written as a linear combination of these basis vectors, which can be arranged in a matrix $\mathbf{G}$ known as *generator matrix*. It is worth noting that the coefficients of these linear combinations are given by the information bits of the message to be encoded.

Remark 3.1. *There are k basis vectors in an (n, k) linear block code, since it is composed by 2^k codewords.*

---

[1]Since there is a one-to-one mapping (given by the encoder) between messages $\mathbf{u}$ and codewords $\mathbf{v}$, it is easy to see that a linear block code has 2^k codewords.

The generator matrix $\mathbf{G}$ is then rectangular with k rows and n columns and has the form

$$\mathbf{G} = \begin{bmatrix} \mathbf{g_0} \\ \mathbf{g_1} \\ \vdots \\ \mathbf{g_{k-1}} \end{bmatrix} = \begin{bmatrix} g_{0,0} & g_{0,1} & g_{0,2} & \cdots & g_{0,n-1} \\ g_{1,0} & g_{1,1} & g_{1,2} & \cdots & g_{1,n-1} \\ \vdots & \vdots & \vdots & \vdots & \vdots \\ g_{k-1,0} & g_{k-1,1} & g_{k-1,2} & \cdots & g_{k-1,n-1} \end{bmatrix}, \tag{3.1}$$

where $\mathbf{g_0}\ \mathbf{g_1}\ \cdots\ \mathbf{g_{k-1}}$ are the basis vectors. For example, for a (6,3) block code, a generator matrix $\mathbf{G}$ can be written as

$$\mathbf{G} = \begin{bmatrix} 1\ 0\ 1\ 1\ 0\ 0 \\ 1\ 1\ 0\ 0\ 1\ 0 \\ 0\ 1\ 1\ 0\ 0\ 1 \end{bmatrix}. \tag{3.2}$$

Hence, to generate a codeword $\mathbf{v}$, we use the message $\mathbf{u}$ and the generator matrix $\mathbf{G}$. Assume the message $\mathbf{u} = 101$ has to be transmitted to the receiver. The codeword will then be given by $\mathbf{v} = \mathbf{uG}$. Thus,

$$\mathbf{v} = \mathbf{uG},$$
$$= u_0\mathbf{g_1} + u_1\mathbf{g_1} + \cdots + u_{k-1}\mathbf{g_{k-1}},$$
$$= u_0\mathbf{g_0} + u_1\mathbf{g_1} + u_2\mathbf{g_2}. \tag{3.3}$$

As all codes presented in this chapter are binary, additions and multiplications are computed in the binary field. Thus, addition and multiplication operators hereinafter are computed as modulo-2 addition and modulo-2 multiplication, respectively.

Remark 3.2. *It is worth noting that modulo-2 addition is equivalent to bitwise exclusive-OR operation, and a modulo-2 multiplication is the same as a bitwise AND operation.*

Therefore, by applying modulo-2 operations to (3.3), we obtain

$$\mathbf{v} = 1 \cdot \mathbf{g_0} + 0 \cdot \mathbf{g_1} + 1 \cdot \mathbf{g_2}$$
$$\mathbf{v} = 1(101100) + 0(110010) + 1(011001)$$
$$\mathbf{v} = 101100 + 000000 + 011001$$
$$\mathbf{v} = 110101 \tag{3.4}$$

As in this example $k = 3$, there are $2^3 = 8$ possible messages. The respective codewords are shown in Table 3.1.

Note that the matrix $\mathbf{G}$ in (3.2) can be segmented into two parts: a submatrix $\mathbf{P}$ consisting of the parity-check equations responsible to generate the $(n - k)$ redundant bits and an identity matrix, which "copies" the k message bits to the end of the codeword. Linear block codes, whose generator matrices present the same form as (3.2) are known as systematic codes.[2] Codes whose generator matrices do not possess the

---

[2] In fact, a systematic code is one that copies the information bits directly to the codeword.

Table 3.1   Codewords of the code generated by (3.2)

Message	v = uG	Codeword
000	v0 = 0 (101100) + 0 (110010) + 0 (011001)	000000
100	v1 = 1 (101100) + 0 (110010) + 0 (011001)	101100
010	v2 = 0 (101100) + 1 (110010) + 0 (011001)	110010
110	v3 = 1 (101100) + 1 (110010) + 0 (011001)	011110
001	v4 = 0 (101100) + 0 (110010) + 1 (011001)	011001
101	v5 = 1 (101100) + 0 (110010) + 1 (011001)	110101
011	v6 = 0 (101100) + 1 (110010) + 1 (011001)	101011
111	v7 = 1 (101100) + 1 (110010) + 1 (011001)	000111

structure of (3.2) and all their coded bits are computed as linear combinations of information bits are known as nonsystematic.

Thus, the generator matrix $\mathbf{G}$ of a systematic code can be written in canonical form as

$$
\mathbf{G} = \left[\mathbf{P}_{k \times (n-k)} \middle| \mathbf{I}_{k \times k} \right]
$$

$$
= \begin{bmatrix}
p_{00} & p_{01} & p_{02} & \cdots & p_{0,n-k-1} & 1 & 0 & \cdots & 0 \\
p_{10} & p_{11} & p_{12} & \cdots & p_{1,n-k-1} & 0 & 1 & \cdots & 0 \\
\vdots & \vdots & \vdots & \vdots & \vdots & \vdots & \vdots & \vdots & \vdots \\
p_{k-1,0} & p_{k-1,1} & p_{k-1,2} & \cdots & p_{k-1,n-k-1} & 0 & 0 & \cdots & 1
\end{bmatrix}.
\tag{3.5}
$$

Remark 3.3. *In fact, the generator matrix* $\mathbf{G} = \left[\mathbf{I}_{k \times k} \middle| \mathbf{P}_{k \times (n-k)} \right]$ *is also in canonical form.*

3.1.1   Parity check matrix

At the receiver side, the received coded vector may have been contaminated by some noise in the communication channel. Thus, it is the decoder's task to check if the received vector is a valid codeword, and if not, to try to correct it. This task is performed by the parity-check matrix $\mathbf{H}$, which is orthogonal to the generator matrix $\mathbf{G}$ [2,15], i.e.,

$$
\mathbf{GH}^T = \mathbf{0}.
\tag{3.6}
$$

The matrix $\mathbf{H}$, shown below, belongs to the null space of $\mathbf{G}$ and has dimensions $(n-k) \times n$. In fact, matrix $\mathbf{H}$ can be seen as a generator matrix for a linear block code, whose codewords are all orthogonal to the each codeword generated by $\mathbf{G}$. This code generated by $\mathbf{H}$ is known as *dual code* with respect to $\mathbf{G}$.

$$
\mathbf{H} = \begin{bmatrix}
h_{00} & h_{01} & h_{02} & \cdots & h_{0,n-1} \\
h_{10} & h_{11} & h_{12} & \cdots & h_{1,n-1} \\
\vdots & \vdots & \vdots & \vdots & \vdots \\
h_{n-k-1,0} & h_{n-k-1,1} & h_{n-k-1,2} & \cdots & h_{n-k-1,n-1}
\end{bmatrix}
\tag{3.7}
$$

If the generator matrix is in canonical form (3.5), then the parity-check matrix is given by

$$\mathbf{H} = \left[\, \mathbf{I}_{\times(n-k)} \, \middle| \, \mathbf{P}^T \, \right] \tag{3.8}$$

For the generator matrix in (3.2), the matrix $\mathbf{H}$ can be written as

$$\mathbf{H} = \begin{bmatrix} 1 & 0 & 0 & 1 & 1 & 0 \\ 0 & 1 & 0 & 0 & 1 & 1 \\ 0 & 0 & 1 & 1 & 0 & 1 \end{bmatrix} \tag{3.9}$$

To verify that a given received block is a valid codeword, one can just use the orthogonality between $\mathbf{G}$ and $\mathbf{H}$. By projecting the received block (which can be seen as a vector in an n-dimensional space) onto the space spanned by $\mathbf{H}$, a valid codeword will produce a null vector. Otherwise, a nonzero bit sequence is obtained. Mathematically, this is written as

$$\mathbf{v}\mathbf{H}^T = \mathbf{0}. \tag{3.10}$$

As an example, consider the sequence $\mathbf{v} = (101011)$. By projecting $\mathbf{v}$ onto the space spanned by $\mathbf{H}$, we have

$$\mathbf{v}\mathbf{H}^T = (101011) \begin{bmatrix} 1 & 0 & 0 \\ 0 & 1 & 0 \\ 0 & 0 & 1 \\ 1 & 0 & 1 \\ 1 & 1 & 0 \\ 0 & 1 & 1 \end{bmatrix} \tag{3.11}$$

$$\mathbf{v}\mathbf{H}^T = 1(100) + 0(010) + 1(001) + 0(101) + 1(110) + 1(011) \tag{3.12}$$

$$\mathbf{v}\mathbf{H}^T = (000) \tag{3.13}$$

Hence, $\mathbf{v} = (101011)$ is a valid codeword, as can be seen in Table 3.1.

3.1.1.1   Syndrome computation and error detection

Let $\mathbf{r} = (r_1, r_2, \ldots, r_n)$ be a received vector from a noisy channel, i.e.,

$$\mathbf{r} = \mathbf{v} + \mathbf{e}, \tag{3.14}$$

where $\mathbf{e}$ is an error vector. The syndrome of a received block is a vector with $(n - k)$ bits defined as

$$\mathbf{S} = \mathbf{r}\mathbf{H}^T. \tag{3.15}$$

In an algebraic perspective, the syndrome is just the projection of the received vector onto to space spanned by $\mathbf{H}$, which in its turn is orthogonal to the space spanned by $\mathbf{G}$. Therefore, the syndrome can be used to detect if an error has occurred during transmission.

By using (3.14) into (3.15) , the syndrome can be written as

$$\mathbf{S} = (\mathbf{v} + \mathbf{e})\mathbf{H}^T = \mathbf{v}\mathbf{H}^T + \mathbf{e}\mathbf{H}^T, \tag{3.16}$$

and since $\mathbf{v}\mathbf{H}^{\mathrm{T}} = \mathbf{0}$, the syndrome can also be expressed as

$$S = eH^{\mathrm{T}}. \tag{3.17}$$

This equation shows that the syndrome of a received block depends only on the error vector. Thus, if the syndromes for all correctable error patterns are *a priori* known, the original message can be recovered by comparing these syndromes with that computed from the received block $\mathbf{r}$. This method of decoding is known as syndrome decoding.

For instance, suppose the received block is $\mathbf{r} = (000001)$. The syndrome for this block is given by

$$S = rH^T = (000001) \begin{bmatrix} 1 & 0 & 0 \\ 0 & 1 & 0 \\ 0 & 0 & 1 \\ 1 & 0 & 0 \\ 1 & 1 & 0 \\ 0 & 1 & 1 \end{bmatrix}, \tag{3.18}$$

resulting in

$$S = rH^T = 0(100) + 0(010) + 0(001) + 0(101) + 0(110) + 1(011) \tag{3.19}$$

$$S = 011, \tag{3.20}$$

demonstrating that there is an error in this block (specifically in the last bit). More details on linear block codes and syndrome decoding can be found in [2,15].

Once the basic concepts of block codes were presented, the next subsection is devoted to LDPC codes, which are some of the most important block codes in use nowadays.

3.1.2 Low-density parity-check

The LDPC is a linear block code known for its performance, very close to the Shannon limit for the AWGN channel [2,15]. It is characterized by providing a matrix $\mathbf{H}$ containing a small number of 1 compared to the number of 0 (around 1% of the entries of $\mathbf{H}$ are 1s). This matrix $\mathbf{H}$ can be called a sparse or low-density matrix. If all rows and columns have the same number of ones, we have a regular code. Otherwise, we have an irregular code.

LDPC codes were proposed by Gallager in 1961 [16], but because of the high computational effort required for their implementation, they were forgotten for years, until 1981, when Tanner [17] generalized LDPC codes and introduced a graphical representation, facilitating their implementation. This graphical representation has come to be known as Tanner Graph. Subsequently, in the 1990s, LPDC studies were resumed by [18].

3.1.2.1 Tanner graphs

Tanner in 1981 introduced the now-known Tanner Graph [17]. Basically it is a graphical representation that can be used for any block codes, then can also be used for LDPCs.

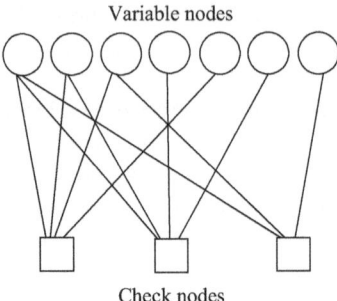

Figure 3.3   Tanner graph

Tanner Graph has 2 sets of nodes: on one side we have the variable nodes and on the other, the check (parity) nodes. We have n variables nodes and $(n - k)$ check nodes. In a Tanner graph, a variable node is connected to a parity node whenever the element ij of the matrix $\mathbf{H}$ equals 1, i.e., when $h_{i,j} = 1$.

For example, consider that for an LDPC code, the parity-check matrix $\mathbf{H}$ is given by

$$\mathbf{H} = \begin{bmatrix} 1 & 0 & 0 & 1 & 1 & 0 & 1 \\ 0 & 1 & 0 & 1 & 0 & 1 & 1 \\ 0 & 0 & 1 & 0 & 1 & 1 & 1 \end{bmatrix} \tag{3.21}$$

The Tanner graph associated to the matrix $\mathbf{H}$ is represented in Figure 3.3.

Through the Tanner graph, it is possible to propose algorithms for iterative decoding of LDPC based on the exchange of messages between nodes. These methods use probabilistic analysis and will be briefly seen in the next section.

3.1.2.2   LDPC decoding

The LDPC decoding algorithms are iterative algorithms based on message exchanges between Tanner's graphs. These algorithms use estimation methods based on probability, such as the maximum a posteriori probability (MAP) estimator [2,15].

To make the correction, the LDPC uses a posteriori likelihoods or log likelihood ratios (LLRs). For the first iteration, the algorithm accepts as input the probability values of each received bit, which are known as a priori probability. Then, the LLR below is computed iteratively, where

$$l_i = \ln \frac{Pr(b_i = 0|\mathbf{r})}{Pr(b_i = 1|\mathbf{r})}, \tag{3.22}$$

$Pr(b_i)$ is the probability of a bit i and $\mathbf{r}$ is the received sequence.

After a number of iterations, a decision is taken based on the LLR values. For a value of LLR above a threshold, a bit 1 is chosen and below that threshold, one chooses bit 0.

There are many algorithms for decoding LDPC Codes, being the sum-product and belief-propagation algorithms the most known [2,15].

3.1.3 Reed–Solomon codes

Besides linear block codes and LDPC codes, briefly described in previous sections, there are some other subclasses of block codes. Among them, one of the most important and widely used in practice is that of Reed–Solomon (RS) codes, which was discovered in 1960 [2,15]. They are typically encountered in digital communications and storage systems, with applications ranging from CDs, DVDs and barcodes to wireless, satellite, deep-space and PLC communications [2,5,15]. The success of RS codes comes from their ability to correct multiple errors in a data block, especially when these errors occur in bursts, and to the existence of efficient coding and decoding algorithms.

Differently from the binary linear block codes shown in Section 3.1, whose messages and codewords are vectors of binary symbols, RS codes are q-ary codes. This means that each element of message and codeword vectors comes from an alphabet of q symbols. Hence, for a message of k information symbols, a q-ary (n, k) RS code consists of a set of q^k codewords of length n.

Since an RS code is a linear block code, all concepts and properties developed for binary codes in the previous sections of this chapter could still be applied with few modifications. Particularly, all operations (additions and multiplications) must now be carried out modulo-q. Also, in all practical applications of RS codes, q is chosen as a power of 2, i.e., $q = 2^s$, where s represents the number of bits that is grouped into a symbol on the q-ary alphabet.

Example 3.1. *In optical communications, data-storage systems, and hard-disk drives, it is usual to employ the* $(255, 239)$ *RS code, with* $s = 8$. *This code is constructed over a 256-symbols alphabet and is capable of correcting eight or fewer symbol errors. Another important RS code is the* $(255, 233)$ *code, which can correct up to 16 symbol errors and is used by NASA as a standard code for deep-space and satellite communications.*

One of the remarkable properties of RS codes is that their error-correction capability is tightly related to the number of parity symbols added by the encoder. Specifically, for a q-ary (n, k) RS code, the maximum number of symbol errors that can be corrected is given by $t = \frac{n-k}{2}$. In other words, the number of redundant symbols added by the encoder of an RS code is twice the maximum number of symbol errors that it can correct. Also, for this code, $n = q - 1$ and $k = q - 1 - 2t$ [2,15].

In general, to correct a symbol error in a received sequence, the decoder must be able to find not only the position of this error in the block but also its value. This last task is equivalent to determine which of the q symbols in the code alphabet must replace the erroneous symbol. As the RS encoder adds two redundant symbols for each correctable symbol error, it is possible to intuitively associate one of these parity symbols to the location of an error (i.e., detection), and the other, to the determination of the error value.

Remark 3.4. *For binary linear block codes, this general decoding process is easier, since each symbol can assume only one of two possible values. Thus, once the location of the error is found, the decoder must simply flip the bit in that position to correct it.*

Although syndrome decoding, as presented previously, could be used to decode RS codes, this is not the most efficient decoder, due to both the RS codewords' lengths and their multiple error-correction capability. Fortunately, there are efficient algorithms to perform RS decoding, such as the Berlekamp–Massey and the Euclidean algorithms, whose details can be found in [2,15] and references therein.

Finally, it is also worth noting that the ability of RS codes to deal with burst errors is due to their nonbinary nature. In fact, during decoding, an incorrect symbol is replaced with a correct one, whether the corruption was found in 1 bit or in all bits of that symbol. Therefore, the decoding of a received sequence is based on replacing an incorrect symbol, regardless how many bits within that symbol have been corrupted, which provides robustness against bursty noise.

Remark 3.5. *RS codes are often concatenated with convolutional codes, which will be described in the sequel, to repair any convolutional decoding errors (which typically occur in bursts).*

A more detailed presentation of RS codes, including their algebraic structure, coding and decoding algorithms, can be found in [2,15].

3.2   Convolutional codes

Convolutional codes have been widely used in a variety of applications, such as wireless and satellite communications. Their popularity comes from their simple encoder structure and the possibility of practical implementation of maximum likelihood (ML) decoding methods [2,4]. As stated in the beginning of this chapter, one of the differences between convolutional and block codes is the presence of memory in the encoder. Another difference is that convolutional codes accept as inputs not only message blocks but also running sequences of information bits.

As with block codes, the encoder of a convolutional code with k inputs and n outputs, with $n > k$, produces a coded bit sequence known as *codeword*. The smallest Hamming distance[3] between any two finite-length codewords of a given convolutional code is known as *free distance* or d_{free} [2]. In general, the higher the free distance, the better the ability of the code to detect and/or correct errors that may occur during the transmission. The ratio $R = k/n$ is called coding rate and indicates the amount of redundancy introduced into the message.

Due to the presence of memory elements in the encoder, the outputs depend not only on the inputs at a certain time instant but also on the inputs in μ previous instants. Therefore, the encoder has a memory of order μ. For a fixed rate R, more

---

[3]The Hamming distance is defined as the number of bits in which two binary sequences differ.

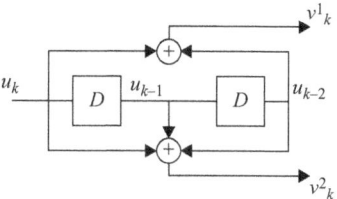

Figure 3.4   A simple convolutional encoder

robust codes can be constructed, up to a certain limit, by increasing the number of memory elements [2]. An example of a rate $R = 1/2$ convolutional encoder can be seen in Figure 3.4, where the blocks marked with D are the memory elements and the operator $\oplus$ indicates modulo-2 addition. In Figure 3.4, the output bit v^1 at instant k, v_k^1, is computed by modulo-2 addition between input bits u_k and u_{k-2}, while output v^2 is formed by summing u_k, u_{k-1} and u_{k-2}.

Example 3.2. *Suppose the message* $\mathbf{u} = (1\,0\,0\,1\,0)$ *has to be encoded by the encoder shown in Figure 3.4. Considering that the memory elements are set to an initial value of zero, one has at the first time instant* $u_k = 1$ *(i.e., the first bit of the message) and* $u_{k-1} = u_{k-2} = 0$. *This produces as outputs* $v^1 = 1$ *and* $v^2 = 1$.

In the next time instant, the inputs of the memory elements, which can be implemented by shift registers, are shifted to their outputs and the second information bit, which is a 0, arrives. Therefore, the outputs are given by $v^1 = u_k \oplus u_{k-2} = 0 \oplus 0 = 0$ *and* $v^2 = u_k \oplus u_{k-1} \oplus u_{k-2} = 0 \oplus 1 \oplus 0 = 1$. *In the third time instant, another message bit arrives and the memory elements shift their inputs, producing* $v^1 = 0 \oplus 1 = 1$ *and* $v^2 = 0 \oplus 0 \oplus 1 = 1$. *Continuing this procedure up to the end of the message, outputs* $v^1 = (1\,0\,1\,1\,0)$ *and* $v^2 = (1\,1\,1\,1\,1)$ *are generated.*

Finally, to obtain the encoder output, $\mathbf{v}^1$ *and* $\mathbf{v}^2$ *must be multiplexed in time, such as* $\mathbf{v} = (v_0^1\,v_0^2\,v_1^1\,v_1^2\cdots)$, *where* v_j^i *represents the encoder output* v^i *at time j. Hence, the codeword for the message* $\mathbf{u} = (1\,0\,0\,1\,0)$ *is given by* $\mathbf{v} = (1\,1\,0\,1\,1\,1\,1\,1\,0\,1)$.

Since modulo-2 addition is a linear operation, the convolutional encoder is a linear system. Consequently, each output sequence ($\mathbf{v}^1$ and $\mathbf{v}^2$ in Figure 3.4) is computed by the convolution[4] between the input sequence $\mathbf{u}$ and their respective "impulse responses." These impulse responses are obtained by observing the coded bits generated by the input $\mathbf{u} = (1\,0\,0\,\cdots)$. For the encoder in Figure 3.4, the impulse responses are $\mathbf{g}_1 = (1\,0\,1)$ and $\mathbf{g}_2 = (1\,1\,1)$, or $\mathbf{g}_1 = 5$ and $\mathbf{g}_2 = 7$ in octal.

Taking linearity into account, the discrete-time convolution could be replaced, as in the Z-transform, by a product in a transformed domain. Then, by representing the input sequence as a polynomial of the form $\mathbf{u}(D) = u_0 + u_1 D + u_2 D^2 + \cdots$ and the impulse response as $\mathbf{g}(D) = g_0 + g_1 D + \cdots + g_\mu D^\mu$, the output sequence is given

---

[4]Hence the name *convolutional* codes.

by $\mathbf{v}(D) = \mathbf{u}(D)\mathbf{g}(D)$. In these expressions, D may be interpreted as a delay operator and the powers of D denote the number of time units a bit is delayed with respect to the first bit of the message. Thus, for the encoder in Figure 3.4, it is possible to write $\mathbf{g_1}(D) = 1 + D^2$ and $\mathbf{g_2}(D) = 1 + D + D^2$. These polynomials provide information on the structure of the code and are known as *generator polynomials*.

As convolutional encoders are sequential circuits, their operation can be well described by state diagrams, where the states are defined by the contents of the memory elements. Each of the 2^μ states has 2^k input branches and 2^k output branches, being k the number of inputs to the encoder.

Example 3.3. *The encoder in Figure 3.4 has two memory elements ($\mu = 2$) and one input ($k = 1$). Hence, it has $2^\mu = 2^2 = 4$ states and $2^k = 2^1 = 2$ branches entering and leaving each state. The states correspond to the possible combinations of the outputs of the memory elements. In this way, the state S_0 can be associated with both memory outputs in 0, the state S_1 with the first memory element in 0 and the second, in 1, and so on for the other states.*

By looking at the encoder in Figure 3.4 and by considering the memory elements are set to 0, it is easy to see that when a bit 0 arrives at the input, both outputs will be 0 and the next state will be again S_0. On the other hand, if a bit 1 arrives, both outputs will be one and the next state will be S_2.

Suppose now the encoder is at state S_1. For an incoming bit of 0, the state at the next time instant will be S_0 and $v^1 = v^2 = 1$. However, if the incoming bit is 1, $v^1 = v^2 = 0$ and the next state will be S_2. To obtain the complete state diagram, this procedure must be repeated to all states.

Hence, the state diagram of the convolutional code represented in Figure 3.4 is shown in Figure 3.5. In this diagram, the arrows indicate transitions between two states, while the first symbol of the set u/v^1v^2 defines the input (information) bit and the next two symbols, the coded output.

Remark 3.6. *It is important to observe that the state diagram can also be used to encode a message. In fact, a message corresponds to a path through the state diagram. Consider, for instance, the message $\mathbf{u} = (1\ 0\ 0\ 1\ 0)$ and the state diagram in Figure 3.5. By starting at S_0, the sequence of states reached by $\mathbf{u}$ is $S_0 \rightarrow S_2 \rightarrow S_1 \rightarrow S_0 \rightarrow S_2 \rightarrow S_1$ and the codeword is $\mathbf{v} = (1\ 1\ 0\ 1\ 1\ 1\ 1\ 0\ 1)$, which is exactly the same as that of Example 3.2.*

As can be readily seen from Figure 3.5, it is possible to determine the current state by knowing only the previous state and the input bit that caused the transition between these states. Therefore, the encoding process can be seen as a *Markov process* and its temporal evolution represented by a trellis diagram, such as that of Figure 3.6, which corresponds to a message block of length 3 coded by the encoder in Figure 3.4 followed by two more bits that force the encoder to initial state S_0.[5]

---

[5]The bits that force the encoder back to the initial state are called *tail bits*. In general, μ tail bits are appended to the end of the message.

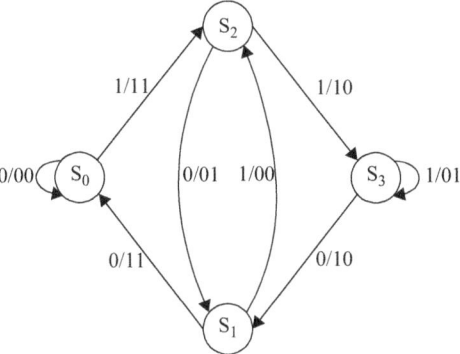

Figure 3.5   State diagram for the code (5,7)

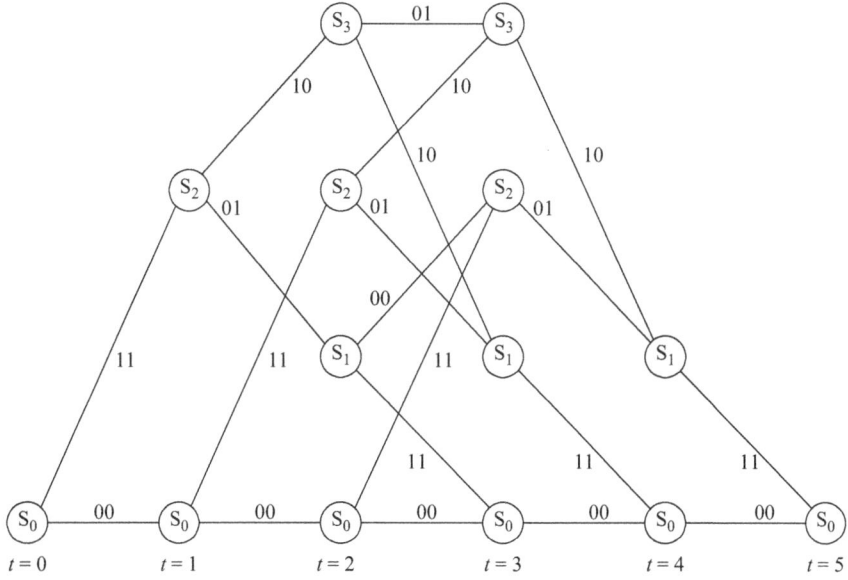

Figure 3.6   Trellis diagram for code (5,7) supposing a message composed by 3 information bits and 2 tail bits

In a trellis, the branches leaving a state are associated with each possible input to the encoder. Thus, for the code with generators (5,7) shown so far, the lower branch leaving a state is associated with the information bit "0," while the upper branch is associated with the input bit "1." The labels of each output branch show the coded bits generated by the respective information bit. In addition, each codeword is represented by a unique path through the trellis, that is, each codeword is obtained through concatenation of the encoder output bits at each state transition.

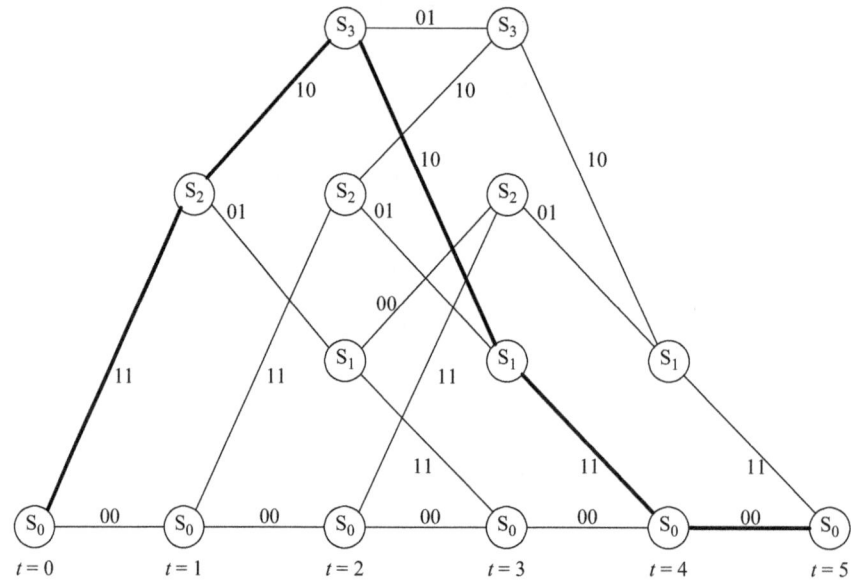

Figure 3.7   Path associated to the message $\mathbf{u} = (1\ 1\ 0\ 0\ 0)$

Example 3.4. *Suppose the message* (1 1 0) *has to be encoded by the code (5,7), whose trellis diagram is shown in Figure 3.6. Before starting the encoding process, $\mu = 2$ tail bits are appended to* **u**, *generating the "new" message* **u** $= (1\ 1\ 0\ 0\ 0)$. *Then, considering S_0 as the initial state in $t = 0$, the first incoming bit (bit 1) triggers the state transition to S_2 and the encoder outputs $v^1 = 1$ and $v^2 = 1$. When the second bit 1 arrives at the encoder input, at $t = 1$, the encoder produce as coded outputs $v^1 = 1$ and $v^2 = 0$. In this case, the next state at $t = 2$ can only be S_3. The third information bit (which is a bit 0), in its turn, causes the transition from S_3 to S_1, generating $v^1 = 1$ and $v^2 = 0$. Finally, the two tail bits force the encoder back to S_0, causing the transitions from S_1 to S_0 at $t = 4$ and from S_0 to S_0 at $t = 5$. The resulting codeword is obtained by concatenating the coded outputs at each trellis branch and is given by* **v** $= (1\ 1\ 1\ 0\ 1\ 0\ 1\ 1\ 0\ 0)$. *Figure 3.7 shows the path of the message* **u** $= (1\ 1\ 0\ 0\ 0)$ *through the trellis.*

Depending on the way the codewords are computed from the input message bits, the convolutional codes can be classified as systematic or nonsystematic, recursive or nonrecursive. In a systematic code, k outputs are replicas of the k inputs (information bits). The other outputs of the encoder correspond to the *parity bits*. In a recursive systematic code, the parity bits are generated through a feedback loop. In nonsystematic codes, none of the outputs is exactly the same as any of the inputs. Figure 3.8 illustrates the encoder of a systematic convolutional code with generators $\mathbf{g}_1 = 20$ and $\mathbf{g}_2 = 37$ (in octal), while Figure 3.9 shows the encoder of the recursive systematic

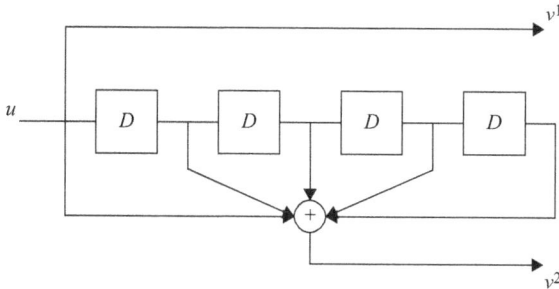

Figure 3.8 Systematic convolutional code (20,37)

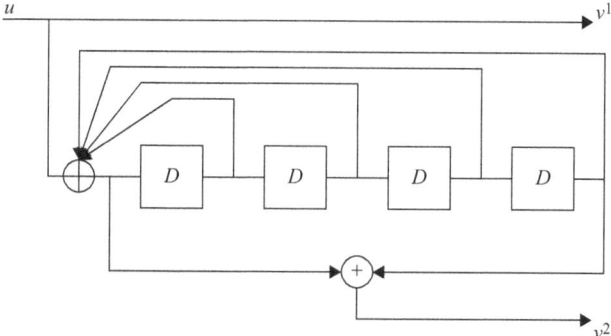

Figure 3.9 Recursive systematic convolutional code (37,21)

convolutional code with $\mathbf{g}_1 = 37$ and $\mathbf{g}_2 = 21$. Finally, a non-systematic convolutional code with $\mathbf{g}_1 = 37$ and $\mathbf{g}_2 = 21$ can be seen in Figure 3.10. It is worth noting that the recursive systematic and the nonsystematic codes shown in Figures 3.9 and 3.10, respectively, have the same generator polynomials, that is, they can be described by the same trellis diagram, differing only in the input and output bits associated with each branch of the trellis.

To apply channel-coding techniques in practical communication systems, the receiver must be able to recover unambiguously the original message from the encoded information, i.e., there must be a one-to-one correspondence, given by the generator polynomials, between message bits and codewords. Hence, the task of the decoder is to produce an estimate of the original message based on the received noisy signal. In other words, the decoder must find the path of the trellis that generated the codeword sent through the channel.

A *decoding rule* is a strategy for choosing a codeword for each received sequence. A decoding error occurs when the estimated message differs from the original one. A decoding rule frequently used in the decoding of convolutional codes is the one

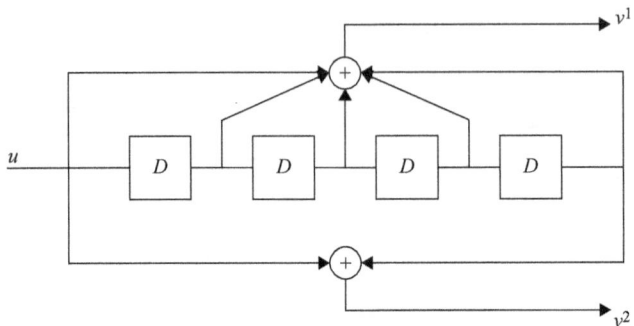

Figure 3.10   Nonsystematic convolutional code (37,21)

that minimizes the sequence error probability, i.e., $P(\hat{\mathbf{v}} \neq \mathbf{v})$. It can be shown [2–4] that the minimization of $P(\hat{\mathbf{v}} \neq \mathbf{v})$ is accomplished by choosing the codewords $\hat{\mathbf{v}}$ that maximize the *a posteriori* probability of $\mathbf{v}$, that is,

$$\hat{\mathbf{v}} = \arg \max_{\mathbf{v}} P(\mathbf{v}|\mathbf{r}), \tag{3.23}$$

where $\mathbf{r}$ is the noisy received sequence. This decoding rule is known as MAP.

By using the well-known Bayes rule, the *a posteriori* probability $P(\mathbf{v}|\mathbf{r})$ can be written as

$$P(\mathbf{v}|\mathbf{r}) = \frac{p(\mathbf{r}|\mathbf{v})P(\mathbf{v})}{P(\mathbf{r})}. \tag{3.24}$$

In this equation, $P(\mathbf{v})$ is the *a priori* probability of codeword $\mathbf{v}$. Since $P(\mathbf{r})$ does not depend on $\mathbf{v}$, (3.23) can be rewritten as

$$\hat{\mathbf{v}} = \arg \max_{\mathbf{v}} p(\mathbf{r}|\mathbf{v})P(\mathbf{v}). \tag{3.25}$$

The function $p(\mathbf{r}|\mathbf{v})$ is known as likelihood function. By comparing (3.25) and (3.23), one can realize that a decoder that maximizes the likelihood is equivalent to the MAP decoder when the codewords are equally likely. In many systems, the probability of occurrence of each codeword is not previously known by the receiver, which makes the MAP decoding impossible. In these cases, the ML decoding is usually the best available strategy. In practice, ML decoding is performed by the Viterbi algorithm [2–4], whose implementation is simpler than that of the MAP decoder.

Basically, the Viterbi algorithm is an efficient way to find the path through the trellis that is most likely to the received bit sequence. It does that by moving sequentially through the trellis and comparing a metric, such as the Hamming distance, between the received sequence and all paths entering a state. When two or more paths enter the same state, the one having the smallest Hamming distance is chosen, while the others are removed from the possible candidate paths. The chosen path is known as the *survival*. This selection of surviving paths is performed for all states until the

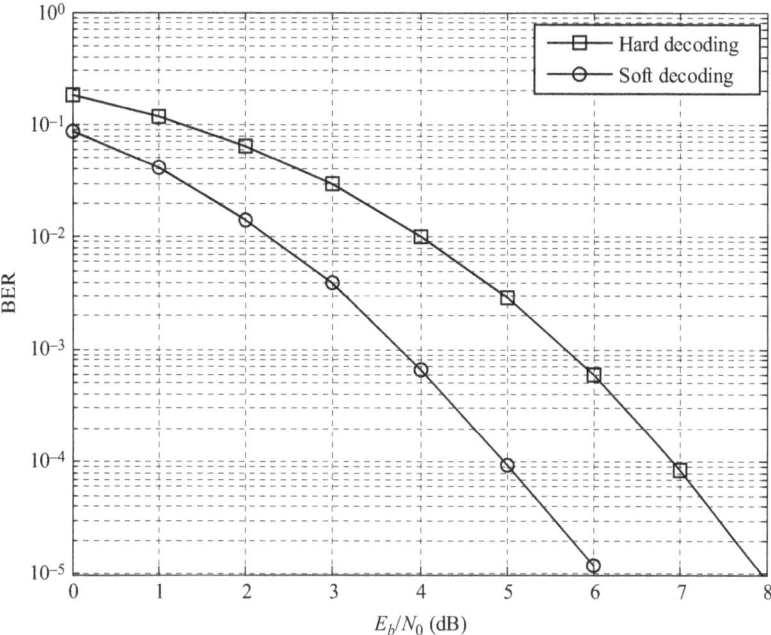

*Figure 3.11 Performance comparison between hard- and soft decoding for the
code of Figure 3.4*

end of the trellis, when the path with the smaller distance is chosen as the decoded
sequence. More details on the Viterbi algorithm can be found in [2–4].

Another point in most digital communication systems using binary encoding is
that the received signal is generally quantized before the decoding process. Thus, the
decoder input is composed of a sequence of "0"s and "1"s, and the device performs
the so-called *Hard Decoding*. On the other hand, when the decoder input has more
than two levels or it is not quantized at all, the decoder performs a *Soft Decoding*.
Although hard decoding is simpler to implement, its performance is not as good as
that of the soft decoding, since hard decoding does not use all the information available in the received signal. This is evident from Figure 3.11, which shows the bit error
rate (BER) performances obtained by simulation of communication systems transmitting blocks of 150 information bits a through an AWGN (Additive White Gaussian
Noise) channel. These information bits were coded by the rate $R = 1/2$ convolutional
code of Figure 3.4 using both hard- and soft-decoding at the receiver side, which used
the Viterbi algorithm. This results match the ones presented in [3], indicating there is
a 2 dB difference, in terms of BER, between hard- and soft-decoding.

The properties and characteristics presented thus far are valid for any convolutional code, including the *turbo codes*, which are some of the most powerful codes,

in terms of error correction capability, in use in current communication systems. The next section is then devoted to provide an overview on turbo codes.

3.3 Turbo codes

In 1993, a new class of convolutional codes, whose performance approaches the Shannon limit, was presented to the scientific community [19]. These new codes, called turbo codes, were one of the major advances in channel-coding theory in the last decades, evolving rapidly due to their performances and their unique combination of some new ideas with "forgotten" concepts and algorithms.

As will be seen in the sequel, it is the iterative decoding procedure, which has some similarities with the turbo engines working cycles, that gives the name to the codes. The first encoder of a turbo code [19] was composed by the parallel concatenation of two identical recursive systematic convolutional codes separated by an interleaver, as is illustrated in Figure 3.12. However, the recursive systematic convolutional codes can also be serially concatenated [20], as indicated in Figure 3.13.

The interleaver is of fundamental importance in the turbo codes structure. During the encoding step, the interleaving reduces the number of codewords with low Hamming weights, i.e., the number of words near to d_{free} [20]. In addition, the permutation of the bits that form a codeword allows to "create" long codes from short convolutional codes, that is, from codes with few memory elements. This is important since only long codes may approach the Shannon limit [21].

The decoding step, in its turn, is carried out by the serial concatenation of two decoders, each of which is associated with one of the encoders. As the message

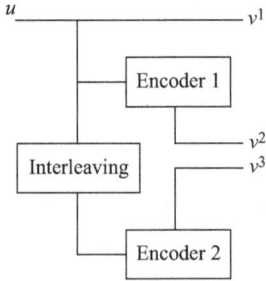

Figure 3.12   Turbo code with parallel concatenation

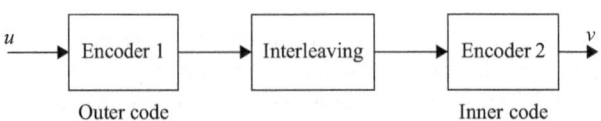

Figure 3.13   Turbo code with serial concatenation

is encoded by more than one encoder prior to transmission, the receiver can refine its estimates by iteratively exchanging information between the decoders. This means that the same received sequence passes many times through the decoding processes before a final decision on the message can be taken. Figure 3.14 illustrates the iterative decoder for a parallel concatenated encoder, while Figure 3.15 illustrates the decoder for a serial concatenated encoder.

As mentioned in Section 3.2, soft decoding has a better performance than hard decoding. Therefore, the two decoders of a turbo decoder must be able to exchange soft information between them. This soft information (also known as soft decisions) is usually expressed by means of the logarithm of the likelihood ratio (LLR), defined as follows:

$$L(u_k) \triangleq \ln \frac{P(u_k = 1|\mathbf{r})}{P(u_k = 0|\mathbf{r})}, \tag{3.26}$$

where u_k is the message bit at instant k, and $\mathbf{r}$ is the noisy received sequence. It is important to notice that the hard decisions taken from $L(u_k)$ provide the MAP estimation of the message bits, while the absolute value of $L(u_k)$ indicates the reliability that the decision taken is correct [21,22].

As can be seen in Figures 3.14 and 3.15, the output of one decoder is used by the other one as an *a priori* information on the received signal. This information, also known as *extrinsic* information, corresponds to an incremental information on bit u_k obtained by the decoding process of all bits of the data block, except the systematic bit received at instant k. In other words, each decoder jointly processes the received bits around the systematic bit u_k and the *a priori* information on u_k to generate a new LLR over u_k, taking into account the restrictions imposed by the code structure.

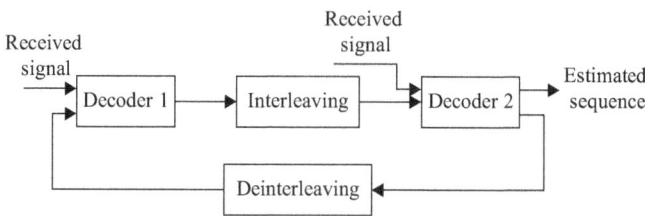

Figure 3.14   Iterative receiver for a parallel concatenated turbo code

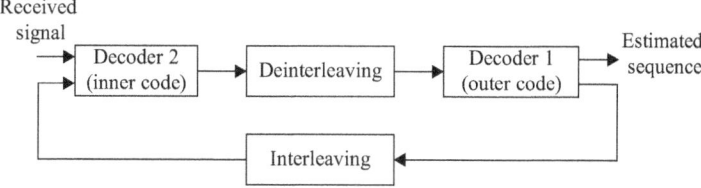

Figure 3.15   Iterative receiver for a serial concatenated turbo code

One of the keys elements to the excellent performance of the iterative decoding performed by turbo codes is the fact that the decoders only exchange extrinsic information between them. This ensures that each decoder use as *a priori* information just the values that has not been generated directly by itself in the previous iteration, preventing a positive feedback, which could cause numerical instability to the turbo decoder. It is still possible to show [15,19,21–23], from (3.26) and the Bayes rule, that the LLR at the output of each decoder can be written as the sum of three terms: one originating from the channel (received sequence), another from the *a priori* bit probabilities and the last one corresponding to the extrinsic information generated during the previous decoding step. Hence, once the full LLR's $L(u)$ are computed, the extrinsic information can be obtained simply by subtracting from $L(u)$ the terms corresponding to the channel and to the *a priori* probabilities.

Decoding algorithms that accept as input estimates of the *a priori* probability of message bits and compute new estimates of these probabilities given the reception of the entire sequence are called soft-input, soft-output decoders. Examples of such decoders are the BCJR-MAP [2,15,19,24] and SOVA (*soft-output Viterbi algorithm*) [2,15,25]. One difference between them lies in the fact that the states estimated by SOVA form a feasible path through the code trellis, while the states estimated by the BCJR algorithm are not, necessarily, connected.

Given the importance of turbo codes for PLC systems, as could be seen in [7,10, 11] and in other chapters of this book, the MAP-decoding algorithm usually employed in turbo decoders will be briefly described in the sequel. It is worth noting that this algorithm minimizes the bit error probability, while the Viterbi decoder minimizes the sequence error probability.

The symbol-to-symbol MAP algorithm, better known as BCJR (BCJR stands for Bahl, Cocke Jelinek and Raviv, which are the authors who proposed this decoding method) [24], calculates the *a posteriori* probability for each state transition as well as for the message bits of a Markov process given a noisy observed sequence and then computes the LLR (3.26), as [2,15,19,24]

$$L(u_k) = \ln \frac{P(u_k = 1|\mathbf{r})}{P(u_k = 0|\mathbf{r})} = \ln \frac{\sum_{\{(s',s):u_k=1\}} p(s',s,\mathbf{r})}{\sum_{\{(s',s):u_k=0\}} p(s',s,\mathbf{r})}, \tag{3.27}$$

where s' and s represent the trellis states at instants $k-1$ and k, respectively. The sums in the numerator and in the denominator are performed for all existing transitions between states s' and s triggered by $u_k = 1$ and $u_k = 0$, respectively.

By using the properties of a Markov process, it is possible to show [2,15,24] that the joint probabilities $p(s',s,\mathbf{r})$ can be written as the product of three independent terms, i.e.,

$$\begin{aligned} p(s',s,\mathbf{r}) &= p(s',\mathbf{r}_{j<k}) \cdot p(s,\mathbf{r}|s') \cdot p(\mathbf{r}_{j>k}|s) \\ &= \underbrace{p(s',\mathbf{r}_{j<k})}_{\alpha_{k-1}(s')} \cdot \underbrace{P(s|s') \cdot p(r_k|s',s)}_{\gamma_k(s',s)} \cdot \underbrace{p(\mathbf{r}_{j>k}|s)}_{\beta_k(s)} \cdot \end{aligned} \tag{3.28}$$

In this equation, $\mathbf{r}_{j<k}$ represents the sequence of symbols received from the beginning of the block to instant $k-1$, while $\mathbf{r}_{j>k}$ is the sequence between time instant $k+1$ and the end of the trellis.

The terms $\alpha_k(s)$ and $\beta_k(s)$ can be computed recursively as

$$\alpha_k(s) = \sum_{s'} \gamma_k(s', s) \cdot \alpha_{k-1}(s'), \tag{3.29}$$

$$\beta_{k-1}(s') = \sum_{s} \gamma_k(s', s) \cdot \beta_k(s). \tag{3.30}$$

Considering a finite trellis starting and ending in the zero state, (S_0), $\alpha_k(s)$ and $\beta_k(s)$ are initialized as follows:

$$\alpha_{\text{beginning}}(0) = 1 \quad \text{and} \quad \alpha_{\text{beginning}}(s) = 0, \quad \forall s \neq 0$$
$$\beta_{\text{end}}(0) = 1 \quad \text{and} \quad \beta_{\text{end}}(s) = 0, \quad \forall s \neq 0 \tag{3.31}$$

The BCJR algorithm computes the values of α_k by moving through the code trellis from its beginning to its end (forward recursion). On the other hand, β_k's are obtained by starting at the end of the trellis and going back to its beginning (backward recursion).

The term $\gamma_k(s', s)$ in (3.29) and (3.30) depends on the channel's current output and, as can be seen in (3.28), is expressed in terms of the state transition probability $P(s|s')$ and the conditional probability $p(r_k|s', s)$, where

$$P(s|s') = P(u_k) = \begin{cases} \dfrac{e^{La_k}}{1 + e^{La_k}}, & u_k = 1 \\[3mm] \dfrac{1}{1 + e^{La_k}}, & u_k = 0 \end{cases} \tag{3.32}$$

and

$$p(r_k|s', s) = \prod_{l=1}^{n} \frac{1}{\sigma_n\sqrt{2\pi}} e^{\left(-(1/2\sigma_n^2)(r_{kl} - av_k^l)^2\right)}. \tag{3.33}$$

In these equations, La_k is the *a priori* information generated by the previous decoder, n is the number of coded bits at the output of the encoder at instant k, $v_k^l, l = 1, \ldots, n$, represents the outputs of the encoder originated from transition between states s' and s, $r_{k,l}$ is the received symbol corresponding to v_k^l, σ_n^2 denotes the noise variance and a is the amplitude of the channel fading ($a = 1$ for an AWGN channel).

Remark 3.7. *It is important to highlight that for serially concatenated turbo codes, the outer decoder has no access to the channel outputs and therefore $p(r_k|s', s)$ reduces to a constant. Hence, $\gamma_k(s', s)$ depends only on the* a priori *information.*

The output of the BCJR-MAP decoder can then be rewritten as

$$L(u_k) = \ln \frac{\sum_{\{(s',s):u_k=1\}} \alpha_{k-1}(s') \cdot \gamma_k(s',s) \cdot \beta_k(s)}{\sum_{\{(s',s):u_k=0\}} \alpha_{k-1}(s') \cdot \gamma_k(s',s) \cdot \beta_k(s)}. \qquad (3.34)$$

A more detailed discussion on the BCJR algorithm and its implementation aspects can be found in [2,15,24].

3.4 Final remarks

This chapter briefly described the fundamentals of error-correcting codes, trying to provide an intuitive understanding of how these codes work, instead of an in-depth mathematical formulation and analysis. The interested readers can find invaluable information on this subject in many references, such as [2,15], which cover both classical (block and convolutional codes) and modern (LDPC and turbo codes) channel-coding theories.

References

[1]   Shannon CE. A mathematical theory of communication. Bell System Technical Journal. 1948;27(3):379–423.

[2]   Lin S, Costello DJ. Error Control Coding: Fundamentals and Applications. 2nd ed. Upper Saddle River, NJ: Prentice-Hall; 2004.

[3]   Lee EA, Messerschmit DG. Digital Communication. 2nd ed. New York, NY: Kluwer Academic Press; 1994.

[4]   Vucetic B, Yuan J. Turbo Codes – Principles and Applications. Chichester: Kluwer Academic Publishers; 2000.

[5]   Chuah TC. On Reed–Solomon coding for data communications over power-line channels. IEEE Transactions on Power Delivery. 2009;24(2):614–620.

[6]   Ouahada K. Nonbinary convolutional codes and modified M-FSK detectors for power-line communications channel. Journal of Communications and Networks. 2014;16(3):270–279.

[7]   Wada T. A study on the performance of turbo coding for noise environments in power lines. In: Communications, 2003. ICC'03. IEEE International Conference on. vol. 5; 2003. p. 3071–3075.

[8]   Wada T. A study on performance of LDPC codes on power line communications. In: 2004 IEEE International Conference on Communications; 2004. p. 109–113.

[9]   Andreadou N, Assimakopoulos C, Pavlidou FN. Performance evaluation of LDPC codes on PLC channel compared to other coding schemes. In: 2007 IEEE International Symposium on Power Line Communications and Its Applications; 2007. p. 296–301.

[10]   Prasad G, Latchman HA, Lee Y, *et al.* A comparative performance study of LDPC and Turbo codes for realistic PLC channels. In: 18th IEEE International Symposium on Power Line Communications and Its Applications; 2014. p. 202–207.

[11]   Tseng DF, Tsai TR, Han YS. Robust turbo decoding in impulse noise channels. In: 2013 IEEE 17th International Symposium on Power Line Communications and Its Applications; 2013. p. 230–235.

[12]   Asghari H, Newman RE, Latchman HA. Bandwidth-efficient forward-error-correction-coding for high speed powerline communications. In: 2006 IEEE International Symposium on Power Line Communications and Its Applications; 2006. p. 355–360.

[13]   Al-Hinai N, Hussain ZM, Sadik AZ. Performance of BCH coding on transmission of compressed images over PLC channels. In: 2009 International Conference on Advanced Technologies for Communications; 2009. p. 39–42.

[14]   Vinck AJH. Coding for a terrible channel. In: EU-COST289 2nd Workshop, Special Topics on 4G Technologies; 2005. p. 6–7.

[15]   Ryan W, Lin S. Channel Codes: Classical and Modern. Cambridge: Cambridge University Press; 2009.

[16]   Gallager RG. Low-density parity-check codes. IRE Transactions on Information Theory. 1962;8(1):21–28.

[17]   Tanner R. A recursive approach to low-complexity codes. IEEE Transactions on Information Theory. 1981;27(5):533–547.

[18]   MacKay DJC, Neal RM. Near Shannon limit performance of low-density parity-check codes. Electronics Letters. 1996;32(18):1645.

[19]   Berrou C, Glavieux A, Thitimajshima P. Near Shannon limit error-correcting coding and decoding: turbo-codes. In: Proceedings IEEE International Conference on Communications. ICC'93. vol. 2/3. Geneva, Switzerland; 1993. p. 1064–1070.

[20]   Benedetto S, Divsalar D, Montorsi G, *et al.* Serial concatenation of interleaved codes: analysis, design and iterative decoding. IEEE Transactions on Information Theory. 1998;44(3):909–926.

[21]   Woodard JP, Hanzo L. Comparative study of turbo decoding techniques: an overview. IEEE Transactions on Vehicular Technology. 2000;49(6):2208–2233.

[22]   Hagenauer J. The turbo principle: tutorial introduction and state of the art. In: Proceedings of the International Symposium on Turbo Codes and Related Topics. Brest, France; 1997. p. 1–11.

[23]   Valenti MC. Iterative Detection and Decoding for Wireless Communications [doctoral thesis]. Virginia State University (EUA); 1999.

[24]   Bahl LR, Cocke J, Jelinek F, *et al.* Optimal decoding of linear codes for minimizing symbol error rate. IEEE Transactions on Information Theory. 1974;20(2):284–287.

[25]   Hagenauer J, Hoeher P. A Viterbi algorithm with soft-decision outputs and its applications. In: Proceedings IEEE Global Telecommunications Conference. GLOBECOM'89. vol. 3; 1989. p. 1680–1686.

Chapter 4

Principles of orthogonal frequency division multiplexing and single carrier frequency domain equalisation

Cristiano Panazio[1] and Renato Lopes[2]

This chapter provides fundamental principles of orthogonal frequency division multiplexing and single carrier (SC) modulation schemes, with a focus on frequency-domain equalisation. We begin with basic concepts, followed by the description of the cyclic prefix (CP) technique, which allows for a low-complexity equaliser. We then show how to optimally calculate the equaliser coefficients, and then we present some bit error rate (BER) results that illustrate the performance differences among the possible modulation schemes. Finally, we show some peak-to-average power ratio differences among the two modulation schemes.

4.1 Introduction

With the use of more bandwidth to satisfy the need for higher data rates, the received signal becomes more sensitive to the channel time dispersion, which generates inter-symbol interference (ISI) and can considerably degrade the receiver performance. Such impairment is usually mitigated through channel equalisation at the receiver, but as the distortions increase, so does the complexity of the equalisers. To reduce this complexity, frequency domain equalisation (FDE) has become the technique of choice due to its simple and efficient implementation.

Historically, before becoming an important technique for SC modulation schemes, FDE first became popular with orthogonal frequency domain multiplexing (OFDM) schemes, which were introduced by Chang [1] in 1966. The idea was to transmit data symbols with periods much larger than the channel time dispersion. As a result, each stream would experience a flat frequency response channel, needing only a simple phase and gain correction. In contrast, SC modulations conventionally required a complex time-domain equaliser. Nevertheless, in order to achieve the same data rates as SC modulation, this basic concept would need to transmit multiple

[1]Department of Telecommunications and Control Engineering (PTC), Escola Politécnica of the University of São Paulo (EPUSP), Brazil
[2]Department of Communications (DECOM), University of Campinas (UNICAMP), Brazil

symbols in parallel, using frequency division multiplexing. Chang's main contribution was a method that, through some appropriate frequency shaping, allowed the carriers to have minimum frequency separation. In fact, he showed that the subchannels associated to each carrier presented some frequency overlapping, achieving maximum spectral efficiency, and yet could easily be demultiplexed.

In the following year of 1967, Saltzberg [2] extended the work of Chang to use quadrature amplitude modulation (QAM), proposing what is nowadays called OFDM/OQAM (offset QAM). However, a more practical digital implementation of OFDM came only in 1971, with Weinstein and Ebert [3] when they showed that the Cooley–Tukey fast-Fourier transform (FFT) algorithm [4] could be used to generate and demodulate the parallel orthogonal frequency streams. However, in the presence of ISI, the approach in [3] was not able to attain perfect orthogonal channels at the receiver, which made it hard to demultiplex the streams.

A solution to the ISI problem was only devised by Peled and Ruiz [5] in 1980, when they proposed the present form of OFDM. They introduced the concept of the CP and showed that, as long as the CP length is larger than the channel impulse response, the received OFDM symbol will not suffer from ISI and can preserve the orthogonality of the subchannels. They also showed that, in this case, equalisation can be perfectly achieved by a single-tap phase and gain compensation on each subchannel.

SC modulation has been used for a longer time than OFDM and provides lower peak-to-average power ratio (PAR) and superior robustness to carrier frequency offsets [6–9]. However, it traditionally uses time–domain equalisation [10] to deal with ISI, which tends to be more complex than the equalisation strategy of OFDM. This is only exacerbated by the modern large-bandwidth systems, which makes the time–domain equalisation approach extremely computationally expensive.

To circumvent the complexity of time–domain equalisation, SC-FDE has become a popular solution. The origin of SC-FDE can be traced back to 1973 with Walzman and Schwartz [11]. However, only more recently [6,9,12–14], it has become a serious alternative. One of the main advantages of FDE is that, through the introduction of a CP, the equaliser can be implemented in the frequency domain, thus benefiting from the low complexity of the FFT. As a result, the equaliser is as computationally efficient as the OFDM counterpart. Nowadays, SC-FDE is used in the uplink of the 4G Long-Term Evolution (LTE) [9] and is considered for 5G systems [15,16].

In the sequel of this chapter, we show the fundamental concepts needed to implement FDE for OFDM and SC system. We first describe a mathematical model that allows us to show how to achieve the simple but nonetheless efficient frequency domain equaliser for both OFDM and SC modulation schemes. Then, we present some simulation results that show how some choices impact the performance of these schemes. Finally, some PAR comparisons are shown that are followed by the chapter conclusions.

4.2   Mathematical preliminaries and basic concepts

Before we define the system model and explain the frequency domain equaliser concept, let us introduce some notation and basic concepts. Let us start by defining the N-point discrete Fourier transform (DFT) as a matrix $\mathbf{F}$, where the element $[\mathbf{F}]_{k+1,n+1} = (1/\sqrt{N})e^{-j2\pi kn/N}$, with $k, n = 0, 1, \ldots, N-1$ denoting a frequency and a time shift, respectively, and $j = \sqrt{-1}$. The inverse DFT (IDFT) is given by $\mathbf{F}^{-1} = \mathbf{F}^H$, where $(\cdot)^H$ is the transpose-conjugate operator and $[\mathbf{F}^{-1}]_{k+1,n+1} = (1/\sqrt{N})e^{j2\pi kn/N}$.

We suppose that the channel can be modelled by a finite impulse response (FIR) filter with L coefficients defined by the vector $\mathbf{h} = \begin{bmatrix} h_0 & h_1 & \cdots & h_{L-1} \end{bmatrix}^T$, where $(\cdot)^T$ is the transpose operator. Given a sequence $\mathbf{x} = \begin{bmatrix} x_0 & x_1 & \cdots & x_{M-1} \end{bmatrix}^T$ to be transmitted, the received sequence at the channel output is

$$\mathbf{y} = \mathbf{h} * \mathbf{x}, \tag{4.1}$$

where the symbol $*$ represents the linear convolution operator, and each element of $\mathbf{y}$ is given by

$$y_n = \sum_{k=0}^{L+M-1} x_k h_{n-k}, \tag{4.2}$$

where $n = 0, \ldots, L + M - 1$. This convolution can also be expressed in matrix form as

$$\mathbf{y} = \underbrace{\begin{bmatrix} h_0 & 0 & \cdots & 0 & \cdots & 0 \\ h_1 & h_0 & & & & 0 \\ \vdots & & \ddots & & & \\ h_{L-1} & & & h_0 & & \\ 0 & \ddots & & & \ddots & \\ & h_{L-1} & & & & h_0 \\ \vdots & & & h_{L-1} & & h_1 \\ & & & & \ddots & \vdots \\ 0 & & \cdots & & 0 & h_{L-1} \end{bmatrix}}_{\mathcal{H}} \mathbf{x}, \tag{4.3}$$

where $\mathcal{H}$ is called the convolution matrix and has dimension $L + M - 1 \times M$.

4.2.1 The basics of OFDM

Weinstein and Ebert proposed in [3] to efficiently synthesise the subcarriers using the IFFT (for modulation) and FFT (for demodulation), which are the fast implementations of the (I)DFT for sequences of length N, where N is usually a power of two.[1] Ignoring the presence of ISI, and thus the need for the CP, a single OFDM symbol in baseband is formed in the transmitter, in the discrete–time domain, as

$$\mathbf{s} = \mathbf{F}^H \breve{\mathbf{a}}, \tag{4.4}$$

---

[1]There are (I)FFT for non-power of two lengths, as long as the length can be factored in small primes, such as two and three.

where $\breve{\mathbf{a}} = \begin{bmatrix} \breve{a}_0 & \breve{a}_1 & \cdots & \breve{a}_{N-1} \end{bmatrix}^T$ is the vector containing the digital modulation symbols [e.g. binary phase shift keying (BPSK) or QAM symbols] and the breve over the letter means that vector represents a signal in the frequency domain. In other words, each symbol $\breve{a}_k$ is modulated by a complex sinusoid of frequency k/N, i.e. $(\breve{a}_k/\sqrt{N})e^{j2\pi kn/N}$, and their sum along the time n is represented in (4.4), i.e.

$$s_n = \frac{1}{\sqrt{N}} \sum_{k=0}^{N-1} \breve{a}_k e^{j2\pi kn/N}. \tag{4.5}$$

As seen before, $\mathbf{x}$ denotes the transmitted signal, which, in OFDM, is formed by the introduction of a CP to $\mathbf{s}$. However, as already mentioned, at this point, we consider an ideal channel, which does not introduce ISI. Hence, we have no need to use the CP, so that the transmitted signal is $\mathbf{x} = \mathbf{s}$. We also assume that there is no noise, so that the received signal is equal to the transmitted signal, i.e. $\mathbf{r} = \mathbf{x}$. Then, in order to demodulate, we first down-convert the signal component in frequency m/N. To that end, we multiply the received signal r_n by $e^{-j2\pi mn/N}$. Then, we need a matched filter. To that end, note that each symbol is modulated be the carrier multiplied by an N-sample discretised rectangular pulse. The filter should be matched to the rectangular pulse, so that it is simply the sum of N consecutive samples of the down-converted signal, i.e.

$$\breve{u}_m = \frac{1}{\sqrt{N}} \sum_{n=0}^{N-1} r_n e^{-j2\pi mn/N}. \tag{4.6}$$

Clearly, the combination of down-conversion and matched filtering corresponds to a DFT operation on the received signal, which can be represented in matrix form by

$$\breve{\mathbf{u}} = \mathbf{Fr}. \tag{4.7}$$

Finally, using the fact that $\mathbf{r} = \mathbf{x} = \mathbf{s}$ and that the IDFT operation in (4.4) and the DFT operation in (4.7) are inverses of each other, we conclude that, in the absence of ISI and noise, $\breve{\mathbf{u}} = \breve{\mathbf{a}}$. In other words, in this ideal case, the output of the matched filters are equal to the transmitted symbols.

4.2.2   The basics of SC

For simplicity, but without loss of generality, consider a baseband model sampled at the symbol rate. Again, we assume that there is no ISI, so no CP is introduced. Thus, $\mathbf{x} = \mathbf{s}$. Furthermore, in SC, the symbols are transmitted directly in the time domain. Thus, instead of $\breve{\mathbf{a}}$, which represented the information symbols in the frequency domain for OFDM, we assume that we want to transmit the vector $\mathbf{a} = \begin{bmatrix} a_0 & a_1 & \cdots & a_{N-1} \end{bmatrix}^T$. Finally, since $\mathbf{a}$ is already in the time domain, we have that $\mathbf{s} = \mathbf{a}$.

As in the previous section on OFDM, besides an ideal channel we also assume the absence of noise, so that the received signal is equal to the transmitted signal. It is worth noting that this model implicitly assumes that the transmission pulse and its matched filter at the receiver results in a Nyquist filter.

4.3 Frequency domain equalisation

Time-dispersive channels introduce ISI, which may heavily affect the performance of OFDM and SC. In this section, we show how to efficiently compensate for these distortions in OFDM by using the CP technique. For the SC scheme, we describe a system model that allows us to easily understand the SC-FDE approach by using the same structure used in the cyclic prefixed OFDM. However, as we will show later, the SC-FDE performance is sensitive to the equalisation criterion used to calculate the equaliser coefficients.

Before showing the FDE schemes for OFDM and SC-FDE, we describe how ISI may be represented as a simple entrywise product in the frequency domain for a channel that performs cyclic convolution. We then show how this can be achieved in practical channels, which actually perform linear convolutions, through the use of cyclic prefixes.

4.3.1 The channel distortion as a simple entrywise product

In order to achieve an entrywise product representation for the ISI, the first step is to transform the linear convolution represented in (4.2) in a circular one, represented by

$$r_n = \sum_{k=0}^{N-1} s_k h_{n-k \bmod N}, \tag{4.8}$$

where $m \bmod N$ is the remainder of the division of m by N, and r_n is the result of the circular convolution of the signal s_n with the channel coefficients h_n.

Now, before showing how we can emulate a circular convolution through the CP, let us demonstrate how this circular property can simplify the equalisation by writing the circular convolution in matrix form:

$$\mathbf{r} = \underbrace{\begin{bmatrix} h_0 & 0 & \cdots & h_{L-1} & \cdots & h_1 \\ h_1 & h_0 & & & \ddots & \vdots \\ \vdots & & \ddots & & & h_{L-1} \\ h_{L-1} & & & h_0 & & 0 \\ 0 & \ddots & & \vdots & \ddots & \vdots \\ \vdots & & h_{L-1} & h_{L-2} & \cdots & h_0 \end{bmatrix}}_{\mathcal{H}_c \; N \times N} \begin{bmatrix} s_0 \\ s_1 \\ s_2 \\ s_3 \\ \vdots \\ s_{N-1} \end{bmatrix}, \tag{4.9}$$

where $\mathcal{H}_c$ is an $N \times N$ circular convolution matrix. Note that we can write $\mathcal{H}_c$ as

$$\mathcal{H}_c = h_0 \mathbf{I}_N + h_1 \mathbf{Q} + h_2 \mathbf{Q}^2 + \cdots + h_{L-1} \mathbf{Q}^{L-1}, \tag{4.10}$$

where $\mathbf{I}_N$ is an identity matrix with dimension $N \times N$, and

$$\mathbf{Q} = \begin{bmatrix} 0 & 0 & \cdots & 0 & 1 \\ 1 & 0 & \ddots & 0 & 0 \\ 0 & 1 & & \vdots & \vdots \\ \vdots & \ddots & \ddots & 0 & 0 \\ 0 & \cdots & & 1 & 0 \end{bmatrix} \tag{4.11}$$

is also a circular convolution matrix that corresponds to a unity delay channel, i.e. $\mathbf{h} = [0 \ 1]^T$.

Now, we want to equalise in the frequency domain, so we take the DFT of the received signal, resulting in

$$\check{\mathbf{u}} = \mathbf{Fr}. \tag{4.12}$$

We also consider that the transmitted signal is generated as an OFDM signal, as described in (4.4). Thus, substituting (4.4) and (4.12) into (4.9), we can write

$$\check{\mathbf{u}} = \mathbf{F}\mathcal{H}_c\mathbf{F}^H\check{\mathbf{a}}. \tag{4.13}$$

Now, we show that the matrix

$$\mathbf{\Lambda} = \mathbf{F}\mathcal{H}_c\mathbf{F}^H \tag{4.14}$$

is actually a diagonal matrix, whose diagonal elements correspond to the DFT of $\mathbf{h}$. To that end, we use (4.10) to write

$$\mathbf{\Lambda} = \left(h_0\mathbf{I} + h_1\mathbf{FQF}^H + h_2\mathbf{FQ}^2\mathbf{F}^H + \cdots + h_{L-1}\mathbf{FQ}^{L-1}\mathbf{F}^H\right). \tag{4.15}$$

We know that any vector multiplied by $\mathbf{Q}$ will be circularly shifted, so that is not difficult to show that [17]

$$\mathbf{D} = \mathbf{FQF}^H = \text{diag}\left\{\left[1 \ e^{-j2\pi/N} \ \cdots \ e^{-j2\pi k/N} \ \cdots \ e^{-j2\pi(N-1)/N}\right]\right\}, \tag{4.16}$$

where diag $\{\mathbf{x}\}$ represents a diagonal matrix where its diagonal are the elements of $\mathbf{x}$, and all other elements are zero.

Note that the diagonal of $\mathbf{D}$ corresponds to the DFT of a unity delay, which is exactly the DFT time-shift property. Also, note that $\mathbf{D}^n = \mathbf{FQ}^n\mathbf{F}^H$, since, from (4.16), $\mathbf{Q} = \mathbf{F}^H\mathbf{DF}$. Thus, we can write (4.15) as

$$\mathbf{\Lambda} = \left(h_0\mathbf{D}^0 + h_1\mathbf{D} + h_2\mathbf{D}^2 + \cdots + h_{L-1}\mathbf{D}^{L-1}\right) = \sum_{n=0}^{N-1} h_n\mathbf{D}^n$$

$$= \sum_{n=0}^{N-1} \text{diag}\left\{h_n\left[1 \ e^{-j2\pi n/N} \ \cdots \ e^{-j2\pi kn/N} \ \cdots \ e^{-j2\pi(N-1)n/N}\right]\right\}, \tag{4.17}$$

where $h_n = 0$ for $n > L - 1$ and the $(k+1)$th element of the diagonal is

$$[\mathbf{\Lambda}]_{k+1,k+1} = \check{h}_k = \sum_{n=0}^{N-1} h_n e^{-j2\pi kn/N}, \ k = 0, 1, \ldots, N - 1, \tag{4.18}$$

which is clearly the DFT of $\mathbf{h}$, i.e. $\check{\mathbf{h}} = \sqrt{N}\mathbf{Fh}$.

Therefore, we can rewrite (4.12) as

$$\breve{\mathbf{u}} = \boldsymbol{\Lambda}\breve{\mathbf{a}} = \breve{\mathbf{h}} \circ \breve{\mathbf{a}} = \left[\breve{h}_0\breve{a}_0 \ \breve{h}_1\breve{a}_1 \ \cdots \ \breve{h}_{N-1}\breve{a}_{N-1}\right]^T. \tag{4.19}$$

In conclusion, the kth channel output, $\breve{u}_k$, is the entrywise product of the DFT of the channel impulse response and the original signal. In vector notation, we represent the entrywise product by the $\circ$ operator.

It is worth noting that one fundamental aspect to make $\boldsymbol{\Lambda}$ a simple diagonal matrix is that the channel must be static during the N samples of an OFDM symbol.[2] Otherwise, off-diagonal entries will appear in the matrix $\boldsymbol{\Lambda}$ and the more time-varying the channel is, the larger these entries will be with respect to the diagonal. In this scenario, equalisation is still possible, but at a much higher computational cost, even with some simplifications and clever additional manipulations, due to the need to deal with a full matrix (e.g. [18,19]). Such time-varying channels may arise, for instance, as a result of mobility in wireless communication systems. In order to avoid additional complexity, the system must be designed considering that each received block of N samples experiences a quasi-static channel, which can be achieved if such block period is less than the channel coherence time [20].

Finally, in order to achieve such entrywise product of the transmitted signal and the channel discrete frequency response, we must transform the linear convolution into a circular one for at least N samples. Fortunately, this can be achieved by the CP technique, as we show next.

4.3.2 The cyclic prefix technique

In this section, we show how the CP can be used to transform the linear convolution performed by the channel into a circular convolution. We begin with a simple example, which also introduces the topic of FDE, which will be discussed in full generality in the next section.

Thus, assume we want to transmit four information symbols, denoted by $\breve{a}_0$, $\breve{a}_1$, $\breve{a}_2$ and $\breve{a}_3$. Assume also that the Z-transform of the channel is $H(z) = 1 + \alpha z^{-1}$. Finally, we will use an OFDM symbol length of $N = 2$.

Our first task is to compute the IFFT, to generate the symbols x_n that will be input to the channel. Since $N = 2$, we must group the transmitted symbols in blocks of $N = 2$ symbols. In the first block, and using $N = 2$ in (4.5), this yields the time-domain symbols

$$
\begin{aligned}
s_0 &= \frac{1}{\sqrt{2}}\left(\breve{a}_0 + \breve{a}_1\right) \\[2mm]
s_1 &= \frac{1}{\sqrt{2}}\left(\breve{a}_0 - \breve{a}_1\right).
\end{aligned}
\tag{4.20}
$$

---

[2]As we are going to see, the circular convolution is achieved through the use of CP. In this case, the channel must be static during $N + L - 1$ samples.

For the second block, the resulting symbols are

$$s_2 = \frac{1}{\sqrt{2}} (\breve{a}_2 + \breve{a}_3)$$

$$s_3 = \frac{1}{\sqrt{2}} (\breve{a}_2 - \breve{a}_3) .$$

(4.21)

Now, normally we would transmit the four symbols in a sequence, i.e. the channel input would be s_0, s_1, s_2 and s_3. Assume, however, that we insert a CP to the second block, transmitting its last symbol before actually transmitting the block. This would yield the following channel inputs: $x_0 = s_0$, $x_1 = s_1$, $x_2 = s_3$, $x_3 = s_2$ and $x_4 = s_3$. After transmitting these symbols, and ignoring noise, this would result in the following received symbols:

$$r_2 = x_2 + \alpha x_1 = s_3 + \alpha s_1$$

$$r_3 = x_3 + \alpha x_2 = s_2 + \alpha s_2$$

$$r_4 = x_4 + \alpha x_3 = s_3 + \alpha s_2.$$

(4.22)

Note that r_3 and r_4 depend on symbols coming only from the second block. Furthermore, it is not hard to see that r_3 and r_4 are the circular convolution between the channel and the transmitted symbols $x_3 = s_2$ and $x_4 = s_3$. Thus, by transmitting $x_2 = s_3$ before actually transmitting the second block, we turned the linear convolution performed by the channel on the symbols x_n into a circular convolution performed on the symbols s_n. Finally, note that the circular convolution does not depend on the fact that the symbols s_n are the outputs of an IDFT operator. It only depends on the insertion of the CP.

Before generalising the CP, and studying its impact on some system parameters, we continue with the example to illustrate the entrywise product model for the channel in (4.19). To that end, let $\breve{u}$ be the length-2 DFT of r_3 and r_4:

$$\breve{u}_0 = \frac{1}{\sqrt{2}} (r_3 + r_4)$$

$$\breve{u}_1 = \frac{1}{\sqrt{2}} (r_3 - r_4) .$$

(4.23)

Now, using (4.22) and (4.21), it can be seen that

$$\breve{u}_0 = (1 + \alpha) \breve{a}_2$$

$$\breve{u}_1 = (1 - \alpha) \breve{a}_3.$$

(4.24)

Note that the terms multiplying the transmitted symbols in the equation above are exactly the DFT of the channel impulse response, as predicted by (4.19).

Also note some important consequences of the CP in some system parameters. To that end, let T_s be the symbol period at the channel input. Note that each OFDM block corresponds to three symbols: one for the CP and two for the actual information symbols. Thus, the period for an OFDM block is $3T_s$.

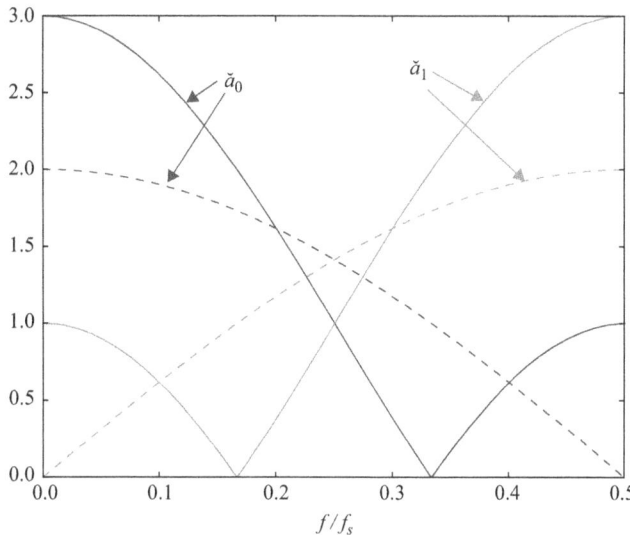

Figure 4.1 *Frequency content of a length-2 OFDM symbol at the transmitter. The solid lines correspond to the symbols without cyclic prefix, the dashed lines correspond to symbols with a length-one CP. The spectra centred at frequency 0 correspond to the first carrier in the OFDM symbol, while the spectra centred at $f_s/2$ correspond to the second carrier*

The frequency content of the OFDM symbols at the channel input in this example, both with and without CP, is depicted in Figure 4.1. In both cases, the length-2 DFT samples the spectrum at the frequencies 0 and $f_s/2$ [3], so the symbols a_2 and a_3 are centred at these frequencies. Without CP, these symbols are modulated and multiplied by a length-2 rectangular window, and the resulting spectra are shown as the solid lines in Figure 4.1. However, with the length-one CP used in this example, the transmitted OFDM symbol actually lasts three time samples. This results in the spectra were shown as dashed lines in Figure 4.1. Note that, with CP, the symbols are no longer orthogonal at the transmitter. However, as seen before, after the CP is removed, the symbols become orthogonal even in the presence of ISI.

Generalising the concept of the CP is now straightforward. Its first task is to guarantee that neighbouring OFDM blocks do not interfere with each other and to make each block look like a circular convolution. Both these goals can be achieved as long as the CP is at least as large as the memory of the channel, $L - 1$. All we have to do, then, is to copy the last $L - 1$ values of the OFDM symbol to the beginning, as illustrated in Figure 4.2. This figure illustrates the OFDM symbols at the channel input, i.e. the values of x_n in (4.5). These were obtained under the assumption that only one value of a_k, namely a_2 was non-zero. In other words, the figures show a symbol with only one carrier. An actual OFDM symbol would then be obtained as the sum of N figures like Figure 4.2, one for each value of k in (4.5).

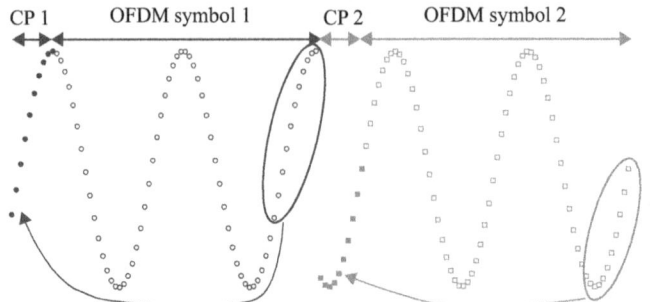

The symbols at the end (empty marks)   Again, the symbols at the end are copied
are copied at the beginning (solid marks)   at the beginning.
forming the cyclic prefix.

Figure 4.2   Illustration of an OFDM symbol with cyclic prefix. This plot shows the
time-domain OFDM symbol, i.e. the values of x_n in (4.5). The regular
OFDM symbol, without the cyclic prefix, is shown as the empty
markers. The cyclic prefix is shown as the solid markers, and its values
are copied from the end of the OFDM symbols. In this figure, we show
the time-domain OFDM symbol with a single non-zero frequency
domain symbol, in this case $ă_2$. An actual OFDM symbol would be the
sum of several plots like this one, each plot corresponding to a
frequency domain symbol, i.e. a value of k in (4.5)

The main reason for choosing a single value of k when plotting Figure 4.2 is to illustrate the fact that there is no discontinuity between the CP and the beginning of the corresponding symbol. This is because each value of k in (4.5) generates a complex exponential that has k periods within an OFDM symbol. For instance, in Figure 4.2, we used $k = 2$, which results in two periods for the sinusoid. This periodicity ensures that there is no discontinuity between the beginning and the end of the symbol. Note, however, that there may be severe discontinuity between adjacent OFDM symbols, as seen in Figure 4.2. This results in poor spectral containment in OFDM [8]. A large body of literature exists that tries to improve this characteristic of OFDM [21].

Note that using a CP decreases the system spectral efficiency, since we now transmit N symbols over $N + L - 1$ samples. Note, however, that this rate penalty decreases as N increases, which means that long OFDM symbols may be desirable. On the other hand, the channel has to remain constant during the transmission of an OFDM symbol, so that N cannot be too long in some applications with time-varying channels.

Finally, we note a common misconception of OFDM systems: to ensure that each symbol $ă_k$ sees a one-tap channel, the frequency response of the channel is *not* required to remain constant for each subchannel. All we need is a CP of appropriate length. To see this, consider again our example with $N = 2$ symbols and assume $\alpha = 1$. The second carrier uses the frequencies from $f_s/4$ to $f_s/2$. Over this range, the frequency response varies, in magnitude, from $\sqrt{2}$ to 0, which is certainly not a

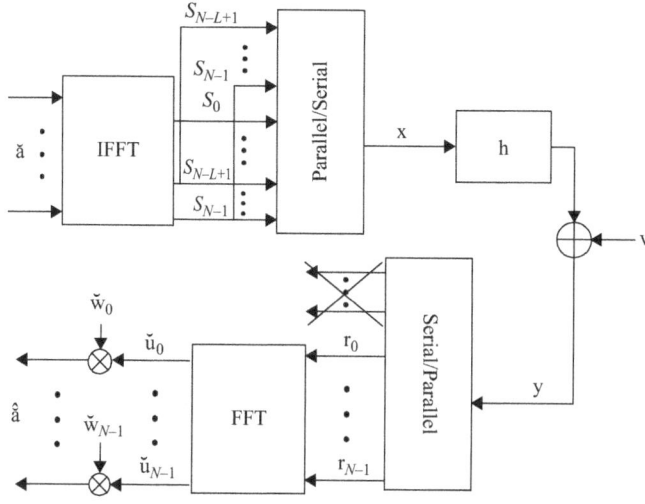

Figure 4.3　OFDM system model

negligible variation. Still, as shown in (4.24), the channel model for the demodulated signal is a pair of one-tap channels, and there is no ISI between $\breve{a}_2$ and $\breve{a}_3$.

4.3.3　OFDM equalisation

Consider the OFDM system model presented in Figure 4.3. In it, we assume that the CP length is longer than the channel impulse response, perfect synchronisation and that the channel is static for each OFDM symbol. Besides, we also consider the presence of additive white Gaussian noise (AWGN) with zero mean and variance σ_v^2, so that the received signal can be written as

$$\breve{\mathbf{u}} = \mathbf{Fr} = \mathbf{F}(\mathcal{H}_c\mathbf{s} + \mathbf{v}) \tag{4.25}$$

$$\breve{\mathbf{u}} = \mathbf{F}\mathcal{H}_c\mathbf{F}^H\breve{\mathbf{a}} + \mathbf{Fv}$$

$$= \mathbf{\Lambda}\breve{\mathbf{a}} + \breve{\mathbf{v}}$$

$$= \left[\breve{h}_0\breve{a}_0 \;\; \breve{h}_1\breve{a}_1 \;\; \cdots \;\; \breve{h}_{N-1}\breve{a}_{N-1}\right]^T + [\breve{v}_0 \;\; \breve{v}_1 \;\; \cdots \;\; \breve{v}_{N-1}]^T ,$$

where $\mathbf{v} = [v_0 \;\; v_1 \;\; \cdots \;\; v_{N-1}]^T$ is the AWGN noise vector and $\breve{v}_k$ is also an AWGN noise with zero mean and variance σ_v^2, due to the orthonormal characteristics of the N-point DFT.

From (4.25), we can see each subcarrier as an AWGN channel where the symbol $\breve{a}_k$ is rotated and scaled by $\breve{h}_k$ and a Gaussian white noise $\breve{v}_k$ is added. Therefore, a simple correction to this distortion is to multiply it by a scalar, i.e. the one-tap equaliser

$$\hat{\breve{\mathbf{a}}} = \breve{\mathbf{w}} \circ \breve{\mathbf{u}} \tag{4.26}$$

$$= \left[\breve{w}_0 \breve{h}_0 \breve{a}_0 \ \breve{w}_1 \breve{h}_1 \breve{a}_1 \ \cdots \ \breve{w}_{N-1} \breve{h}_{N-1} \breve{a}_{N-1} \right]^T + \left[\breve{w}_0 \breve{v}_0 \ \breve{w}_1 \breve{v}_1 \ \cdots \ \breve{w}_{N-1} \breve{v}_{N-1} \right]^T,$$

where $\breve{w}_k$ is the kth equaliser coefficient and $\hat{\breve{\mathbf{a}}} = \left[\hat{\breve{a}}_0 \ \hat{\breve{a}}_1 \ \cdots \ \hat{\breve{a}}_{N-1} \right]^T$ is the vector with the estimated values of $\breve{\mathbf{a}}$. At this point, we can recover $\breve{a}_k$ by making

$$\breve{w}_k = \frac{1}{\breve{h}_k}. \tag{4.27}$$

The signal-to-noise ratio (SNR) at the equaliser output is given by

$$SNR_{\mathrm{OFDM}_k} = |\breve{h}_k|^2 \frac{\sigma_{\breve{a}_k}^2}{\sigma_v^2}, \tag{4.28}$$

where $\sigma_{\breve{a}_k}^2$ is the variance of the symbol on subcarrier k. The value of SNR_{OFDM_k} is important in order to allow optimal channel decoding of the information bits encoded in the symbols. Also note that any scalar multiplication of the received signal will not change the SNR, so any other criterion to determine the equaliser coefficients will yield the same SNR. From (4.28), we can also calculate the theoretical average BER, considering an M-QAM modulation:

$$\mathrm{BER}_{\mathrm{OFDM}} = \frac{1}{N} \sum_{k=0}^{N-1} P_{e_{M\text{-QAM}}}(SNR_{\mathrm{OFDM}_k}), \tag{4.29}$$

where the bit error probability for M-QAM, considering Gray mapping, is

$$P_{e_{M\text{-QAM}}}(SNR) = \left(1 - \frac{1}{\sqrt{M}}\right) \frac{4}{\log_2 M} Q\left(\sqrt{\frac{3}{M-1} SNR}\right), \tag{4.30}$$

where $Q(x) = (1/\sqrt{2\pi}) \int_x^\infty e^{(-u^2/2)} du$ is the Q-function of a normal Gaussian distribution with zero mean and unitary variance.

4.3.4   SC frequency domain equalisation

In order to make it easier to understand the SC-FDE scheme, and to put it in the same framework we have been using, we can see the SC modulation as a linear precoded OFDM scheme. This approach is commonly adopted in the literature (e.g. [14,22]) and is depicted in Figure 4.4. Such precoding is achieved by making the OFDM signal to be transmitted as

$$\breve{\mathbf{a}} = \mathbf{F}\mathbf{a}, \tag{4.31}$$

where the vector $\mathbf{a}$ is formed with digital modulation symbols to be transmitted. Recall that the time-domain signal $\mathbf{s}$ in our framework is formed as $\mathbf{s} = \mathbf{F}^H \breve{\mathbf{a}}$, which, for SC, results in $\mathbf{s} = \mathbf{a}$, since $\mathbf{F}^H \mathbf{F} = \mathbf{I}$. In other words, in SC, the digital modulation

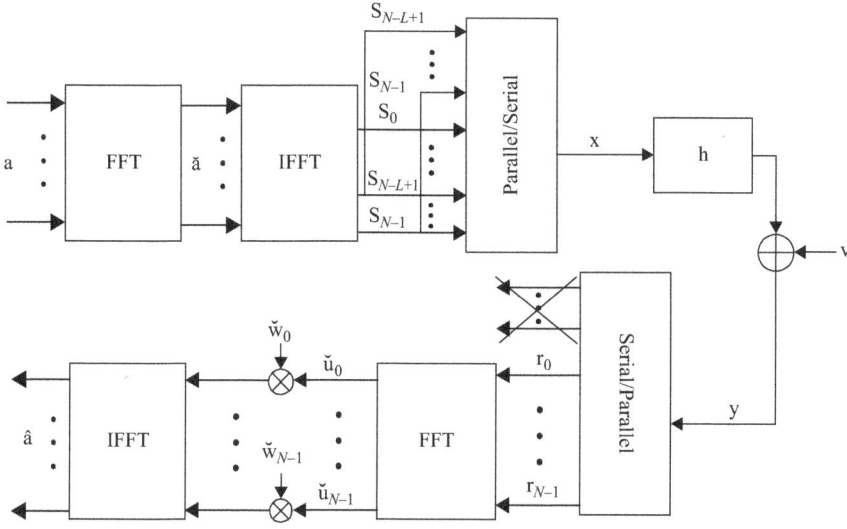

Figure 4.4   SC-FDE as linear precoded OFDM system

symbols **a** are transmitted directly in the time-domain signal **s**. In contrast, in OFDM, the information symbols are used to form the frequency-domain vector **ă**.

It is important to note that the frequency domain OFDM symbols **ă** in this model are no longer statistically independent, since they are a linear combination of the symbols **a**. This characteristic, as we will show, impacts the equaliser coefficients design and also can help exploit the channel frequency diversity.

As the received signal is the same of the OFDM scheme (4.25), we can use the OFDM one-tap equaliser in (4.26) to estimate the *frequency-domain* symbols **ă**: $\hat{\text{ă}} = \text{w̌} \circ \text{ǔ}$. We can then recover the time-domain symbols by multiplying the equaliser output by the inverse of the precoding matrix, i.e.

$$\hat{\mathbf{a}} = \mathbf{F}^{H}(\text{w̌} \circ \text{ǔ}). \tag{4.32}$$

However, in contrast with OFDM, the resulting SNR depends on several factors. To see this, we expand (4.32) and, after some algebraic manipulation, write the *k*th estimated symbol as a combination of the desired symbol, the remaining ISI and noise:

$$\hat{a}_k = \underbrace{\frac{1}{N}\sum_{n=0}^{N-1}\left[\text{w̌}_n\text{ȟ}_n\right]a_k}_{\text{desired signal}} \tag{4.33}$$

$$+ \underbrace{\frac{1}{N}\sum_{\substack{n=0\\n\neq k}}^{N-1}\left[\sum_{m=0}^{N-1}\text{w̌}_n\text{ȟ}_n e^{-j2\pi(k-n)m/N}\right]a_m}_{\text{intersymbol interference}} + \underbrace{\frac{1}{\sqrt{N}}\sum_{n=0}^{N-1}\text{w̌}_n\text{v̌}e^{j2\pi kn/N}}_{\text{noise}},$$

From (4.33), the SNR of the SC-FDE can be written as

$$SNR_{SC-FDE} = \frac{P_{signal}}{P_{ISI} + P_{noise}}, \tag{4.34}$$

where P_{signal} is the desired signal power, given by

$$P_{signal} = \left| \frac{1}{N} \sum_{n=0}^{N-1} \left[\breve{w}_n \breve{h}_n \right] \right|^2 \sigma_a^2, \tag{4.35}$$

P_{ISI} is the ISI power, given by

$$P_{ISI} = \frac{1}{N} \left[\sum_{n=0}^{N-1} \left| \breve{w}_n \breve{h}_n \right|^2 - \frac{1}{N} \left| \sum_{n=0}^{N-1} \breve{w}_n \breve{h}_n \right|^2 \right] \sigma_a^2, \tag{4.36}$$

and finally, P_{noise} is the noise power

$$P_{noise} = \frac{\sigma_v^2}{N} \sum_{n=0}^{N-1} |\breve{w}_n|^2 . \tag{4.37}$$

To design the equaliser, consider again (4.25), which we rewrite here as

$$\breve{u} = \breve{h} \circ \breve{a} + \breve{v}, \tag{4.38}$$

where, as before, $\breve{h}$ is the vector with the N-point DFT of the channel impulse response, and $\breve{a}$ and $\breve{v}$ represent the transmitted signal and the noise, both in the frequency domain. Then, we may write the equaliser output as

$$\hat{a} = \mathbf{F}^H (\breve{w} \circ \breve{h} \circ \breve{a} + \breve{w} \circ \breve{v}), \tag{4.39}$$

Now, assume that we want to mitigate ISI by inverting the channel, which is the so-called zero-forcing (ZF) solution. To that end, we may simply choose $\breve{w}_i = 1/\breve{h}_i$. Therefore, with the aid of the CP that turns the channel linear convolution performed by the channel into a circular one, it is possible to perfectly invert a FIR channel with another FIR filter.[3] This is an advantage over the conventional SC linear equaliser (LE) with linear convolution, and it allows us to completely eliminate ISI and make $P_{ISI} = 0$.

Note that we may need larger values of $\breve{w}_k$, where the channel frequency response magnitude is small ($|\breve{h}_k| \ll 1$). However, in contrast to OFDM, as stated in (4.28), the SC-FDE noise power (P_{noise}) depends on the equaliser coefficients as seen in (4.37). Furthermore, the larger the equaliser coefficients, the larger is the noise power after equalisation. This is known as noise enhancement, and it may considerably degrade the SC-FDE performance if not taken into account. Also, note that using the ZF solution (4.27) and (4.34), we have [23]

$$SNR_{SC-FDE}^{ZF} = \frac{1}{(1/N) \sum_{k=0}^{N-1} (\sigma_v^2 / (\sigma_a^2 |\breve{h}_k|^2))} \tag{4.40}$$

---

[3]As long the frequency response of this channel satisfies $\breve{h}_k \neq 0, \forall k$.

which corresponds to the harmonic mean of $(\sigma_a^2/\sigma_n^2)|\check{\mathbf{h}}|^2$. As a consequence, the presence of spectral nulls, i.e. small values of $|\check{h}_k|$, will incur in a very high performance penalty.

The equalisation literature shows that a good compromise for the trade-off between ISI mitigation and the noise enhancement which comes with it can be achieved through the use of a minimum mean square error (MMSE) criterion, which chooses the equaliser coefficients as the solution to the following optimisation problem:

$$\underset{\check{w}}{\operatorname{argmin}} = \mathrm{E}\left\{\|\mathbf{Fa} - \check{\mathbf{w}} \circ \check{\mathbf{u}}\|^2\right\}, \tag{4.41}$$

where $\|\cdot\|$ is the norm of the vector and $\mathrm{E}\{\cdot\}$ is the expectation operator.

In order to find the optimal value of $\check{w}_k$ for the MMSE criterion, we differentiate (4.41) with respect to $\check{w}_k^*$ and set it to zero, obtaining

$$\check{w}_k^{MMSE} = \frac{\check{h}_k^*}{|\check{h}_k|^2 + (\sigma_v^2/\sigma_a^2)}. \tag{4.42}$$

By substituting (4.42) into (4.34), we have [23]

$$SNR_{SC-FDE}^{MMSE} = \frac{1}{(1/N)\sum_{k=0}^{N-1}(\sigma_v^2/(\sigma_v^2 + \sigma_a^2|\check{h}_k|^2))} - 1, \tag{4.43}$$

which can be seen as the harmonic mean of $\left(1 + ((\sigma_a^2/\sigma_n^2)|\check{\mathbf{h}}|^2)\right) - 1$. Note that in contrast to (4.40), in (4.43) a small $|\check{h}_k|$ will not incur in a large SNR loss, giving better performance results as we are going to see in the next section.

Also, contrary to a traditional, time-domain MMSE LE [24], the MMSE SC-FDE coefficients can be easily calculated and do not need to use a training delay as the LE does.[4] The optimal value of this delay depends on the received SNR and the channel frequency response, making the search for the optimal value costly. In practice, a conservative value is guessed, which will likely lead to suboptimal performance. This problem does not exist when using the CP, since the one-tap frequency domain equaliser corresponds to a circular convolution and, thus, the training delay is irrelevant as it just circularly shifts the combined equaliser-channel impulse response. Besides, even when compared to the LE with the optimal delay, the MMSE SC-FDE solution will result in lower MMSE since the one-tap frequency domain equaliser, which is able to perfectly invert an FIR channel, corresponds to an infinite length LE. Such advantage is going to be shown in the next section.

4.4 Simulation results

In this section, we show some simulations to illustrate the performance and some interesting behaviours of OFDM and SC-FDE schemes under time-dispersive channels.

---

[4]By training delay, we mean the fact that the MMSE LE minimises $\mathrm{E}\left\{|a_n - \hat{a}_{n-\delta}|^2\right\}$, for some delay δ.

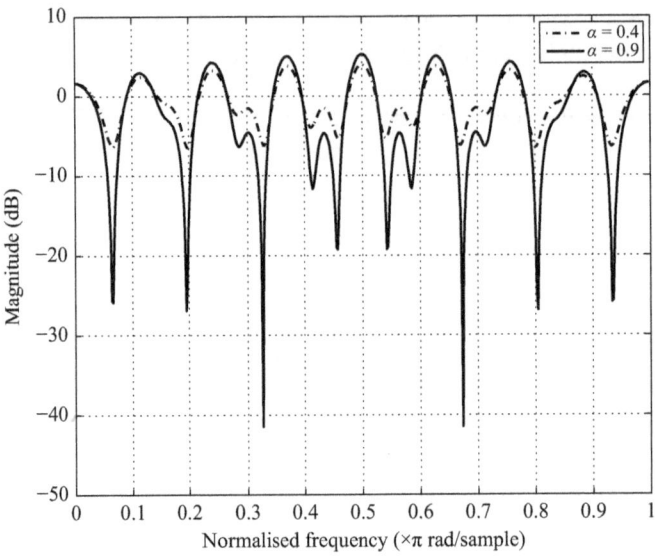

Figure 4.5   Frequency response for $H(z) = 1 + \alpha z^{-16} - (\alpha/2)\, z^{-30} + (\alpha/3)\, z^{-32}$, $\alpha = 0.4$ and 0.9

In our simulations, for simplicity, the channel is deterministic and time-invariant, with four multipath components, defined as $H(z) = 1 + \alpha z^{-16} - (\alpha/2)z^{-30} + (\alpha/3)z^{-32}$, where $\alpha = 0.4$ or 0.9. These channels represent a three-branch power line communication channel, and their frequency response is shown in Figure 4.5. Note that the larger the parameter α, the more frequency-selective is the channel, as seen in Figure 4.5. We also assume that the CP length is equal to $L - 1$. For all simulations, the modulation is 4-QAM, the DFT length is $N = 256$, and the CP length is $L = 33$, unless otherwise indicated.

First, in Figure 4.6, we show the OFDM BER with respect to the SNR of the received signal for both channels, i.e. $\alpha = 0.4$ and 0.9. The figure shows both the theoretical and the simulated BER performances, and they clearly match. Furthermore, we see that larger values of α result in more performance degradation, which is a consequence of the increased frequency selectivity shown in Figure 4.5. This degradation results from the fact that the data sent in subchannels with highly attenuated values $|\check{h}_k|$ have higher error probability, due to the low SNR_{OFDM_k} values. The BER in these channels will dominate the overall BER.

Next, in Figure 4.7, the BER performance of the ZF and MMSE SC-FDE is shown for the same SNR and α values. We also show the performance of an MMSE LE without CP, using 256 coefficients and optimal training delay for each SNR and channel frequency response. From the results, the simulated values match the theoretical values and it is clear that the MMSE solutions outperform, as expected, the ZF. As mentioned before, this is due to the larger noise enhancement of the ZF equaliser, especially when the channel is more frequency-selective, i.e. for larger values of α. Also, for higher SNR values, the MMSE SC-FDE is superior to the MMSE LE

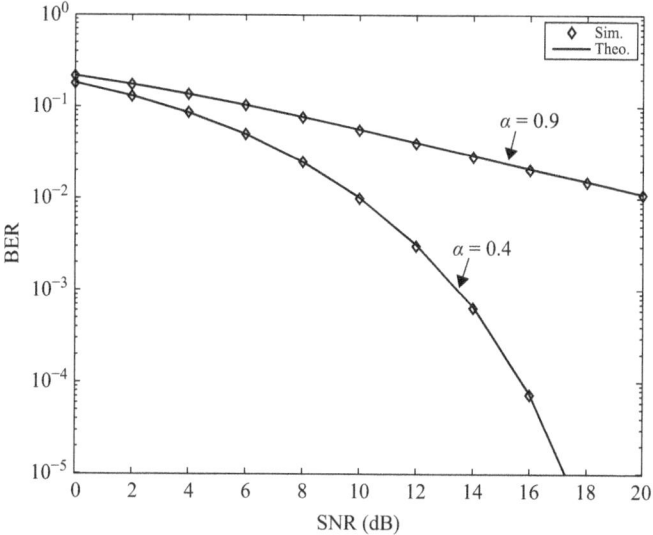

Figure 4.6   Bit error rate of OFDM for α = 0.4 and 0.9

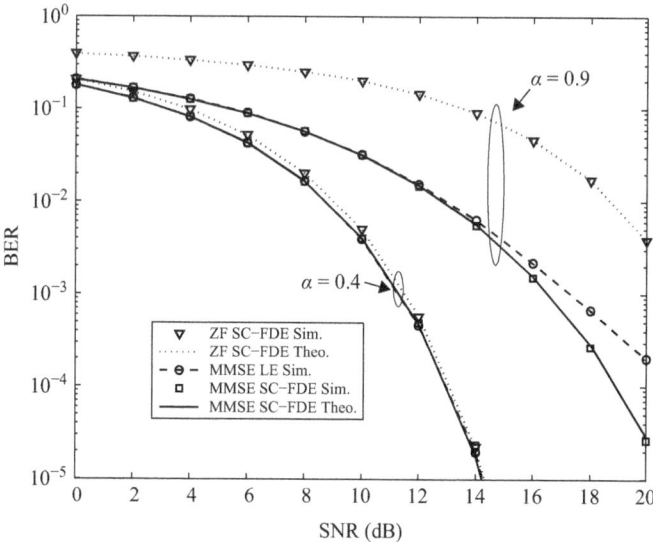

Figure 4.7   Bit error rate of the linear time domain MMSE equaliser (LE), ZF
SC-FDE and MMSE SC-FDE for α = 0.4 and 0.9

equaliser without CP, since, for these SNR values, the SC-FDE benefits from the possibility to perform a better (and even perfect) channel inversion even with a finite number of coefficients. The difference is more notable when α is larger, that is when the ISI is more severe.

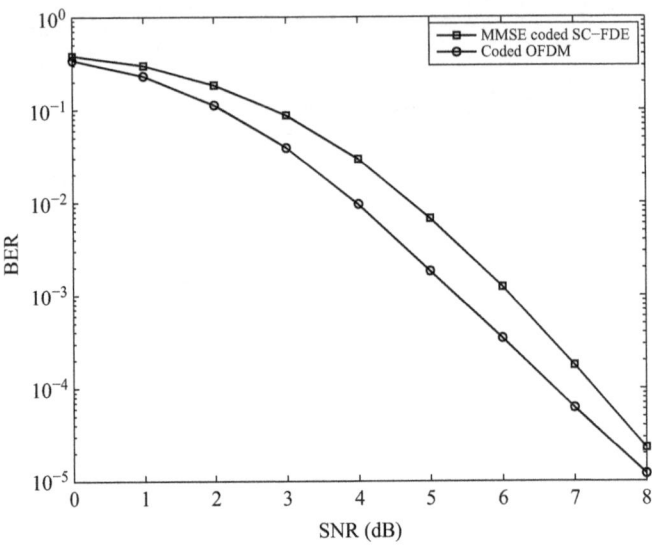

Figure 4.8   Bit error rate for 4-QAM and (171 133)$_{octal}$ code and $\alpha = 0.9$

Finally, when comparing the results in Figures 4.6 and 4.7, we can see that OFDM is considerably outperformed by the MMSE SC-FDE, particularly when $\alpha = 0.9$. This comes from the fact that, in OFDM, the BER at the attenuated subcarriers is close to 0.5, and the values dominate the average BER. This does not happen with the SC-FDE since the precoding averages the SNR, allowing the SC solution to exploit the channel frequency-diversity and achieve much better results than OFDM. In fact, the SNR of all the symbols is the same.

Fortunately, it is possible to overcome the lack of frequency diversity in OFDM using two solutions. The first and optimal solution is to make use of waterfilling/bit-loading [8] when the channel is known at the transmitter side. Using waterfilling, we do not transmit on the highly attenuated subcarriers; instead the available power is used to maximise the capacity on the best subcarriers.

However, if the channel is not known at the transmitter, we must use channel coding and interleave the coded bits in the frequency domain. This combination exploits the frequency diversity by correcting errors generated in low SNR subcarriers using the more reliable subcarriers. In order to show this, in Figure 4.8, we present the BER performance of an OFDM system with 4-QAM, a rate-1/2 convolutional code (171 133)$_{octal}$, and random interleaving. As can be seen, the coded OFDM performance is not only much better than the non-coded one shown in Figure 4.6, but it also outperforms the coded MMSE SC-FDE solution. This comes from the fact that the combination of channel coding and OFDM is more efficient to exploit the channel frequency-diversity than the coded SC-FDE, which relies solely on linear precoding.

It is worth noting that OFDM is less effective at higher coding rates than SC-FDE. Also, the latter can be further improved using decision feedback equalisation, but further details and comparisons are out of the scope of this chapter. Additional results comparing coded OFDM and SC-FDE, using linear and non-linear equalisation schemes, can be found in [13,23,25,26]. The general conclusion is that the use of non-coded OFDM is not of practical interest, as observed by [6].

4.5 Peak-to-average power

The more a power amplifier (PA) works close to its saturation, the more efficient it is. One way to measure this is to evaluate the PAR of the baseband transmitted signal

$$\text{PAR} = \frac{\max |s_n|^2}{\text{E}\left\{|s_n|^2\right\}}. \tag{4.44}$$

A PAR close to one indicates that the envelope of s_n is close to a constant, and then the PA can work almost all the time close to the saturation. On the other hand, a large PAR value indicates the opposite, i.e. the envelope fluctuates over a wide range, indicating that the PA cannot work most of the time close to saturation; otherwise, it could not accommodate the (large) signal excursion without unacceptable distortions.

In this sense, it is interesting to compare both SC and OFDM schemes in terms of PAR. However, if we took the PAR defined in (4.44), the OFDM will achieve very large values when compared to SC. For instance, consider a BPSK signal in baseband and at Nyquist rate. Using (4.44), the SC PAR would be equal to 1, but the PAR for OFDM would be N. The problem is that achieving this PAR is unlikely, since it occurs only in four out of 2^N possible OFDM symbols: $\breve{\mathbf{a}} = \pm[1 \; \cdots \; 1]$ or $\breve{\mathbf{a}} = \pm[1 \; -1 \; \cdots \; 1 \; -1]$. Then, to make a more fair comparison, we define the PAR considering the length of an OFDM symbol

$$\text{PAR}_m = \frac{\max_{(m-1)N \leq n < mN} |s_n|^2}{\text{E}\left\{|s_n|^2\right\}}, \tag{4.45}$$

where PAR_m becomes a random variable and m indicates the mth OFDM symbol. Also, it is worth noting that the addition of the CP will not change the PAR.

As a last remark, it is important to note that the oversampled and interpolated signal will have a peak power value larger than the Nyquist sampled one, since the interpolating filter will make the value overshoot near the peak value of the Nyquist sampled signal. Then, by considering s'_n the oversampled by a factor K and interpolated version of s_n, we have[5]

$$\text{PAR}_m = \frac{\max_{(m-1)KN \leq n < mKN} |s'_n|^2}{\text{E}\left\{|s'_n|^2\right\}}. \tag{4.46}$$

---

[5]The upconverted signal will have the PAR doubled, but since this will happen for both SC and OFDM signals, we can ignore this in the comparison.

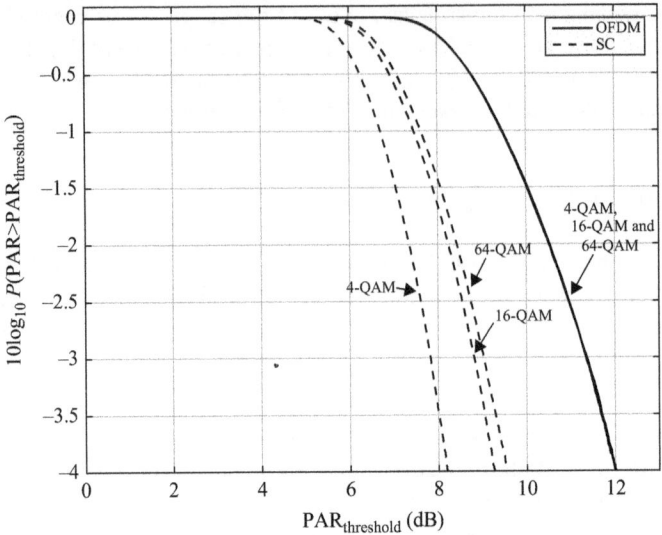

Figure 4.9   *PAR CCDF of SC and OFDM for N = 256, 8 times oversampling and 4, 16 and 256-QAM*

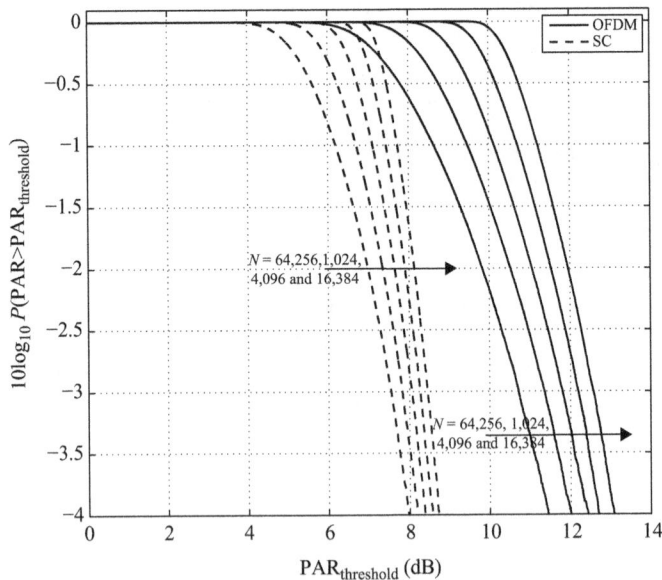

Figure 4.10   *PAR CCDF of SC and OFDM for N = 64, 256, 1,024, 4,096 and 16,384, 8 times oversampling and 4-QAM*

The literature (e.g. [8]) indicates that an oversampling factor of $K = 4$ already is close to the $K \gg 1$ scenario.

Then, using the SC as a precoded OFDM, and $K = 8$ oversampling factor and interpolating both SC and OFDM signals in the frequency domain by using a zero-padded IDFT, we show in Figures 4.9 and 4.10 the complementary cumulative distribution function (CCDF) of the random variable PAR_m, i.e. the probability of PAR_m being larger than a certain threshold $PAR_{threshold}$.

In Figure 4.9, we show that the OFDM PAR is insensitive to the cardinality of the chosen modulation, whereas the SC PAR gets decreasingly larger with the increasing cardinality. Also, as expected, the SC provides lower PAR than OFDM and the OFDM PAR is far away from the worst case scenario defined in (4.44), which is equal to 24 dB.

Finally, in Figure 4.10, using only 4-QAM, we present the PAR for $N = 64, 256,$ 1,024, 4,096 and 16,384. As N grows, it slightly increases the PAR values, but it has a smaller impact on the SC scheme than OFDM. For instance, considering a probability of 10^{-3}, going from $N = 64$ to $N = 16,384$ (a 256 fold increase) makes the SC and OFDM PAR increase by 0.91 and 1.84 dB, respectively.

It is worth noting that the OFDM PAR can be reduced through signal processing techniques [8], but they are complex and many of them require the transmission of side information.

4.6 Concluding remarks and further considerations

This chapter has presented fundamental concepts of OFDM and SC-FDE. Both schemes are today the standard techniques for modern wire and wireless systems and future technique being discussed are usually derived from them. The main principle of the techniques presented here is the use of the CP in order to obtain a simple yet efficient equaliser. We have also shown how to calculate the equaliser coefficients, as well some performance limitations of both schemes and their PAR behaviour. We believe that, together with the previous chapters, the reader has the basic concepts to understand these systems and perform some simulations.

References

[1]   Chang RW. Synthesis of band-limited orthogonal signals for multichannel data transmission. The Bell System Technical Journal. 1966;45(10):1775–1796.

[2]   Saltzberg B. Performance of an efficient parallel data transmission system. IEEE Transactions on Communication Technology. 1967;15(6):805–811.

[3]   Weinstein S, Ebert P. Data transmission by frequency-division multiplexing using the discrete Fourier transform. IEEE Transactions on Communication Technology. 1971;19(5):628–634.

[4]   Cooley JW, Tukey JW. An algorithm for the machine calculation of complex Fourier series. Mathematics of Computation. 1965;19(90):297–301.

[5]   Peled A, Ruiz A. Frequency domain data transmission using reduced computational complexity algorithms. Acoustics, Speech, and Signal Processing, IEEE International Conference on ICASSP'80. 1980;5:964–967.

[6]   Sari H, Karam G, Jeanclaude I. Transmission techniques for digital terrestrial TV broadcasting. IEEE Communications Magazine. 1995;33(2):100–109.

[7]   Wang Z, Ma X, Giannakis GB. OFDM or single-carrier block transmissions? IEEE Transactions on Communications. 2004;52(3):380–394.

[8]   Li YG, Stüber GL, editors. Orthogonal Frequency Division Multiplexing for Wireless Communications. 1st ed. Springer, New York, NY; 2006.

[9]   Myung HG, Lim J, Goodman DJ. Single carrier FDMA for uplink wireless transmission. IEEE Vehicular Technology Magazine. 2006;1(3):30–38.

[10]   Proakis J, Salehi M. Digital Commun. 5th ed. McGraw-Hill, New York, NY; 2008.

[11]   Walzman T, Schwartz M. Automatic equalization using the discrete frequency domain. IEEE Transactions on Information Theory. 1973;19(1):59–68.

[12]   Shynk JJ. Frequency domain and multirate adaptive filtering. IEEE Signal Processing Magazine. 1992;9(1):14–37.

[13]   Falconer D, Ariyavisitakul SL, Benjamin-Seeyar A, *et al.* Frequency domain equalization for single-carrier broadband wireless systems. IEEE Communications Magazine. 2002;40(4):58–66.

[14]   Pancaldi F, Vitetta G, Kalbasi R, *et al.* Single-carrier frequency domain equalization. IEEE Signal Processing Magazine. 2008;25(5):37–56.

[15]   Filho JCM, Panazio C, Abrão T. Uplink performance of single-carrier receiver in massive MIMO with pilot contamination. IEEE Access. 2017;5: 8669–8681.

[16]   da Silva MM, Dinis R. A simplified massive MIMO implemented with pre or post-processing. Physical Communication. 2017; Available from: http://www.sciencedirect.com/science/article/pii/S187449071630218X.

[17]   Horn RA, Johnson CR. Matrix Analysis. 1st ed. Cambridge University Press, New York, NY; 1987.

[18]   Mostofi Y, Cox DC. ICI mitigation for pilot-aided OFDM mobile systems. IEEE Transactions on Wireless Communications. 2005;4(2):765–774.

[19]   Barhumi I, Leus G, Moonen M. Equalization for OFDM over doubly selective channels. IEEE Transactions on Signal Processing. 2006;54(4):1445–1458.

[20]   Yacoub MD. Foundations of Mobile Radio Engineering. 1st ed. CRC Press, Inc., Boca Raton, FL; 1993.

[21]   Banelli P, Buzzi S, Colavolpe G, *et al.* Modulation formats and waveforms for 5G networks: who will be the heir of OFDM?: an overview of alternative modulation schemes for improved spectral efficiency. IEEE Signal Processing Magazine. 2014;31(6):80–93.

[22]   Lin YP, Phoong SM, Vaidyanathan PP. Filter Bank Transceivers for OFDM and DMT Systems. 1st ed. Cambridge University Press, New York, NY; 2011.

[23]   Paula A, Panazio C. A comparison between OFDM and single-carrier with cyclic prefix using channel coding and frequency-selective block fading channels. Journal of Communication and Information Systems. 2011;1(26):19–29.

[24]   Haykin S. Adaptive Filter Theory. 4th ed. Prentice Hall, Upper Saddle River, NJ; 2002.

[25]   Carmon Y, Shamai S, Weissman T. Comparison of the achievable rates in OFDM and single carrier modulation with I.I.D. inputs. IEEE Transactions on Information Theory. 2015;61(4):1795–1818.

[26]   Paula AS, Panazio CM. Comparison of OFDM and ideal SC-DFE achievable rates and performances without channel knowledge at the transmitter. Journal of Communication and Information Systems. 2017;32(1):23–28.

Chapter 5
Modern power line communication technologies

Samuel C. Pereira[1], Ivan R. S. Casella[2],
and Carlos E. Capovilla[2]

Power line communication (PLC) technologies allow the transmission of data through the electrical cables employed in power transmission, distribution and consumption networks, or in any other electrified environment, taking advantage of these important existing branched and interconnected infrastructures to reduce the time and costs of deploying communication networks.

PLCs have been used for a long time, but only with recent advances in digital communication, signal processing and circuit design techniques have become sufficiently reliable and secure for applications in data networks and smart grids (SGs).

PLC technologies have been designed for both narrowband and broadband applications and, as a result, several standards were developed to regulate their use. These standards specify several techniques to overcome the challenge of transmitting data through a medium (electrical cables) that was not originally designed for this purpose. Currently, PLCs are able to compete or complement wireless technologies in a variety of applications, presenting, under certain conditions, a number of advantages over them.

5.1 Introduction to PLC technologies

PLCs can be defined as any technology that transmits data through electrical power cables. Such data transmission may be carried out either in indoor environments (e.g., industrial, residential and vehicular electrical networks) or outdoor (e.g., power transmission and distribution networks).

Basically, the data to be transmitted, after being properly modulated (in baseband or bandpass), is injected into the electrical cables by means of a coupling circuit. The resulting PLC signal (modulated data) can coexist with the 50/60-Hz power line signal waveform, by operating at a significantly higher frequency, as shown in Figure 5.1. Moreover, PLC signals can also be injected into direct current (DC) power grids [1–3].

[1] Department of Automation and Process Control, Federal Institute of São Paulo (IFSP Suzano), Brazil
[2] Center for Engineering, Modeling and Applied Social Sciences (CECS), Federal University of ABC (UFABC), Brazil

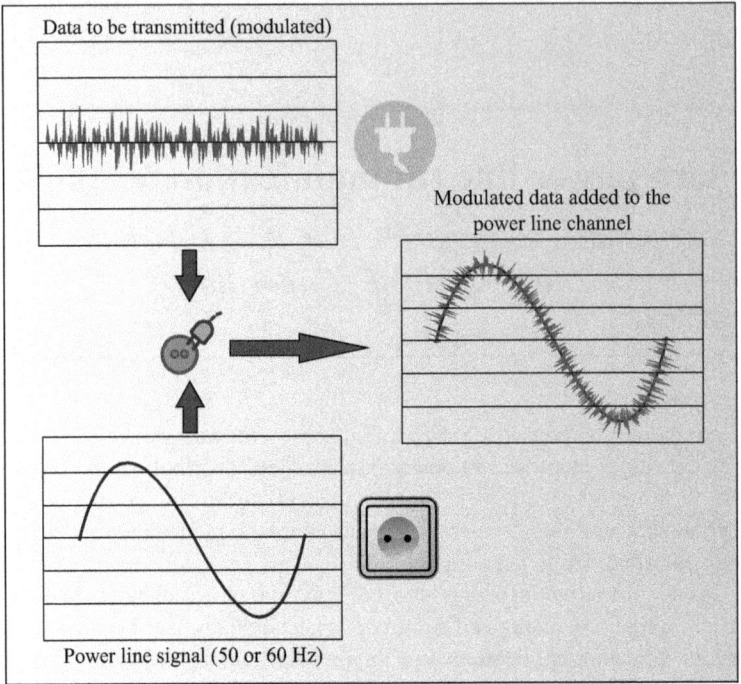

Data to be transmitted (modulated)

Modulated data added to the power line channel

Power line signal (50 or 60 Hz)

Figure 5.1   PLC and 50/60 Hz power line signals

Figure 5.2 shows a generic block diagram of a digital PLC modem (modulator/demodulator). In the transmitter, several types of external devices, with different communication protocols, can be connected to the PLC modem through a suitable data interface, which should be compatible with the PLC standard used. The information to be transmitted, in binary digital format, is encoded by a forward error correction (FEC) scheme and interleaved to protect the information against errors caused by the communication channel (i.e., power lines). The modulator then maps the digital (encoded) bitstream into an analog baseband or bandpass signal to be efficiently transmitted through the power line cables. The modulated signal is injected into the electrical cables by means of a coupling circuit which is generally an analog baseband or bandpass filter.

At the receiver, the decoupling circuit, which is also a baseband or bandpass filter, can mitigate the 50/60-Hz power line signal and unwanted electrical noises by selecting only the frequency range of the modulated information signal. The demodulator demaps the analog baseband or bandpass signal from the decoupling circuit to the interleaved (encoded) bit stream, which may contain errors introduced by the communication channel and by different types of noise in the power line. Finally, the bit stream is deinterleaved and directed to the decoder that detects and corrects the errors encountered, delivering the correct data information to the external devices.

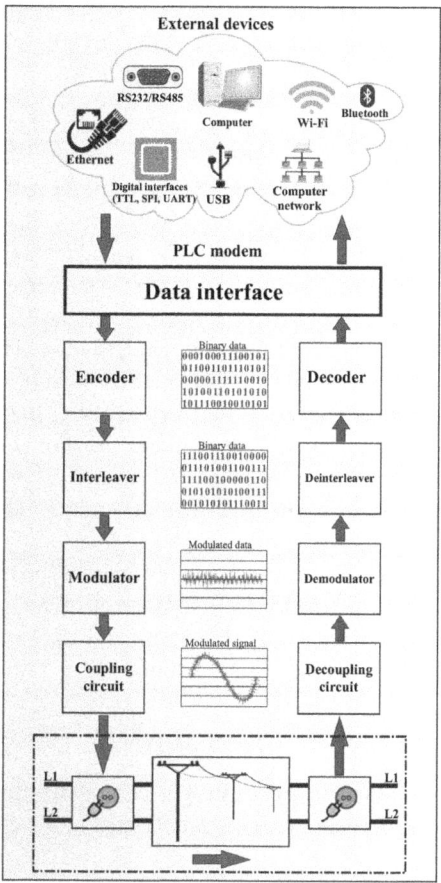

Figure 5.2   Generic block diagram of a PLC modem

5.2   Advantages and disadvantages of PLC technologies

As 84 percent of the world population has access to electricity [4], it is possible to conclude that the power grid is by far the largest physical network in the world. Therefore, the use of this enormous high capillarity infrastructure to convey information seems to be very advantageous. When compared to other wired solutions, this advantage is mainly related to the reduction of time and cost for the implantation of new communication infrastructures. Since this same advantage is offered by wireless solutions, it is necessary to analyze the advantages and disadvantages of PLC technologies more deeply before choosing them, rather than wireless technologies.

5.2.1   *Advantages of PLC technologies*

The main advantages of PLC technologies are as follows:

- Use of an existing infrastructure (i.e., the power grid), which reaches the majority of the world's population [4], covering entire countries, including the interior of buildings, homes and industrial facilities and remote electrified regions (reduction in time and cost of deploying the communication infrastructure);
- Ability to penetrate places where wireless signals have difficulties, such as tunnels, basements of buildings and houses, concrete and metal shielded structures;
- Physical contact with the electrical grid is required to gain access to PLC signals, which improves information and system security;
- Long-range communications. Ultra-narrowband PLC (UNB-PLC) systems can establish a low data rate (LDR) communication link over power lines with receivers located up to 200 km from the transmitter [5], while narrowband PLC (NB-PLC) systems up to 6 km over medium voltage (MV) lines (from 1 to 35 kV) [6];
- High data rate (HDR) communications. Modern broadband PLC (BB-PLC) systems offer maximum (theoretical) data rates of up to 1 Gbps, operating in single-input–single-output (SISO) mode, and up to 2 Gbps operating in multiple-input–multiple-output (MIMO) mode [7].

5.2.2   *Disadvantages of PLC technologies*

The power line cannot be considered a proper communication channel at all. The disadvantages of PLC technologies are mainly related to the channel characteristics, such as:

- The parameters of the power line operating as a communication channel vary over time according to the location and characteristics of the loads connected to the grid [6];
- The topology of the electrical network and its components are not designed to transmit data [high frequency (HF) signals]. The power line, mainly in low voltage (LV) networks (up to 1,000 V), has many branches with different characteristics (e.g., different cables, loads, electrical components, open ends). These ramifications and consequent impedance mismatches cause reflections of the PLC signals at various points, a phenomenon known as multipath propagation, which can degrade PLCs considerably [8];
- Under certain conditions, the noise power may be greater than the power of the PLC signal, resulting in negative signal-to-noise ratios (SNRs) [9]. In addition, in the LV power lines, there may be an SNR variation of the order of 10 dB abruptly due to the connection of a load in the power line [10];
- PLC signals can suffer severe attenuation. For example, a typical 200-m point-to-point PLC link over an LV power line may suffer signal attenuations of the order of 30 dB [6]. Furthermore, depending on the used PLC technology, voltage

transformers in the network can cause signal attenuations of the order of 50 dB or more [11];

- Power line cables are not shielded, so portions of the radio frequency (RF) energy they carry can be radiated. Thus, PLC systems operating in the range of 3.8–30 MHz may interfere with licensed radio communication services using the same frequencies, such as amateur radio and emergency bands [12,13].

5.3   History of PLC technologies

The first mention of a technology in which information is sent through power cables goes back to 1838 [14]. At that time, the English scientist Edward David (1806–55) created a system, known as the "Electrical Renewal," in which batteries were installed on the transmission lines to improve the signal range [15]. In order to remotely monitor the voltage level of these batteries, he developed a communication system through the power supply line of the London-Liverpool telegraph system [16].

Almost 60 years later, in 1897, Joseph Routin and C. E. Brown, in Zurich (Switzerland), patented a power meter controlled by signals sent from the power utility via power line. It was rudimentary and probably infeasible, but it started a concept that became reality decades later, i.e., energy meters remotely monitored by the utilities [17]. Following the same line, Chester H. Thordarson, in 1905, created a power meter that could be read remotely via power line plus an additional wire for signaling, in the United States of America (USA) [16].

In 1918, AT&T inaugurated commercial carrier frequency (CaF) over telephone lines. CaF technology, developed in 1911, allowed the transmission of multiple telephone channels over single telephone circuit (each channel using its own transmission frequency) [18]. At that time, the telephone lines were unreliable, expensive and fragile to weather conditions and mountainous terrain. As the power lines were more robust and already available, it was decided to use them for transmitting voice by means of CaF [19]. That is the reason, PLC technologies are also known as "power line carrier." Then, still in 1918, the first test and commercial operation of telephone signal over power lines were performed in Japan when, in May of that year, voice was transmitted over a 144-km power line (Kinugawa Hydro-Electric Co.). In the same year, commercial operation of this system was done over a three-phase power line of 22 kV (Fuji Hydroelectric Co.) [20].

In the USA, in July 1920, a communication system operating above 5 kHz was successfully tested, transmitting voice over a power line of 11 kV and 33 km in length (American Gas and Electric Co.). In late 1920, the telephone signal via the power grid, known as "wireless wired," was already widespread in the USA and Europe. In 1929, there were approximately 1,000 similar systems installed in Europe and the USA, operating in the frequency range between 50 and 150 kHz [20]. Due to the few branches, high voltage power lines (35–230 kV) proved to be excellent carriers of RF energy, to the point that a signal transmitted with a power of 10 W can reach distances of 500 km [21].

The CaF technology was also used for radio broadcasting through power lines. In 1923, in State Island (New York), the first radio system transmitted by the power grid was tested by the Wired Radio Service Company (controlled by The North American Company, a public utility). Radio signals within the amplitude modulation (AM) frequency band were inserted into the MV lines so that subscribers could receive and tune the signals radiated by the electrical lines (they were charged for this service directly in the electricity bill). This system, known as carrier current (CaC) was simpler and easier to tune than conventional radio systems that required more sensitive receivers and antennas and, therefore, more expensive and susceptible to reception problems [22]. CaC systems were also used in Switzerland, Germany, and Norway and were very common in the Soviet Union in the 1930s because of their low cost and accessibility [23]. In the USA, CaC was widely used to transmit local radios within the university campus and to provide news and music broadcast services to subscribers [24]. Nowadays, they are still used in drive-in cinemas and in travelers information systems inside bus stations [25].

Due to the intrinsic characteristics such as large number of branches, open terminations and different types of conductors that results in impedance mismatch and high attenuation of the HF signals, the LV and MV power lines were not initially intended for the deployment of communications systems operating at relatively high frequencies. Initially, in 1930 in Germany, low frequency (LF) unidirectional communication systems (below 3 kHz) were employed in LV and MV power lines for load-management applications [21]. These systems became quite popular in the 1950s when the Brown Boveri Electric, one of ABB's predecessor companies, along with other companies, developed a PLC system to manage the connection or disconnection of loads to the power grid [26] (the first patent for a similar communication system was issued by the French Cesar René Loubery in 1899 [27]). This system, known as the ripple control system (RCS), was later deployed in several European countries. In the RCS, the power utility sends electrical control signals at frequencies below 3 kHz, which are detected by the receivers connected to the LV power line at the customer's premises. These signals remotely switch high-power loads or groups of loads (machines, boilers, electric heating and public lighting) allowing the power utility to control peak loads [26]. These low frequencies signals with high power can cover a wide area, being able to cross MV/LV transformers in order to reach customers connected to the LV power lines [28].

In the end of 1972, substations with total capacity of 120 GVA, normally, had RCSs communicating with approximately five million receivers. At that time, these systems had been installed in most of the major European countries, Australia, New Zealand and in parts of Africa and South America [29].

Although the PLC technology available in the 1940s was commercially used in baby alarms [30], the first relevant commercial project involving PLCs emerged in 1975 with the X10 from Pico Electronics (a Scottish company), proposed to transmit signals to control lights and household appliances through power lines. In 1978, the X10 standard was introduced in the US market, kicking off the home automation industry [31], and today, even with the most modern communication technologies available, it is still popular all over the world [32]. In X10 systems, digital data (consisting of an address and a command) are sent from a controller device to a controlled device, by modulating a 120 kHz carrier and transmitting the modulated signal in bursts during the zero crossing of the 50/60-Hz signal (one bit per crossover) of the power grid [33]. These systems can achieve data rates of 20 bps (usable rate) [34].

In the 1980s to the early 1990s, available hardware technologies were able to implement FSK (frequency shift keying) modems using FEC and spread spectrum (SS) techniques and enabled the development of PLC systems with data rates of 1–144 kbps (operating at frequencies below 200 kHz and covering distances of 300–500 m) [21,28,35–37]. At that time, PLC systems above 200 kHz were not feasible due to distortions caused by branched power lines. However, in 1991, Nortel (a Canadian telecommunications company) and United Utilities (British power utility) began to investigate together the use of power lines to provide telecommunications services. At the outset, these companies have proved for the first time that a typical LV power line could propagate HF signals (over 1 MHz) by means of suitable coupling circuits. Then, employing appropriate compression, coding and modulation techniques, these HF signals could be used for modern telecommunications services [38].

In 1995, in Manchester, these companies started trials with 25 residential customers on the power line-based telephone system, providing a 32 kbps digital voice communication channel for each customer. In 1998, Nor.Web Communications was created, a joint venture between the two companies (Nortel and United Utilities). At that time, with its PLC technology, Nor.Web was able to offer a bidirectional digital communication link through the power line with data rates above 1 Mbps and with capacity for more than 200 subscribers. In addition to the United Kingdom, tests to provide massive, low-cost Internet access were successfully completed in the USA, Australia and Germany. However, it was found that the size and shape of UK street lamps made them capable of irradiating PLC signals (such as antennas), causing serious interference to British Broadcasting Corporation services and also affecting amateur radios and emergency bands. Because of this, Nor.Web was closed in 1999 even though it had received investments around 15–20 million of pounds sterling [10,13,39].

A major annual conference, the International Symposium on PLCs (ISPLC) and Its Applications, was created in 1997 by communications researchers in Europe and Asia seeking a forum for discussions on the challenges and achievements of PLC. As of 2006, this conference became financially and technically sponsored by the Institute of Electrical and Electronics Engineers (IEEE) Communications Society (ComSoc) [40].

In 2000, the HomePlug Power Alliance was founded, a nonprofit industry association that also aims to create standards for BB-PLC applications. In 2001, this alliance launched its first standard, the HomePlug 1.0, suitable for indoor applications (e.g., sharing broadband Internet within households) and operating in the frequency band of 4.5–21 MHz [41]. This standard, based on orthogonal frequency division multiplexing (OFDM), a multicarrier (MC) modulation technique and with carrier sense multiple access with collision avoidance (CSMA/CA) mechanism, presents data rates from 1.3 to 5.3 Mbps [42] with a typical maximum distance between devices of approximately 300 m [43]. In 2004, the Open PLC European Research Alliance (OPERA) project was created, consisting of 35 participants (utilities, universities, telecommunications operators and PLC technologies companies) and research funding from the European Community (EC). The objective of this project was the improvement of existing PLC technologies, standardization and development of solutions for connection of PLC networks to backbones, aiming to offer low-cost broadband access service to all European citizens through power lines [44].

Currently, new NB-PLC technologies operating below 500 kHz such as power-line intelligent metering evolution (PRIME) and G3-PLC, proposed respectively by PRIME Alliance and G3-PLC Alliance, can offer theoretical data rates of 150 kbps to 1 Mbps [45,46], while recent BB-PLC technologies operating at frequencies of up to 100 MHz, such as HomePlug AV2, can achieve theoretical data rates of up to 2 Gbps in indoors environments [7].

PLC technologies operating below 500 kHz have been of particular interest for use in SGs because they meet the data rates required by a large number of SG applications and can be used in outdoor environments where low frequencies are needed to reduce the level of interference in licensed radiocommunication bands, as well as providing a broader coverage [47].

Since the beginning of 2010, IEEE and the International Telecommunication Union (ITU) have been publishing a series of international standards for PLC technologies [48,49], which has been very important for the promotion of these technologies worldwide. Figure 5.3 shows the time line of PLC technologies history, highlighting the most relevant achievements in its development.

5.4   PLC classification, frequency bands and standards

PLC systems are generally classified according to their data transmission bandwidth, and can be divided into three main groups [50]:

- UNB-PLC
- NB-PLC
- BB-PLC

Each group has its own frequency bands, data rates, applications and standards.

5.4.1   UNB-PLC systems and its applications

PLC systems operating in the super LF (30–300 Hz) and ultra LF (0.3–3 kHz) bands are classified as UNB-PLC. Because of their LF, these systems provide long-range communication even through MV/LV transformers, which allow most customers to connect to LV/MV networks without the need for signal repeaters and/or additional coupling circuits (reduction of infrastructure costs). Their main disadvantage is the LDRs (120 bps maximum) [50].

Currently, UNB-PLC technologies are mainly used as communication systems for load management. In these systems, known as RCSs, power utilities send control

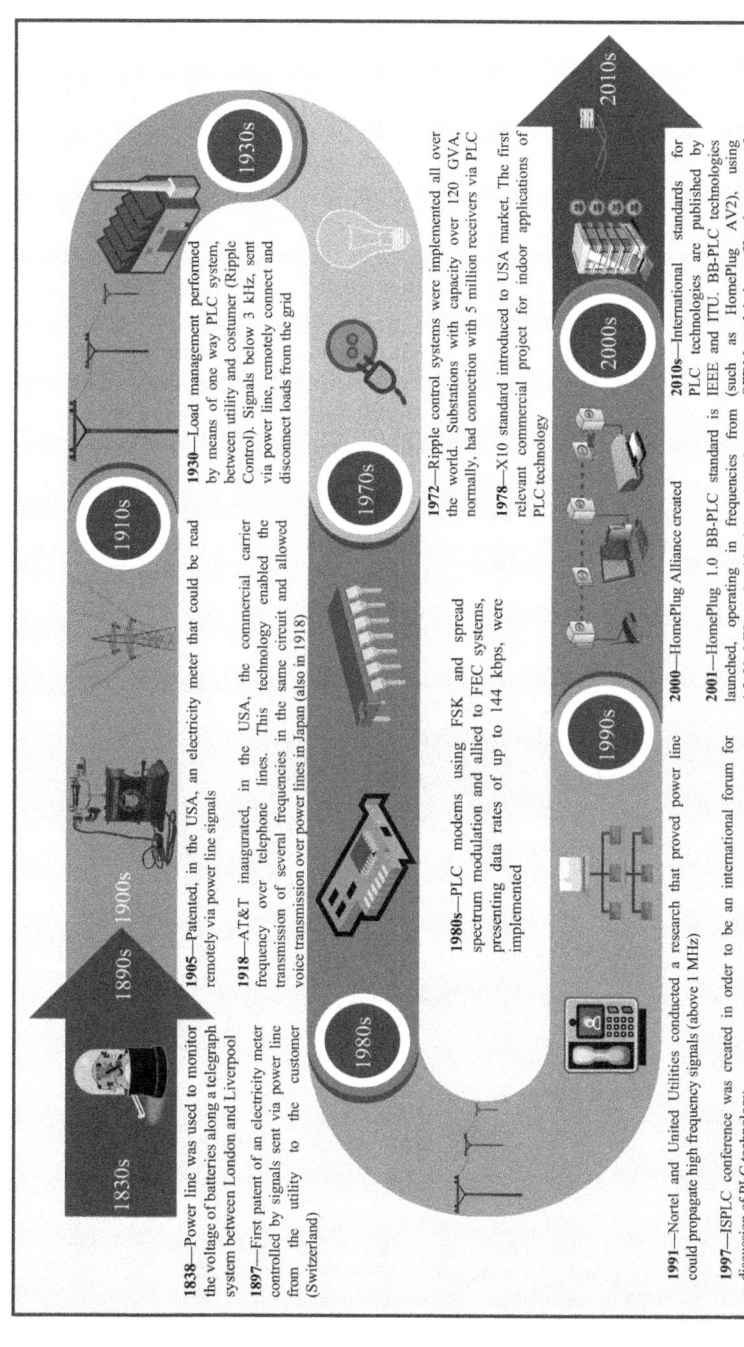

Figure 5.3   PLC technologies time line

signals through the electrical grid to remotely connect and/or disconnect loads or generators from the grid so that electricity can be managed mainly at peak times of demand. RCS signals occupy the frequency band between 100 Hz and 3 kHz (typically between 150 and 1,350 Hz) and are injected by power utilities into MV power lines where the control center (CC) is connected. They can reach clients located 150–200 km from the CC, in industrial and residential facilities connected to the MV or LV (by crossing transformers) power lines. At the customers side, there are devices installed by the power utilities and connected to specific loads or group of loads (e.g., hot water boilers, electric heating, household appliances, street lighting) and generators, making possible to connect or disconnect them from/to the power grid, according to the commands received. Customers who allow the installation of these devices can receive rebates on the electricity bill [5,26,28].

By allowing adequate control of power demand and reducing the problem of electricity supply during periods of peak demand, RCSs have been widely used in many different countries and continents, such as the USA, Europe and Oceania [51]. As shown in Figure 5.4, they typically use a one-way communication channel, but more modern systems can provide bidirectional communication, additionally enabling the transmission of the current status of the loads and generators, which is an important feedback on the success or otherwise of the remote switching operation. RCSs with bidirectional communication capability can also provide remote measurement of customer power consumption, known as automatic meter reading (AMR) [52].

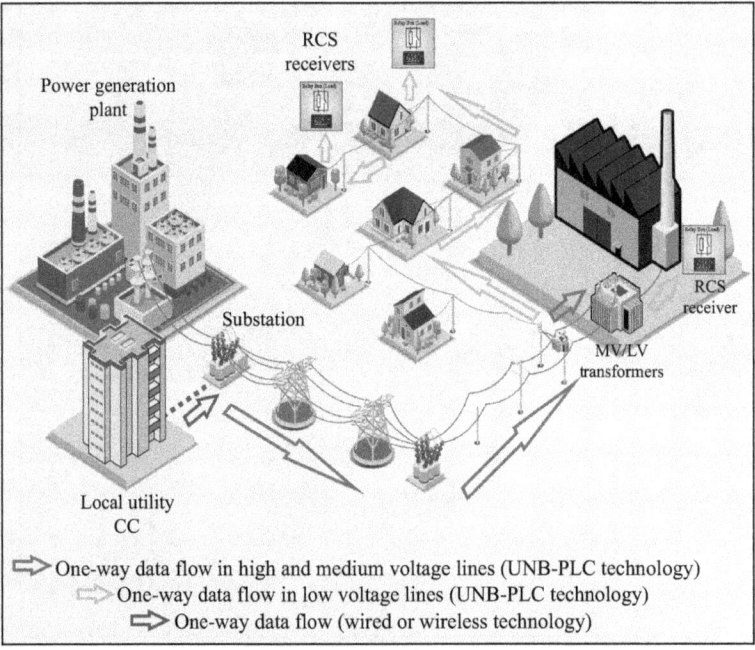

Figure 5.4   RCS example

5.4.2 Standards and frequency bands for UNB-PLC systems

The RCSs are tailor made for companies in the electrical sector, so there is neither standardization nor license fees for this type of service. However, for RCSs dedicated only for AMR applications, there are two main UNB-PLC standards: two-way automatic communication system (TWACS) and Turtle [50].

TWACS is a standard of communication through power line created in 1982 and widely used today. It was developed to provide bidirectional communication for AMR applications to the power utilities within their service areas [53,54]. It can also be used for load-management purposes and is mainly employed in small towns and rural areas, where residences are located farther from urban centers, since TWACS provides long range communication (over 300 km in rural areas) [55]. It relies on modifying the 50/60-Hz signal waveform of the power line. For transmissions from the CC to customers, voltage pulses are injected into the power line during zero crossing of the 50/60-Hz signal (these voltage pulses are superimposed on the power line signal as minor distortions) [53]. In the transmissions of the customers to the CC, current pulses are used [56]. The information to be transmitted is modulated by the polarity and quantity of these voltage/current pulses [53] which are within the range of 200–600 Hz [57]. TWACS systems can achieve data rates of up to 120 bps in 60 Hz power lines and 100 bps in 50 Hz (2 bits per cycle) [58].

On the other hand, created in 1995, the Turtle system was developed for AMR applications. This technology offers one-way communication through power lines and enables thousands of devices to transmit digital information continuously. Each power meter uses a specific frequency within the range of 5–10 Hz with a bandwidth of 0.0015 Hz. This very narrow bandwidth provides a data rate of only 0.0005 bps, so that the simple reading of a meter usually takes a whole day, which makes this system unsuitable for load management applications (response time for connecting or disconnecting loads is impractical). However, this LDR has the advantage of requiring very low transmit power (some milliwatts allow the PLC signal to reach hundreds of kilometers and also to cross MV/LV transformers), which enables the development of low cost power meters with built-in Turtle technology transmitters. In addition to AMR applications, the Turtle system can provide warnings of power failure (when a device stops sending information, disregarding a potential device failure, it indicates a power outage in your area) [59,60].

5.4.3 NB-PLC systems and its applications

NB-PLC systems operate in the frequency band between 3 and 500 kHz, which covers the very LF (3–30 kHz) and LF (30–300 kHz) and part of the medium frequency band (MF – 300 kHz to 3 MHz) [50]. Like UNB-PLC systems, the use of low frequencies allows the use of NB-PLC technologies in external applications without the risk of interfering with licensed radio communication services.

The first relevant NB-PLC system was the X10 protocol, launched in 1978 and still popular today. It is designed for residential automation (e.g., lights and appliances control) through the power lines, operating at 120 kHz and offering usable data rates of 20 bps [31,32,34].

In addition to the home automation application, NB-PLC systems can also be used in home energy management systems (HEMS), dedicated to monitoring, control and analysis of energy in homes, which allows the monitoring of energy received from the power utility (energy meter), monitoring and control of household appliances, heating system and lighting, integration of photovoltaic panels and batteries, and even communication with plug-in electric vehicles. HEMS can improve the efficiency of energy consumption through the automatic management of loads and power generators [50,61].

However, the need to develop a system to read energy meters and also to connect devices for monitoring and control the conditions of the outdoor power lines that integrate the distribution network were the main factors that motivated the development of NB-PLC systems with bidirectional capacity and high rate data transmission [28].

In this context, NB-PLC systems can be used to develop the advanced metering infrastructure (AMI), which allows utilities to exchange information with consumers energy meters (which may include electricity, gas and water readings), allowing customers to monitor their energy consumption in real time, check the value of current power rates and even to remotely connect or disconnect loads and generators.

For AMI, NB-PLC systems can be the last mile communication solution connecting the energy meters, known as smart meters, to the local LV power lines (the range of NB-PLC systems in LV power lines can be of the order of 200–300 m) [6,37], as shown in Figure 5.5. Smart meters, after receiving remote commands from the power utility, send the read-in information to a local data concentrator that forwards them to the CC, usually via broadband communication technologies (e.g., BB-PLC).

PLC is the main communication technology used in smart meters, with 60 percent of the market share [62]. As modern NB-PLC systems offer higher data rates, it is possible for the power utility to take advantage of AMI to add new services, such as distribution automation (DA), responsible for monitoring and control of distribution networks, street lighting control and local loads and generators management (integration between the CC and HEMS at the customer's facilities), i.e., in the direction on the SGs.

5.4.4   Frequency bands for NB-PLC systems

In 1991, the European Committee for Electrotechnical Standardization (CENELEC), responsible for standardization in the field of electrotechnical engineering in Europe, published the standard EN50065-1, which regulates the frequency band between 3 and 148.5 kHz for signaling on LV power lines. This frequency band was divided in four groups referred to as CENELEC A, B, C and D bands, each one dedicated to different users and applications [63,64]:

- CENELEC A (3–95 kHz), reserved for energy providers and their utilities;
- CENELEC B (95–125 kHz), reserved for customers of power utilities (all applications), without the need for any access protocol;
- CENELEC C (125–140 kHz), reserved for customers of power utilities with the purpose of home networking. In this band, the use of access protocol,

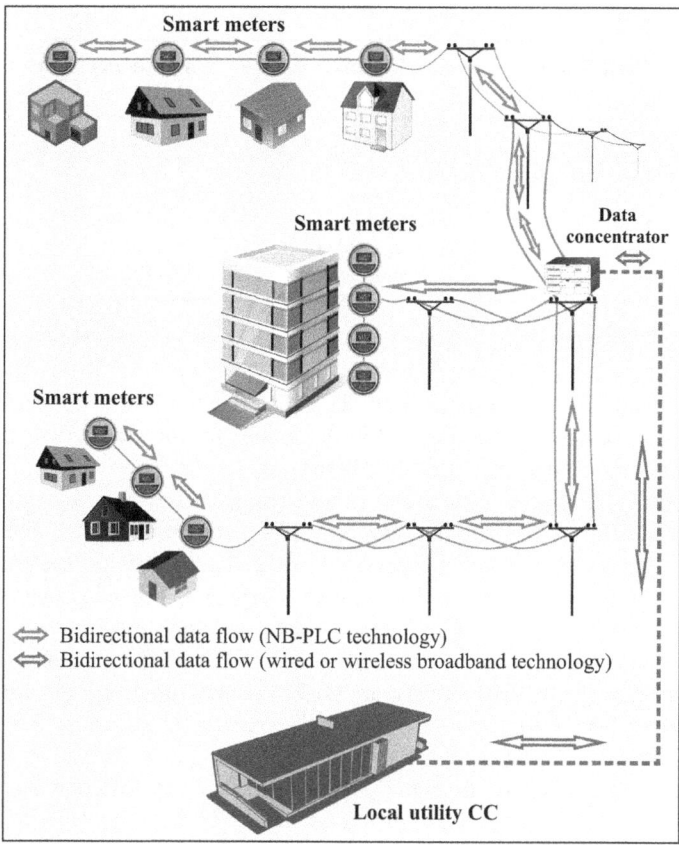

Figure 5.5   AMI last mile solution based on NB-PLC technologies

i.e., CSMA/CA, is mandatory which allows the coexistence of different and/or incompatible systems operating in this band;

• CENELEC D (140–148.5 kHz), reserved for customers of power utilities (alarms and security systems, without access protocol).

In the USA, the Federal Communications Commission (FCC) is responsible to regulate the telecommunications in all 50 states, District of Columbia and in the US territories. This agency allocated the frequency band of 9–490 kHz for electric utilities to operate NB-PLC systems in all applications [65]. This wider frequency band allows PLC systems with higher data rates in comparison with CENELEC band, because any device can use the whole band.

In Japan, the Association of Radio Industries and Businesses (ARIB), an agency established to promote the research and development of radio systems and to standardize telecommunication activities in Japan [66], allocated the frequency band of

Table 5.1   *Frequency bands for NB-PLC systems*

Organization	Region	Frequency band (kHz)
ARIB	Japan	10–450
CENELEC	Europe	3–148.5
CEPRI	China	3–90
FCC	USA	9–490
ITU	Worldwide	9–490
IEEE	Worldwide	10–490

10–450 kHz for NB-PLC systems [67]. The China Electric Power Research Institute (CEPRI), government agency focused in researches for the electric sector, defined the frequency band between 3 and 90 kHz [68].

In addition to government agencies that regulate telecommunications locally (regional regulation), international organizations recommend frequency bands and standards worldwide to promote interoperability and compatibility between devices from different manufacturers and countries. These recommendations are usually adopted in most countries, making them mandatory in their territories.

For instance, the ITU recommended an NB-PLC standard operating from 9 to 490 kHz [49], while the IEEE created an NB-PLC standard operating in the 10–490-kHz frequency band [69]. Both standards can be configured to operate in the regional frequency bands.

Table 5.1 describes the frequency bands established by different organizations around the world. It is important to mention that CENELEC is considering to extend its frequency band to 500 kHz in order to harmonize with FCC and ARIB bands [50]. NB-PLC systems operating in the 148.5–500-kHz frequency band (i.e., outside of CENELEC band) may have access to the European market as long as they comply with the electromagnetic compatibility (EMC) rules of the EC [70].

5.4.5   Standards for NB-PLC systems

NB-PLC systems can be divided in two groups: LDR and HDR [50].

5.4.5.1   LDR NB-PLC

LDR NB-PLC systems employ single carrier (SC) modulation techniques and offer data rates below 10 kbps. The X10 can be considered the first relevant LDR NB-PLC standard, being released in 1978 with a data rate of 20 bps (useful rate). In the following decades, new standards were developed with FEC and SS techniques that resulted in more reliable communication systems with higher data rates, enabling Internet Protocol (IP) operation. LDR NB-PLC standards have been developed for residential automation (indoor applications), remote metering and public lighting control (outdoor applications). Some examples of popular LDR NB-PLC standards are:

- LonWorks (ISO/IEC 14908): Introduced in 1990 by Echelon Corporation, Lon-Works (Local Operating Network) was approved by the International Organization for Standardization (ISO) and the International Electrotechnical Commission (IEC), in 2008, as an international standard (ISO/IEC 14908), covering all seven layers of the Open System Interconnection (OSI) model. At that time (2008), there were already over 100 million devices with LonWorks standard installed around the world, being used in different markets, like transportation, utilities, process control and home automation [71,72]. LonWorks uses binary phase shift keying modulation with differential Manchester encoding, operating in CENELEC bands (A and C). In addition to power lines, it also works over twisted pair cables and wireless medium. Over power line cables, in CENELEC A band, LonWorks has a data rate of 3.6 kbps and operates in the frequency band of 75–86 kHz. In CENELEC C band, the data rate is 5.4 kbps operating in the frequency band of 115–132 kHz [73].
- IEC 61334: This is an international standard, known as "distribution automation using distribution line carrier systems," which offers LDRs over power line for remote energy metering and SG applications. IEC 61334 covers all seven layers of OSI model and in the Physical (PHY) layer operates in the frequency band of 20–95 kHz with spread FSK (S-FSK) modulation and without FEC schemes, providing an effective data rate of 2.4 kbps [68]. NB-PLC systems operating with FSK and S-FSK modulation are the most commonly deployed technology for AMR infrastructure (widely used by power utilities) [72].
- HomePlug C&C: The HomePlug Command and Control (HomePlug C&C) standard was defined by HomePlug Powerline Alliance, being the PHY and medium access control (MAC) layers ratified in 2007. This NB-PLC standard, which covers all seven layers of OSI model, is designed for command and control purposes in home appliances, home automation and SG applications. HomePlug C&C devices are plug and play, being able to create a secure network automatically (just connecting the device to the power line), and have CSMA/CA mechanism. The PHY layer specifies data interleaving processes, FEC coding and an SS modulation called differential code shift keying, for operation in the FCC, ARIB and CENELEC frequency bands (A and B). HomePlug C&C offers a maximum data rate of 7.5 kbps [74].

Other examples of NB-PLC standards: ISO/IEC 14543-3-5 (KNX), CEA-600.31 (CEBus), Insteon, SITRED, Ariane Controls, BacNet [50].

5.4.5.2 HDR NB-PLC

HDR NB-PLC systems are based on OFDM and can achieve theoretical data rates of up to 1 Mbps (at the PHY layer) [46]. HDR NB-PLC systems are designed to be the last mile communication solution for the AMI and meet the requirements of other applications planned for the SGs in addition to the AMI, such as DA, public lighting control, load management, which are not supported by UNB-PLC and LDR NB-PLC systems.

The first HDR NB-PLC systems, PRIME and G3-PLC, were not created by standard development organizations (SDOs), but they were later ratified as international standards. The IEEE and the ITU also published their HDR NB-PLC international standards, respectively, IEEE 1901.2 and ITU-T G.hnem. These standards only specify the two lower layers of OSI model (PHY and MAC layers) [72].

PRIME

The PRIME initiative was started at the end of 2006 by Iberdrola, a power utility from Spain, aiming to develop an open and royalty free PLC standard, covering PHY and MAC layers, in order to offer a solution that attends either AMI or SG needs. In 2007, the Spanish government created a directive to oblige all power utilities to replace, by 2018, their meters under 15 kW (10 million devices only for Iberdrola) with smart meters [75]. The field tests with PRIME standard also started in 2007 and several companies joined to the PRIME Alliance, including utilities and semiconductor companies [28]. In 2018, more than 18 million smart meters with PRIME standard were deployed in over 15 countries worldwide [76].

In its latest version (v1.4 from 2014), PRIME uses the frequency band from 41.992 to 471.679 kHz (divided in eight frequency channels that can be selected and combined in different ways), being able to operate in CENELEC, ARIB and FCC bands (the previous version, v1.3.6, works only on CENELEC bands). It is based on OFDM, using differential phase shift keying (DPSK) (carrying one, two or three bits per symbol) mapping for the subcarriers. Besides convolutional (CONV) coding and data interleaving for protecting the information, PRIME offers an optional robust operation (ROBO) mode in which the information inside the data frame is repeated by a factor of four, which allows its operation even in conditions with negative SNR. Combining the eight channels, PRIME v1.4 achieves a theoretical data rate ranging from 5.4 to 1,028.8 kbps [46,77].

The convergence layer of PRIME MAC, responsible for interfacing the information with upper layers of OSI model, is divided in two sublayers. The common part convergence sublayer (CPCS) provides generic services (only data segmentation and reassembly) and the service specific convergence sublayer (SSCS) with specific services for different communication profiles. SSCS allows transference of IPv4 and IPv6 packets over PRIME networks and also supports data packets in the format specified in IEC 61334-4-32 standard, which allows to replace the PHY and MAC layers of old IEC 61334 systems with PRIME standard. PRIME works with 128-bit Advanced Encryption Standard (AES-128), which guarantees a secure PLC network even at lower levels. The channel access is made by means of CSMA/CA mechanism [78].

In 2012, PRIME was approved, by the ITU, as an international standard, being the ITU-T G.9904 recommendation [79].

G3-PLC

In 2008, Maxim Integrated won a contract to develop a new PLC specification for implementing AMI to the Electricite Reseau Distribution France (ERDF), a French power utility with over 35 million customers. This partnership between Maxim and ERDF, also with participation of Sagem Communications, resulted the G3-PLC standard which was released in 2009 [80,81]. In 2011, 12 companies of the SG sector (utilities, equipment and semiconductor companies) created the G3-PLC Alliance in order to promote G3-PLC worldwide as an open international standard and also its applications in the SG and home automation scenarios [82].

The G3-PLC standard specifies the PHY and MAC layers, and similarly to PRIME, it is also based on OFDM with DPSK mapping for the subcarriers. However, in addition to data interleaving and CONV coding, G3-PLC also employs a Reed–Solomon (RS) coding in a concatenated way (before CONV coding) to improve the robustness of data transmission. The G3-PLC also provides a ROBO mode by means of a repetition coding with factor four (applied after the CONV coding and only with DPSK mapping for the subcarriers). There is also a super ROBO mode in which the information of frame control header (FCH), part of the frame, is repeated six times (the encoded data is repeated four times either for ROBO or super ROBO mode). G3-PLC operates in CENELEC, ARIB and FCC bands, offering a theoretical data rate of up to 300 kbps in FCC band and in its less ROBO mode when associated with differential 8-phase shift keying (D8PSK) mapping.

The G3-PLC has an adaptive tone mapping (ATM) function in which, based on the channel estimation, it is possible to carry data only on the subcarriers located in parts of the frequency spectrum that have better conditions, that is, higher SNRs. This function allows the adaptation of the communication system to the dynamic behavior of the PLC channel [83].

Due to its robustness, the G3-PLC signal has the ability to traverse MV/LV transformers without the need for coupling circuits [6]. This feature is especially important in the deployment of metropolitan area networks for Internet access or neighborhood area networks (NANs) for smart metering in countries such as the USA where an MV/LV transformer serves only six to eight households (in countries of Asia and Europe, a transformer can serve 200–300 households) [10]. Thus, for instance, to form a NAN between several smart meters, the PLC signal will probably have to traverse a couple of transformers and G3-PLC will simplify the overall deployment of the system. This reliability and robustness of G3-PLC can also provide long range communication in MV power lines. During the G3-PLC field tests in the USA, a communication link of 8 km over MV power lines and of 1.5 km was established when crossing transformers (MV to LV network) [84].

The G3-PLC MAC layer is based on IEEE 802.15.4, with AES-128 and, similarly to PRIME, the access to the medium is also performed by means of CSMA/CA mechanism. The convergence layer is based on 6LoWPAN (low-power wireless personal area network/IETF RFC 4944) which accommodates IPv6 packets at the top of G3-PLC MAC layer [85] and becomes G3-PLC compatible with small and simple devices (low power), inside the internet of things (IoT) concept.

In 2012, G3-PLC was approved, by the ITU, as an international standard, being the ITU-T G.9903 recommendation [83].

IEEE 1901.2

The IEEE working group which developed IEEE 1901.2 standard started in the beginning of 2009, aiming to create a PLC solution for automotive industry. In November of this same year, during a meeting sponsored by the NIST (National Institute of Standards and Technology), IEEE agreed to sponsor a new PLC standard for SG applications below 500 kHz, suitable for alternating and DC power lines, indoor and outdoor applications (either for LV or MV power lines), with ability to cross MV/LV transformer and data rate of up to 500 kbps. This standard would define PHY and MAC layers and should have mechanism to guarantee coexistence (i.e., to share the same frequency band) between the different NB-PLC technologies available [86]. In November of 2013, IEEE 1901.2 "Standard for LF (less than 500 kHz) Narrowband PLCs for Smart-Grid Applications" was approved by the IEEE, being published in the next month [87].

IEEE 1901.2 is based on G3-PLC standard, having similar PHY layer and MAC based on IEEE 802.15.4 with support to IPv6 packets (convergence layer not specified). Security and access mechanism are also similar to G3-PLC. Despite these similarities, there are some differences that make G3-PLC and IEEE 1901.2 noninteroperable technologies [88]. These differences are related to [88]

- size of the data unit on MAC layer;
- lack of specification of collision avoidance mechanism for fragmented packets and frame counter for anti-replay protection in IEEE 1901.2;
- use of information elements (IE) in the frames of IEEE 1901.2 (IE are optional and variable in length);
- possibility of the super ROBO mode (repetition by a factor of six) for the payload (user data) part of the IEEE 1901.2 PHY frame (G3-PLC applies super ROBO mode only for the FCH).

IEEE 1901.2 also has ATM function, operates in the CENELEC (A and B), ARIB and FCC bands and can coexist with other NB-PLC technologies [88].

ITU-T G.hnem

In January 2010, the ITU started the G.hnem project in order to create a worldwide NB-PLC standard, for SG applications, which aggregates features of PRIME and G3-PLC, with new functions and improvements for a better performance. In 2011, the ITU-T recommendations G.9955 and G.9956, respectively, PHY and MAC specifications of the ITU-T NB-PLC standard were approved and published [89].

The ITU-T G.hnem PHY layer is similar to G3-PLC, with RS and CONV encoder. However, the interleaver splits the data into blocks, and in the ROBO mode, these blocks can be repeated 2, 4, 6 or 12 times (before interleaving). G.hnem is also based on OFDM, but unlike G3-PLC and PRIME, it does not use DPSK mapping for the subcarriers but quadrature amplitude modulation (QAM), and each QAM symbol can be configured to carry 1, 2 and 4 bits. QAM mapping needs a coherent receiver, i.e., the subcarrier phase is used as reference for demodulation, which increases the receiver complexity, but improves the SNR gain (less degradation of PLC signal). G.hnem also operates in CENELEC (A, B and CD) and FCC bands with ATM function [89].

G.hnem MAC layer defines CSMA/CA mechanism for coexistence with other NB-PLC technologies and its convergence layer supports multiple protocols, including IPv4, IPv6 and Ethernet. Security is obtained by means of AES-128 code [49,89].

5.4.6 BB-PLC systems and its applications

BB-PLC systems operate in the frequency band between 1.8 and 250 MHz, which cover part of MF (300 kHz to 3 MHz), HF (3–30 MHz) and part of very HF (30–300 MHz) bands [50]. Because of their operation in high frequencies, these systems are suitable for applications involving data rates of tens or hundreds of Mbps. Today, BB-PLC systems can reach data rates of up to 2 Gbps (over PHY layer) [7].

In 1991, it was proved that power line could propagate HF signals (above 1 MHz) [38], which opened the doors for the implementation of BB-PLC systems. The initial purpose of this technology was to deliver broadband telecommunication services to residential customers, but it was noticed that in outdoor applications, it could affect licensed radio communication services that uses the same frequency band (similarly for antennas, power line is not shielded and emit RF radiation related to the PLC signal) [12,13]. In 2002, the Japan's government, based on studies that monitored radiation emitted by BB-PLC systems, prohibited PLC systems operating between 2 and 30 MHz due to harmful effects on licensed radio services operating in HF band [90]. In the USA, BB-PLC systems, in outdoor applications, are not prohibited; however, FCC rules regulate the transmission power, and consequently the signal range, so that a BB-PLC network would require a large number of signal repeaters, increasing the deployment costs [50].

Although the OPERA project (conducted by the EC in 2004) which promoted research to develop technologies for providing low-cost broadband telecommunication services via power line to all European citizens [44], BB-PLC systems became restricted to indoor applications (sharing Internet connection and multimedia data in an in-home scenario) [50]. In this type of application (shown in Figure 5.6), which started in 2001 with HomePlug 1.0 standard from HomePlug Alliance [41], BB-PLC modems connected to the power sockets create an in-home local area network (LAN) in order to share Internet connection, peripherals (printers, scanners and so on) and multimedia data (audio and video). It is an alternative to the popular Wi-Fi 802.11 with the advantage of reaching all the rooms of the house (Wi-Fi signals have difficult to penetrate infrastructure). Today, there are over 200 million HomePlug BB-PLC devices, from different manufacturers, in the market [91].

5.4.7 Frequency bands for BB-PLC systems

In Europe, CENELEC EN 50561-1 standard, from 2013, regulated the frequency band of 1.6–30 MHz for BB-PLC systems. This standard demands a specific dynamic protection for radio broadcast services so that BB-PLC systems need to be equipped with a detector of radio signals in order to not transmit in the frequencies these signals are detected (if no signal is detected, the whole frequency band available can be used for transmitting PLC data). Although this dynamic control, which must be performed

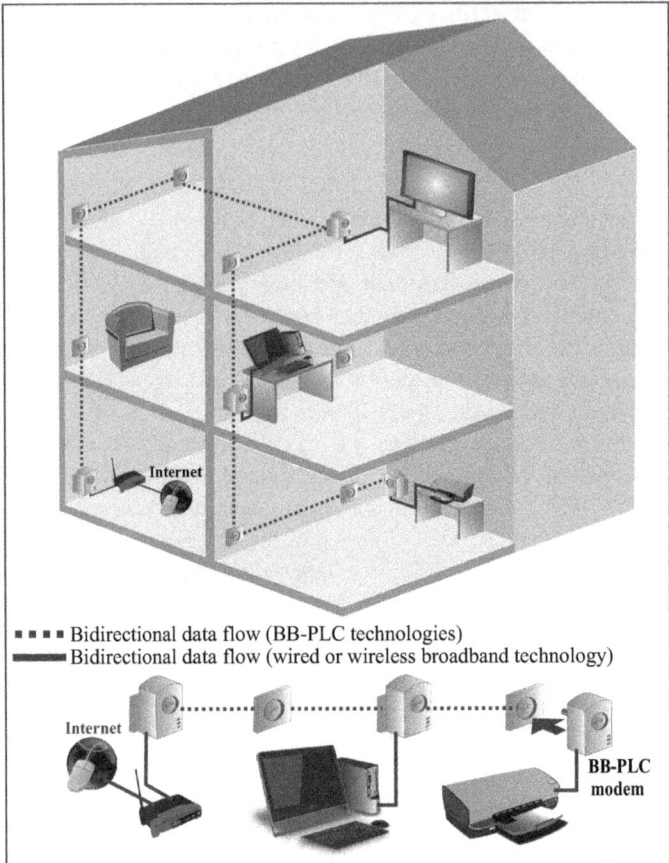

Figure 5.6   In-home LAN with BB-PLC technologies

automatically by the BB-PLC devices, CENELEC EN 50651-1 also mentions some frequencies bands, inside the frequency band of 1.6–30 MHz, permanently excluded because of their use in radio communication services. BB-PLC standards operating above 30 MHz, such as IEEE 1901-2010 (2–50 MHz) and ITU-T G.hn (2–100 MHz), can have access to the European market by complying with the EMC regulation established by the EC [70,72,92,93].

In the USA, there is no regulated frequency band for BB-PLC systems. In 2004, in order to cover these systems, FCC published a "Reporter and Order" (FCC 04-245) which implemented changes to part 15 of its rules. FCC considers, from a regulatory standpoint, that BB-PLC systems are an unlicensed and unintentional emitter of RF energy and requires that this type of device must not cause interference to licensed radio services, having to comply with the FCC EMC rules which limit the power of their RF radiation. These systems must also accept any interference caused by normal activity of these same radio services [94]. However, differently from CENELEC, FCC does not specify any excluded frequency band and/or mechanism of protection for radio services.

5.4.8   Standards for BB-PLC systems

In the 2000s, industry associations were created in order to develop BB-PLC standards for in-home applications, e.g., the HomePlug Power Alliance (HomePlug), the Universal Powerline Association, the High-Definition PLC (HD-PLC) Alliance and the HomeGrid Forum [50]. The standards created by these associations specify PHY and MAC layers and are based on OFDM. Later, the IEEE and the ITU published their international standards, respectively, IEEE 1901-2010 and ITU-T G.hn.

5.4.8.1   HomePlug

The HomePlug Power Alliance was created in 2000 in order to develop standards for home powerline networking products. Its first standard, the HomePlug 1.0, was launched in 2001 and operates in the frequency band of 4.5–21 MHz. The PHY layer employs a concatenated FEC scheme based on RS and CONV coding, interleaving, DPSK mapping and OFDM modulation with 84 subcarriers. The MAC layer, with CSMA/CA mechanism and security scheme based on the 56-bit Data Encryption Standard (56-bit DES), was developed to work with IEEE 802.3 frame formats, which allows integration with the widely deployed Ethernet standard. The theoretical maximum data rate over PHY layer is 13.78 Mbps and field tests in a typical residential environment presented data rates of 1.6–5.3 Mbps (at application layer of OSI model) [41,42]. In 2008, the Telecommunications Industry Association (TIA) published the TIA-1113 international standard which is based on HomePlug 1.0 [95].

In 2005, the HomePlug AV (audio and video) standard was launched. Different from HomePlug 1.0, that was designed for sharing broadband Internet connection in a home environment, the HomePlug AV was designed to distribute high definition video and audio, e.g., high-definition TV (HDTV) and voice over IP (VoIP), also in a home environment. For that, it provides a theoretical data rate of 200 Mbps over PHY layer and 150 Mbps of information rate, operating in the frequency band of 2–30 MHz. Data at PHY layer are protected by a turbo convolutional coding and interleaving; then, they are mapped to DPSK or QAM symbols and transmitted by means of OFDM modulation with 1,155 subcarriers (of which 917 carry data). The MAC layer supports CSMA/CA with optional time division multiple access (TDMA) scheme, and the security is based on AES-128 (the convergence layer supports Ethernet format packets so that all IP based protocols are compatible). HomePlug AV standard also offers ROBO mode, supports ATM function and coexistence and optional interoperability with HomePlug 1.0 [96].

In 2010, a simplified version of HomePlug AV in order to offer a low cost and low power solution that is suitable for SG and energy applications in home area networks was introduced. This version, called HomePlug Green PHY, uses the same PHY layer of HomePlug AV but supports only differential quaternary phase shift keying (DQPSK) subcarrier mapping, which limits the data rate over PHY layer to 10 Mbps. HomePlug Green PHY does not support ATM and its MAC layer has only CSMA/CA mechanism (differently from HomePlug AV, it does not support TDMA). Moreover, it has a power save mode for saving energy when the device is not transmitting or receiving data. HomePlug Green PHY is interoperable with HomePlug AV, AV2 and the IEEE 1901-2010 [97].

In the future, the integration of several multifunction devices inside a home environment (HDTV, VoIP, interactive gaming, broadband Internet, home automation, security monitoring, IoT and so on) will demand higher data rates from network devices. For covering this demand, in 2012, it was launched the HomePlug AV2 that can achieve theoretical data rates of 1012 Mbps (over PHY layer). It uses the same architecture of HomePlug AV and HomePlug Green, which allows interoperability between them. However, it has an extended frequency band, from 1.8 to 86.16 MHz, which allows the use of 3,455 subcarriers and, consequently, a higher data rate in comparison to HomePlug AV. HomePlug AV2 supports MIMO operation, which allows to transmit data in two pairs of wires (line-neutral and line-ground or neutral-ground inside a three wire configuration). The second pair of wires allows using a second independent transmitter, which doubles the data rate, achieving a theoretical data rate of 2,024 Mbps (over PHY layer). In premises that do not have a three wire configuration to implement MIMO, the HomePlug AV2 device automatically changes the configuration to standard SISO operation. The MAC layer is similar to HomePlug AV except that it was added a power save mode, like in HomePlug Green [7,98].

5.4.8.2   HD-PLC

HD-PLC is a BB-PLC standard (MAC and PHY layers) created in 2005 by Panasonic Corporation [99]. In order to promote the world wide operation and interoperability of its standard, Panasonic founded in 2007 the HD-PLC Alliance, a nonprofit association that currently has 22 members, including industries and SDOs [100,101]. The efforts of this alliance led HD-PLC to be used as part of the international standard IEEE 1901-2010 [99].

HD-PLC PHY layer is based on the wavelet OFDM technique (W-OFDM) with pulse width modulation (PAM) mapping in each subcarrier. In comparison to the conventional fast Fourier transform OFDM (FFT-OFDM), W-OFDM does not require guard intervals between OFDM symbols. Due to this feature, the W-OFDM provides higher transmission efficiency so that HD-PLC PHY can achieve a theoretical data rate of 240 Mbps over a 26-MHz bandwidth (frequency range 2–28 MHz). Concatenated RS and CONV FEC coding and interleaving are employed to increase the robustness of the system to transmission errors caused by the communication channel. Optionally, the low density parity check CONV (LDPC-CONV) coding may also be employed [102].

MAC layer supports two types of media access mechanism, CSMA/CA and dynamic virtual toking passing (DVTP). The security uses AES-128 encryption code and the convergence layer supports Ethernet packets. HD-PLC also supports ATM function which allows, after channel estimation, the physical layer to use an optimized configuration according to channel conditions and consequently to improve network throughput [102].

For avoiding interference with licensed radio communication services, HD-PLC supports dynamic notch (DN) and dynamic power control (DPC) functions. The HD-PLC device with DN function periodically senses the presence of short wave radio signals in the medium, and once it is detected, the device will add notches to the PLC signal in the regions of the frequency spectrum in which the radio signals were detected. The DPC function reduces the power of the PLC signal when the HD-PLC device has a good communication link that does not need maximum power transmission [102].

5.4.8.3 IEEE 1901-2010

In 2005, the IEEE P1901 Working Group, sponsored by the IEEE ComSoc and aiming to develop an international BB-PLC standard with data rates over 100 Mbps (over PHY layer) and frequencies below 100 MHz, was created [103]. At the end of 2010, the standard created by this working group (IEEE 1901-2010) was published [104].

IEEE 1901-2010 BB-PLC standard defines two PHY/MAC layers, one based on HomePlug AV and another based on HD-PLC standards. The choice for one of the layers is optional, but IEEE 1901-2010 specifies a mandatory Intersystem Protocol (ISP) that guarantees the coexistence between both PHY layers (but no interoperability) and other BB-PLC standards. In addition to operating in the frequency band of HomePlug AV and HD-PLC (up to 30 MHz), IEEE 1901-2010 standard can optionally operate in frequencies above 30 MHz (1.8–50 MHz), which enables theoretical data rate of 500 Mbps (over PHY layer and operating in the whole frequency band) [105].

The reason for the use of two PHY/MAC layers is related to the divergence of technologies in the current PLC market. This can be considered a step toward convergence of PLC technologies which will occur in the future once it is established a solid market demand for BB-PLC systems [50].

IEEE 1901-2010 is designed to either in-home or access networks. Access networks involves several nodes (hundreds or thousands) in a larger area and with centralized control, offering broadband communication services (Internet or VoIP) and power utilities services (SG applications and building/factory control). On the other hand, in-home networks consist of a few devices close to each other in a residential environment. For covering both scenarios, MAC layer has different algorithms which allow its use either in a typical in-home network (with one or two hops) and in an access network with over 1,000 nodes (multihop environment), also supporting dynamic topologies [106].

5.4.8.4 ITU-T G.hn

In 2006, the ITU, later supported by the HomeGrid Forum (industry alliance created in 2008), started the G.hn project in order to develop an international standard for wired broadband communication (up to 1 Gbps) in an in-home environment (up to 250 nodes), being able to operate over power lines, phone lines, coaxial and Cat-5 cables [107]. In 2010, all three components of ITU-T G.hn, the PHY layer (G.9960), MAC layer (G.9961) and the coexistence protocol (G.9972), were complete and approved by the ITU [107].

ITU-T G.hn PHY layer (G.9960) works with three different profiles. One of these profiles is designed for SG applications and works in the frequency band of 2–25 MHz, with data rates of up to 20 Mbps, low complexity and low power consumption (features required by SGs). The other two profiles work in the frequency bands of 2–50 and 2–100 MHz, with data rates of up to 1 Gbps (over PHY layer) [49]. It employs a quasi-cyclic LDPC (QC-LDPC) mandatory coding, interleaving, coherent phase shift keying (PSK) or QAM mapping, and OFDM modulation [108].

ITU-T G.9972 defines ISP for G.hn standard which allows its coexistence with IEEE 1901-2010 and consequently with HomePlug and HD-PLC devices, ever since the same ISP is implemented in all devices. The ISP allows these devices to share the same frequency band by means of time division multiplexing, i.e., each device uses the same medium and frequency band, but in different times [109].

5.5 Conclusion remarks

This chapter presented the concept behind PLC technologies, its history, classification and finally an overview of the current PLC standards present in the market. While old PLC technologies are still used by electric utilities in RCSs or for home automation purposes, the modern PLCs clearly have advantages that allows its use as a complement or even substitute for the current wired and wireless communication technologies (in either narrowband or broadband applications).

Today, NB-PLCs are strongly present in smart metering applications, being the last mile communication solution that connects millions of consumer power meters to the CC of electric utilities. These applications can be considered one of the first steps towards SGs, as this communication infrastructure can be used for implementation of other services and functions inside SG concept.

On the other hand, BB-PLCs are able to replace or work together with wireless technologies for sharing Internet and multimedia data in in-home environments, overcoming the wireless signal issues to cross structures such as walls and floors. For SGs applications in outdoor environments, the possibility of BB-PLCs interfering with licensed radio and communication services is a concern, but precautions, such as using low transmit power and devices capable of detecting radio signals (avoiding transmitting at the frequencies where they are detected), can alleviate such problem.

The lack of standardization of PLCs is not an issue anymore since international standards of both IEEE and ITU are available for both NB-PLC and BB-PLC.

References

[1] Nguyen TV, Petit P, Sawicki JP, *et al.* DC power-line communication based network architecture for HVDC distribution of a renewable energy system. Energy Procedia. 2014;50(1):147–154.

[2] Wade ER, Asada HH. Design of a broadcasting modem for a DC PLC scheme. IEEE/ASME Transactions on Mechatronics. 2006;11(5):533–540.

[3] Tsuzuki S, Areni IS, Yamada Y. A feasibility study of 1 Gbps PLC system assuming a high-balanced DC power-line channel. In: IEEE International Symposium on Power Line Communications and Its Applications; 2012. p. 386–391.

[4] IEA. World Energy Outlook 2016. Paris (France): International Energy Agency, Organisation for Economic Co-operation and Development; 2016. ISBN:9789264264946.

[5] Spoor DJ, Browne N, Crisafulli CA. Mitigation of ripple control signal amplification with LV harmonic filters. In: IEEE International Conference on Harmonics and Quality of Power; 2008. p. 1–5.

[6] Razazian K, Umari M, Kamalizad A, *et al.* G3-PLC specification for powerline communication: overview, system simulation and field trial results. In: IEEE International Symposium on Power Line Communications and Its Applications; 2008. p. 313–318.

[7] Yonge L, Abad J, Afkhamie K, *et al.* An overview of the HomePlug AV2 technology. Journal of Electrical and Computer Engineering. 2013;(892628): 1–20.

[8] Gotz M, Rapp M, Dostert K. Power line channel characteristics and their effect on communication system design. IEEE Communications Magazine. 2004;42(4):78–86.

[9] Lin J, Nassar M, Evans BL. Impulsive noise mitigation in powerline communications using sparse Bayesian learning. IEEE Journal on Selected Areas in Communications. 2013;31(7):1172–1183.

[10] Clark D. Powerline communications: finally ready for prime time? IEEE Internet Computing. 1998;2(1):10–11.

[11] LeClare J. Overcoming Smart Grid Communications Challenges with Orthogonal Frequency Division Multiplexing (OFDM) and IEEE 1901.2. Maxim Integrated; 2012. Application Note 5356.

[12] Tsuchiya F, Misawa H, Nakajo T, *et al.* Interference measurements in HF and UHF bands caused by extension of power line communication bandwidth for astronomical purpose. In: IEEE International Symposium on Power Line Communications and Its Applications; 2003. p. 265–269.

[13] Wheen A. The last mile. In: Webb S, editor. Dot-Dash to Dot.Com: How Modern Telecommunications Evolved from the Telegraph to the Internet. New York, NY: Springer; 2011. p. 103–111.

[14] Ahola J. Applicability of Power-Line Communications to data transfer of Online Condition Monitoring of Electrical Drives [dissertation]. Lappeenranta University of Technology. Lappeenranta (Finland); 2003.

[15] Windelspecht M. Groundbreaking Scientific Experiments, Inventions, and Discoveries of the 19th Century. 1st ed. Krebs RE, editor. Westport, CT: Greenwood Press; 2003.

[16] Brown P. Power line communications – past, present and future. In: IEEE International Symposium on Power Line Communications and Its Applications; 1999. p. 1–8.

[17] Durham T. Is Big Brother reading your meter. New Scientist. 1983; 98(1354):168–171.

[18] Schwartz M. The origins of carrier multiplexing: Major George Owen Squier and AT&T. IEEE Communications Magazine. 2008;46(5):20–24.

[19] Schwartz M. Carrier-wave telephony over power lines – early history. In: IEEE Conference on the History of Electric Power; 2007. p. 244–254.

[20] Schwartz M. Carrier-wave telephony over power lines: early history [History of Communications]. IEEE Communications Magazine. 2009;47(1):14–18.

[21] Dostert K. Telecommunications over the power distribution grid – possibilities and limitations. In: IEEE International Symposium on Power Line Communications and Its Applications; 1997. p. 1–9.

[22] Harris Jr W. Giving the public a light-socket broadcasting service. Radio Broadcast Magazine. 1923;3(6):465–470.

[23] Pusung FF, editor. Smart Grid. 1st ed. Mainz (Germany): PediaPress; 2010.

[24] Baumgarten L. Elevator Going Down: The Story of Muzak [homepage on the Internet]. Red Bull Music Academy Daily; 2012 [cited 2018 May 09]. Available from: http://daily.redbullmusicacademy.com/2012/09/history-of-muzak.

[25] Vernon T. Found in the Attic: LPB Carrier Current AM Transmitter [blog on the Internet]. The Telos Alliance; 2014 [cited 2018 May 09]. Available from: http://blogs.telosalliance.com/found-in-the-attic-february-2014.

[26] Switchgear Manual, Chapter 14.6: Load Management, Ripple Control. ABB Calor Emag; 10th revised edition, 2001.

[27] Sixth generation of ripple control system receivers [homepage on the Internet]. Metering and Smart Energy International; 2006 [cited 2018 May 09]. Available from: https://www.metering.com/regional-news/europe-uk/sixth-generation-of-ripple-control-system-receivers/.

[28] Dzung D, Berganza I, Sendin A. Evolution of powerline communications for smart distribution: from ripple control to OFDM. In: IEEE International Symposium on Power Line Communications and Its Applications; 2011. p. 1–9.

[29] Morgan M, Talukdar S. Electric power load management: some technical, economic, regulatory and social issues. Proceedings of the IEEE. 1979;67(2):241–312.

[30] Broadbridge R. Power line modems and networks. In: IEEE National Conference on Telecommunications; 1989. p. 294–296.

[31] Rye D. My life at X10. Hometoys Inc.; 1999 [cited 2018 May 09]. Available from: http://www.hometoys.com/content.php?url=/htinews/oct99/articles/rye/rye.htm.

[32] Fritz R. Is X-10 an Obsolete Technology?. Lifewire.com; 2017 [cited 2018 May 09]. Available from: https://www.lifewire.com/is-x-10-obsolete-818408.

[33] Dzung D, Berganza I, Sendin A. Effects of power lines on performance of home control system. In: IEEE International Conference on Power Electronics, Drives and Energy Systems; 2006. p. 1–6.

[34] Vasseur JP, Dunkels A. Chapter 12: Communication mechanisms for smart objects. In: Adams R, Bevans D, editors. Interconnecting Smart Objects with IP: The Next Internet. Burlington (MA): Morgan Kaufmann (Elsevier Science); 2010. p. 147–165.

[35] Hagmann W. A spread spectrum communication system for load management and distribution automation. IEEE Transactions on Power Delivery. 1989;4(1):75–81.

[36] Pavlidou N, Vinck H, Yazdani J, Honary B. Power line communications: state of the art and future trends. IEEE Communications Magazine. 2003;41(4):34–40.

[37] Tang L, So PL, Gunawan E, *et al.* Characterization of power distribution lines for high-speed data transmission. In: IEEE International Conference on Power System Technology; 2000. p. 445–450.

[38] Brown P. Directional coupling of high frequency signals onto power networks. In: IEEE International Symposium on Power Line Communications and Its Applications; 1997. p. 19–24.

[39] The end of the line? [homepage on the Internet]. Modern Power Systems (Global Trade Media); 2000 [cited 2018 May 09]. Available from: http://www.modernpowersystems.com/news/newsthe-end-of-the-line-.

[40] International Symposium on Power Line Communications and Its Applications (ISPLC) [homepage on the Internet]. IEEE; 2013 [cited 2018 May 09]. Available from: http://www.isplc.org/.

[41] HomePlug 1.0 Technology White Paper. San Ramon (CA): HomePlug Powerline Alliance; 2001.

[42] Lee MK, Newman RE, Latchman HA, *et al.* HomePlug 1.0 powerline communication LANs protocol description and performance results. International Journal of Communication Systems. 2003;16(5):447–473.

[43] Stephenson P. Real World Performance of Solwise HomePlug Powerline Data Transfer Devices. Solwise Ltd.; 2013 [cited 2018 May 09]. Available from: https://www.solwise.co.uk/net-powerline-real-world-performance.htm.

[44] Sanz JV, Martín EG. Open PLC European Research Alliance Project. In: IEEE International Symposium on Power Line Communications and Its Applications; 2004. p. 1–5.

[45] ITU-T. Narrowband Orthogonal Frequency Division Multiplexing Power Line Communication Transceivers for G3-PLC Networks. Geneva (Switzerland): International Telecommunication Union; 2014. G.9903.

[46] Berganza I, Bois S, Brunschweiler A, *et al.* PRIME v1.4 White Paper. Brussels (BE): PRIME Alliance; 2014.

[47] DeLisle JJ. What's the Difference Between Broadband and Narrowband RF Communications? Microwaves and RF; 2014 [cited 2018 May 09]. Available from: http://www.mwrf.com/systems/what-s-difference-between-broadband-and-narrowband-rf-communications.

[48] Pereira SC, Caporali AS, Casella IRS. Power line communication technology in industrial networks. In: IEEE International Symposium on Power Line Communications and Its Applications; 2015. p. 216–221.

[49]   Oksman V. New ITU-T recommendations for smart grid in-home and access communications. In: ETSI Smart Grid Workshop; 2015. p. 1–32.

[50]   Galli S, Scaglione A, Wang Z. For the grid and through the grid: the role of power line communications in the smart grid. Proceedings of the IEEE. 2011;99(6):998–1027.

[51]   Crossley D. Review of Load Management and Demand Response in Australia. Hornsby Heights (NSW): Energy Futures Australia Pty Ltd.; 2006.

[52]   Elster LCR-212 Single Phase Load Control Ripple Receiver: Installation and Use. Essential Energy; 2011. CEOP8086 – Issue 3.

[53]   Mak ST, Reed DL. TWACS, a new viable two-way automatic communication system for distribution networks. Part I: Outbound communication. IEEE Transactions on Power Apparatus and Systems. 1982;PAS-101(8): 2941–2949.

[54]   TWACS PLC [homepage on the Internet]. Aclara Technologies LLC; 2017 [cited 2018 May 09]. Available from: http://www.aclara.com/products-and-services/communications-networks/twacs-plc/.

[55]   Inside Aclara TWACS Technology [homepage on the Internet]. Aclara Technologies LLC; 2017 [cited 2018 May 09]. Available from: http://www.aclara.com/wp-content/uploads/2017/02/TWACS_Metrum-Cellular-Advantage.pdf.

[56]   Mak ST, Moore TG. TWACS, a new viable two-way automatic communication system for distribution networks. Part II: Inbound communication. IEEE Transactions on Power Apparatus and Systems. 1984;PAS-103(8): 2141–2147.

[57]   Mak ST. Propagation of transients in a distribution network. IEEE Transactions on Power Delivery. 1993;8(1):337–343.

[58]   Rieken DW, Hessling Jr JB, inventors; ACLARA Technologies Inc., Assignee. TWACS Transmitter and Receiver. 2015 Apr 2.

[59]   Santini Jr LC. Sistema de telemedição rural "turtle" – Experiência da Enersul. Enersul; 2008.

[60]   Hunt PC, Hunt LR. Using ultra narrow bandwidth to overcome traditional problems with distribution line carrier. In: IEEE Rural Electric Power Conference; 1995. p. 1–7.

[61]   Davis A. Energy Management System: Strategies and Benefits [blog on the internet]. Lawrence Berkeley National Laboratory; 2015 [cited 2018 May 09]. Available from: http://homeenergypros.lbl.gov/profiles/blogs/energy-management-system-strategies-and-benefits.

[62]   Moscatelli A. From smart metering to smart grids: PLC technology evolutions. In: IEEE International Symposium on Power Line Communications and Its Applications; 2011.

[63]   Who we are [homepage on the internet]. CENELEC (European Committee for Electrotechnical Standardisation); 2018. [cited 2018 May 09]. Available from: https://www.cenelec.eu/aboutcenelec/whoweare/index.html.

[64]   Standard: CENELEC EN 50065-1 – Signalling on low-voltage electrical installations in the frequency range 3 kHz to 148.5 kHz. CENELEC

(European Committee for Electrotechnical Standardisation); 1991. EN 50065-1.

[65] FCC Online Table of Frequency Allocations – 47 C.F.R.. FCC (Federal Communications Commission); 2017 [cited 2018 May 09]. Available from: https://transition.fcc.gov/oet/spectrum/table/fcctable.pdf.

[66] Establishment of ARIB [homepage on the internet]. ARIB; [cited 2018 May 09]. Available from: https://www.arib.or.jp/english/arib/about_arib.html.

[67] Power Line Communication Equipment (10 kHz – 450 kHz). ARIB. STD-T84; 2002.

[68] Sato T, Kammen D M, Duan B, *et al*. Communications in the smart grid. In: Smart Grid Standards: Specifications, Requirements, and Technologies. Singapore: Wiley; 2015. p. 247–294.

[69] Logvinov O. Netricity PLC and the IEEE P1901.2 Standard. HomePlug Powerline Alliance.

[70] Koch M. EU Regulation of High-Speed Powerline Communication in the spectrum 150–500 kHz. Aachen (Germany): Devolo AG; 2016.

[71] LonWorks Technology Achieves ISO/IEC Standardization [homepage on the internet]. LonMark International; 2008 [cited 2018 May 09]. Available from: http://www.lonmark.org/news_events/press/2008/1208_iso_standard.

[72] Berger LT, Schwager A, Galli S, *et al*. Current power line communication systems: a survey. In: Berger LT, Schwager A, Pagani P, *et al*., editors. MIMO Power Line Communications: Narrow and Broadband Standards, EMC, and Advanced Processing. Boca Raton (FL): CRC Press; 2014. p. 253–270.

[73] Introduction to the LonWorks Platform (Revision 2.0). San Jose (CA): Echelon Corporation. 078-0183-01B.

[74] Bradbury S. HomePlug Command & Control (C&C) Overview White Paper. HomePlug Powerline Alliance; 2008.

[75] Berganza I, Sendin A, Arriola J. PRIME: powerline intelligent metering evolution. In: CIRED Seminar SmartGrids for Distribution; 2008. p. 1–3.

[76] Interoperable Standard for Advanced Meter Management and Smart Grid [homepage on the internet]. PRIME Alliance; 2018. [cited 2018 May 09]. Available from: http://www.prime-alliance.org/.

[77] Sendin A, Kim IH, Bois S, *et al*. PRIME v1.4 evolution: a future proof of reality beyond metering. In: IEEE International Conference on Smart Grid Communications; 2014. p. 332–337.

[78] Specification for Powerline Intelligent Metering Evolution. PRIME Alliance; 2014. v1.4-20141031.

[79] ITU-T. Recommendation ITU-T G.9904. International Telecommunication Union; 2012. E38120.

[80] Sagem Communications and Maxim Join Forces to Offer an AMM Solution Based on OFDM Technology [homepage on the internet]. Market Wired (Nasdaq); 2008 [cited 2018 May 09]. Available from: http://www.marketwired.com/press-release/sagem-communications-maxim-join-forces-offer-amm-solution-based-on-ofdm-technology-902553.htm.

[81]   Maxim and ERDF Release the First OFDM-Based Powerline-Communication Specification Supporting the IPv6 Internet Protocol for Smart Grids [homepage on the internet]. Globe Newswire; 2009 [cited 2018 May 09]. Available from: http://www.globenewswire.com/news-release/2009/09/01/403823/172371/en/Maxim-and-ERDF-Release-the-First-OFDM-Based-Powerline-Communication-Specification-Supporting-the-IPv6-Internet-Protocol-for-Smart-Grids.html.

[82]   G3-PLC Alliance Formed to Drive Communications Standard for Smart Grid Development [homepage on the internet]. Business Wire; 2011 [cited 2018 May 09]. Available from: http://www.businesswire.com/news/home/20110930005068/en/G3-PLC-Alliance-formed-drive-communications-standard-smart.

[83]   ITU-T. Recommendation ITU-T G.9903. International Telecommunication Union; 2014. E39134.

[84]   Razazian K, Kamalizad A, Umari M, *et al.* G3-PLC field trials in U.S. distribution grid: initial results and requirements. In: IEEE International Symposium on Power Line Communications and Its Applications; 2011. p. 153–158.

[85]   Razazian K, Niktash A, Loginov V, *et al.* Enhanced 6LoWPAN Ad hoc routing for G3-PLC. In: IEEE International Symposium on Power Line Communications and Its Applications; 2013. p. 137–142.

[86]   LeClare J, Niktash A, Levi V. An Overview, History, and Formation of IEEE P1901.2 for Narrowband OFDM PLC. Maxim Integrated; 2013. Application Note 5676.

[87]   Kelly V. IEEE Approves Standard Designed to Support Low-frequency, Narrowband Power-line Communications and Smart Grid Applications. Business Wire; 2013 [cited 2018 May 09]. Available from: https://www.businesswire.com/news/home/20131120005492/en/IEEE-Approves-Standard-Designed-Support-Low-Frequency-Narrowband.

[88]   Galli S, Lys T. Next generation narrowband (under 500 kHz) power line communications (PLC) standards. China Communications. 2015;12(3):1–8.

[89]   Oksman V, Zhang J. G.HNEM: the new ITU-T standard on narrowband PLC technology. IEEE Communications Magazine. 2011;49(12):36–44.

[90]   JARL. Japan's Government concluded it is not suitable to allow HF band for PLC. The JARL News. 2002;15(4):1.

[91]   "Simply Connect" with HomePlug networking. HomePlug Powerline Alliance; [cited 2018 May 09]. Available from: http://www.homeplug.org/media/filer_public/6b/05/6b05d6a2-0e29-4b35-bb06-d9610a896711/homeplug_brochure.pdf.

[92]   HD-PLC is Ready for New Approved EMC Standard for Broadband Power Line Communications. HD-PLC Alliance; 2012.

[93]   Power line communication apparatus used in low-voltage installations – Radio disturbance characteristics – Limits and methods of measurement – Part 1: Apparatus for in-home use. CENELEC (European Committee for Electrotechnical Standardisation); 2013. EN 50561-1.

[94]   FCC Report and Order 04-245. FCC (Federal Communications Commission); 2004.

[95]   HomePlug 1.0 Technology Integrated into New TIA-1113 International Standard by Telecommunications Industry Association [homepage on the internet]. HomePlug Powerline Alliance; 2008 [cited 2018 May 09]. Available from: http://www.homeplug.org/news/member-pr/160/.

[96]   HomePlug AV White Paper. HomePlug Powerline Alliance; 2015. HPAVWP-050818.

[97]   HomePlug Green PHY 1.1 White Paper. HomePlug Powerline Alliance; 2012.

[98]   HomePlug AV2 White Paper. HomePlug Powerline Alliance; 2012.

[99]   HD-PLC Alliance Certifies Powerline Network Products Complying to Recently Ratified IEEE 1901 Standard [homepage on the internet]. MegaChips; 2010 [cited 2018 May 08]. Available from: http://www. megachips.com/announcements/hd-plc-alliance-certifies-powerline-network-products-complying-to-recently-ratified-ieee-1901-standa.

[100]   Welcome to HD-PLC Alliance [homepage on the internet]. HD-PLC Alliance; 2015 [cited 2018 May 08]. Available from: http://www.hd-plc.org/modules/alliance/message.html.

[101]   HD-PLC Alliance Members [homepage on the internet]. HD-PLC Alliance; 2018 [cited 2018 May 08]. Available from: http://www.hd-plc.org/modules/alliance/members.html.

[102]   Alliance HP. IEEE 1901 HD-PLC (High Definition Power Line Communication. HD-PLC Alliance; 2009.

[103]   Galli S, Logvinov O. Recent developments in the standardization of power line communications within the IEEE. IEEE Communications Magazine. 2008;46(7):64–71.

[104]   Latchman HA, Katar S, Yonge L, *et al.* Introduction. In: Homeplug AV and IEEE 1901: A Handbook for PLC Designers and Users. Piscataway (NJ): Wiley-IEEE Press; 2013. p. 1–9.

[105]   IEEE. IEEE Standard for Broadband over Power Line Networks: Medium Access Control and Physical Layer Specifications. Institute of Electrical and Electronics Engineers; 2010. 978-0-7381-6472-4.

[106]   Goldfisher S, Tanabe S. IEEE 1901 access system: an overview of its uniqueness and motivation. IEEE Communications Magazine. 2010;48(10): 150–157.

[107]   United Nations ITU-T's G.hn Approved as Global Standard for Wired Home Networking [homepage on the internet]. Market Wired (Nasdaq); 2010 [cited 2018 May 08]. Available from: http://www.marketwired.com/press-release/united-nations-itu-ts-ghn-approved-as-global-standard-for-wired-home-networking-1274797.htm.

[108]   Oksman V, Galli S. G.hn: the new ITU-T home networking standard. IEEE Communications Magazine. 2009;47(10):138–145.

[109]   Stark B. Coexistence and interoperability in a mixed vendor PLC environment. The HomeGrid Forum; 2017. v.2.

Chapter 6
Power line communication channel models
Ricardo Suyama[1]

Advances in communications system have been achieved through the use of reliable models able to capture essential characteristics that are present in real environments. In order to obtain a representative model for the signal degradation observed in PLC communication systems, it is important to characterize the distortions imposed by the propagation of the signals in the wires, in terms of the channel frequency response (or, equivalently, the channel impulse response), and correctly model the noise, which may be even more stringent than in other communication systems, such as in ADSL or wireless communications systems.

Several works have proposed models for the PLC channel, essentially trying to capture the propagation effects due to multipath propagation in the physical medium and to obtain realistic noise models, following two main approaches to accomplish this task: a bottom-up and a top-down approach.

In the bottom-up approach, the channel transfer function is modeled using transmission line theory and requires prior knowledge about the network topology and loads connected to it [1–4]. Nevertheless, statistical extensions of the model were proposed in [5–7], so that channels with similar characteristics could be generated without the knowledge about a specific network configuration.

On the other hand, in the top-down approach, the channel model is obtained by considering a general multipath propagation model for the signals, with parameters that can be fitted from real measurements [8,9] or statistically modeled [10,11], allowing the generation of channels similar to those found in practice.

In this chapter, rather than providing a broad view about the existing models found in the literature, we follow the top-down approach, aiming to provide a concise description of simple models that can be used in simulations to assess the performance of PLC communications systems and smart grid applications. In this sense, the chapter has been structured as follows: first, the analytical multipath propagation model is presented, which is the basis for generating synthetic frequency responses, for both broadband and narrowband PLC (NB-PLC); then, we present usual models for the noise present in the PLC communication channel. In the sequence, we present the general procedure for generating channel frequency responses for simulation scenarios

[1]Center for Engineering, Modeling and Applied Social Sciences (CECS), Federal University of ABC (UFABC), Brazil

for broadband and NB-PLC, and, finally, we comment about recent extensions of such models to MIMO PLC systems.

6.1 Multipath propagation model

Due to the very nature of low voltage and medium voltage distribution network, signal propagation in such medium does not occur in a direct "line-of-sight" path between transmitter and receiver. Since there may exist several connections and derivations in the network, the received signal is a composition of the transmitted signal and additional echoes (due to the propagation in additional paths) [9]. This situation is illustrated in a simple example in Figure 6.1, which depicts a hypothetical communication channel with a branch somewhere between "A" and "B."

Let us assume that both transmitter and receiver ends are matched with the transmission lines, i.e., the loads impedances Z_A and Z_B are equal to the lines characteristic impedances Z_{L1} and Z_{L2}, respectively. In this case, signal transmitted from A to B is possibly subject to reflections in the branch connection point (denoted in the figure by point "X"), and in the end of the branch (point "C" in the figure), due to impedance mismatch. Hence, one can define the reflection factors r_{1X}, r_{3X} and r_{3C} to represent how much of the impinging signal is reflected in points X and C, and transmission factors t_{1X} and t_{3X} to model how much of the impinging signal is transmitted through X.

It is not difficult to imagine different possible propagation paths that may arise due to multiple reflections, for example, paths like $A \rightarrow X \rightarrow B$, which can be considered a "line-of-sight" path, $A \rightarrow X \rightarrow C \rightarrow X \rightarrow B$, $A \rightarrow X \rightarrow X \rightarrow X \rightarrow C \rightarrow X \rightarrow B$, and so on. Thus, in order to obtain the channel model, it is necessary to consider the attenuation and delays associated with each propagation path.

The attenuation observed in a particular path is due to the reflection and transmission factors along the path—represented by a gain g_p which is the product of the reflection and transmission factors (denoted by r_{ij} and t_{ij} in Figure 6.1) along the path—and also due to the attenuation $A(f, d)$ related to losses of the cables, which increases with the path length and frequency. On the other hand, the propagation delay

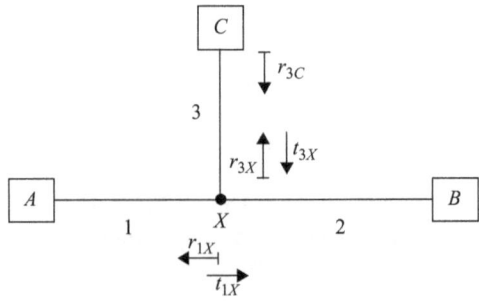

Figure 6.1　Multipath signal propagation in a transmission line with a single tap

associated with a particular path p depends on the propagation velocity in the cables and is given by

$$\tau_p = \frac{d_p\sqrt{\varepsilon_r}}{c_0} = \frac{d_p}{v}, \tag{6.1}$$

where ε_r denotes the dielectric constant of the insulating material, c_0 is the speed of light and d_p denotes the length of the path. Hence, considering these factors, the frequency response of the channel connecting A to C can be expressed as

$$H(f) = \sum_{p=1}^{N_{\text{paths}}} g_p \cdot A(f, d_p) \cdot e^{-j2\pi f \tau_p}. \tag{6.2}$$

The attenuation due to cable losses can depends on physical characteristics of the cable [12], but based on results obtained from real channel measurements, it was found that $A(f, d_p)$ can be approximated by

$$A(f, d) = e^{-\alpha(f)d} = e^{-(\alpha_0 + \alpha_1 \cdot f^k)d}, \tag{6.3}$$

where α_0 and α_1 are attenuation parameters that can be fitted from real measurements and k is a constant that has typical value between 0.5 and 1 [9]. Hence, using (6.3) in (6.2), one obtains that the channel frequency response can be modeled by[1]

$$H(f) = A \sum_{p=1}^{N_{\text{paths}}} g_p \cdot e^{-(\alpha_0 + \alpha_1 \cdot f^k)d_p} \cdot e^{-j2\pi f(d_p/v)}, \tag{6.4}$$

where v is the propagation velocity and A is a normalization constant.

The signal propagation model presented in (6.4) has been used as an essential building block for PLC channel models [9–11]. Figure 6.2 illustrates the frequency responses of two different channels that were obtained by fitting (6.4) to measurements. In the first case, the channel is modeled with $N = 5$ dominant paths, while in the second, $N = 4$. Despite the simplicity of the model and the reduced number of dominant paths, the resulting frequency response is able to capture the main characteristics observed in real channels [9].

In order to cope with possible variations found in real channel responses, several studies in the literature were devoted to explore statistical information about PLC channel from worldwide measurement campaigns, from which it was possible to develop methods to generate channel models that exhibit the same statistical properties of the real measured channels. It is worth mentioning that it is similar to the approach followed in the development of wireless communications systems [13–15], obtaining statistical reference models for the propagation of wireless signals in different scenarios. In Sections 6.3 and 6.4, we discuss in more detail such models, presenting a methodology to generate PLC channels for simulations of broadband and NB-PLC communications systems.

---

[1]A more general model would consider that the reflection and transmission coefficients present complex values and be frequency dependent—in which case g_p should be replaced by $|g_p|e^{\phi_{g_p}(f)}$.

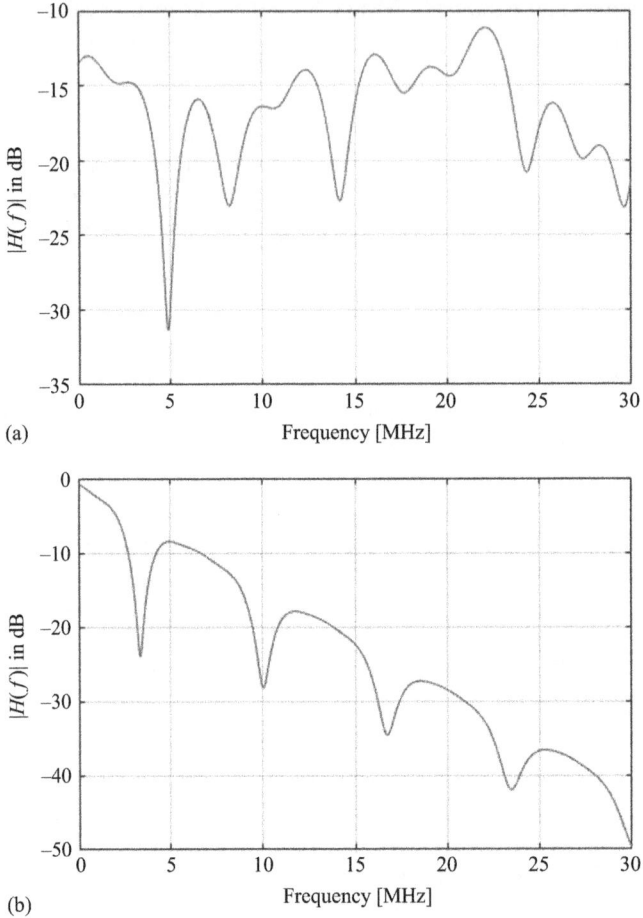

Figure 6.2   Frequency response magnitude obtained with (6.4) using parameters presented in (a) [8] and (b) [9]

6.2 Noise in PLC channels

An important part of PLC channel modeling is related to the noise model to be adopted. Since the noise in such environment is, in fact, a composition of noise sources connected to the network, such as electrical motors, dimmers, consumer electronics, and from external sources, such as radio broadcast, it is not easy to obtain a single model that can encompass the characteristics of all distinct sources.

A usual approach is to consider different models, one for each particular type of noise. In this chapter, we discuss three common types of noise that can be observed in the PLC channel, as indicated in Figure 6.3: colored background noise, which can be understood as the result of a superposition of several noise sources connected to the network; narrowband noise, commonly associated with radio broadcast; and

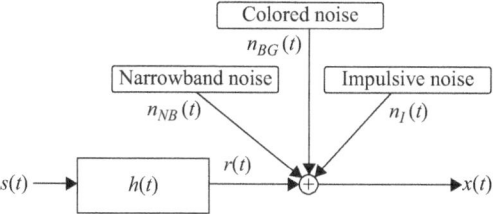

Figure 6.3 Different types of noise present in PLC channels

impulsive noise, due, for instance, to the connection or disconnection of electrical equipment in the network [16].

6.2.1 Colored background noise

Due to the contribution of several sources, the disturbances in the frequency range used in PLC systems give rise to a colored background noise, which varies with frequency [17]. In the literature, different models for the background noise were proposed, but share a similar power spectral profile, with a low density in almost all frequency range, which, however, significantly increases for lower frequencies [7,9].

A simple model that exhibit such characteristics considers that the background noise has a Gaussian distribution with power spectral density given by

$$S_{n_{BG}}(f) = a + b \left| \frac{f}{1 \times 10^6} \right|^c , \tag{6.5}$$

where a, b and c are parameters that can be obtained from measurements, and f is given in Hz. For example, in [17], the authors proposed, based on measurements, that a, b and c be set to -140, 38.75 and -0.72, respectively, for a "good" channel condition, and -145, 53.23 and -0.337, respectively, for a "bad" channel condition. Both situations are depicted in Figure 6.4(a), and the basic difference between each other is related to the noise floor level, since both scenarios maintain the same decay profile.

Another model for the background noise was proposed in the context of the OMEGA project [18], a European research project involving several companies and universities that made several contributions to the characterization of the PLC channel. Based on experimental measurements, the proposed model expresses the power spectral density of the background noise as

$$S_{n_{BG},OMEGA}(f) = 10 \log_{10} \left(\frac{1}{f^a + 10^b} \right) , \tag{6.6}$$

where a and b are the model parameters. The noise level profile, for $a = 2$ and $b = -15.5$, as originally proposed in [18], is shown in Figure 6.4(b), indicating the same frequency dependent behavior observed in the previous model.

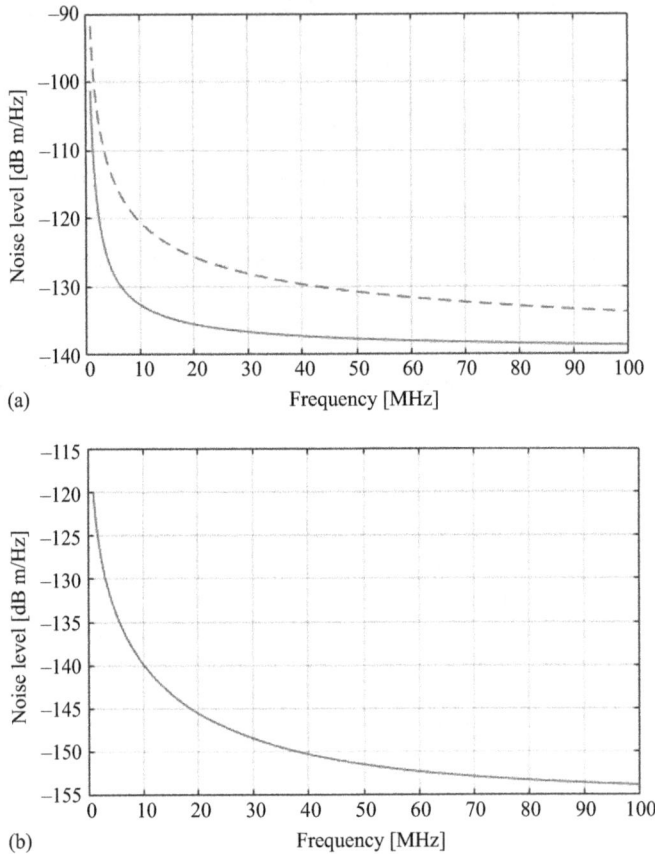

*Figure 6.4   Background noise models (a) proposed in [17] (b) Omega Project [18].
In (a) both best- and worst-case scenarios are depicted*

6.2.2   Narrowband noise

In general, there are two main sources of narrowband interference in PLC systems.
First, devices connected to the power network that operate switching the input current
(such as dimmers and switching power supplies), which originates a noise associated
with the switching frequency, that can be as low as 50 Hz, in the case of dimmer
devices, up to hundreds of kHz, for switching power supplies. The second group of
noise sources is associated with radio frequency transmission, from radio broadcast-
ers in long, medium and short wave range, as well as amateur radio. In the short and
medium wave band, transmissions are mainly based on narrow band analog modulation techniques, with bandwidths around 9 kHz, which may be seen, in broadband
OFDM communication schemes, as a single-tone interference [19].

For those reasons, in simulations, the narrowband noise is often modeled as a sum of multiple sinusoidal terms, i.e.,

$$n_{NB}(t) = \sum_{k=1}^{N_{\text{interf}}} A_k \sin(2\pi f_{0,k} t + \phi_k), \tag{6.7}$$

where N_{interf} is the number of disturbers, each one with a different central frequency $f_{0,k}$, amplitude A_k and phase ϕ_k. In the absence of prior information about possible disturbers, an usual approach is to consider that the frequencies are randomly distributed in the frequency spectrum, with arbitrary values for their amplitudes and phases.

A more elaborated model, based on real noise measurements, was proposed in [20]. According to the model, developed in the frequency domain, the power spectral density associated with the narrowband interference can be represented by [21]

$$S_{n_{NB}}(f) = \sum_{k=1}^{N_{\text{interf}}} A_k e^{-(f-f_{0,k})^2/(2B_k^2))}, \tag{6.8}$$

where B_k denotes the bandwidth of each interferer. Generation of noise samples exhibiting such power spectral density can be carried out by filtering a white Gaussian noise.

An interesting point in the model presented in [20] is that it also provides statistical models for each of the parameters in (6.8), which were obtained from measurements made in residential and office buildings. For example, in a "typical" residential building, the author proposes that N_{interf} should follow a normal distribution (for positive values) with a mean value $\mu = 0.88$ and standard deviation $\sigma = 5.47$ for frequencies below 10 MHz, $\mu = 0.35$ and $\sigma = 3.94$ for frequencies between 10 and 20 MHz, and $\mu = 0.72$ and $\sigma = 2.53$ for frequencies in in the 20–30-MHz range. By generating random numbers following such distributions (considering only the positive values), the following can be observed: for the frequency range of 0–10 MHz, there will be an average of $N_{B1} = 5$ disturbers; for the 10–20 MHz range, the average is $N_{B2} = 3$; and for the 20–30 MHz range, the average is $N_{B3} = 2$.

Similar statistical models for A_k and B_k are presented in [20], and in Figure 6.5, the power spectral density in the 0–30-MHz frequency range, for a particular realization, is depicted. Using the parameters presented in [20], a number of narrowband disturbers were generated, leading to the narrowband noise level illustrated in Figure 6.5(a). In Figure 6.5(b), the narrowband noise was added to the background noise model presented in (6.6), and a general overview of the noise level in the channel is obtained.

6.2.3 *Impulsive noise*

Impulsive noise in the PLC channel is mainly associated to the occurrence of disturbances that present a high level of interference in a short period of time, a characteristic that is not easily captured by simple models as the ones used to represent background

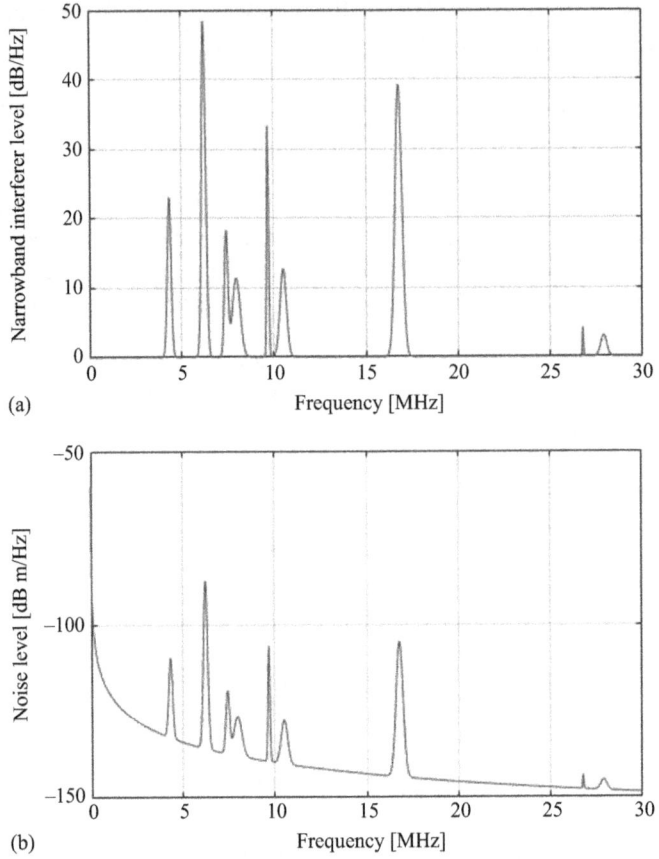

*Figure 6.5   Background noise models (a) proposed in [17] (b) Omega Project [18].
In (a) both best- and worst-case scenarios are depicted*

noise. In the sequence, some of the most common impulsive noise models are presented: the Bernoulli–Gaussian model, the Middleton Class A model, the α-stable model and the Markov–Gaussian model.

6.2.3.1   Bernoulli–Gaussian model

Considered to be one of the simplest impulsive noise models, the Bernoulli–Gaussian model represents the impulsive characteristic by means of the product of two independent random sequences: a real Bernoulli $\varepsilon(t)$ and a complex Gaussian process $n_1(t)$, i.e.,

$$n_I(t) = \varepsilon(t)n_1(t). \tag{6.9}$$

The occurrence of a high level of interference is controlled by the Bernoulli distribution, and the amplitude of this disturbance is associated with the value drawn from the Gaussian distribution. Hence, if the background noise is also assumed to be a

Gaussian process, with standard deviation σ_{BG}, the overall model for the background and impulsive noise can be expressed by a mixture of Gaussians, i.e.,

$$n(t) = \varepsilon n_1(t) + (1 - \varepsilon)n_{BG}(t) \tag{6.10}$$

where $n_{BG}(t)$ is the background noise. In this case, the noise at any given time instant will be characterized by a random value with probability density function given by

$$p(n_{BG}(t)) = (1 - P)\mathcal{N}(0, \sigma_{BG}^2) + P\mathcal{N}(0, K\sigma_{BG}^2), \tag{6.11}$$

where P denotes the probability that an impulsive noise occurs, $\mathcal{N}(\mu, \sigma^2)$ denotes the Gaussian distribution with mean value μ and variance σ^2, and $K\sigma_{BG}^2$ denotes the variance of the impulsive noise (K times higher than the background noise). Figure 6.6 depicts two different noise scenarios, with $\sigma_{BG}^2 = 0.01$ and $K = 400$, to illustrate the effect of the impulsive noise under different impulse probabilities – $P = 0.01$ in Figure 6.6(a), a mild noise scenario, and $P = 0.1$ in Figure 6.6(b), a scenario heavily disturbed by impulsive noise.

6.2.3.2 Middleton Class A

The Middleton Class A model has been widely used and studied noise model [21–24], that still preserves the simplicity of Bernoulli–Gaussian model, but has a greater flexibility. Mathematically, noise samples following Middleton's Class A noise model follows the distribution given by

$$p(n_I(t)) = \sum_{m=0}^{\infty} P_m \mathcal{N}(0, \sigma_m^2), \tag{6.12}$$

where

$$P_m = \frac{\Delta^m e^{-\Delta}}{m!}, \tag{6.13}$$

and

$$\sigma_m^2 = \sigma_I^2 \frac{m}{\Delta} + \sigma_{BG}^2, \tag{6.14}$$

with σ_I^2 and σ_{BG}^2 denoting, respectively, the impulse noise and background noise variance. In such model, the parameter Δ is associated with the density of impulses occurring in an observation period.

From (6.12) one can observe that the distribution is also given by a mixture of Gaussian distributions, as in the Bernoulli-Gaussian model, but considering an infinite number of terms. Nonetheless, an approximation of (6.12) taking only the first terms in the summation is often enough to model the impulsive noise [25]. In this case, truncating (6.12) to the first N terms results in

$$p(n_I(t)) = \sum_{m=0}^{N-1} P_m' \mathcal{N}\left(0, \sigma_m^2\right), \tag{6.15}$$

where

$$P_m' = \frac{P_m}{\sum_{m=0}^{K-1} P_m}. \tag{6.16}$$

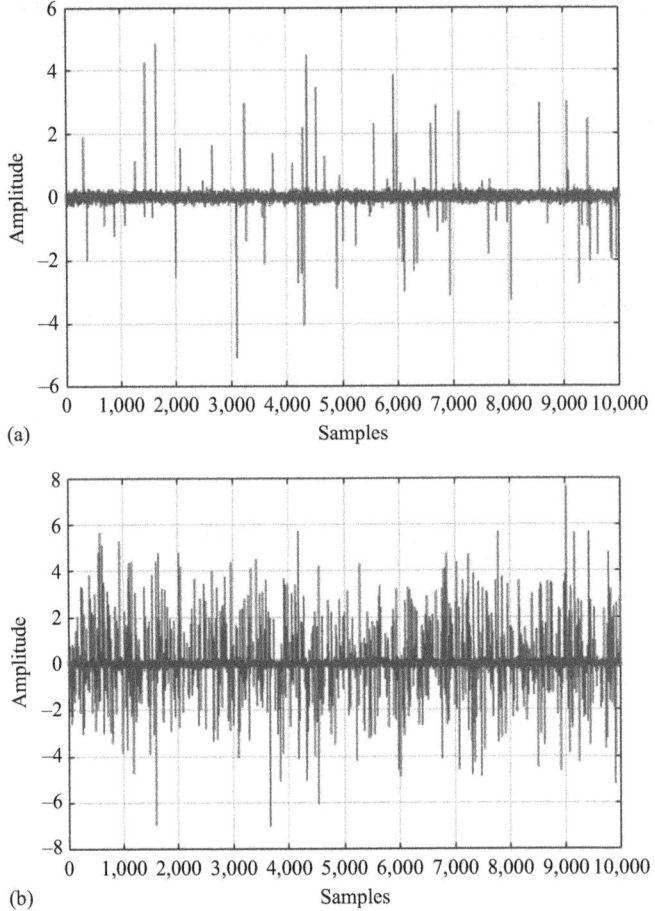

Figure 6.6   Impulsive noise model based on a Bernoulli–Gaussian distribution, for $\sigma_{BG}^2 = 0.01$, $K = 400$ and two different scenarios (a) $P = 0.01$ (b) $P = 0.1$

In Figure 6.7, two different scenarios with Middleton's noise model are depicted, in similar conditions to that presented in Figure 6.6. In the simulation, only the first 5 terms of (6.12) were used. The difference in Figure 6.7(a) and 6.7(b) is due to different values of the parameter Δ (0.01 and 0.1, respectively).

6.2.3.3 α-Stable distributions

α-Stable distributions have been considered as an appropriate distribution to model the impulsive characteristic of noise in PLC systems [25,26], and arise from a generalized form of the Central Limit Theorem from probability theory, which states that the sum of independent and identically distributed (i.i.d.) random variables with

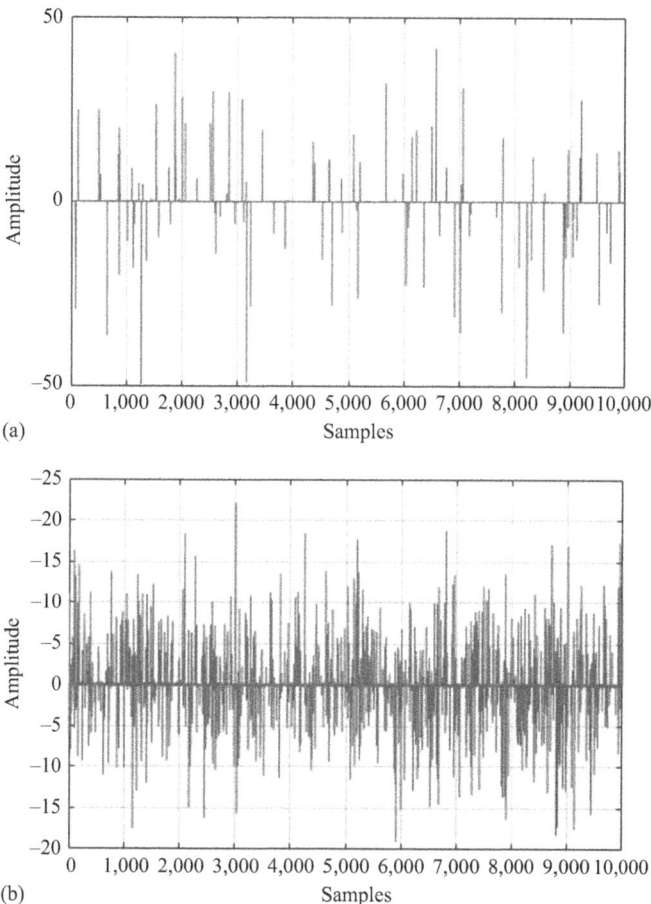

*Figure 6.7   Impulsive noise model based on Middleton's Class A distribution,
for $\sigma_g^2 = 0.01$, $\sigma_I^2 = 4$ and two different scenarios (a) $\Delta = 0.01$
(b) $\Delta = 0.1$*

finite or infinite variance converges in distribution to the α-stable distribution [27].
If only finite variances are allowed in the sum, then the normalized sum converges in
distribution to the Gaussian distribution.

The α-stable distribution, denoted by $S_\alpha(\sigma, \beta, \mu)$, is defined by four parameters:

α represents an index of stability $(0 < \alpha < 2)$ and is related to the tail decay of the
distribution.

σ is a scale parameter $(\sigma \geq 0)$.

β is the skewness parameter $(-1 \leq \beta \leq 1)$, being equal to 0 for a symmetrical
distribution and assuming other values for skewed variables.

μ represents a shift parameter $(\mu \in \Re)$, controlling the location of the variable.

For $\alpha = 2$, the distribution is equivalent to the Gaussian distribution, with the distinction that σ^2 is not equivalent to the variance—in fact, $\sigma^2 = (variance/2)$.

In general, an α-stable random variable $X \sim S_\alpha(\sigma, \beta, \mu)$ has no closed form for the probability density function (pdf), hence being usually defined in terms of its characteristic function

$$\Phi_X(\theta) = \begin{cases} \exp\{j\mu\theta - \sigma^\alpha|\theta|^\alpha(1 + j\beta \operatorname{sign}(\theta)\tan(\pi\alpha/2))\} & \text{if } \alpha \neq 1 \\ \exp\{j\mu\theta - \sigma|\theta|(1 - j\beta(2/\pi)\operatorname{sign}(\theta)\ln|\theta|)\} & \text{if } \alpha = 1 \end{cases}, \quad (6.17)$$

where $\theta \in \Re$ and

$$\operatorname{sign}(\theta) = \begin{cases} 1 & \text{if } \theta > 0 \\ 0 & \text{if } \theta = 0 \\ -1 & \text{if } \theta < 0 \end{cases}. \quad (6.18)$$

Even though it is not possible to obtain a closed-form expression for the probability density function (pdf) from (6.17), one can obtain it numerically [26], which is depicted in Figure 6.8 for different values of α. From the pdf, by numerical integration, it is also possible to obtain the cumulative distribution function, which can then be used to produce samples following the distribution by the method of inverse transformation [27]. Figure 6.8(b) and 6.8(c) shows a set of noise samples drawn from the α-stable distribution for two distinct values of α. In Figure 6.8(c), the samples correspond to the distribution for $\alpha = 2$, which is equivalent to the normal distribution, whereas in Figure 6.8(b), the noise was generated with $\alpha = 1.5$, thus exhibiting a more impulsive behavior.

The summarized procedure to generate samples from an α-stable distribution, as indicated in [26], comprises the following steps:

1. Sample $\Phi_X(\theta)$ using increasing values for θ, obtaining a numerical representation of the characteristic function;
2. Evaluate the inverse discrete Fourier transform of the resulting sequence, obtaining a numerical representation of the desired pdf. It is worth mentioning that the samples of the characteristic function should maintain Hermitian symmetry before applying the inverse Fourier transform;
3. Obtain an estimate of the cumulative density function (cdf) by numerical integration of the pdf;
4. Once the cdf is estimated, it is possible to generate α-stable random numbers by using a pdf transformation method

6.2.4 Markov–Gaussian noise model

One major drawback of the previous models, despite their simplicity, is that they are unable to model possible temporal structure that may be present in the noise. For instance, the impulsive noise often occurs in bursts [9], a characteristic that is not captured by a simple distribution. In such scenario, a simple model based on

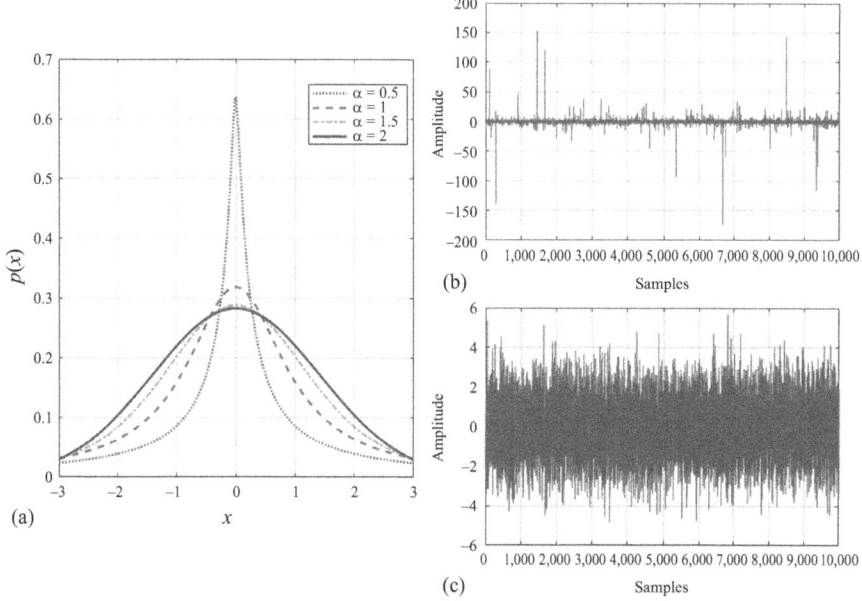

Figure 6.8   *α-Stable distribution for different values of parameter α, with β = 0,*
μ = 0 and σ = 1. (a) Distributions for different values of α, (b) samples
generated with α = 1.5 and (c) for α = 2

an association of a Markov process and the Gaussian distribution, denoted here as
Markov–Gaussian model [28], can be interesting to describe this burst behavior of
the noise.

This model can be considered a simplified version of the model presented in [9]
and is based on the approach described in [29,30] to model channel events that occur
in bursts. In the Markov–Gaussian model, the noise is a complex circularly symmetric
Gaussian random variable with variance depending on the channel state s_k—which
can be either G, i.e., a "good" channel, or "B," i.e., a "bad" channel, as indicated in
Figure 6.9, where P_{IJ} denote the transition probability from state I to state J. The pdf
of n_k in each case is defined as

$$p(n_k|s_k = G) = \mathcal{N}(0, \sigma_{BG}^2),\tag{6.19}$$

and

$$p(n_k|s_k = B) = \mathcal{N}(0, K\sigma_{BG}^2),\tag{6.20}$$

where the parameter $K \geq 1$ represents the ratio between the average noise power
present in the bad channel and that in the good channel.

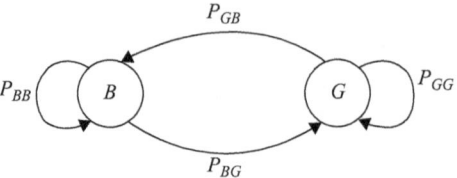

Figure 6.9   Markov chain used in the Markov–Gauss noise model

In order to completely describe the noise model, it is necessary to define the statistical description of the state process $\{s_k\}$, which, in this model, is assumed to be a stationary first-order Markov process [27], i.e.,

$$P(\{s_k\}) = P(s_1) \prod_{k=1}^{N-1} P(s_{k+1}|s_k), \tag{6.21}$$

for each realization of the process $\{s_k\}$. Thus, a complete description of the model is given by the state transition probabilities $P_{GB} = P(s_{k+1} = I|s_k = J)$ for $I, J \in \{G, B\}$. In this simple case, with only two states, it is only necessary to define the transition probabilities $P_{BG} = P(s_{k+1} = G|s_k = B)$ and $P_{GB} = P(s_{k+1} = B|s_k = G)$. With these values, the probabilities of being in a given state are

$$P_G = P(s_k = G) = \frac{P_{BG}}{P_{GB} + P_{BG}}, \tag{6.22}$$

and

$$P_B = P(s_k = B) = \frac{P_{GB}}{P_{GB} + P_{BG}}. \tag{6.23}$$

In [28], the authors define the parameter

$$\xi = \frac{1}{P_{GB} + P_{BG}}, \tag{6.24}$$

which controls the behavior of the noise. For $\xi > 1$, the channel tends to stay in a state for a period that is longer than that observed in the memoryless case, while for $\xi < 1$, the channel tends to stay in a state for a period that is, on average, shorter than the memoryless case. Thus, for typical scenarios with bursts affecting the channel, one should consider values of $\gamma \geq 1.0$.

In order to illustrate the different behaviors of the model, in Figure 6.10, the samples generated with the Gauss–Markov model are depicted for $\xi = 1, \xi = 5$ and $\xi = 10$. It is clear that for $\xi = 1$, the noise samples profile is similar to that observed in the noise models without memory, whereas for $\xi = 5$ and $\xi = 10$, the impulses appear in bursts, as was originally expected.

6.2.5   Noise in narrowband systems

The noise in NB-PLC systems is mainly composed of continuous colored background noise and continuous tone jammers, together with periodic impulses [31]. In the

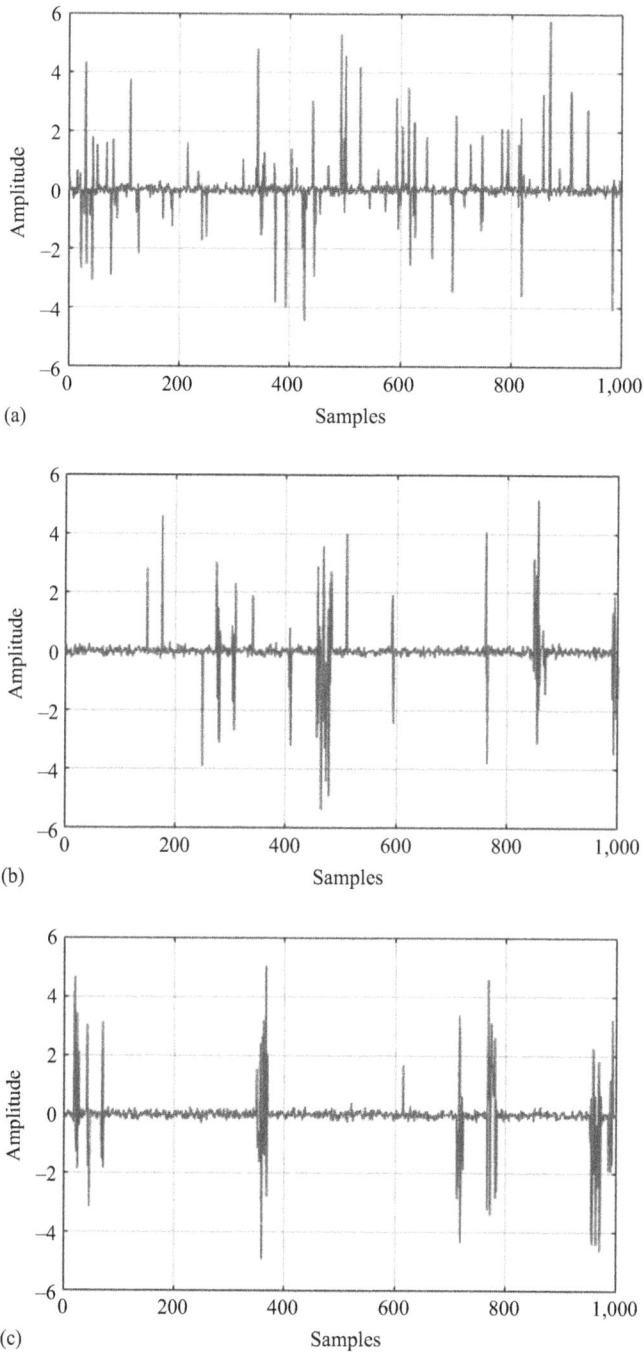

Figure 6.10 *Noise samples generated with the Markov–Gaussian model, for three*
different values of ξ, with σ = 0.1, P$_B$ = 0.1 and R = 400. (a) ξ = 1,
(b) ξ = 5 and (c) ξ = 10

literature, two main noise models have been used to assess the performance of NB-PLC systems, denoted here as Katayama model [31] and LPTV Model [32]. In this chapter, we focus our attention on the first model.

The Katayama model was developed taking into account a set of measurements obtained in Japan. In the measurements, the authors found that the noise presents a cyclic behavior and thus could be modeled as a cyclostationary process. In this model, the distribution of the noise samples at a particular time instant is given by a Gaussian distribution, with zero mean and variance that varies synchronously to the AC voltage of the mains. In other words, the probability density function of the noise at time instant $t = iT_s$ can be expressed by

$$p(n_{NB}(t)) = \mathcal{N}(0, \sigma_{NB}^2(t)), \tag{6.25}$$

where $\sigma_{NB}^2(t)$ is the instantaneous variance of the noise and is a periodic time function, with period that equals to half of the mains frequency, i.e., $T/2$.

The form of $\sigma^2(t)$ can be estimated considering an average of the instantaneous noise power, taken at every $T/2$ seconds, for $t = 0, \ldots, T/2$. However, the authors propose the use of a simple function to approximate the behavior of $\sigma^2(t)$, thus reducing the number of parameters to be estimated. The periodic function to approximate the variance is expressed by

$$\hat{\sigma}_{NB}^2(t) = \sum_{l=0}^{L-1} A_l |\sin(2\pi t/T + \theta_l)|^{n_l}, \tag{6.26}$$

where A_l, θ_l and n_l are the parameters that characterize the variance of the noise.

In general, $L = 3$ is sufficient to capture most of the noise characteristics in NB-PLC systems, where the term for $l = 0$ represent the background noise (and, hence, $n_0 = 0$ and θ_0 is an arbitrary constant value), the term associated with $l = 1$ is related to continuous periodic noise, and the term for $l = 2$ is used to mode the periodic impulsive noise [31].

The resulting model is able to generate uncorrelated noise samples, and in order to introduce a spectral profile—which is typically assumed to have an exponential decay—the Gaussian noise should be filtered by a linear time invariant system in order to shape its power spectral density. The shaping filter is defined in the frequency domain as

$$\delta(f) = \frac{a}{2} \exp(-a|f|), \tag{6.27}$$

with a being estimated from measurements. The procedure for generating noise samples according to the model can be summarized as follows:

1. Determine the set of parameters for $\sigma^2(t)$
2. Generate uncorrelated noise samples with instantaneous variance given by $\sigma^2(t)$
3. Filter the noise samples with an LTI system with frequency response given by $\sqrt{\delta(f)}$, as defined in (6.27).

In Figure 6.11, an example with $a = 1.2 \times 10^{-5}$, as indicated in [31], is presented.

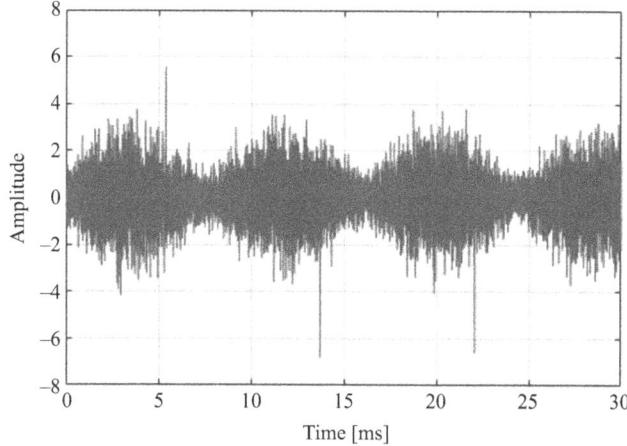

Figure 6.11   Noise samples from the Katayama model, with parameters indicated in [31]

A further refinement in the noise model for NB-PLC systems was presented in [32], where the authors introduce the possibility of variation of the noise spectral content. In the Katayama model, the noise samples produced by a cyclostationary process is filtered by a single filter in order to produce a decaying spectral profile. In [32], a new model denoted linear periodically time-varying (LPTV) noise model is presented, in which the authors consider that a given observation period can be divided into intervals where the noise spectral shape remains unchanged. As in other approaches, the parameters for each filter in the LPTV noise model is obtained by fitting the model with real measurements.

6.3   Generating channels for broadband PLC

The analytical model for the PLC channel transfer function in (6.4) depends on a set of parameters, which can be fitted to represent a particular scenario [8]. However, in general, rather than obtaining the results for a particular channel, it is interesting to have a more flexible description of the scenario—e.g., a model that represents a channel in a "common" in-home scenario or in a "common" industrial scenario. This flexibility can be achieved considering that some of the parameters in (6.4) are, in fact, random variables, and the parameters are obtained after fitting the model with real measurements.

Following the model presented in [10], let us assume that reflection and transmission coefficients in a network are uniformly distributed in $[-1, 1]$ or, equivalently, are the result of the product between a uniform random variable between $[0, 1]$ and a random sign flip. Thus, since g_p is a product of these transmission and reflection coefficients, its distribution will approach a log-normal distribution [27] multiplied by a random sign flip. It is further assumed that g_p is zero mean, with variance equal to $\sigma_g^2 = 1$, without loss of generality since constant A can be properly adjusted afterwards.

The path lengths d_p can also be modeled a random variable, and unless more information is provided, a usual practice is to consider that it is uniformly distributed in the interval $[0, L]$, for a particular length L. Finally, regarding the number of paths, it is assumed that it is a Poisson random variable, i.e.,

$$N_{\text{paths}} \sim \mathcal{P}(n_{\text{paths}}) \quad \rightarrow \quad P(N_{\text{paths}} = k) = \frac{e^{-\Lambda L}}{1 - e^{-\Lambda L}} \frac{(\Lambda L)^k}{k!} \tag{6.28}$$

where $k \geq 1$ and ΛL define the mean value of the distribution. In other words, Λ can be interpreted as the number of paths per unit length (paths/m). In [10], for example, it is assumed that $\Lambda = 0.2$ (paths/m).

Plausible values for the remaining parameters in (6.4) are $v = 2 \times 10^8$, $A = 6.7 \times 10^{-4}$, $\alpha_0 = -0.015$, $\alpha_1 = 3.7 \times 10^{-5}$, $k = 0.35$ and $L = 320$. A complete set of parameter values corresponding to real channel measurements can be found in [10].

Based on these statistical models, the procedure for generating the channels can then be summarized as follows:

1. Determine the number of paths in the channel, according to the distribution defined in (6.28);
2. Path lengths should be uniformly sampled in the interval $[0, L]$;
3. Obtain the path gains g_p from a lognormal distribution, with zero mean and $\sigma_g^2 = 1$, multiplied by a random sign flip;
4. Determine the channel frequency response using the obtained parameters in (6.4).

6.4 Generating channels for narrowband PLC

Following a similar approach as in the broadband PLC case, in [11], the authors propose a channel generator for NB-PLC, considering the channel frequency response below 500 kHz. The proposed model is also based on the multipath propagation model (6.4), but, in this case, the model explicitly includes a frequency dependent path gain $g_p(f)$ in the channel frequency response, and it is assumed that it can be modeled as

$$H(f) = A \sum_{p=1}^{N_{\text{paths}}} g_p(f) \cdot e^{-(\alpha_0 + \alpha_1 \cdot f) v \tau_p} \cdot e^{-j2\pi f \tau_p}, \tag{6.29}$$

where τ_p is the path delay and $v = 2 \times 10^8$ m/s is the signal propagation velocity in the cables.

As in the previous case, the application of this model to the NB-PLC case depends on the proper selection of its parameters, which can also be described in terms of statistical models obtained from field measurements [33]. A simplification in (6.29) was proposed to approximate the behavior of the path gain $g_p(f)$ and obtain a compact and simpler expression for the channel frequency response, given by

$$H(f) = A \sum_{p=1}^{N_{\text{paths}}} a_p \cdot e^{b_p f} \cdot e^{j\phi_p f} \cdot e^{-j2\pi f \tau_p}, \tag{6.30}$$

in which the path gain and the attenuation in the cable are modeled by parameters a_p, b_p and ϕ_p.

The procedure for generating the synthetic channels is very similar to that described for the broadband channels, and can be summarized as follows [11]:

1. Obtain the number of paths N_{paths}, which should be drawn from a Poisson distribution, with parameter $\lambda = 25.04$;
2. Generate the delays τ_p by generating samples from a Beta distribution, i.e.,

$$X \sim Beta(\alpha, \beta) \quad \rightarrow \quad p(x) = \frac{x^{\alpha-1}(1-x)^{\beta-1}}{B(\alpha, \beta)}, \tag{6.31}$$

where

$$B(\alpha, \beta) = \frac{\Gamma(\alpha)\Gamma(\beta)}{\Gamma(\alpha + \beta)} \tag{6.32}$$

with $\Gamma(x)$ representing the Gamma function [27]. In this case, the shape parameters α_τ and β_τ given by 0.43 and 0.45. Since the values are normalized, τ_p is obtained by multiplying the random value by $\tau_{max} = 0.808$ ms;

3. For each path, generate ϕ_p, a_p and b_p as random samples from the following distributions:
   ϕ_p Uniform distribution in the interval $[-\pi, \pi]$;
   a_p Lognormal distribution with mean μ and standard deviation σ.
   b_p Gamma distribution with shape parameter a and scale parameter b.

Plausible parameter values for generating a_p and b_p are $\mu = -90$, $\sigma = 2.5$, $a = 1 \times 10^{-5}$ and $b = 2$. A detailed description of the parameter values obtained from real channel measurements is given in [11].

6.5 Extensions to MIMO PLC

The availability of three wires—phase (P), neutral (N) and protective earth (PE)—in powerline infrastructure in many countries worldwide, can be explored to implement a MIMO PLC system, which may help increase the achievable data rates in the communication system. In Figure 6.12, the possible subchannels that can be used involving the existing wires in the powerline infrastructure are illustrated.

A first approach for generating a MIMO channel model was presented in [34], where the authors present a methodology for generating a 2×4 MIMO channel. The key idea explored by the authors was to obtain the MIMO channel model from a single-input–single-output (SISO) channel generated using the model proposed in the context

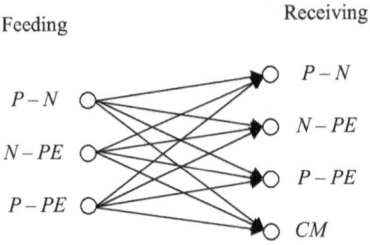

Figure 6.12   Individual physical paths in a MIMO PLC channel. CM denotes the common mode

of the European OPERA project [35]. Assuming that a reference SISO channel was generated, subchannels that compose the MIMO model are obtained by multiplying the taps of the SISO channel by random phase factors. Since all subchannels are derived from the same reference SISO channel, the larger the variation introduced by these phase factor, the more uncorrelated the channels will become. This same idea was used in [36] to generate a 3×3 MIMO channel. Nevertheless, with this approach, it is not possible to have total control about the correlation between the subchannels, an important aspect to be considered in the development of strategies for MIMO communication systems designs.

In this sense, the work presented in [37] tackles the problem of channel modeling from a different perspective, developing a model able to capture the spatial correlations observed in real measurements corresponding to a 2×3 MIMO channel. Considering that the estimated channel coefficients for each subcarrier at frequency f are organized in a matrix $\mathbf{H}(f)$, of size $N_r \times N_t$, where N_r and N_t represent the number of receiving and transmitting ports, it can be factored as

$$\mathbf{H}(f) = K \cdot \mathbf{R}_r^{1/2} \cdot \mathbf{H}'(f) \cdot \left(\mathbf{R}_t^{1/2}\right)^T \tag{6.33}$$

where K is a constant, $\mathbf{R}_r$ and $\mathbf{R}_t$ are the transmitter and receiver correlation matrices, respectively, and $\mathbf{H}'(f)$ is a channel matrix with taps corresponding to i.i.d. variables with a complex Gaussian distribution (zero mean and unit variance). Hence, if one is able to obtain the correlation matrices that represent a particular scenario, it is easy to generate several MIMO channels presenting the same spatial correlation structure.

This idea was further explored in [38] and a more elaborated model was derived, using the formulation previously described based on the multipath propagation model (6.4), but introducing the correlation between subchannels using (6.33).

6.6   Concluding remarks

In this chapter, we presented some simple models for the channel frequency response and noise in the context of PLC systems.

Using an analytical formulation for the signal propagation in the powerlines, and results from real measurements, it was possible to obtain statistical models that can be used to generate random channels—both for broadband and narrowband PLC. In the MIMO case, the same propagation model can be used, but some modifications are necessary to model possible correlations between sub-channels. In addition to that, different models for the three main categories were described, aiming to emulate the behavior of signals observed in channel measurements.

The statistical models described in this chapter were adjusted based on real measurements and provide faithful channel and noise models that can be helpful in the design and assessment of PLC communication systems and smart grid applications.

Acknowledgement

The author would like to thank the support from the National Council for Scientific and Technological Development – CNPq.

References

[1]   Hooijen OG. On the relation between network-topology and power line signal attenuation. In: Proc. International Symposium on Power Line Communications; 1998.

[2]   Galli S, Banwell TC. A deterministic frequency-domain model for the indoor power line transfer function. IEEE Journal on Selected Areas in Communications. 2006;24(7):1304–1316.

[3]   Anatory J, Theethayi N, Thottappillil R. Power-line communication channel model for interconnected networks Part I: Two-conductor system. IEEE Transactions on Power Delivery. 2009;24(1):118–123.

[4]   Cañete FJ, Cortés JA, Díez L, *et al.* A channel model proposal for indoor power line communications. IEEE Communications Magazine. 2011;49(12): 166–174.

[5]   Tonello AM, Zheng T. Bottom-up transfer function generator for broadband PLC statistical channel modeling. In: 2009 IEEE International Symposium on Power Line Communications and Its Applications. IEEE; 2009. p. 7–12.

[6]   Tonello AM, Versolatto F. Bottom-up statistical PLC channel modeling – Part II: Inferring the statistics. IEEE Transactions on Power Delivery. 2010;25(4):2356–2363.

[7]   Tonello AM, Versolatto F. Bottom-up statistical PLC channel modeling – Part I: Random topology model and efficient transfer function computation. IEEE Transactions on Power Delivery. 2011;26(2):891–898.

[8]   Philipps H. Modelling of powerline communication channels. In: IEEE Intl. Symp. on Power Line Commun. and Its Appl. (ISPLC); 1999. p. 14–21.

[9]   Zimmermann M, Dostert K. A multipath model for the powerline channel. IEEE Transactions on Communications. 2002;50(4):553–559.

[10] Tonello AM, Versolatto F, Bejar B, *et al.* A fitting algorithm for random modeling the PLC channel. IEEE Transactions on Power Delivery. 2012;27(3):1477–1484.

[11] Gassara H, Rouissi F, Ghazel A. Top-down random channel generator for the narrowband power line communication. AEU—International Journal of Electronics and Communications. 2018;89(March):146–152.

[12] Wentworth SM. Applied Electromagnetics: Early Transmission Lines Approach. Hoboken, NJ: Wiley; 2007.

[13] Rappaport TS. Wireless Communications: Principles and Practice (2nd Edition). Upper Saddle River, NJ: Prentice Hall; 2002.

[14] Goldsmith A. Wireless Communications (1st Edition). Cambridge: Cambridge University Press; 2005.

[15] Tse D, Viswanath P. Fundamentals of Wireless Communication. Cambridge: Cambridge University Press; 2005.

[16] Zimmermann M, Dostert K. An analysis of the broadband noise scenario in powerline networks. In: International Symposium on Powerline Communications and its Applications; 2000. p. 5–7.

[17] Esmailian T, Kschischang FR, Gulak PG. In-building power lines as high-speed communication channels: channel characterization and a test channel ensemble. International Journal of Communication Systems. 2003;16(5): 381–400.

[18] Tlich M, Pagani P, Avril G, *et al.* OMEGA Deliverable 3.2: PLC Channel Characterization and Modelling. OMEGA; 2009 Jan.

[19] Galda D, Rohling H. Narrow band interference reduction in OFDM-based power line communication systems. In: Proc. of IEEE International Symposium on Power Line Communications and Its Applications; 2001. p. 345–351.

[20] Benyoucef D. A new statistical model of the noise power density spectrum for powerline communication. In: Proc. 7th International Symposium on Power-Line Communications and Its Applications (ISPLC). Kyoto, Japan; 2003. p. 136–141.

[21] Andreadou N, Pavlidou FN. Modeling the noise on the OFDM power-line communications system. IEEE Transactions on Power Delivery. 2010;25: 150–157.

[22] Middleton D. Statistical-physical model of electromagnetic interference. IEEE Transactions on Electromagnetic Compatibility. 1977;19(3):106–126.

[23] Berry LA. Understanding Middleton's canonical formula for Class A noise. IEEE Transactions on Electromagnetic Compatibility. 1981;EMC-23(4): 337–344.

[24] Miyamoto S, Morinaga N, Katayama M. Performance analysis of QAM systems under Class A impulsive noise environment. IEEE Transactions on Electromagnetic Compatibility. 1995;37(2):260–267.

[25] Shongwey T, Vinck AJH, Ferreira HC. A study on impulse noise and its models. In: Power Line Commun and its Applicat (ISPLC), 2014 18th IEEE Int Symp on. 2014; p. 12–17.

[26]   Laguna-Sanchez G, Lopez-Guerrero M. On the use of alpha-stable distri-
       butions in noise modeling for PLC. IEEE Transactions on Power Delivery.
       2015;30(4):1863–1870.

[27]   Papoulis A, Pillai SU. Probability, Random Variables and Stochastic Processes.
       Boston, MA: McGraw-Hill; 2002.

[28]   Fertonani D, Colavolpe G. On reliable communications over channels
       impaired by bursty impulse noise. IEEE Transactions on Communications.
       2009;57(7):2024–2030.

[29]   Gilbert EN. Capacity of a Burst-Noise channel. Bell System Technical Journal.
       1960;39(5):1253–1265.

[30]   Elliott EO. Estimates of error rates for codes on Burst-Noise channels. Bell
       System Technical Journal. 1963;42(5):1977–1997.

[31]   Katayama M, Yamazato T, Okada H. A mathematical model of noise in nar-
       rowband power line communication systems. IEEE Journal on Selected Areas
       in Communications. 2006;24(7):1267–1276.

[32]   Nassar M, Dabak A, Kim IH, *et al.* Cyclostationary noise modeling in nar-
       rowband powerline communication for smart grid applications. In: 2012
       IEEE International Conference on Acoustics, Speech and Signal Processing
       (ICASSP); 2012. p. 3089–3092.

[33]   Gassara H, Rouissi F, Ghazel A. Empirical modeling of the narrowband
       power line communication channel. In: 2016 5th International Conference
       on Multimedia Computing and Systems (ICMCS); 2016. p. 570–575.

[34]   Canova A, Benvenuto N, Bisaglia P. Receivers for MIMO-PLC channels:
       throughput comparison. In: IEEE ISPLC 2010 – International Symposium
       on Power Line Communications and its Applications; 2010. p. 114–119.

[35]   Babic M, Hagenau M, Dostert K, Bausch J. Theoretical Postulation of the PLC
       Channel Model OPERA. IST Integrated Project No. 507667 funded by EC,
       OPERA Deliverable D4; 2005.

[36]   Hashmat R, Pagani P, Zeddam A, *et al.* A channel model for multiple input
       multiple output in-home power line networks. In: 2011 IEEE International
       Symposium on Power Line Communications and Its Applications. IEEE;
       2011. p. 35–41.

[37]   Tomasoni A, Riva R, Bellini S. Spatial correlation analysis and model for
       in-home MIMO power line channels. 2012 IEEE International Symposium
       on Power Line Communications and Its Applications, ISPLC 2012; 2012.
       p. 286–291.

[38]   Pagani P, Schwager A. A statistical model of the in-home MIMO PLC channel
       based on European field measurements. IEEE Journal on Selected Areas in
       Communications. 2016;34(7):2033–2044.

Chapter 7

Narrowband power line communication systems

Samuel C. Pereira[1], Ivan R. S. Casella[2], and Carlos E. Capovilla[2]

Advanced metering infrastructure (AMI) can be considered as one of the first steps towards smart grids (SGs) that provide direct access to consumers [1]. In a complementary way, power line communications (PLCs) appear as one of the main used technologies today for smart metering applications (60 per cent of market share [2]), demonstrating the importance of studying these technologies for the evolution of the communication infrastructures of power systems.

Modern narrowband PLC (NB-PLC) technologies are a very attractive solution for last-mile communication of the AMI [2,3] and a great promise to integrate future SGs applications. These technologies can provide two-way communication with the quality needed for neighbourhood area networks (NANs) that interconnect the smart meters, to meet the requirements of the AMIs (smart meters can offer in addition to electricity services, gas and water services). The NANs are connected to the utility's control centre (CC) through field area networks and wide area networks enabling AMI to offer remote metering of power consumption (this information can be accessed in real-time by consumers and/or the utility) and other bidirectional services such as remote load activation/deactivation and power meter tamper detection [1].

Currently, there are several widely used NB-PLC standards such as powerline intelligent metering evolution (PRIME) and G3-PLC, respectively, developed by the PRIME Alliance and G3-PLC Alliance, and the standards IEEE 1901.2 and ITU-T G.hnem, created by standard development organizations (SDOs) of the International Telecommunication Union (ITU) and the Institute of Electrical and Electronic Engineers (IEEE). These standards are essential to deliver transparent, efficient and integrated communication through the power line network. Basically, a standard is a document that provides requirements and guidelines for a product, process or service. Specifically in relation to the mentioned NB-PLC standards, they focus on the specifications of the two lowest layers of the open systems interconnection (OSI) model, i.e. the physical (PHY) and medium access control (MAC) layers [4]. One particularity of the above-mentioned modern NB-PLC standards lies in the fact that the PHY

[1]Department of Automation and Process Control, Federal Institute of São Paulo (IFSP Suzano), Brazil
[2]Center for Engineering, Modeling and Applied Social Sciences (CECS), Federal University of ABC (UFABC), Brazil

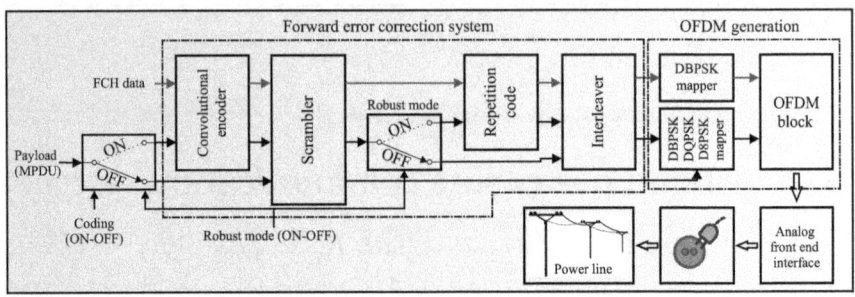

Figure 7.1   Block diagram of the PRIME v1.4 PHY layer transmitter

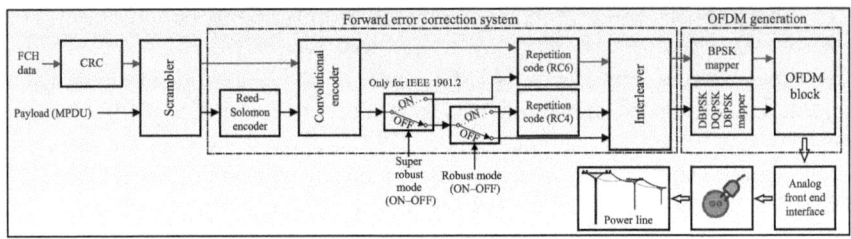

Figure 7.2   Block diagram of G3-PLC/IEEE 1901.2 PHY layer transmitter

layer of all of them uses multi-carrier (MC) modulation systems based on the OFDM (orthogonal frequency division multiplexing) technique.

In this context, this chapter will first present an overview of the PHY layer of the PRIME, G3-PLC, IEEE 1901.2 standards, and then the results obtained from the performance simulations of PHY layer of these standards will be presented and discussed considering multipath fading channels with additive white Gaussian noise (AWGN) and impulsive noise.

7.1   PHY layer description of PRIME, G3-PLC and IEEE 1901.2 standards

In this section, the main features of the PHY layers of the standards PRIME v1.4 [5], G3-PLC [6] and IEEE 1901.2 [7] will be described more deeply. As G3-PLC and IEEE 1901.2 have similar PHY layers, the description of both will be presented together.

The block diagram of the PRIME v1.4 PHY layer (transmitter) is shown in Figure 7.1. The payload coming from the MAC layer and the frame control header (FCH) generated in the PHY layer are encoded by the forward error correction (FEC) system and sent to the OFDM generation block where the encoded bits are mapped into subcarriers and transformed in OFDM symbols. Afterwards, the OFDM symbols are sent to the analogue front end (AFE) interface that conditions the signal in order to be coupled to the power line by means of specific filters.

On the other hand, Figure 7.2 shows the block diagram of a G3-PLC/IEEE 1901.2 PHY layer (transmitter). Again, the FCH and payload data are encoded by the FEC

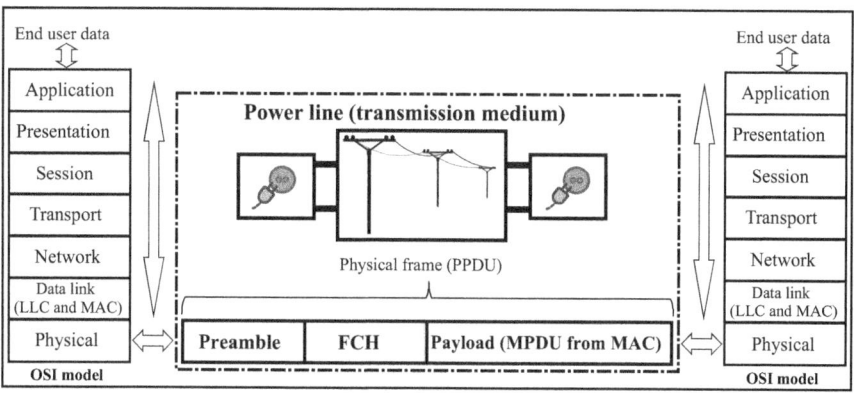

Figure 7.3   PHY frame structure and OSI model

system, transformed in OFDM symbols in the OFDM generation block and finally coupled to the power line by the AFE interface.

7.1.1   PHY frame

All the NB-PLC standards specify only the two lowest layers of the OSI model, PHY and MAC. In this model, shown in Figure 7.3, the information to be transmitted by a source user passes through seven layers (application, presentation, session, transport, network, data link and PHY) until reaching the communication channel, starting at the application layer (layer 7) and ending at the PHY layer (layer 1). After passing through the channel and reaching the receiver device, the information does the inverse path until reaching the destination user application layer. At each layer, additional data (e.g. control information or error detection) can be appended to the data information to better organize the transmission of the information and ensure the reliability of communication between different devices.

The PHY layer is responsible for defining the interfaces for medium access (e.g. signal type, channel characteristics, timing, modulation and coding schemes). On the other hand, the MAC layer is responsible for controlling and organizing access to the transmission medium, which can be shared by several network devices, whether or not based on the same standard (the MAC layer must ensure coexistence between devices sharing the same medium in order to not interfere with each other). The MAC is actually one of the two sublayers that constitute the link layer (MAC and logical link control), but it is widely called the MAC layer.

The MAC layer creates a data frame known as MAC protocol data unit (MPDU) that contains the original information at application layer plus all additional data appended through the upper layers. Before the information is transmitted through the PLC channel, the PHY layer attaches a preamble and an FCH to the MPDU from the MAC layer, creating the PHY layer frame, also known as the physical protocol data unit (PPDU). At PHY layer, the MPDU is called physical service data unit (PSDU) or payload. Each PHY frame starts with a preamble that is a fixed sequence of information known and expected by the receiver and used for detecting and synchronizing the received frame sent by the transmitter (it is also used to estimate channel conditions). The preamble sequence is followed by the FCH which contains the information necessary to demodulate the payload part of the frame [like modulation type and frame length (FL)], including a cyclic redundancy check (CRC), that is an error detecting code used to detect if the received FCH information contains any error caused by characteristics of the PLC channel (like noise and multipath effect). If the FCH CRC processing indicates any error, the information to demodulate the data frame is not reliable and the received frame is excluded. Then, the receiver can request the transmitter to resend the excluded frame, which increases the transmission latency but guarantees the reliability of the data transmission and avoids data loss. The FCH is followed by the payload which contains the data (information) delivered by the MAC layer (MPDU) and that will be transmitted. The payload data is also checked by means of a CRC code; however, it is performed at upper levels.

Figure 7.3 shows the PHY frame structure. Each part of the frame (preamble, FCH and payload) has its length normally specified in amount of OFDM symbols (integer number, except the preamble that can be a non-integer number of symbols and even not to use OFDM symbols).

7.1.1.1   PRIME PHY frame

The PRIME latest version (v1.4) describes two types of PHY frames, A and B. The type A corresponds to the one described in the previous PRIME version (v1.3.6), whereas the type B was introduced in the latest version for supporting robust operation (ROBO) transmission modes. This section will describe only the type B frame (information about type A can be found in [5]).

Preamble

The PRIME preamble is based on linear chirp sequence, i.e. a constant envelope signal (constant amplitude) in which the frequency varies linearly by the time. The frequency range depends on the number of frequency channels used, as PRIME frequency band (41.992–471.679 kHz) is divided in eight frequency channels that can be selected and combined in different ways (some additional details will be presented in Section 7.1.4.2). Each used frequency channel forms a preamble sub-symbol in which the chirp signal ranges from the lower to the upper frequency. The head of each sub-symbol is shaped with a raising roll-off region, whereas the tail has a decreasing roll-off region. The preamble sub-symbols (one for each used frequency channel) are concatenated to form a preamble symbol with duration of 2.048 ms (the shaped head and tail of each sub-symbol are overlapped with the adjacent sub-symbol in order to maintain constant signal envelope and good autocorrelation properties of the preamble symbol).

Figure 7.4 shows an example of PRIME preamble structure when frequency channels 1, 3 and 6 are used (the frequency range of each frequency channel can be found in Annex G of [5]). In the PHY frame, the preamble is repeated four times

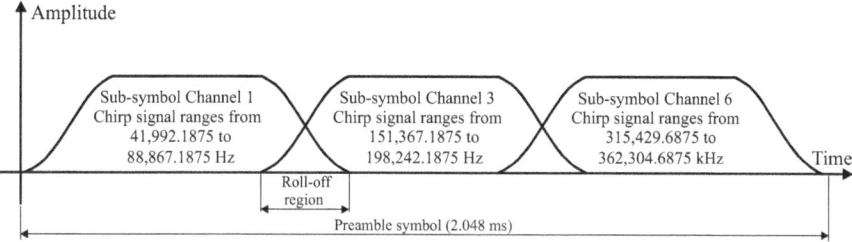

Figure 7.4   PRIME preamble structure when three frequency channels are used

FCH

Protocol	LEN	PAD_LEN	Reserved	CRC_Ctrl	MPDU1	FLUSHING_H	PAD_H
4	8	9	3	12	0 to 288	6	0 to 6

4 symbols (8.96 ms)

PAYLOAD

MPDU2	FLUSHING_P	PAD
$M \times 8$	8	$8 \times$ PAD_LEN

M symbols ($M \times 2.24$ ms)

Figure 7.5   PRIME FCH and payload structure

(in the last repetition, the preamble has a phase inversion that can be used for frame synchronization at the receiver). The total preamble duration is 8.192 ms.

Frame control header and payload

The PRIME FCH is composed of four OFDM symbols (84 data subcarriers plus 13 pilots for each activated frequency channel), just after the preamble and with duration of 8.96 ms (some additional details will be presented in Section 7.1.4.2). For the subcarriers, FCH always use differential binary phase shift keying (DBPSK) with convolutional (CONV) and repetition coding activated. The payload symbols, after the FCH, carry 96 data subcarriers for each active frequency channel (the number of pilots is equal to the number of active carriers) and according to the MAC configuration, the subcarriers can be DBPSK, differential quaternary phase shift keying (DQPSK) and differential 8-phase shift keying (D8PSK) modulated (with or without coding). Figure 7.5 shows the PRIME FCH and payload structure with their respective fields and lengths in bits (before encoding). The maximum number of payload symbols M is 252 when repetition code is activated and 63 for the other settings.

The FCH is composed by the following information fields:

- Protocol: indicates the transmission mode of the payload (subcarrier modulation type and activation of CONV and repetition encoding).
- LEN (Length): indicates the length of the payload (after coding) in number of OFDM symbols.
- PAD_LEN: indicates the length of the PAD field at payload (in bytes);

- Reserved: bits reserved for future features.
- CRC_Ctrl: this field contains the CRC checksum (CRC-12) over protocol, LEN, PAD_LEN and reserved fields (used for detecting errors added in these fields during the transmission over the PLC channel). The CRC for the payload is performed at upper levels.
- MPDU1: first part of the MPDU (payload) transmitted in the FCH. The firsts bits of the MPDU are transmitted in this field, being the amount defined by the number of active frequency channels selected, ranging from 0 (one active frequency channel) to 288 bits (eight active frequency channels).
- FLUSHING_H: bits needed to reset the CONV encoder (all six bits set to zero).
- PAD_H: bits (zeros) added (before coding) for adjusting the FCH length (number of bits) in order to fill the four OFDM symbols (after coding). It depends on the number of active frequency channels and ranges from zero (one active frequency channel) to six (eight active frequency channels).

The payload is composed by the following information fields:

- MPDU2: second part of the MPDU, i.e. all the bits following the MPDU1;
- FLUSHING_P: bits needed to reset the CONV encoder (all six bits set to zero). Only used when CONV encoder for payload is active;
- PAD: bits (zeros) added (before coding) for adjusting the payload length (number of bits) in order to fill an integer number of OFDM symbols (after coding). It depends on the number of active frequency channels and ranges from zero (one active frequency channel) to six (eight active frequency channels).

7.1.1.2   G3-PLC and IEEE 1901.2 PHY frame

The G3-PLC and IEEE 1901.2 NB-PLC specification supports two types of PHY frames: data frame and acknowledge (ACK) frame. The second one does not carry data, only preamble and FCH, and it is used for signalling purposes (e.g. frames correct reception verification). The PHY frames have two types of structures: one for the CENELEC (European Committee for Electrotechnical Standardization) regulated frequency band (3–148.5 kHz) and the other for the FCC (Federal Communications Commission) frequency band (9–490 kHz). This section will describe only the data frame with FCC structure (information about ACK frame and CENELEC structure can be found in [6,7]).

Preamble
The preamble consists of eight identical specific synchronization symbols called SYNCP symbols followed by one and a half identical specific synchronization symbols denoted as SYNCM symbols, transmitted before the FCH (IEEE 1901.2 can have eight or twelve SYNCP symbols). The duration of each preamble symbol is 213 μs for FCC band plan and 640 μs for CENELEC. The SYNCP symbols are used for synchronization, automatic gain control, initial phase reference and channel estimation. It is obtained setting the subcarriers of the symbol with the phase vectors described in [6,7], according to the frequency band used. The frequency bands available (CENELEC, FCC and ARIB) have different number of subcarriers so that there is a

specific SYNCP symbol for each band. The SYNCM is identical except that all sub-carriers are inverted (π phase shifted) in relation to SYNCP symbol, being the phase distance between SYNCP and SYNCM symbols used for frame synchronization.

Frame control header and payload

For FCC band plan, the FCH is composed of 12 OFDM symbols (232 µs each one) with 72 data subcarriers just after the preamble, being all bits protected by a CONV coding. The FCH subcarriers are mapped to coherent binary phase shift key-ing (BPSK) modulation with Super ROBO (S-ROBO) mode in which the data, after coding, is repeated six times. For G3-PLC, only the FCH can use the S-ROBO mode, but IEEE 1901.2 allows applying S-ROBO mode to the payload of the PHY frame.

The FCH has information for demodulating the payload frame, including the tone map (TM). G3-PLC and IEEE 1901.2 have adaptive tone mapping (ATM) function that allows the devices to adapt to the PLC channel conditions of an specific commu-nication link, using PHY layer parameters optimized for the channel conditions and consequently to provide maximum throughput. For that, the receiver estimates the SNR of each subcarrier of the received signal in order to select which subcarriers will carry data (the ones with better conditions in terms of SNR), the most appropriate modulation type and code rate (ROBO or S-ROBO modes) and the power level of the transmitter. The ATM function is handled by the MAC in conjunction with the PHY layer (more details in Section 7.1.5).

The payload has only one field because the MAC layer is responsible for padding the data in order to fit it to the PHY layer requirements. Differently from the FCH, the payload data is protected by the concatenation of Reed–Solomon (RS) and CONV encoders. It also has a selectable ROBO mode in which the encoded bits are repeated by a factor of four. In IEEE 1901.2, the implementation of S-ROBO mode (encoded bits repeated by a factor of six) for the payload is mandatory for FCC and ARIB band plans and optional for CENELEC bands (G3-PLC does not support S-ROBO mode for the payload).

The subcarriers payload can be mapped on the DBPSK, DQPSK and D8PSK (mandatory) modulations and, optionally, G3-PLC (ITU-T specification) and IEEE 1901.2 support coherent modulations such as BPSK, quaternary phase shift keying (QPSK), 8-phase shift keying (8-PSK) and 16-quadrature amplitude modulation (16-QAM). The maximum number of payload symbols is 252 for CENELEC band plans and 511 for FCC.

Figure 7.6 shows the G3-PLC and IEEE 1901.2 FCH and payload structures with their respective fields and lengths in bits (before encoding).

The FCH is composed by the following information fields, common for G3-PLC and IEEE 1901.2:

*   PDC (phase detection counter): indicates the phase difference between the trans-mitter and the receiver. For that, the devices must have an internal counter synchronized to the zero crossing of the power line signal. This counter ranges from 0 to 255 (8 bits) in one period of the power line signal. At the time the frame is transmitted, the value of this counter is stored in the PDC field. Then, the receiver

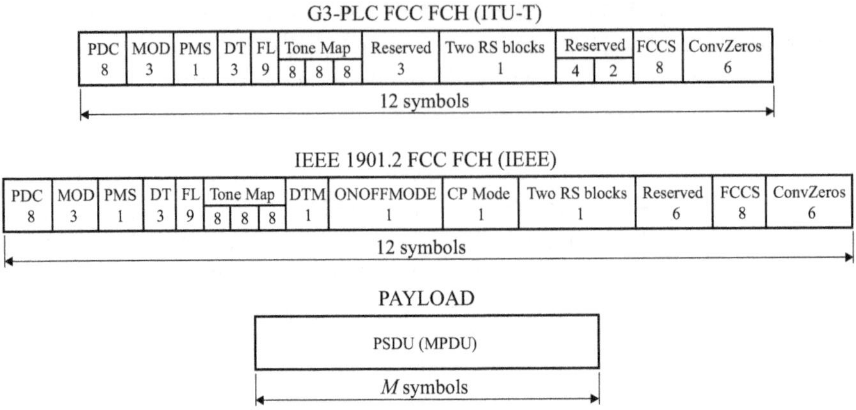

G3-PLC FCC FCH (ITU-T)

PDC	MOD	PMS	DT	FL	Tone Map			Reserved	Two RS blocks	Reserved	FCCS	ConvZeros
8	3	1	3	9	8	8	8	3	1	4	2	8	6

|← 12 symbols →|

IEEE 1901.2 FCC FCH (IEEE)

PDC	MOD	PMS	DT	FL	Tone Map			DTM	ONOFFMODE	CP Mode	Two RS blocks	Reserved	FCCS	ConvZeros
8	3	1	3	9	8	8	8	1	1	1	1	6	8	6

|← 12 symbols →|

PAYLOAD

PSDU (MPDU)

|← M symbols →|

Figure 7.6　G3-PLC and IEEE 1901.2 – FCH and payload structure

compares the PDC with the value of its counter at the time of the PDC reading. The difference of these values (transmitter and receiver counters, both synchronized to the zero crossing of the power line signal) indicates the phase difference between the devices, and it is stored in a neighbour table of each device. This information is used for detecting losses on the distribution line and the presence of incorrect neutral and phase installations (its use is optional).

- MOD (modulation): indicates the transmission mode of the payload, i.e. the subcarrier modulation type (DBPSK, DQPSK, D8PSK, ROBO and S-ROBO mode). Optionally, coherent modulation (BPSK, QPSK, 8PSK and 16-QAM) can be used. Only IEEE 1901.2 allows the use of the S-ROBO mode for the payload.
- PMS (payload modulation scheme): indicates if the subcarriers are differentially or coherently modulated.
- DT (delimiter type): indicates whether the frame expects or not a response. IEEE 1901.2 presents additional features, like indication that the frame was rejected or stored for later reading due to a busy device (the busy device sends this indication to the sender via DT field).
- FL: indicates the PHY FL in number of symbols.
- TM: it consists of three fields, each one with 8 bits. These fields are used to indicate which sub-bands of the payload OFDM symbol are carrying data (more details in Section 7.1.5).
- Reserved: it reserves some bits for future features.
- Two RS blocks: G3-PLC and IEEE 1901.2 allow dividing the MPDU (PSDU) in two equal size RS blocks. This field indicates whether or not the MPDU was divided.
- FCCS (frame control check sequence): this field contains the CRC checksum (CRC-8 for FCC band plan) over all previous FCH fields (used for detecting errors added in these fields during the transmission over the PLC channel). The CRC for the payload is performed at upper levels.
- ConvZeros: bits needed to reset the CONV encoder (6 bits set to zero).

The FCH can also have the following information fields, exclusive for IEEE 1901.2:

- DTM (data tone mask): The tone mask indicates the subcarriers used, being applied to preamble, FCH and payload. However, IEEE 1901.2 presents the multi-tone mask function which allows the use of a specific tone mask for the data and other to the preamble and FCH. The DTM field indicates if the single or multi-tone mask is used (the TM field can be used to indicate the subcarriers used on the payload).
- ONOFFMODE: indicates if the subcarriers inactivated by the ATM function are turned off (no energy transmitted) or carry dummy bits.
- CP mode (cyclic prefix mode): IEEE 1901.2 supports a long CP mode in which the number of samples of the CP is 52 (rather than 30 samples from the standard CP mode). The long CP can only be applied to the payload, and it must be indicated on the CP mode field of the FCH.

The payload is composed by the following information field:

- PSDU: this field carries the MPDU coming from the MAC layer, i.e. the data portion. Ever since the MAC layer is responsible for padding the data in order to fit it to the PHY layer requirements, no additional field is needed.

7.1.2   Scrambling schemes

Scrambling is a digital data processing that can be employed to randomize the data information in order to redistribute bits, encrypt data (in a simple concept) or aid in data synchronization. It avoids a long sequence of zeros or ones that can increase the crest factor (relationship between peak and root mean square power) at the output of the OFDM block and promote energy dispersal on the subcarriers of the OFDM symbol.

7.1.2.1   PRIME and G3-PLC scrambler

Both PRIME and G3-PLC use the same scrambling scheme. The only difference is that PRIME applies the scrambler to the encoded data whereas G3-PLC before encoding. Scrambling is always applied to FCH and payload, even when coding (in the case of PRIME) or ROBO mode is not activated.

The output data of the scrambler block is obtained by means of the XOR (exclusive OR) function of each bit of the data stream and a pseudo noise (PN) sequence. This PN sequence is obtained by the generator polynomial $P(x)$ described in (7.1) which results in a sequence of 127 bits that is cyclically applied to the data stream by means of the XOR function.

$$P(x) = x^7 + x^4 + 1 \tag{7.1}$$

Figure 7.7 shows the PRIME scrambler structure, detailing the PN sequence generator obtained by means of a linear-feedback shift register (LFSR) that must be initialized with "all ones" state. The XOR function between each bit of the PN

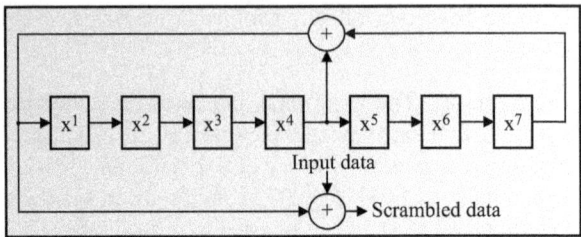

*Figure 7.7   Scrambler structure of both PRIME and G3-PLC/IEEE 1901.2
PHY layer*

sequence (cyclically applied to the XOR block) and the data bit stream results the
scrambled data.

7.1.3   Forward error correction system

The use of power lines as a communication channel is very challenging. The conditions and parameters of power lines vary dynamically according to the types of loads
connected, their locations within the network and their behaviour over time [8]. For
low-voltage (LV) networks, in less than one second, the signal-to-noise ratio (SNR)
may vary by 10 dB due to only a single load connected [9]. Moreover, impedance
mismatches, due to the presence of several branches with different characteristics,
cause propagation delays and reflections that may significantly degrade the received
signal (multipath effect) [10].

Power lines can also greatly attenuate the PLC signal (typically 30 dB for a 200 m
link in LV networks [8]) and, taking into account the presence of different sources of
noise and interference, the resulting SNR may be negative [2], which means that the
noise and/or interference power may be higher than the PLC signal power.

In order to overcome all these issues, an FEC scheme is used. FEC encodes the
message with redundant data so that errors introduced by the communication channel
can be corrected at the receiver. The more redundant data is added, the more robust is
the system, however, the lower is the effective data rate (proportionally reduced by the
redundant data added). The performance of the FEC can be improved by concatenating
two or more different coding schemes as it is done in G3-PLC.

7.1.3.1   PRIME forward error correction system

The FEC system of PRIME, shown in Figure 7.8, processes the FCH and payload
differently. The FCH uses a fixed configuration in which it passes through a CONV
encoder, scrambler, repetition encoder and data interleaver (following this order).
The subcarriers will be always DBPSK modulated. The payload can have different
configurations:

- Coding OFF: in this configuration, the payload is processed only by the scrambler.
  The subcarriers can be DBPSK, DQPSK or D8PSK modulated. This is the less
  robust mode and offers the highest data rates.

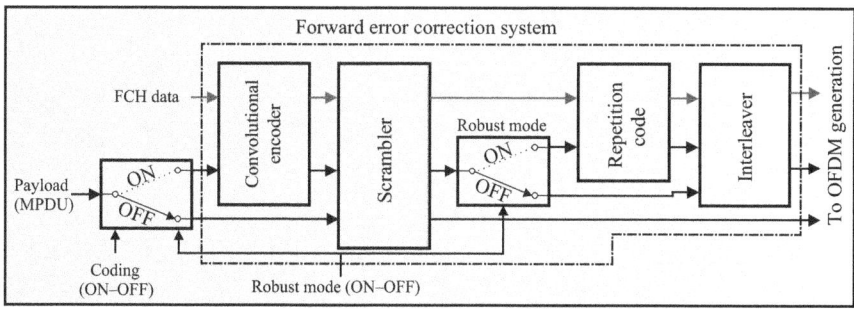

Figure 7.8 Block diagram of PRIME v1.4 FEC system

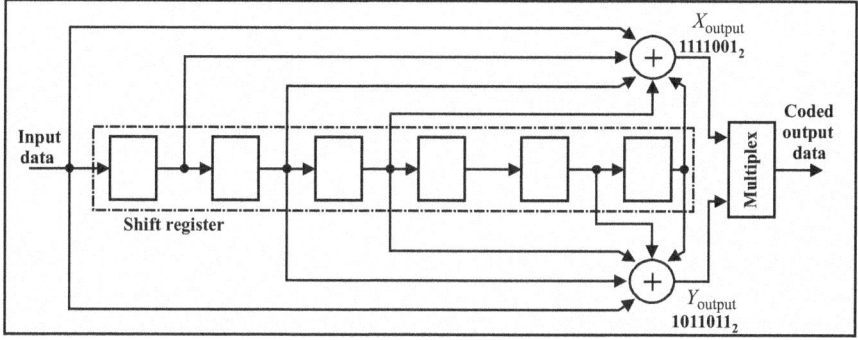

Figure 7.9 PRIME CONV encoder structure

- Coding ON: the payload is processed by a sequence of CONV encoder, scrambler and data interleaver. The subcarriers also can be DBPSK, DQPSK or D8PSK modulated.
- ROBO mode: the payload is processed similarly to the FCH, i.e. the payload passes sequentially through CONV encoder, scrambler, repetition encoder and data interleaver. The only difference is that the subcarriers can be either DBPSK or DQPSK modulated.

Convolutional encoder

The payload (when coding or ROBO are activated) and FCH data employ a CONV encoder (most significant bit first) which has code rate of $1/2$ (for each input bit, there will be two output bits), constraint length $K = 7$ (seven memory elements) and generator polynomials $(171, 133)_8$. The generator polynomials correspond to the tap connections of the LFSRs that integrate the encoder, represented in binary by the polynomials $(1111001)_2$ for X_{output} and $(1011011)_2$ for Y_{output}. Figure 7.9 shows the PRIME CONV encoder structure. At the start and end of each frame, the encoder state is always set to zero (the error probability at the CONV decoder is lower when the starting and ending states are known) by means of the flushing zeros appended to the end of the FCH and payload, as shown in Figure 7.5 from Section 7.1.1.1.

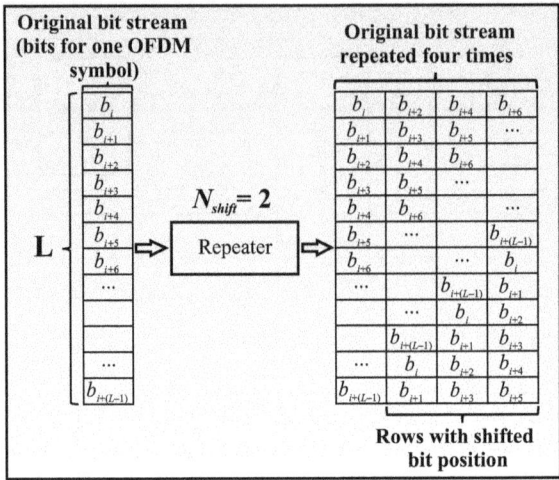

Figure 7.10   PRIME repetition code with a shift value of 2

Repetition code

The repetition code is always applied to the FCH data and, when the ROBO mode is activated, to the payload. The repetition code consists in repeating the bit data stream four times in a way that time and frequency diversity are introduced. For that, the bit data stream will be divided and organized in rows whose number of lines L will be equal the number of bits carried by each OFDM symbol, according to the chosen subcarrier mapping (DBPSK or DQPSK) and number of selected frequency channels N_{ch}. As explained previously, the PRIME frequency band is divided in eight frequency channels that can be selected and combined in different ways (some additional details will be presented in Section 7.1.4.2). Each channel has 84 data subcarriers for the FCH and 96 for the payload (each carrying 1 or 2 bits depending on the choice of DBPSK and DQPSK mapping). Then, L is equal $84 \cdot N_{ch}$ for the FCH and $96 \cdot N_{ch}$ and $192 \cdot N_{ch}$ for the payload when using, respectively, DBPSK and DQPSK subcarrier mapping.

Once the rows are organized, each one will be repeated in order to form a matrix of four rows. For introducing time and frequency diversity, there will be a cyclic shift of the bits position. The first row maintains the original bit position (most significant bit first), and the next three rows have their bit position shifted N_{shift} lines in relation to the previous one so that the last row (fourth) will be $3 \cdot N_{shift}$ shifted in relation to the first one. N_{shift} is two for DBPSK subcarrier modulation (FCH and payload) and four for DQPSK (payload). Figure 7.10 shows an example of PRIME repetition code with $N_{shift} = 2$. The original bit stream contains the amount of bits for one OFDM symbol (one bit per line), and after the repeater, there will be four OFDM symbols. The lines of the matrix represent the frequency domain (subcarriers inside OFDM symbol), whereas the rows represent the time domain (OFDM symbols).

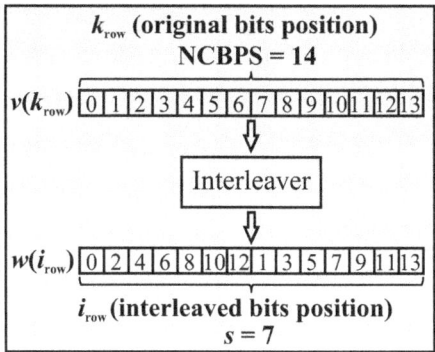

Figure 7.11   Example of interleaving performed by PRIME interleaver block

Interleaver

Interleaving is applied only when coding or ROBO mode is activated. Power line channel presents frequency selective fading, in which only a specific group of subcarriers inside the used frequency spectrum is affected, which causes burst bit errors. For avoiding these burst errors, the interleaver performs permutation of coded bits so that adjacent bits are separated from each other by several bits, which ensures bit errors are randomly scattered before decoding, improving the error correction capability of the decoder.

The encoded bits of each OFDM symbol are interleaved separately so that the interleaver block must have the size corresponding to the number of coded bits per OFDM symbol N_{CBPS}. The bit stream of each OFDM symbol is organized in a row vector $v(k_{row})$, with $k_{row} = 0, 1, \ldots, N_{CBPS} - 1$ (k_{row} represents the row positions of each element of the vector, which in this case each element is a bit) and applied to the input of the interleaver. The output of the interleaver will have a row vector $w(i_{row})$, with $i_{row} = 0, 1, \ldots, N_{CBPS} - 1$ (i_{row} represents the row positions of each element of the interleaved vector). The elements of each position of vector $w(i_{row})$ is determined by (7.2), for $k_{row} = 0, 1, \ldots, N_{CBPS} - 1$ and $s = 7$ for the FCH. For the payload, s is determined by (7.3), where N_{CBPSC} is the number of encoded bits per subcarrier (1 for DBPSK, 2 for DQPSK and 3 for D8PSK). The *rem*() function finds the remainder after division of k_{row} by s and the function *floor*() rounds the division of k_{row} by s to the nearest integer less than or equal to that result.

$$w(i_{row}) = v\left(\frac{N_{CBPS}}{s} \cdot rem(k_{row}, s) + floor\left(\frac{k_{row}}{s}\right)\right) \tag{7.2}$$

$$s = 8 \cdot \left(floor\left(\frac{N_{CBPS}}{2}\right) + 1\right) \tag{7.3}$$

Figure 7.11 shows an example of the bit permutation performed by the interleaver block with $N_{CBPS} = 14$ and $s = 7$; however, for the FCH , $N_{CBPS} = 84 \cdot N_{CBPSC} \cdot N_{ch}$ and for the payload, $N_{CBPS} = 96 \cdot N_{CBPSC} \cdot N_{ch}$, where N_{ch} is the number of frequency channels used for the transmission.

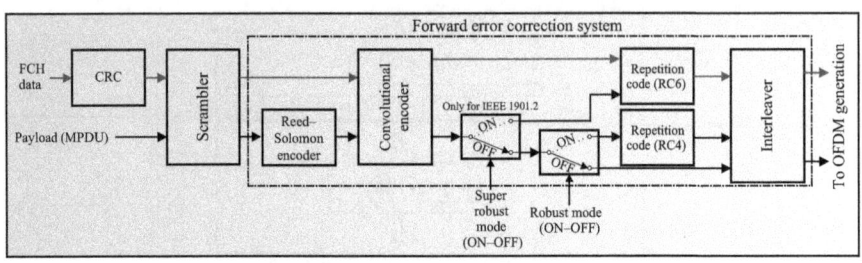

Figure 7.12   Block diagram of G3-PLC/IEEE 1901.2 FEC system

At the receiver, the received interleaved vector $w'(i_{row})$ will be applied to the deinterleaver block that, by means of (7.4), for $i_{row} = 0, 1, \ldots, N_{CBPS} - 1$, will restore bits to their original position inside the vector $v'(k_{row})$ so that it can be decoded.

$$v'\left(\frac{N_{CBPS}}{s} \cdot rem(i_{row}, s) + floor\left(\frac{i_{row}}{s}\right)\right) = w'(i_{row}) \tag{7.4}$$

7.1.3.2   G3-PLC/IEEE 1901.2 forward error correction system

Similarly to PRIME, G3-PLC and IEEE 1901.2 FEC system (shown in Figure 7.12) also processes the FCH and payload differently. The FCH uses a fixed configuration in which, after the scrambling, it passes through a CONV encoder, repetition code for S-ROBO mode (encoded data bits are repeated six times) and data interleaver (following this order).

Differently from PRIME, the FEC system for the payload in G3-PLC/IEEE 1901.2 consists of the concatenation of the RS and CONV coding schemes (it is not possible to disable it). There is also a selectable ROBO mode in which the encoded data bits are repeated four times. IEEE 1901.2 also offers an S-ROBO mode in which the encoded data bits are repeated six times (G3-PLC uses the S-ROBO mode only for the FCH and without RS encoding).

G3-PLC/IEEE 1901.2 CONV encoder
G3-PLC and IEEE 1901.2 use exactly the same PRIME's CONV encoder described in Section 7.1.3.1.

G3-PLC/IEEE 1901.2 Reed–Solomon encoder
The RS encoder is applied only to the payload of the PHY frame, after the scrambler block and before the CONV encoder. The scrambled data is encoded using a RS code with the following parameters:

- Normal mode (without repetition code): RS ($n = 255, k = 239, t = 8$).
- ROBO and S-ROBO modes (with repetition code): RS ($n = 255, k = 247, t = 4$).

Where n represents the codeword length in bytes (data bits plus eight bytes of parity bits), k the message length in bytes (input data bits must be organized in blocks with this length) and t is the number of symbols in which the encoder is able to correct (i.e. errors in up to 8 bytes of any part of the codeword can be corrected). Only one

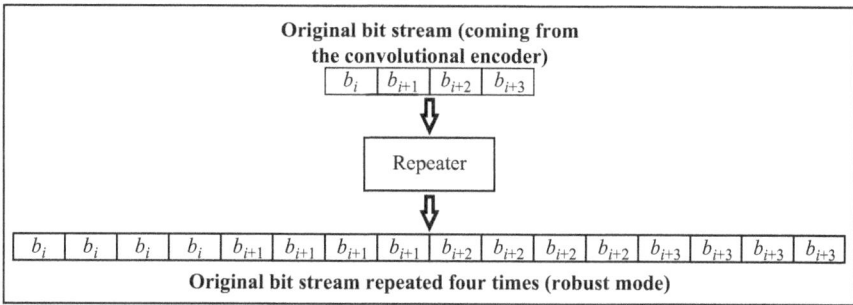

Figure 7.13   G3-PLC/IEEE 1901.2 repetition code for ROBO mode

RS block is used for each PHY frame so that the n and k parameters are adjusted according to the amount of data to be transmitted that will depend on the subcarrier modulation, coding rate and available subcarriers over the frequency band. The RS block sizes for different PHY parameters are defined in [6,7].

   The RS codeword is generated using the generator polynomial presented in (7.5). The RS parity symbols are given by the polynomial shown in (7.6), representing a Galois Field $GF(2^8)$.

$$g(x) = \prod_{i=0}^{2t} (x - \alpha^i) \tag{7.5}$$

$$P(x) = x^8 + x^4 + x^3 + x^2 + 1 \tag{7.6}$$

Repetition code

The repetition code is always applied to the FCH and, when the ROBO or S-ROBO modes are activated, to the payload as well. The repetition coding is always applied to the FCH and also to the payload when the ROBO or S-ROBO modes are activated. It consists of repeating each bit at the CONV encoder output four times for ROBO (RC4) mode and six times for S-ROBO (RC6). The repetition is performed bit by bit sequentially as shown in Figure 7.13. The FCH will always use the S-ROBO mode. For the payload, IEEE 1901.2 allows to select ROBO and S-ROBO modes, and in G3-PLC only, ROBO mode is supported.

Interleaver

G3-PLC/IEEE 1901.2 interleaving is done in a way that both time and frequency diversities are added, offering protection against burst errors that affect consecutive OFDM symbols (time) and frequency deep fade that affects adjacent subcarriers (frequency). The coded bits of each PHY frame are jointly interleaved. For this, the bit data stream of each PHY frame, coming from the CONV encoder or from the repetition block (in case of ROBO modes), is organized in a matrix whose number of columns m_{col} is equal to the number of active subcarriers per OFDM symbol and the number of rows n_{row} is equal to the number of OFDM symbols of the PHY frame times the modulation size N_{bps}. The number of rows n_{row} is found by means of (7.7), where *ceil*() is a function that returns a number equal to the nearest integer greater than or

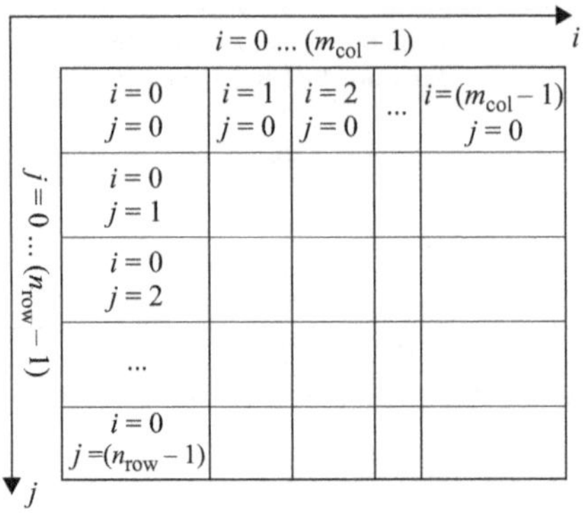

Figure 7.14   G3-PLC/IEEE 1901.2 interleaver matrix structure

equal to its input, N_{bits} is the total number of encoded bits to be transmitted in one PHY frame and N_{bps} is the modulation size, i.e. the number of bits per active subcarrier that depends on the selected mapping (DBPSK $=1$, DQPSK $=2$, D8PSK $=3$ and 16QAM $=4$).

$$n_{row} = ceil \left(\frac{N_{bits}}{m_{col} \cdot N_{bps}} \right) \cdot N_{bps} \tag{7.7}$$

Figure 7.14 shows the structure of the G3-PLC/IEEE 1901.2 interleaver matrix, where i and j represent, respectively, column and row positions. Once the matrix is organized, interleaving is performed in two steps. In the first step, each column is circularly shifted a different number of times. In the second one, each row is also circularly shifted. The number of circular shifts for the rows is determined by (7.8) whereas the columns by 7.9, where I and J represent, respectively, column and row interleaved positions, n_j is the first coprime of n_{row} larger than 2 and n_i the second one (I obtained from J). Similarly, m_i is the first coprime of m_{col} larger than 2 and m_j the second one (if it is not possible to find two coprimes larger than 2, one of the coprimes must be set to 1 with $n_i > n_j$ and $m_j > m_i$). The $rem(a, b)$ function finds the remainder after division of a by b.

$$J = rem(j \cdot n_j + i \cdot n_i, n_{row}) \tag{7.8}$$

$$I = rem(i \cdot m_i + J \cdot m_j, m_{col}) \tag{7.9}$$

The spreading behaviour of the G3-PLC/IEEE 1901.2 interleaver for a 10×8 matrix is shown in Figure 7.15. As $n_{row} = 8$, then $n_j = 3$ and $n_i = 5$ (coprime numbers for 8, except 1 and 2, are 3, 5 and 7) and as $m_{col} = 10$, then $m_i = 3$ and $m_j = 7$ (coprime numbers for 10, except 1 and 2, are 3, 7 and 9). With these parameters, (7.8) and (7.9) can be applied to the original matrix in order to find, respectively, I and J coordinates in the interleaved matrix.

When DBPSK subcarrier modulation is used, the interleaved matrix must not be modified prior to the mapping block, because each position already has the number of bits in which each DBPSK symbol carries, i.e. 1 bit per symbol. However, if higher

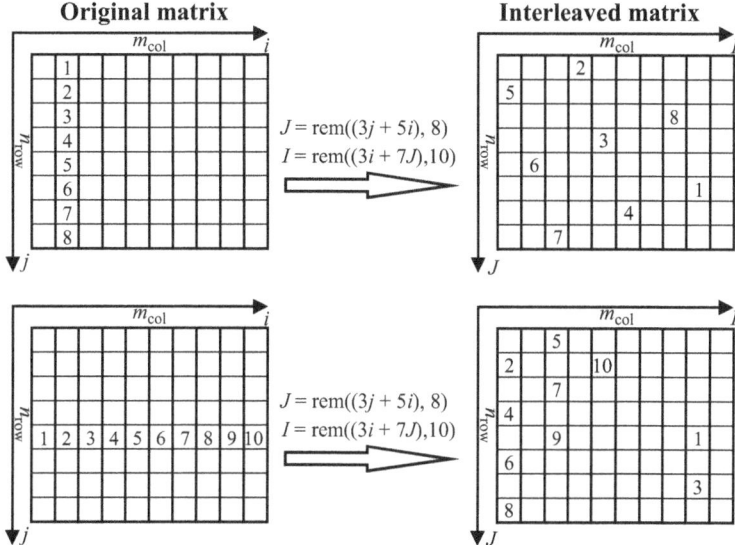

Figure 7.15   Example of the spreading behaviour of G3-PLC/IEEE 1901.2 interleaver

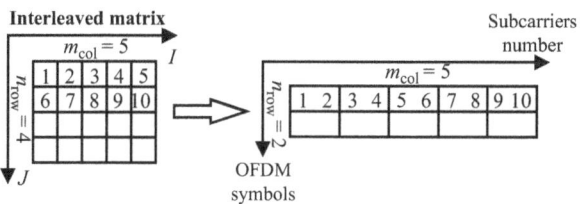

Figure 7.16   Example of interleaved matrix organization of G3-PLC/IEEE 1901.2 for DQPSK mapping

order subcarrier modulation, such DQPSK and D8PSK, is used, the data bits in the interleaved matrix must be organized so that each position has the number of bits carried by each DPSK symbol. For example, if each DQPSK symbol carries 2 bits, then each interleaved matrix position must be organized to carry 2 bits, reducing the number of rows by a factor of two (after that, each row will represent one OFDM symbol), as shown in Figure 7.16. The same applies to D8PSK which carries 3 bits in each symbol.

7.1.4   OFDM generation

Modern NB-PLC systems are based on MC modulation with OFDM. When compared to single carrier techniques, OFDM presents a higher spectrum efficiency and more robustness against frequency selective fading and interference, because the whole

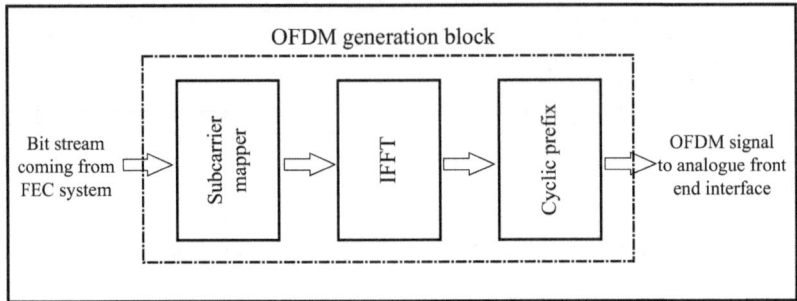

Figure 7.17   PRIME OFDM generation block diagram

spectrum is divided in multiple narrowband signals (subcarriers) that are individually affected by flat fading and interference. It allows the FEC system to recover the corrupted data on the affected subcarriers or even not to use parts of the spectrum that presents higher level of interference or fading. Moreover, due to their orthogonality to each other, the subcarriers are closely spaced, which allows a higher spectrum efficiency when compared to other MC modulation techniques.

For both PRIME and G3-PLC/IEEE 1901.2, the OFDM block receives the data coming from the FEC system (the bits are organized in a matrix so that the columns have the same number of bits carried on each OFDM symbol and the rows have the number of OFDM symbols on each PHY frame). The bits of this matrix are mapped on subcarriers that are DPSK modulated (G3-PLC/IEEE 1901.2 also have optional coherent modulation), and then the OFDM signal is generated in the next blocks.

7.1.4.1   PRIME OFDM generation

In the PRIME OFDM generation block, the bit stream coming from the interleaver block (coding on) or from the scrambler (coding off) is sent to the subcarrier mapper block where it is differentially modulated (DPSK mapping). Then, it is sent to the inverse fast Fourier transform (IFFT) block for generation of the OFDM symbols. Finally, each OFDM symbol is cyclically extended in the CP block. Figure 7.17 shows the PRIME OFDM generation block diagram.

Subcarrier mapper

The bit stream coming from the data interleaver (when coding is enabled) or from the scrambler block (when coding is not activated) is divided in groups (each group is organized to fill one OFDM symbol) and processed by the subcarrier mapper (the first bit is the MSB) that maps these bits for DPSK modulation. The first DPSK symbol is used as phase reference for demodulation of the first symbol containing data (for the following symbols the previous ones are used as reference). The DPSK symbols are placed on adjacent subcarriers inside the same OFDM symbol, i.e. frequency-domain differential modulation (FDDM) where every first subcarrier of each OFDM symbol must carry the reference symbol needed for demodulation at the receiver.

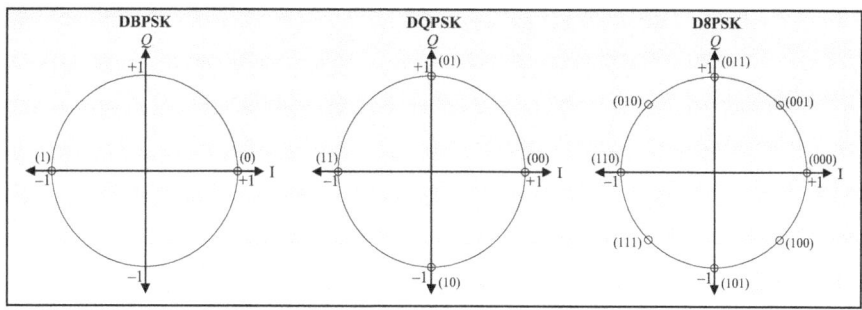

Figure 7.18   Constellation diagram of the modulation schemes used in PRIME

The FCH data is always DBPSK modulated, while the payload can be, according to the selected configuration, DBPSK, DQPSK and D8PSK modulated. Figure 7.18 shows the constellation diagram of the differential modulation schemes applied to PRIME, considering the phase reference at zero degrees.

IFFT block and cyclic prefix

The OFDM symbols are generated by the IFFT block. The data of each OFDM symbol is organized and allocated into the inputs of the IFFT block together with its mirrored complex conjugate (symmetric input) in order to only get the real values as output (the subcarriers mapping for a 2,048-point IFFT block can be found in [5]). After the IFFT block, a CP of 192 samples is appended to the OFDM symbol (192 samples of the end of the symbol delivered by the IFFT block are appended to the beginning of this same symbol). The PRIME OFDM subcarriers use the frequency band from 41.992 to 471.679 kHz that is divided in eight frequency channels that can be used either as single independent channels or combined in different ways. Each frequency channel has 97 equally spaced subcarriers, and it is separated from adjacent frequency channels by a guard interval of 15 subcarriers (7.3 kHz). The allocation of each subcarrier for the eight available frequency channels can be found in Annex G of [5]. The duration of each OFDM symbol is 2.24 ms (192 µs of this period consists of the cyclic prefix appended after the IFFT block). The parameters of PRIME OFDM block are shown in Table 7.1. The sample rate (1 MHz) and fast Fourier transform (FFT) block size (2,048) parameters are not mandatory on the implementation.

For the FCH (four OFDM symbols), each channel contains 13 pilot subcarriers, starting at the first subcarrier and separated from each other by seven data subcarriers so that only 84 subcarriers of each channel will carry data. The pilots from FCH can be used to estimate the sampling start error and the sampling frequency offset. For the payload, only the first subcarrier of each active channel will carry a pilot, and it has the function to provide a phase reference for the DPSK demodulation on frequency domain (96 subcarriers will carry data). Pilot subcarriers are coherent BPSK modulated, and their phases are controlled by a PN sequence of 127 bits obtained by a PN generator similar to the one described for the PRIME scrambler

Table 7.1   PRIME OFDM parameters

Parameter	Value
Sampling frequency	1 MHz
Number of FFT points	2,048
Number of subcarriers per active channel	97
Subcarrier spacing	488.28125 Hz
Number of data subcarriers per active channel	84 (FCH)/96 (payload)
Number of pilot subcarriers per active channel	13 (FCH)/1 (payload)
Symbol duration	2,240 μs (with CP)
Cyclic prefix duration	192 μs (192 samples)

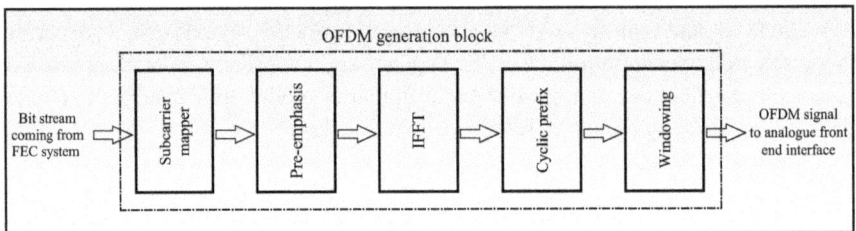

Figure 7.19   G3-PLC/IEEE 1901.2 OFDM generation block diagram

block in Section 7.1.3.1. More details on the distribution of this PN sequence for the pilot subcarriers can be found in [5].

7.1.4.2   G3-PLC/IEEE 1901.2 OFDM generation

In the G3-PLC/IEEE 1901.2 OFDM generation block, the bit stream coming from the interleaver block is sent to the subcarrier modulator block where it is differentially modulated (DPSK and an optional coherent modulation). Then, it is sent to the pre-emphasis block for frequency shaping before the OFDM generation is performed at the IFFT block. After the IFFT block, each OFDM symbol is cyclically extended in the CP block, and, finally, raised cosine shaping is applied in the windowing block. Figure 7.19 shows the G3-PLC/IEEE 1901.2 OFDM generation block diagram.

Subcarrier mapper
The bit stream coming from the data interleaver, already organized in groups to fill one OFDM symbol, is processed by the subcarrier mapper that, similar to PRIME, also maps these bits for DPSK modulation. However, different from PRIME that places the symbols in adjacent subcarriers (FDDM), the symbols are placed in adjacent OFDM symbols, i.e. time-domain differential modulation (in this case, the phase reference is sent at the beginning of each frame, which reduces the overhead in comparison to PRIME FDDM in which the phase reference must be sent in every OFDM symbols). The constellation diagram of the differential modulation schemes

Table 7.2   G3-PLC/IEEE 1901.2 OFDM parameters

Parameter	Value
Sampling frequency (minimum)	1.2 MHz
Number of FFT points (minimum)	256
Number of subcarriers	72
Subcarrier spacing	4.6875 kHz
Symbol duration	232 μs (with CP)
Cyclic prefix duration	18.3 μs (30 samples/8 overlapped)

applied to G3-PLC/IEEE 1901.2, considering the phase reference at zero degrees, is shown in Figure 7.18.

When operating in FCC band, the FCH is always coherently BPSK modulated (the phase reference for its demodulation comes from the last preamble and the payload uses the phase of the last FCH symbol as reference for DPSK demodulation). The payload can be, according to the selected configuration, DBPSK, DQPSK and D8PSK modulated. In ROBO and S-ROBO modes, IEEE 1901.2 uses BPSK modulation, while G3-PLC in ROBO mode uses DBPSK with optional BPSK. G3-PLC and IEEE 1901.2 have also the optional coherent modulation (BPSK, QPSK, 8PSK and 16QAM) for the payload.

Frequency domain pre-emphasis
Frequency domain pre-emphasis is an optional mechanism for frequency shaping of the transmitted signal, e.g. compensation for frequency dependent attenuation and spectral shaping. If implemented, frequency domain pre-emphasis must be applied to all OFDM symbols (preamble, FCH and payload) and in according to parameters TXRES (transmission gain resolution) and *TXCOEF* (transmitter gain for each group of subcarriers) that are related to the neighbour table which has information of PLC devices sharing the same frequency band (more detailed information in [6,7]).

IFFT block and cyclic prefix
The OFDM symbols are generated by the IFFT block. The data of each OFDM symbol is organized into the inputs of a 256-point IFFT block (each subcarrier index must be placed on the corresponding input of the IFFT block with all other inputs set to zero). Only the real part of the time-domain output of the IFFT block must be taken (imaginary parts are discarded). After the IFFT block, a CP of 30 samples is appended to the OFDM symbol. IEEE 1901.2 has an optional longer cyclic prefix (52 samples).

For FCC band, G3-PLC and IEEE 1901.2 use 72 equally spaced subcarriers for transmitting data (FCH and payload) and, inside FCC NBPLC frequency band (9–490 kHz), there are different band plans, found in [7] (IEEE 1901.2) and in [11] (G3-PLC). The duration of each OFDM symbol is 232 μs. Pilot subcarriers are only used when coherent modulation is used in the payload (FCH never uses pilot subcarriers). The parameters of G3-PLC/IEEE 1901.2 OFDM block are shown in Table 7.2. The sample rate (1.2 MHz) and FFT size (256) are the minimum values allowed.

Windowing

Raised cosine shaping is applied to all FCH and payload OFDM symbols with the purpose to reduce out of band emission and the spectral side lobe. Eight samples of each side of the symbol (head and tail) are shaped by the raised cosine function. For the preamble, the same raised cosine function is applied to the head of the first SYNCP symbol (eight samples) and to the tail of the last 1/2 SYNCM symbol (also eight samples). The tail of the SYNCM symbol will be overlapped with the head of the first FCH symbol. Parameters of the raised cosine function are found in [6,7].

7.1.5   *G3-PLC/IEEE 1901.2 ATM function*

The ATM function provides the greatest network throughput for a specific communication link according to the channel conditions. For that, each G3-PLC/IEEE 1901.2 device keeps a neighbour list (at MAC level) which contains all devices within its range. For each neighbour device, the list informs the optimized PHY parameters that must be used for transmitting data frames (only applied to the payload as the FCH uses fixed PHY parameters). These parameters are obtained by means of a channel estimation.

Every time a device needs to send a data frame and there is no valid neighbour list for the receiver, the MAC layer sends a TM request (TMR) frame via the payload of the PHY frame using ROBO or S-ROBO mode, according to the configuration (only IEEE 1901.2 is able to use S-ROBO mode for the payload). After receiving the TMR frame, the receiver performs channel estimation at PHY level and sends back to the transmitter, via its MAC layer, the TMR frame (also using ROBO or S-ROBO mode) which contains optimized PHY parameters (obtained by this channel estimation) for the communication link. These PHY parameters such as TM (usable subcarriers, i.e. the one with better conditions in terms of SNR), modulation type and transmitter power level. These parameters will be stored in the neighbour list of the transmitter device that will use them for the next transmissions to that specific device. The neighbour list is constantly updated as its parameters (or even the whole list) expire according to a configured valid time.

The PHY parameters used in the current transmission are carried by the FCH. The TM field of the FCH indicates the used subcarriers that are divided in sub-bands. Then, the usable subcarriers can only be selected in groups (for FCC band, each sub-band contains six subcarriers and the frequencies for each one can be found in [6,7]). The receiver PHY layer ignores the unused subcarriers according to the TM field. The subcarriers inactivated by the ATM function (i.e. not carrying information) will carry binary values (dummy bits) obtained by a PN sequence generated by the same generator polynomial used for PRIME and G3-PLC scrambler block and introduced in Section 7.1.2.1. The bits of the generator must be initialized to all ones at the start of the FCH and to shift to the next value according to the subcarrier modulation (each modulation carries a different number of bits) and for every disabled or masked subcarrier and pilot. Alternatively, IEEE 1901.2 also allows to turn off the inactive subcarriers, i.e. not to transmit energy on them, rather than transmit dummy bits.

Preamble and FCH will always use all available subcarriers for the specific band plan, except the ones notched by the tone mask parameter predefined by the MAC layer. The tone mask is applied to preamble, FCH and payload, and it does not depend on channel conditions. This is a static function, and the subcarriers are switched off rather than transmitting dummy bits, like in ATM function.

7.2 Simulation of PRIME and G3-PLC/IEEE 1901.2 PHY layers

As explained in Section 7.1.3 and in previous chapters, PLC channels present multipath fading and high levels of noise. In order to evaluate the performance of PRIME and G3-PLC/IEEE 1901.2 PHY layers regarding the data information transmission (payload) over some of these typical characteristics of the PLC channel, some simulations were performed according to the specifications presented in Section 7.1 and analysed for four distinct scenarios:

- AWGN channel;
- Multipath fading channel;
- AWGN channel with periodic impulsive noise;
- Multipath fading channel with periodic impulsive noise.

The PRIME PHY layer was implemented using a 2,048-sample OFDM symbol (corresponding to an effective maximum of 1,024 data symbols due to the symmetrical operation of the IFFT) with duration of 2,048 μs and a fixed guard interval (CP) of 192 samples (192 μs). All eight frequency channels of PRIME v1.4 frequency band specified in [5] were used, totalling 976 data subcarriers. The FEC system consists of $(171, 133)_8$ CONV coding scheme. The data block interleaving was configured as described in Section 7.1.3.1. The subcarrier mapping considered in the simulations was DBPSK with ROBO mode (DBPSK-ROBO), DBPSK, DQPSK and D8PSK. (For comparison purpose, the DQPSK-ROBO was omitted due to the fact that G3-PLC/IEEE 1901.2 does not specify DQPSK ROBO mode.)

On the other hand, G3-PLC/IEEE 1901.2 PHY layer was implemented using a 256-sample OFDM symbol (corresponding to an effective maximum of 128 data symbols due to the symmetrical operation of the IFFT) with duration of 213.3 μs, a fixed guard interval (CP) of 30 samples (due to overlap of eight samples, the guard interval is 18.3 μs) and 72 data subcarriers (specifications for FCC band [7]). The FEC system is obtained by the concatenation of $(255, 239)$ RS with $t = 8$ and $(171, 133)_8$ CONV coding schemes. For simplicity, the same FEC system will be used in the ROBO mode simulations (RS other than the one specified in the G3-PLC/IEEE 1901.2 standards for ROBO mode), by adding only one repetition coding scheme with factor 4. In the same way, the data block interleaving was implemented randomly. The subcarrier mapping considered in the simulations was coherent BPSK-ROBO, DBPSK-ROBO, DBPSK, DQPSK and D8PSK. IEEE 1901.2 uses coherent BPSK subcarrier modulation for ROBO and S-ROBO modes.

For each point of the simulation, 10^4 frames were transmitted. Depending on the subcarrier mapping, bit-filling can be employed to match the FL.

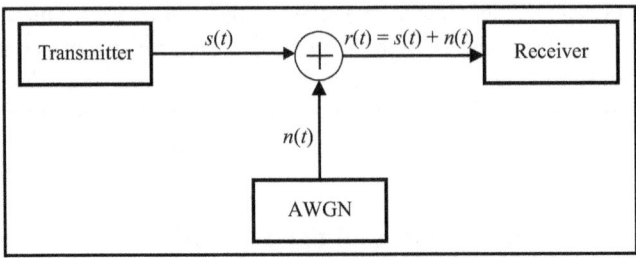

Figure 7.20   Simulation model for an AWGN channel

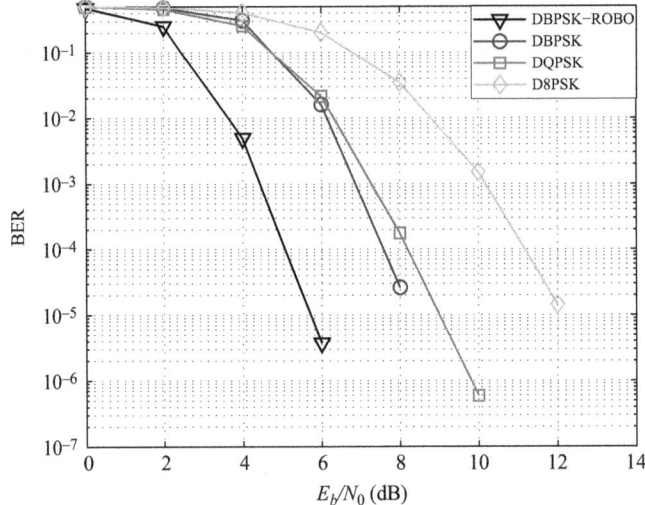

Figure 7.21   BER performance of PRIME PHY layer over an AWGN channel

7.2.1   AWGN channel

In order to first analyse the system performance in the simplest noise scenario, some simulations in an AWGN channel (considering only thermal background noise) were performed [12].

The employed simulation model, shown in Figure 7.20, represents the simple addition operation of AWGN $n(t)$ to the transmitted OFDM signal $s(t)$. The resulting received signal $r(t)$ was processed at the receiver, and the recovered data was compared to the transmitted ones to estimate the bit error rate (BER) of the system.

The simulation results for both PRIME and G3-PLC/IEEE 1901.2 in the AWGN scenario are shown, respectively, in Figures 7.21 and 7.22. The curves express the system performance in terms of BER as a function of the energy per bit to the noise power spectral density ratio (E_b/N_0).

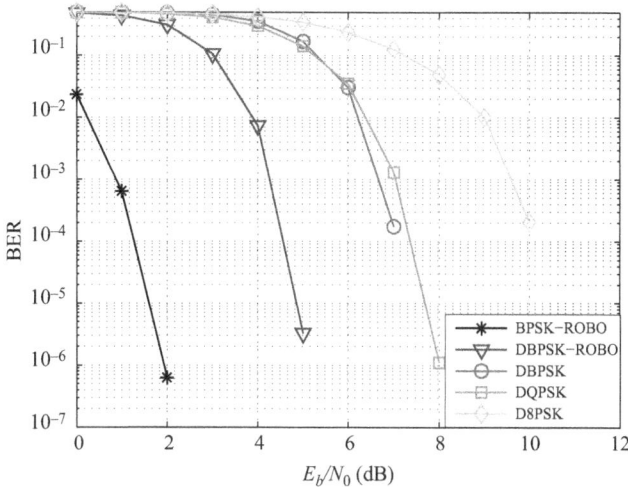

Figure 7.22   BER performance of G3-PLC/IEEE 1901.2 PHY layer over an AWGN channel

For PRIME, it can be verified in Figure 7.21 that, for reaching a BER of $1 \cdot 10^{-4}$, DBPSK-ROBO, DBPSK, DQPSK and D8PSK mapping require, respectively, an E_b/N_0 of approximately 5, 7.6, 8.2 and 11.1 dB.

For G3-PLC/IEEE 1901.2, it can be noted in Figure 7.22 that, for reaching a BER of $1 \cdot 10^{-4}$, BPSK-ROBO, DBPSK-ROBO, DBPSK, DQPSK and D8PSK mapping require, respectively, an E_b/N_0 of approximately 1.2, 4.5, 7.1, 7.3 and 10.2 dB.

Due to the use of a concatenated (RS/CONV) coding scheme, G3-PLC/IEEE 1901.2 clearly presented a better performance in comparison to PRIME (that employs only a CONV coding scheme) in AWGN channels for equivalent modulation mapping schemes (same modulation order). Because of coherent BPSK mapping, the IEEE 1901.2 ROBO mode performed significantly better compared to the G3-PLC ROBO mode using DBPSK.

7.2.2   Multipath fading channel

The multipath fading propagation effects, usually presented in PLC channels, can be depicted by the tapped delay line (TDL) model (corresponding to the channel impulse response). This model has been often used in the literature for modelling different bandlimited communication channels, including PLC channels [13–18], and it represents the multipath fading propagation effects (e.g. reflections, scattering) caused by different resolvable paths (corresponding to distinct symbol time intervals) by varying the complex coefficients of a TDL filter.

In this work, a coded symbol-spaced TDL channel model with 5-taps (five resolvable paths) whose coefficients have Rayleigh distributed magnitudes and uniform distributed phases was considered. An exponential decay profile with factor 2 between coefficients was also employed.

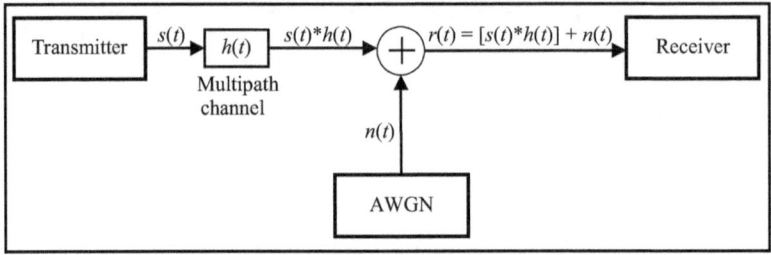

Figure 7.23 Simulation model for multipath fading channel with AWGN

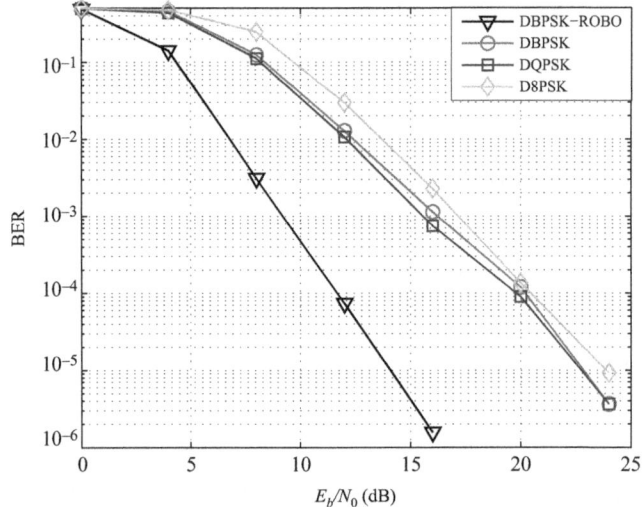

Figure 7.24 BER performance of PRIME PHY layer over multipath fading with AWGN

In the adopted simulation model, described in Figure 7.23, the resulting signal $r(t)$ at the receiver was obtained by the convolution of the transmitted OFDM signal $s(t)$ and the 5-tap multipath channel $h(t)$ and by the addition of AWGN $n(t)$.

In both PRIME and G3-PLC/IEEE 1901.2, the received OFDM symbols were equalized using the minimum mean square error frequency domain equation technique (applied after the FFT operation on the OFDM receiver).

Figures 7.24 and 7.25 present, respectively, the performance results in the multipath fading scenario for both PRIME and G3-PLC/IEEE 1901.2 in terms of BER as a function of E_b/N_0.

For PRIME, it can be verified in Figure 7.24 that, for reaching a BER of $1 \cdot 10^{-4}$, DBPSK-ROBO, DBPSK, DQPSK and D8PSK mapping require, respectively, an E_b/N_0 of approximately 12, 20.5, 20 and 21 dB.

For G3-PLC/IEEE 1901.2, it can be noted in Figure 7.25 that, for reaching a BER of $1 \cdot 10^{-4}$, BPSK-ROBO, DBPSK-ROBO, DBPSK, DQPSK and

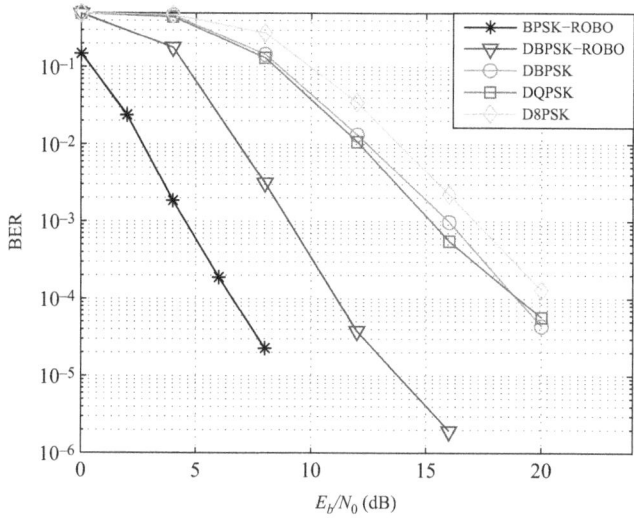

Figure 7.25   *BER performance of G3-PLC/IEEE 1901.2 PHY layer over multipath fading with AWGN*

D8PSK mapping require, respectively, an E_b/N_0 of approximately 6.6, 11, 18.8, 18.8 and 20.2 dB.

Although the multipath channel causes a larger performance degradation compared to the AWGN channel for both standards, the G3-PLC/IEEE 1901.2 also performed better than the PRIME in this scenario for equivalent modulation mapping schemes. Again, IEEE 1901.2 ROBO mode presented a better performance in comparison to G3-PLC ROBO mode.

7.2.3   AWGN channel with periodic impulsive noise

Regarding the noise scenario, it is not possible to analyse the performance of PLC systems considering only AWGN, as in general, the noise in power lines consists of background noise and impulsive noise (very destructive). The impulsive noise occurs generally in bursts, as represented in Figure 7.26, and it can be classified as periodic (occurs at regular time intervals, synchronous or asynchronous to the mains frequency) or aperiodic (random occurrence) [10].

In [19], periodic impulsive noise parameters were evaluated by means of a measuring campaign that established three different scenarios: heavily disturbed, medium disturbed and weakly disturbed. Table 7.3 shows the impulsive noise parameters for these three scenarios, where *IAT* is the inter-arrival time of the impulsive noise and T_{noise} is its average duration.

For evaluating the performance of PRIME and G3-PLC/IEEE 1901.2 PHY layers, the periodic impulsive noise $n_{imp}(t)$, generated according to [20] for the heavily disturbed scenario of Table 7.3, was added to the background noise $n(t)$ (represented

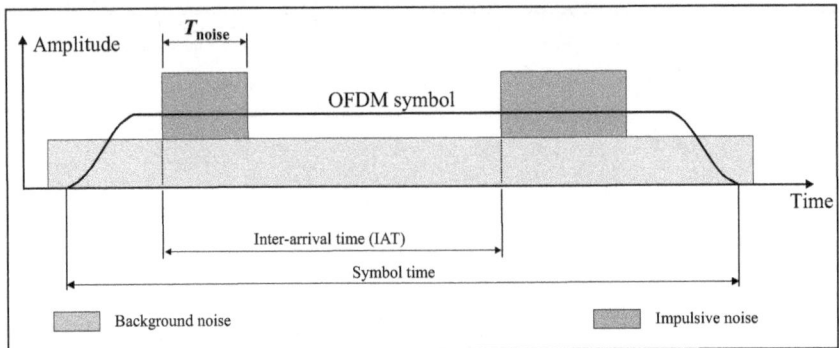

Figure 7.26   Representation of AWGN and impulsive noise over an OFDM symbol

Table 7.3   Periodic impulsive noise parameters

Impulsive noise scenario	IAT (s)	T_{noise} (ms)
Heavily disturbed	0.0196	0.0641
Medium disturbed	0.9600	0.0607
Weakly disturbed	8.1967	0.1107

Source: Reference [20].

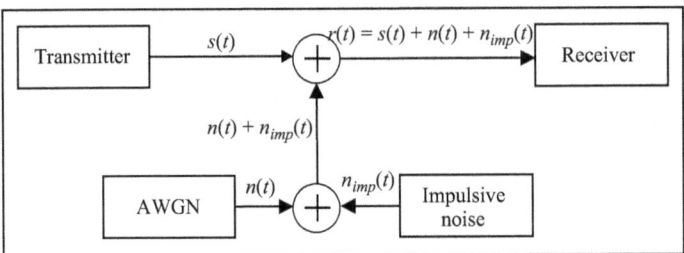

Figure 7.27   Simulation model for AWGN channel with periodic impulsive noise

by the AWGN), as shown in Figure 7.27. The simulations were performed considering two levels of impulsive noise, characterized by the impulsive noise to AWGN power ratio (F_{imp}), one setting $F_{imp} = 10$ and the other setting $F_{imp} = 100$. Although $F_{imp} = 100$ represents an impulsive noise level much higher than $F_{imp} = 10$, which was considered in [19,20], it was also employed here only to investigate more deeply the effects of increasing impulsive noise at extreme levels.

Figures 7.28 and 7.29 present the performance results for PRIME in the AWGN channel with impulsive noise in terms of BER as a function of the E_b/N_0, respectively, for $F_{imp} = 10$ and $F_{imp} = 100$. For comparison purpose, the curves without impulsive noise (only AWGN) were also inserted to the figures.

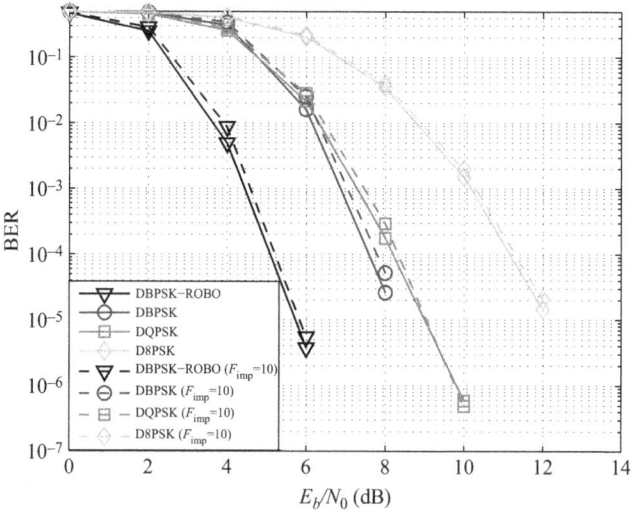

Figure 7.28 BER performance of PRIME PHY layer over AWGN channel with periodic impulsive noise ($F_{imp} = 10$)

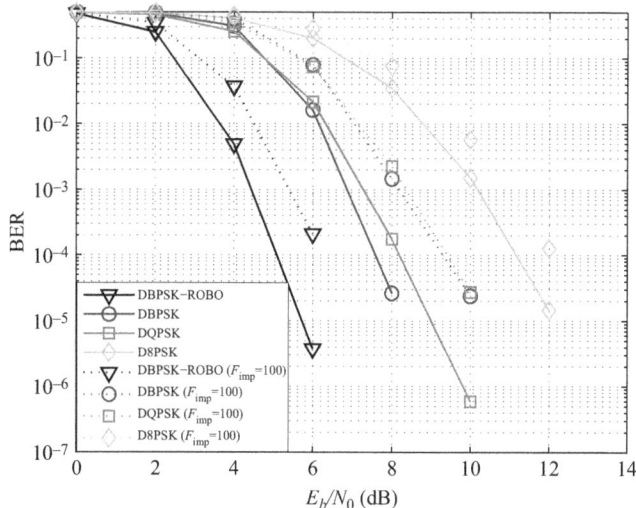

Figure 7.29 BER performance of PRIME PHY layer over AWGN channel with periodic impulsive noise ($F_{imp} = 100$)

It can be verified in Figure 7.28 that, for reaching a BER of 10^{-4}, the E_b/N_0 has to be increased by around 0.2 dB, when compared to the no impulsive noise scenario, independent to the modulation mapping schemes, due to the presence of the periodic impulsive noise. This highlights the robustness of PRIME against periodic impulsive noise.

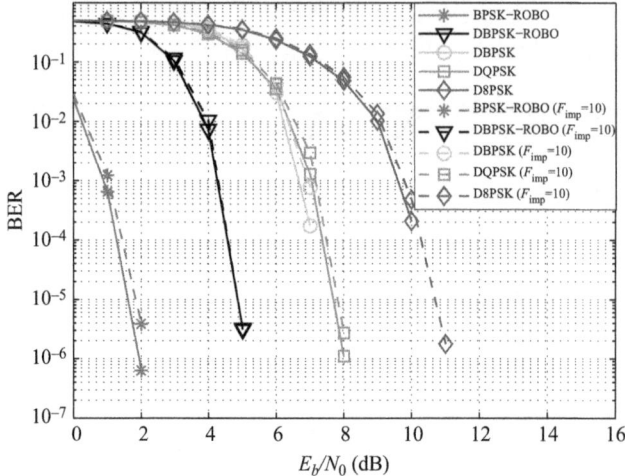

Figure 7.30   BER performance of G3-PLC/IEEE 1901.2 PHY layer over AWGN channel with periodic impulsive noise ($F_{imp} = 10$)

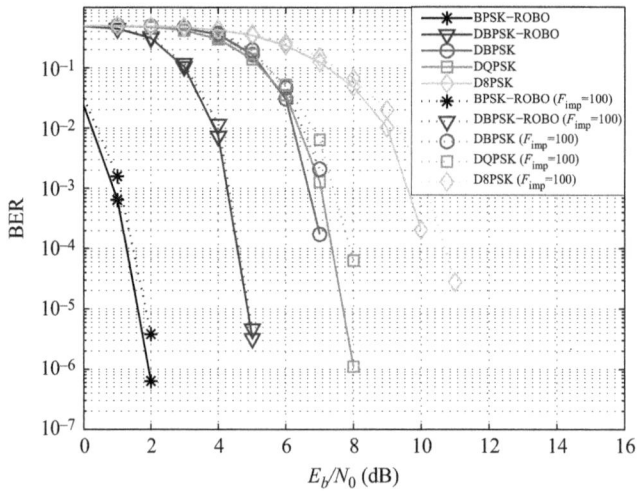

Figure 7.31   BER performance of G3-PLC/IEEE 1901.2 PHY layer over AWGN channel with periodic impulsive noise ($F_{imp} = 100$)

It can be noted in Figure 7.29 that with $F_{imp} = 100$, the higher impulsive noise level reduces the performance of PRIME. For reaching a BER of 10^{-4}, the E_b/N_0 has to be increased between 1 and 2 dB depending on the modulation in relation to the values obtained in the scenario without impulsive noise.

Figures 7.30 and 7.31 show the performance results for G3-PLC/IEEE 1901.2 in the AWGN channel with impulsive noise, respectively, for $F_{imp} = 10$ and $F_{imp} = 100$.

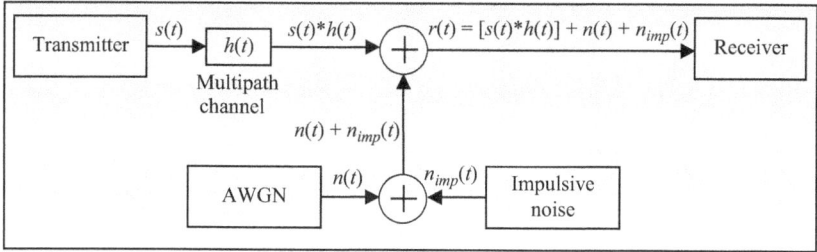

Figure 7.32 *Simulation model for multipath channel with AWGN and periodic impulsive noise*

From Figure 7.30, it can be noted that the presence of impulsive noise with $F_{imp} = 10$ requires an increase in E_b/N_0, when compared to no impulse noise scenario, by around 0.1–0.2 dB for a BER of $1 \cdot 10^{-4}$. It highlights that G3-PLC also presents a good robustness against periodic impulsive noise.

Analysing the results of Figure 7.31, it can be noted that the presence of impulsive noise with $F_{imp} = 100$ require an increase in E_b/N_0, in comparison to the scenario without impulse noise, by around 0.1–0.8 dB for a BER of $1 \cdot 10^{-4}$.

7.2.4 Multipath fading channel with periodic impulsive noise

In this scenario, all PLC channel characteristics previously discussed are combined, as shown in Figure 7.32. The transmitted OFDM signal $s(t)$ pass through the multipath channel $h(t)$ with five resolvable paths, and then AWGN $n(t)$ and periodic impulsive noise $n_{imp}(t)$ are added to the resulting signal at the receiver. The multipath channel was generated according to the parameters shown in Section 7.2.2, and the periodic impulsive noise was configured according to the heavily disturbed scenario presented in Table 7.3, for $F_{imp} = 10$ and $F_{imp} = 100$. Again, the performance with $F_{imp} = 100$ was also analysed for investigation purpose only (verify when impulsive noise levels start to become critical).

Figures 7.33 and 7.34 present the performance results for PRIME in the multipath channel with AWGN and impulsive noise in terms of BER as a function of E_b/N_0, respectively, for $F_{imp} = 10$ and $F_{imp} = 100$. For comparison purpose, the curves without impulsive noise (only multipath fading) were also inserted to the figures.

From Figure 7.33, it can be noted that the presence of impulsive noise with $F_{imp} = 10$ require an increase in E_b/N_0 (when compared to no impulse noise scenario) by around 0.1–0.2 dB for a BER of $1 \cdot 10^{-4}$. The performance degradation due to multipath was much higher than due to the impulsive noise with $F_{imp} = 10$.

Analysing the results of Figure 7.34, it can be noted that the presence of impulsive noise with $F_{imp} = 100$ require an increase in E_b/N_0, when compared to no impulse noise scenario, by around 0.2–0.8 dB for a BER of $1 \cdot 10^{-4}$.

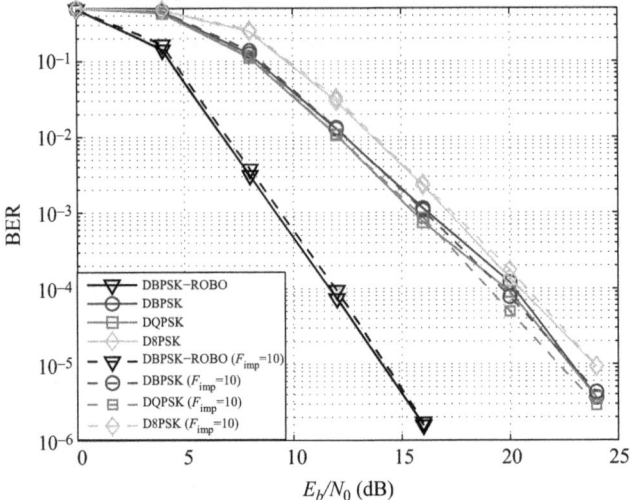

Figure 7.33 *BER performance of PRIME PHY layer over multipath fading channel with AWGN and periodic impulsive noise ($F_{imp} = 10$)*

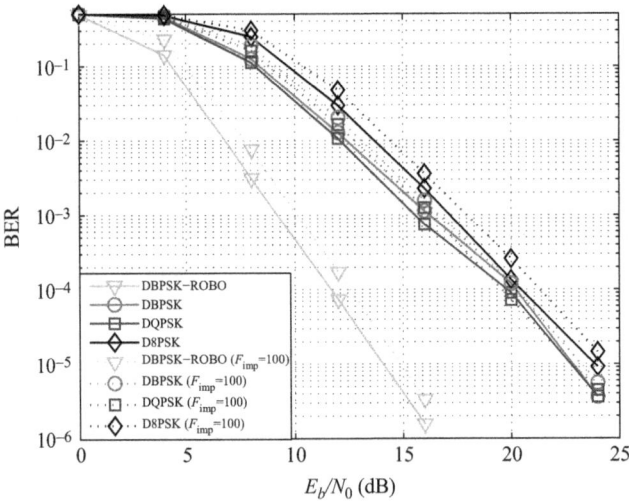

Figure 7.34 *BER performance of PRIME PHY layer over multipath fading channel with AWGN and periodic impulsive noise ($F_{imp} = 100$)*

Figures 7.35 and 7.36 show the performance results for G3-PLC/IEEE 1901.2 in the multipath channel with AWGN and impulsive noise, respectively, for $F_{imp} = 10$ and $F_{imp} = 100$.

From Figure 7.35, it can be noted that the presence of impulsive noise with $F_{imp} = 10$ requires an increase in E_b/N_0, when compared to no impulse noise scenario, by

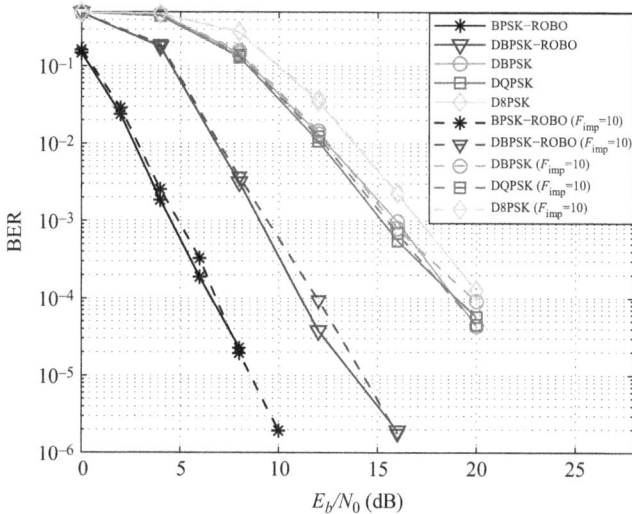

Figure 7.35 BER performance of G3-PLC/IEEE 1901.2 PHY layer over multipath fading with AWGN and periodic impulsive noise ($F_{imp} = 10$)

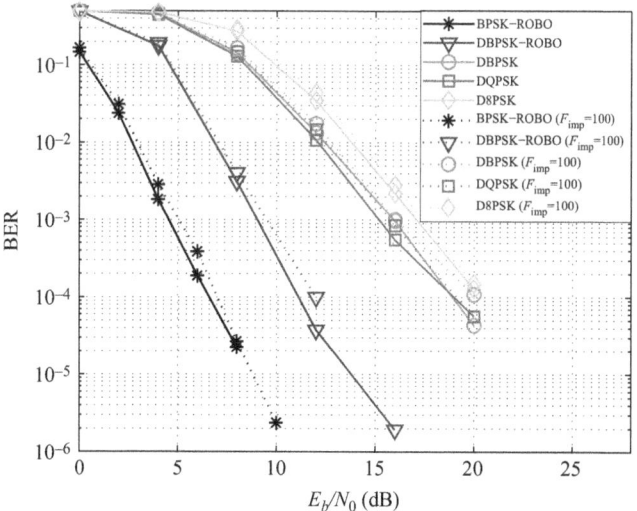

Figure 7.36 BER performance of G3-PLC/IEEE 1901.2 PHY layer over multipath fading with AWGN and periodic impulsive noise ($F_{imp} = 100$)

around 0.05–0.6 dB for a BER of $1 \cdot 10^{-4}$, highlighting the robustness of G3-PLC at this level of periodic impulsive noise. Analysing the results, it can be verified that the performance degradation due to multipath was much higher than due to the impulsive noise with $F_{imp} = 10$.

Analysing the results of Figure 7.36, it can be noted that the presence of impulsive noise with $F_{imp} = 100$ require an increase in E_b/N_0, when compared to no impulse noise scenario, by around 0.2–0.7 dB for a BER of $1 \cdot 10^{-4}$. The results for $F_{imp} = 10$ and $F_{imp} = 100$ were very similar.

7.3   Conclusion remarks

In the analysis conducted in this chapter, it was possible to conclude that both PRIME and G3-PLC presented good robustness against periodic impulsive noise even in very severe scenarios with $F_{imp} = 10$ and $F_{imp} = 100$. The degradation of the system performance caused by multipath propagation was much more critical than the one caused by the periodic impulsive noise (at the investigated levels). It is important to mention that in all scenarios investigated (AWGN and multipath fading channel without and with periodic impulsive noise) G3-PLC outperformed PRIME.

References

[1]   ITU-T. Applications of ITU-T G.9960, ITU-T G.9961 transceivers for smart grid applications: advanced metering infrastructure, energy management in the home and electric vehicles. Geneva (Switzerland): International Telecommunication Union; 2010.

[2]   Lin J, Nassar M, Evans BL. Impulsive Noise Mitigation in Powerline Communications Using Sparse Bayesian Learning. IEEE Journal on Selected Areas in Communications. 2013;31(7):1172–1183.

[3]   Dzung D, Berganza I, Sendin A. Evolution of powerline communications for smart distribution: from ripple control to OFDM. In: 2011 International Symposium on Power-Line Communications and Its Applications (ISPLC); 2011 April 3–6; Udine, Italy. IEEE; 2011. p. 1–9.

[4]   Berger LT, Schwager A, Galli S, *et al.* Current Power Line Communication Systems: A Survey. In: Berger LT, Schwager A, Pagani P, *et al.*, editors. MIMO Power Line Communications: Narrow and Broadband Standards, EMC, and Advanced Processing. Boca Raton (FL): CRC Press; 2014. p. 253–270.

[5]   Specification for Powerline Intelligent Metering Evolution. PRIME Alliance; 2014. v1.4-20141031.

[6]   ITU-T. Narrowband orthogonal frequency division multiplexing power line communication transceivers for G3-PLC networks. Geneva (Switzerland): International Telecommunication Union; 2014. G.9903.

[7]   IEEE. 1901.2-2013 – IEEE standard for low-frequency (less than 500 kHz) narrowband power line communications for smart grid applications. Institute of Electrical and Electronics Engineers; 2013. 978-0-7381-8793-8.

[8]   Razazian K, Umari M, Kamalizad A, *et al.* G3-PLC specification for powerline communication: overview, system simulation and field trial results. In: 2010 IEEE International Symposium on Power Line Communications and Its

Applications (ISPLC); 2010 March 28–31; Rio de Janeiro, Brazil. IEEE; 2010. p. 313–318.

[9]   Clark D. Powerline communications: finally ready for prime time? IEEE Internet Computing. 1998;2(1):10–11.

[10] Gotz M, Rapp M, Dostert K. Power line channel characteristics and their effect on communication system design. IEEE Communications Magazine. 2004;42(4):78–86.

[11] ITU-T. Narrowband orthogonal frequency division multiplexing power line communication transceivers: power spectral density specification. Geneva (Switzerland): International Telecommunication Union; 2014. G.9901.

[12] Liu W, Li C, Dostert K. Emulation of AWGN for noise margin test of powerline communication systems. In: 2011 IEEE International Symposium on Power Line Communications and Its Applications (ISPLC); 2011 April 3–6; Udine, Italy. IEEE; 2011. p. 225–230.

[13] Hoque KMR, Debiasi L, Natale FGB. Performance analysis of MC-CDMA power line communication system. In: IFIP International Conference on Wireless and Optical Communications Networks (WOCN); 2007 July 2–4; Singapore, Singapore. IEEE; 2007. p. 1–5.

[14] Liu L, Cheng T, Yanan L. Analysis and modeling of multipath for indoor power line channel. In: 10th International Conference on Advanced Communication Technology (ICACT); 2008 February 17–20; Gangwon-Do, South Korea. IEEE; 2008. p. 1966–1969.

[15] Tan B, Thompson JS. Powerline Communications Channel Modelling Methodology Based on Statistical Features. CoRR – Computing Research Repository. 2012;1203.3879:1–9.

[16] Lampe L, Vinck AJ. Cooperative multihop power line communications. In: 2012 IEEE International Symposium on Power Line Communications and Its Applications (ISPLC); 2012 March 27–30; Beijing, China. IEEE; 2012. p. 1–6.

[17] Mathur A, Bhatnagar MR, Panigrahi BK. PLC Performance Analysis Over Rayleigh Fading Channel Under Nakagami-m Additive Noise. IEEE Communications Letters. 2014;18(12):2101–2104.

[18] Rjeily CA. Power line communications under Rayleigh fading and Nakagami noise: novel insights on the MIMO and multi-hop techniques. IET Communications. 2018;12(2):184–191.

[19] Zimmermann M, Dostert K. Analysis and Modeling of Impulsive Noise in Broad-Band Powerline Communications. IEEE Transactions on Electromagnetic Compatibility. 2002;44(1):249–258.

[20] Ma YH, So PL, Gunawan E. Performance Analysis of OFDM Systems for Broadband Power Line Communications Under Impulsive Noise and Multipath Effect. IEEE Transactions on Power Delivery. 2005;20(2):674–682.

Chapter 8
Broadband power line communication systems

Ivan R. S. Casella[1], Samuel C. Pereira[2],
and Carlos E. Capovilla[1]

Modern broadband power line communication (BB-PLC) technologies were primarily designed to provide high-speed Internet access and multimedia communication. As they can operate in the frequency range of 1.8–250 MHz (high frequencies), the initial deployments were intended for indoor applications (residential or industrial) in order to avoid electromagnetic interference in licensed radio services that share the same frequency range [1–3]. However, important studies are currently in progress to use these technologies in indoor and outdoor applications of the smart grids [4–6].

BB-PLC can provide theoretical data rates of up to 2 Gbps [7], competing or working in conjunction with wireless technologies in different applications. In addition to offer high data rates, they have the advantage of penetrating facilities where wireless signals have difficulty accessing, such as underground areas, tunnels, basements, different rooms and floors in homes and buildings and remote regions.

Currently, there are two BB-PLC standards developed by industry associations widely spread worldwide, the HomePlug, created by the HomePlug Powerline Alliance (more than 200 million devices sold [8]), and the high definition-PLC (HD-PLC), proposed by the HD-PLC Alliance. Both standards only specify the two lowest layers of the OSI (open systems interconnection) model, that is, the physical layer (PHY) and the medium access control (MAC) layer.

These standards served as the basis for the creation of the international standard IEEE 1901-2010, proposed by the Institute of Electrical and Electronic Engineers (IEEE) and published in 2010. IEEE 1901-2010 defines two PHY layers, one based on HomePlug AV 1.1 standard that uses the FFT-OFDM (fast Fourier transform-orthogonal frequency division multiplexing) technique and the other based on the HD-PLC standard that employs the W-OFDM (wavelet-OFDM) technique. The choice between these layers is optional, but they are not interoperable. On the other hand, coexistence is guaranteed by a mandatory intersystem protocol specified in the standard. The MAC layer defined in IEEE 1901-2010 deals with both PHY layers [9].

[1]Department of Automation and Process Control, Federal Institute of São Paulo (IFSP Suzano), Brazil
[2]Engineering, Modeling and Applied Social Sciences Center (CECS), Federal University of ABC (UFABC), Brazil

In this chapter, an overview of the two PHY layers specified in the IEEE 1901-2010 standard (based on the FFT-ODFM and W-OFDM techniques) will be initially presented. Next, some performance simulations of these PHY layers will be carried out taking into account the effects caused by multipath fading channels, additive white Gaussian noise (AWGN) and impulsive noise. Finally, the results obtained for each case studied will be analyzed.

The similarities between the recent BB-PLC standards, the widespread recognition of the importance of the IEEE 1901-2010 standard, and the fact that it includes the specifications of the two relevant BB-PLC standards (HomePlug and HD-PLC), with some interesting particularities, justifies this chapter to focus only on the IEEE 1901-2010 (favoring objectivity and simplicity of reading).

8.1 Physical layer description of IEEE 1901-2010

In this section, the main features of the PHY layers that integrate the IEEE 1901-2010 standard [9,10], derived from HomePlug AV 1.1 and HD-PLC, will be described more deeply. For simplicity, the specific PHY layer based on HomePlug AV 1.1 will be referred to FFT-OFDM PHY layer, and the specific PHY layer based on HD-PLC will be referred to W-OFDM PHY layer.

The block diagram of the FFT-OFDM PHY layer (transmitter) is shown in Figure 8.1. The header (frame control) and payload fields that came from the MAC layer are encoded by the forward error correction (FEC) scheme, interleaved and sent to the OFDM generation procedure, where the encoded data is mapped into subcarriers and then transformed into OFDM symbols. Subsequently, OFDM symbols are sent to the analog front end (AFE) interface that conditions the signal so that it can be coupled to the power line by means of specific filters (coupling circuits).

On the other hand, Figure 8.2 shows the block diagram of the W-OFDM PHY layer (transmitter). The header and payload (frame body) that came from MAC layer

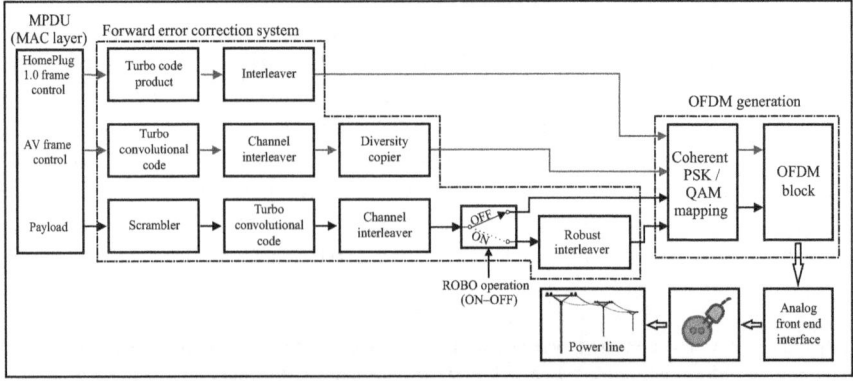

Figure 8.1   Block diagram of the FFT-OFDM PHY layer transmitter

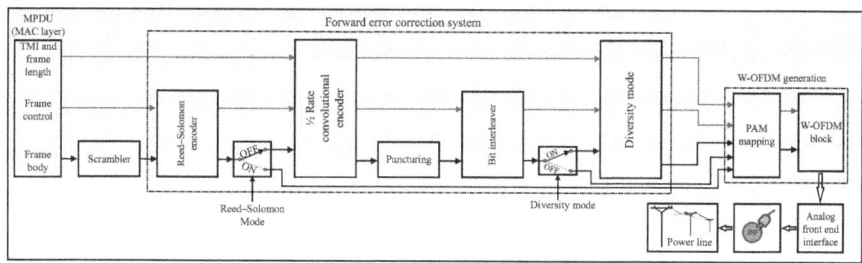

Figure 8.2   Block diagram of the W-OFDM PHY layer transmitter

are processed by the FEC system that encodes and interleave the data and sent it to the W-OFDM generation block where the encoded bits are mapped and transformed into the W-OFDM symbols. The resulting symbols are then sent to the AFE interface, which conditions the signal to be transmitted so that it can be coupled to the power network by means of adequate filters (coupling circuits).

8.1.1   Physical layer frames

IEEE 1901-2010 standard deals with PHY and MAC layers. Its PHY layer is responsible for defining electrical, mechanical and functional interfaces for medium access (e.g., signal type, channel characteristics, timing, modulation and coding schemes). On the other hand, its MAC layer is responsible for controlling and organizing access to the transmission medium, which can be shared by several network devices, whether or not based on the same standard (the MAC layer must ensure coexistence between devices).

IEEE-1901-2010 MAC layer is based on both carrier sense multiple access with collision avoidance and time division multiple access protocols to control the access of transmissions to the PLC channel. For simplicity, this chapter will focus on PHY layer.

The PHY layer of the transmission device receives a data frame known as MAC protocol data unit (MPDU) from the MAC layer. It contains the source user information data entered at the application layer and all additional data appended in each of the upper layers. At the PHY layer, the MPDU is converted in a PHY frame by adding the preamble and header fields. The preamble is a fixed data sequence, known and expected by the receive device, which is used for the detection and synchronization of received frames (can also be used to estimate the PLC channel). The preamble field is followed by a header field containing frame control data required to detect frame type, network identify, presence and length of the payload field (for demodulation at the receiver), and also some other important control functions. In the header field, there is also cyclic redundancy check (CRC) data used to verify the integrity of the received header (if any error is detected, the receiver asks the sender for resending the PHY frame). The last field of the PHY frame is the payload containing variable length broadcast or unicast information from a source user or the network. Figure 8.3 shows the PHY frame structure within the OSI model.

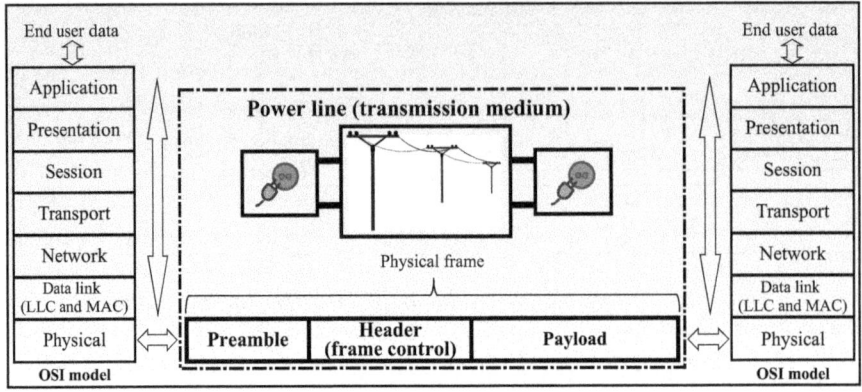

Figure 8.3   PHY frame structure and OSI model

The PHY frame is then encoded by a FEC scheme and interleaved to protect information against errors (possibly generated during transmission). Lastly, it is modulated and conditioned to be transmitted through the PLC channel. After this process, the PHY frame is known as the PHY protocol data unit (PPDU), which is basically the MPDU of the MAC layer converted at the PHY layer to be properly sent through the PLC channel. After crossing the channel and reaching the receiving device, the PPDU makes the reverse path until reaching the destination user's application layer.

8.1.1.1   FFT-OFDM PHY layer frame

The FFT-OFDM PHY layer specification supports two types of PPDU, a short frame only with preamble and frame control and a long frame composed of preamble, frame control and payload. These two types can operate in HomePlug 1.0 (TIA-1113) compatible mode (with preamble and frame control compatible to Home-Plug 1.0) and in AV mode (with preamble and frame control compatible to HomePlug AV 1.1), totaling four possible PPDU formats (two with no payload and two with payload). When operating in the compatible mode, known as hybrid mode, PPDUs are transmitted to prevent HomePlug 1.0 devices from accessing the PLC channel when any IEEE 1901-2010 device is in operation (PPDU in AV mode can be used only when there are no HomePlug 1.0 devices sharing the communication medium).

Figure 8.4 shows the PPDU structure of the FFT-OFDM PHY layer in both modes. In hybrid mode, the PPDU uses a 25-bits HomePlug 1.0 frame control after the preamble, followed by a 128-bits HomePlug AV 1.1 frame control (the presence of the payload field after frame control depends on the type of PPDU). In AV mode, the PPDU does not have the HomePlug 1.0 frame control, and it employs a different preamble. Without loss of generality, this chapter will focus only on the PPDUs in AV mode. It is important to highlight that the payload field of the PPDUs, when present, has variable length, because it depends on the amount of data to transmit.

PPDU structure (hybrid mode)

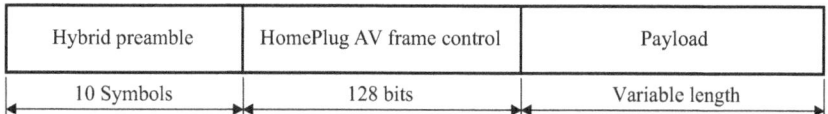

Hybrid preamble	HomePlug 1.0 frame control	HomePlug AV frame control	Payload
9 Symbols	25 bits	128 bits	Variable length

PPDU structure (AV mode)

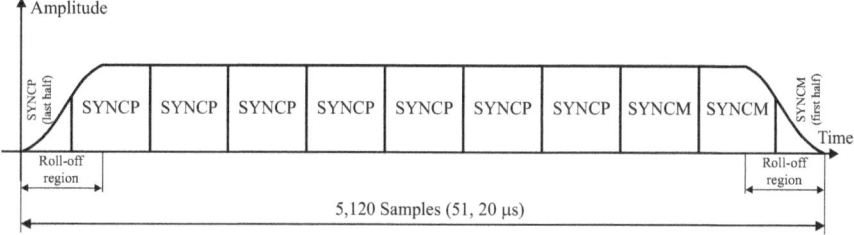

Hybrid preamble	HomePlug AV frame control	Payload
10 Symbols	128 bits	Variable length

Figure 8.4   Frame structure (hybrid and AV modes) of the FFT-OFDM PHY layer

Figure 8.5   Preamble structure of the FFT-OFDM PHY layer (AV mode) in the time domain

Preamble

The preamble of the FFT-OFDM PHY layer starts with the last half of a particular synchronization symbol called SYNCP, followed by seven identical SYNCP symbols and ends with two and a half (first half) identical particular synchronization symbols called SYNCM as shown in Figure 8.5. Each SYNCP symbol is a chirp (frequency sweep) that spans the entire frequency range (1.8–50 MHz) specified for the IEEE 1901-2010 FFT-OFDM standard (excluding masked subcarriers) and with duration of 5.12 μs each (it is obtained setting the subcarriers with the phase vectors described in [10]). As the subcarriers used in the preamble employ the same phases specified for the preambles of the HomePlug 1.0 and HomePlug AV 1.1 standards in their respective frequency ranges (4.5–20.7 and 1.8–30 MHz), due the orthogonality between the subcarriers, the preambles are compatible. It means that HomePlug 1.0 and HomePlug AV 1.1 devices can detect a preamble sent by the FFT-OFDM PHY layer. The process is identical for each SYNCM symbol, except that all phases of the subcarriers are inverted by 90 degrees with respect to the SYNCP symbol.

The roll-off region shown in Figure 8.5 is the part of the waveform arising from the raised cosine windowing process in the time domain (4.96 μs). In this case, the last samples of the roll-off region of the preamble of one frame can overlap the first samples of the roll-off region of the next frame (frame control), reducing the side lobes outside the spectrum used (more details on the preamble specification can be found in [10]).

The sequence of SYNCP and SYNCM symbols provides a repetitive pattern with transition point that can be detected and synchronized by a simple correlation method. Since the preamble signal is known by the receiver, it is also used for channel estimation and equalization.

Frame control (header) and payload

The header of the PPDU of the FFT-OFDM PHY layer is composed of the frame control field. The frame control consists of an FFT-OFDM symbol that carries 128 data bits (generated at the MAC layer) with a duration of 40.96 μs [considering an effective guard interval (GI) of 18.32 μs]. In order to fill one OFDM symbol and also to increase robustness of the transmitted information, the encoded frame control data bits are repeated according to the available subcarriers by the diversity copier block.

The FFT-OFDM symbol is obtained by first applying the data from the frame control field to a concatenated FEC scheme composed of an inner turbo convolutional (TC) encoder with a code rate of 1/2, an interleaver and an outer repeater encoder, which repeats each received bit according to the number of active subcarriers (selected by the mask process discussed later). The resulting encoded bits are then mapped to quaternary phase shift keying (QPSK) symbols (each symbol occupying a subchannel or subcarrier in the frequency domain), which are grouped and transformed into the FFT-OFDM symbol by the inverse fast Fourier transform (IFFT).

In addition to the information related to the MAC layer, the frame control field contains the TMI information, employed to properly demodulate and decode the payload of the PPDU. The TMI is generated by an adaptive tone mapping procedure that is responsible for sending data on the subcarriers located in parts of the frequency spectrum with better conditions in terms of signal-to-noise ratio (SNR) and choosing the most appropriate coding and modulation schemes for this specific scenario.

When establishing a communication link between two devices (unicast transmission), the transmitter first sends a number of sound PPDUs to the receiver. In processing these specific frames, the receiver can adaptively estimate the channel state information of the link (e.g., attenuation, complex gain, delay and SNR of each subcarrier) and determine the best subcarriers and the most appropriate coding and modulation schemes and GI of the OFDM symbol to be used. This information is used to define the TMI that will be employed by the adaptive tone mapping procedure to optimally transmit data PPDUs across the channel in order to maximize the throughput (this procedure will be discussed in more detail later in the chapter).

The transmitter will use the TMI sent by the receiver in its next transmissions (the TMI is informed in the frame control so that the receiver is able to correct, demodulate and decode the payload). The TMI is applied only to the payload field of the PPDUs (the preamble and header will always use all subcarriers available except those excluded by the masking process) and is updated frequently as the receiver continuously monitors the channel, sending a new TMI periodically or when necessary e.g., when bit error rate (BER) is high.

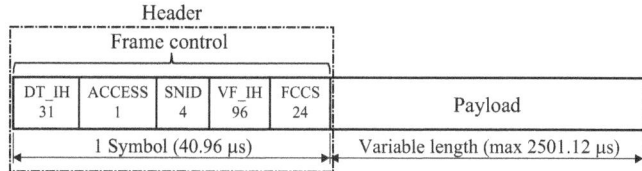

Figure 8.6   Header and payload structure of the FFT-OFDM PHY layer

The payload symbols are present only on the long PPDU, carrying information sent by the MAC layer (source user data or information related to MAC management). The information bits are first divided into blocks of 136 or 520 bytes and then, according to the TMI information, TC encoded, interleaved and mapped to binary phase shift keying (BPSK), QPSK, 8-quadrature amplitude modulation (8-QAM), 16-QAM, 64-QAM, 256-QAM, 1,024-QAM or optionally 4,096-QAM (if the amount of data bits is not sufficient to fill an integer number of symbols, zero padding is provided). Lastly, the resulting mapped symbols are grouped and transformed into a corresponding number of FFT-OFDM symbols.

The first two payload OFDM symbols have a fixed GI of 567 samples (the next ones may have one of the three GIs described in [10] and established by the TMI). The maximum length of the payload field is 2,501.12 μs. Larger frame length (FL) values are optional and the receiver can inform the maximum FL that it can handle by means of the channel estimation procedure (information indicated in the TMI).

Figure 8.6 shows the FFT-OFDM PHY frame control and payload structures with their respective fields and lengths in bits (before TC encoding). The payload field is not described because it has variable length and can carry either end user data or MAC management information (detailed description about its content can be found in [10]).

The header (frame control) of the FFT-OFDM PHY layer is composed of the following information fields:

- DT_IH (delimiter type): indicates the content and format of the variant fields of the frame control. There are six types of variant fields (beacon, start of frame, selective acknowledgment, request to send/clear to send, sound and reverse start of frame);
- ACCESS: this single bit indicates the type of network the MPDU was transmitted (in-home network or access network). In-home networks operate over power lines not owned or managed by an electrical utility, i.e., indoor environment (houses, buildings, industrial facilities and so on). It creates a network for transferring information between electronic devices and to share Internet connection. Access networks are used to interconnect in-home networks to each other and also to their service provider, operating over low voltage (LV) and medium voltage power lines managed by an electrical utility (outdoor environments);
- SNID (short network ID): these 4 bits are used to identify a network and, consequently, to distinguish MPDUs transmitted from different networks that share the same power line medium;

Wavelet-OFDM PPDU structure

Preamble	Tone map index	Frame control	Frame length	Frame body	F-pad
11–17 Symbols	15 bits	278 bits	30 bits	Variable length	Variable

Figure 8.7   Frame structure of the W-OFDM PHY layer

- VF_IH (variant fields): the content of this field depends on the delimiter type. It carries important control information used by both PHY and MAC layers (information for demodulating the payload, like tone map and frame control length are placed in this field). Detailed information about the content of this field can be found in [10];
- FCCS (frame control check sequence): this field contains the CRC checksum (CRC-24) over all previous frame control fields (used for detecting any error added in one or more fields of the frame control during the transmission over the PLC channel).

8.1.1.2   W-OFDM PHY layer frame

Figure 8.7 shows the data frame structure of the W-OFDM PHY layer. It begins with a preamble field of variable length (from 11 to 17 W-OFDM symbols), followed by the TMI field (one symbol), the frame control field (eight symbols) and the FL field (one symbol). Then, there is a variable-length frame body field and, finally, the F-pad field. If there is no payload to be transmitted, the data frame will consist only of the preamble, TMI and frame control (the absence of the payload field is informed in the TMI field). On the other hand, if the data frame has payload, after the frame control there will be the FL indicating the length of the payload (which will also have the CRC of the header) and finally the payload of variable length within the frame body field. The F-pad field, after the frame body, is used for zero padding. The W-OFDM also has an optional post-amble appended to the end of the data frame (similar to the preamble, but using a different phase vector).

In addition to the data frame, there are four other types of frames supported by the W-OFDM MAC layer and transmitted via PHY layer: the acknowledge (ACK) frame (used to inform the transmitter whether the frame was received or dropped, requesting retransmission), request channel estimation (RCE) (used to RCE to the receiver), the channel estimation response (CER) frame and the management frame (different services provided by the W-OFDM MAC layer). Each of these frames has a specific structure that is presented in detail in [10]; however, this chapter will deal only with the data frame.

Preamble

The preamble of the W-OFDM PHY layer consists of a sequence of identical SYNCP symbols followed by a single SYNCM symbol and is used by the receiver for equalization, synchronization and carrier detection. The number of SYNCP symbols can vary from 10 to 16 symbols, so that the preamble can range from 11 to 17 symbols.

The SYNCP symbol is modulated by all ones information, while the SYNCM symbol is equal to the SYNCP symbol multiplied by minus one. The first SYNCP symbol is processed by a ramp function that has zero value for a delay period and then rises linearly (similarly to a ramp) to reach its maximum value. Then, the amplitude at the beginning of the first symbol is zero and rises linearly to reach its maximum value in half the symbol period. The duration of each symbol is $8.192\ \mu s$, but there are two other optional values. In [10], there are detailed information on the preamble structure and its specifications.

Header and payload

Differently from the FFT-OFDM PHY layer, the header of the W-OFDM PHY layer consists of three fields: TMI, frame control and FL. These fields come from the MAC layer as MPDU and become the header of the PPDU by means of the FEC encoding, interleaving and modulation processes.

The TMI and FL fields are, each separately, encoded by a convolutional (CONV) code with code rate of $1/2$ and generator polynomials $(171, 133)_8$. After that, the encoded data is repeated a sufficient number of times so that TMI and FL fills, each one, one W-OFDM symbol (the number of repetitions varies according to the number of active subcarriers defined by tone mask). Then, the bits are mapped into 2-level pulse amplitude modulation (PAM) symbols (each symbols occupying a subchannel or subcarrier in frequency domain). TMI and FL data are not interleaved. Lastly, each resulting block of 2-PAM symbols is transformed into a W-OFDM symbol (one W-OFDM per field) by the inverse discrete wavelet transform (IDWT) operation.

On the other hand, the frame control field is first encoded by a concatenated FEC scheme composed by an outer (50, 34) Reed–Solomon (RS) encoder and an inner CONV encoder with code rate of $1/2$. The encoded bits are then interleaved and repeated in order to fill eight OFDM symbols. Then, these bits are mapped into 2-PAM and transformed into eight W-OFDM symbols by the IDWT operation.

Similar to FFT-OFDM PHY layer, W-OFDM PHY layer also has the adaptive tone mapping procedure. The transmitter sends a RCE frame to the receiver that responds with a CER frame containing the TMI that will be used by the transmit device and informed in the TMI field of its header so that the receiver is able to correctly demodulate the payload field of the frame (this procedure will be discussed in more detail later in the chapter). TMI field also carries 1 bit that informs if the PPDU carries payload (some types of PPDU does not carry payload).

Besides carrying the FL information in number of W-OFDM symbols, the FL field also carries 8-bits CRC information in order to verify the integrity of the header of the PPDU.

The W-OFDM symbols carried in the payload field of the PPDU are obtained by first scrambling the data information and then encoding it by the concatenated FEC scheme (RS and CONV). For the payload, the $1/2$ rate CONV encoder can be punctured for different code rates ($2/3$, $3/4$, $4/5$, $5/6$, $6/7$ and $7/8$) selected according to the channel conditions. Thereafter, the encoded data are organized into groups, and each group is interleaved separately and mapped according to the TMI to 2-PAM, 4-PAM, 8-PAM, 16-PAM or optionally to 32-PAM (each symbol occupying

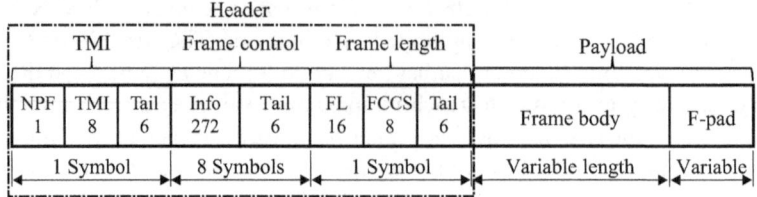

Figure 8.8   Header and payload structure of the W-OFDM PHY layer

a subchannel or subcarrier in the frequency domain). Finally, each mapped symbol group is transformed into a W-OFDM symbol by IDWT operation.

Figure 8.8 shows the W-OFDM header and payload structures with their respective fields and lengths in bits and symbols (as explained, the bits of the header are always repeated in order to fill the indicated number of symbols and also to increase robustness of data transmission).

The header is composed of the following information fields:

- NPF (no-payload flag): this single bit information indicates to the receiver the presence of the payload field.
- TMI (tone map index): this 8-bits information indicates the tone map used in the transmission. When W-OFDM devices communicates with a legacy HD-PLC device, this field contains only 5 bits.
- Tail (for TMI, frame control and FL fields): bits added to reset the CONV encoder (all 6 bits set to zero for constraint length of seven).
- Info: this 272-bits information carries all the frame control information needed to identify the type of frame, network and devices information, device address and other services provided by the MAC layer.
- FL: this 16-bits information indicates to the receiver the length of the payload field in number of symbols. The maximum allowed FL is 1,023 symbols.
- FCCS: this field contains the 8-bits CRC checksum over all previous header fields (used for detecting errors added in these fields during the transmission over the PLC channel).

The payload is composed of the following information fields:

- Frame body: contains the data to be transmitted (payload);
- F-pad: insert a number of zeros to complete an integer number of W-OFDM symbols.

8.1.2   Scrambling schemes

Scrambling is a digital data processing that can be employed to randomize the data information in order to redistribute bits (energy dispersion), encrypt data (in a simple concept) or aid in data synchronization.

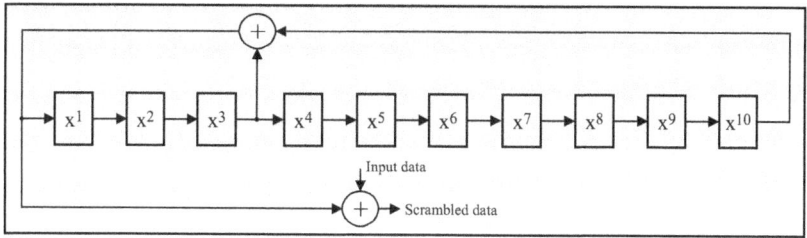

Figure 8.9 Scrambler structure of the FFT-OFDM PHY layer

8.1.2.1 FFT-OFDM PHY layer—scrambler

The scrambler is only applied to the payload field of the PPDU of the FFT-OFDM PHY layer. It randomizes the bits distribution of the data stream, which avoids a long sequence of ones or zeros that can increase the crest factor (peak and root mean square power ratio) at the output of the OFDM block and promotes energy dispersion on the subcarriers of the OFDM symbol. The output data of the scrambler block is obtained by means of the exclusive OR (XOR) function of each bit of the data stream with a pseudo noise (PN) sequence. This PN sequence, represented by the generator polynomial $P(x)$ described in (8.1), is cyclically applied to the data stream by means of the XOR function.

$$P(x) = x^{10} + x^3 + 1 \tag{8.1}$$

Figure 8.9 shows the FFT-OFDM PHY layer scrambler structure, detailing the PN sequence generator obtained by a linear feedback shift register (LFSR) that must be initialized with all ones state. The XOR function between each bit of the PN sequence (cyclically applied to the XOR block) and the data bit stream results the scrambled data.

8.1.2.2 W-OFDM PHY layer—scrambler

The scrambler is only applied to the frame body data of the PPDU of the W-OFDM PHY layer. The output of the scrambler block is obtained by means of the XOR function of each bit of the data stream (frame body) with the PN sequence represented by the generator polynomial $P(x)$ described in (8.2).

$$P(x) = x^7 + x^4 + 1 \tag{8.2}$$

Figure 8.10 presents the scrambler structure of the W-OFDM PHY layer.

8.1.3 Forward error correction system

Data transmission through power lines represents a major challenge. There are many different devices with distinct characteristics connected to them, which may represent different sources of noise, interference and impedance mismatches. In addition, the characteristics of the PLC channels change whenever a new load is connected or disconnected. In LV networks, this dynamic behavior can cause, in less than 1 s,

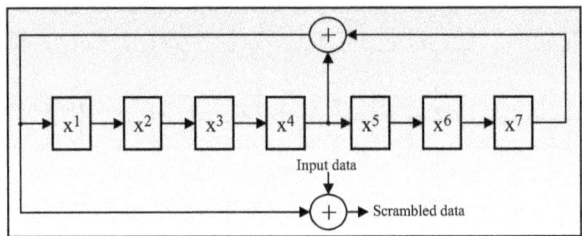

Figure 8.10   Scrambler structure of the W-OFDM PHY layer

variations of more than 10 dB in the SNR [11]. It is common to have negative SNRs on power lines, which means that the PLC signal power is lower than the noise power [12].

The branching characteristic of power lines, with open ends and different loads connected at other ends, causes impedance mismatches and, consequently, signal reflections. The sum of the original signal with its reflections reaching the receiver with different delays, amplitudes and phases, a phenomenon known as multipath propagation, can significantly degrade the PLC signal [13]. The strong attenuation of the PLC signal with the distance is also expected in the power lines, so that a typical connection of 200 m in LV networks can introduce attenuations of the order of 30 dB [14].

FEC techniques are among the most widespread techniques to overcome these problems. The FEC schemes employ the concepts of information theory and digital signal processing to increase the reliability of the system by adding redundant information strongly correlated to the original transmitted data (encoding), so that the errors introduced by the communication channel can be detected and corrected at the receiver (decoding). Thus, FEC schemes play a very important role in the reliability of data transmission by power lines. Two or more FEC schemes may be used in sequence, in a concatenated way, to further increase the robustness of the communication system. The error correction capability of the FEC scheme is usually proportional to the amount of redundant data added to the original data. However, by increasing the amount of redundant data, the usable data rate is reduced (for the same system operating condition). Generally, an interleaving process is associated with the FEC coding techniques to improve the robustness of the communication system against errors in bursts mainly caused by multipath fading channels.

8.1.3.1   FFT-OFDM PHY layer forward error correction coding and interleaving schemes

The FEC scheme of the FFT-OFDM PHY layer processes the header and payload fields differently. In addition, it will also act in a different way over the frame control, depending on whether the PHY layer is operating in hybrid or AV mode.

The 128 bits of the AV frame control pass through a TC encoder with code rate of 1/2, an interleaver and a diversity copier in order to repeat the encoded bits for filling one OFDM symbol. The AV payload field data of the frame passes through a scrambler, a TC encoder and an interleaver (same used in the AV frame control). The

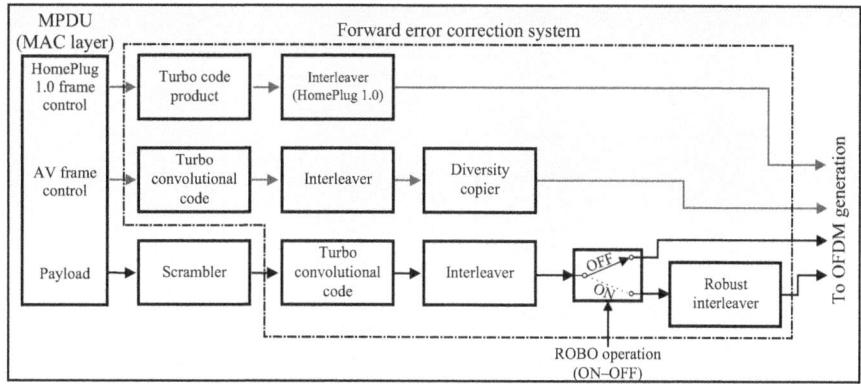

Figure 8.11   Block diagram of the FFT-OFDM PHY FEC system

FEC encoding of the payload is applied on groups of 520 or 136 bytes called PHY blocks. If the robust operation (ROBO) mode is enabled, the interleaver is followed by the ROBO interleaver block.

The hybrid frame control contains an additional frame control compatible with HomePlug 1.0 standard and, as explained previously, it is used only when there are HomePlug 1.0 and FFT-OFDM PHY layer devices sharing the same medium. The FEC system of HomePlug 1.0 frame control employs a turbo product (TP) encoder and an interleaver (this chapter is focused in the AV mode PPDU; however, details about hybrid frame control can be found in [15]).

Figure 8.11 shows the block diagram of the FFT-OFDM PHY layer FEC system. The header (frame control) and payload fields are encoded by the FEC scheme, interleaved and then sent to the OFDM generation block where the data is mapped into subcarriers and transformed in OFDM symbols.

Turbo convolutional encoder
The TC encoder of the FFT-OFDM PHY layer is used in the AV frame control or payload fields of the PPDU. Figure 8.12 shows the block diagram of the TC encoder designed for a code rate of 1/2 and composed of two 8-state recursive systematic convolutional (RSC) encoders, with a 2/3 code rate and generator polynomials $(1, 13/15)_8$, and a turbo interleaver. The outputs of the RSC encoders can be punctured to achieve code rates of 16/21 and, optionally, 16/18.

PHY blocks with sizes of 16, 136 or 520 bytes are supported (the information bits are split into u_1 and u_2 inputs). The output of the TC encoder consists of the original information bits (u_1 and u_2) in their natural order followed by the parity bits generated by the TC encoder (output of the puncturing block) in the order generated by the RSC encoders. Details about the structures of the RSC encoder and turbo interleaver can be found in [10].

Interleaver
Frequency selective fading and impulsive noise, both common characteristics of PLC channels, cause errors in bursts that limit the performance of FEC schemes. In order

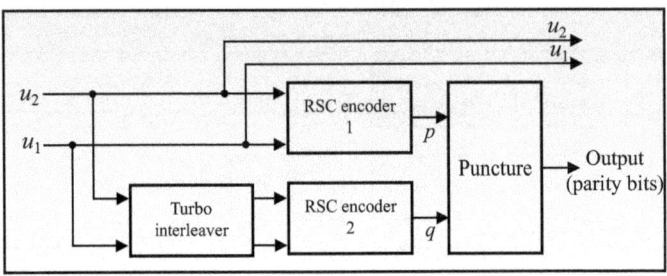

Figure 8.12   TC encoder structure of the FFT-OFDM PHY layer

to prevent the decoder from having to deal with the occurrence of this type of error, an interleaver is normally used after the FEC encoder. By doing this, the interleaver can perform the permutation of the coded bits, so that, adjacent bits are separated from each other by several bits and, consequently, bit errors are randomly spread prior to decoding, which improves the decoder's error correction capability.

The interleaver of the FFT-OFDM PHY layer is applied to the AV frame control (HomePlug 1.0 frame control uses a specific interleaver detailed in [10]) and the payload. In order to improve the decoding process at the receiver, four TC decoders working in parallel are employed. Then, to facilitate this process, interleaving/deinterleaving is performed in four groups of bits so that the four TC decoders can perform simultaneous decoding. At the transmitter, the bits coming from the TC encoder are divided in four subgroups in order to perform interleaving.

As an example, k_{inf} represents the number of information bits and $N - k_{inf}$ the number of parity bits generated by the TC encoder. The information bits are divided in four equal subgroups of $k_{inf}/4$ bits, whereas the parity bits are divided in four equal subgroups of $(N - k_{inf})/4$ bits (for all physical block sizes and code rates available, either the number of information or parity bits are divisible by four). The first $k_{inf}/4$ information bits (in their natural order) form the first information subgroup, the following $k_{inf}/4$ information bits form the second one and so on. Similarly, the first $(N - k_{inf})/4$ bits (in their natural order) form the first parity subgroup, the following $(N - k_{inf})/4$ parity bits form the second one and so on.

Once the subgroups are formed, they are interleaved identically. The four subgroups of information bits are organized into a matrix of $k_{inf}/4$ rows and four columns (each column is a subgroup, with column 0 representing the first, column 1 the second and so on). Equally, the four subgroups of parity bits are organized in a matrix of $(N - k_{inf})/4$ rows and four columns. Then, the rows of both matrices are read in a specific sequence that depends on the code rate of the TC encoder (each reading outputs all 4 bits of the row in their natural order). The readings of the information bits matrix are performed in the following manner:

- The 4 bits of row zero are readout.
- After the reading of row zero, the row number is incremented by factor called *StepSize*, and then the 4 bits of this row are readout, then the row number

is incremented again and so on (row number reading sequence: $0, StepSize$, $2 \cdot StepSize, 3 \cdot StepSize, \ldots, (k_{inf}/4) - StepSize$);

- Once the $(k_{inf}/4) - StepSize$ row is read, the row number is initialized to one and the process is repeated, again the row number is incremented by $StepSize$ factor (row number reading sequence: $1, 1 + StepSize, 1 + 2 \cdot StepSize, 1 + 3 \cdot StepSize, \ldots, (k_{inf}/4) - StepSize + 1$).

- After the last reading, row number is initialized to two and the process is repeated (row number reading sequence in the second run: $2, 2 + StepSize, 2 + 2 \cdot StepSize, 2 + 3 \cdot StepSize, \ldots, (k_{inf}/4) - StepSize + 2$).

- Row number is initialized again (one increment in relation to the last initialization number) and the same process is repeated (the initialization is performed until it reaches the $StepSize$ value and consequently all rows have been read).

The readings of the parity bits matrix (when TC encoder uses code rate of 1/2) are performed in a similar way. The difference is that instead of start reading by the row zero, it starts by a number determined by an *Offset* factor (row number reading sequence in the first run: $Offset, (Offset + StepSize) \bmod T, \ldots, (Offset + T - StepSize) \bmod s$, where T is equal to $(N - k_{inf})/4$). Then, the starting row number is incremented by one (row number reading sequence in the second run: $Offset + 1, (Offset + 1 + StepSize) \bmod T, \ldots, (Offset + 1 + T - StepSize) \bmod T$), so that the process is repeated $StepSize - 1$ more times, until all rows have been read (when the end of matrix is reached, the count continues on row zero until reaching row $Offset - 1$).

For code rates of 16/21 and 16/18, the readings of the parity bits matrix also start with the *Offset* factor; however, the row number is not initialized for each of the $StepSize - 1$ runs. After each row reading, $StepSize$ is added to the row number and a modulo t is performed ($Offset, (Offset + StepSize) \bmod T, (Offset + 2 \cdot StepSize) \bmod T, \ldots$). This process will continue in a sequence until all rows have been read.

Table 8.1 shows *StepSize* and *Offset* factors for different combinations of PHY block sizes and code rates. The transmission sequence of the readings is performed as follows (as explained previously, each reading outputs 4 bits of a specific row):

- Rate 1/2: one reading of information bits followed by one reading of parity bits are transmitted in a sequence (8 bits). This process is repeated until all bits of both matrices are transmitted.

- Rate 16/21: the first three readings of information bits followed by one reading of parity bits (total of 16 bits) are transmitted in a sequence. This is repeated five times, resulting in 20 readings of 4 bits, and then one extra reading of information bits is output. This process is repeated until all bits of both matrices are transmitted.

- Rate 16/18: the first three readings of information bits followed by one reading of parity bits (total of 16 bits) is transmitted in a sequence. Then, five readings of information bits are output. This process is repeated until all bits are transmitted.

There is also an additional interleaving performed by switching bits position of each nibble output by the described rows reading process (information and parity

Table 8.1   Interleaver parameters of the FFT-OFDM PHY layer

PHY block size (bytes)	Code rate	Offset (parity)	StepSize
16	1/2	16	4
520	1/2	520	16
136	1/2	136	16
520	16/21	170	16
136	16/21	40	8
520	16/18	60	11
136	16/18	16	11

Source: Reference [10].

Table 8.2   Interleaving process (switching bits in each nibble) in the FFT-OFDM PHY layer

Output nibble number	Switched bit order
1 or 2	$b_0 b_1 b_2 b_3$
3 or 4	$b_1 b_2 b_3 b_0$
5 or 6	$b_2 b_3 b_0 b_1$
7 or 8	$b_3 b_0 b_1 b_2$
9 or 10	$b_0 b_1 b_2 b_3$

Source: Reference [10].

bits). This interleaving process is shown in Table 8.2 where b_0, b_1, b_2 and b_3 represent the bits output by each row reading of either information or parity bits (the leftmost bit of the nibble is transmitted first). The switching pattern changes after every two nibbles reading and, after a sequence of 10 nibbles, the process is repeated until all nibbles obtained by the first interleaving process have their position switched.

Diversity copier

The 128 bits of the frame control field of the PPDU become 256 bits after the TC encoder with a code rate of 1/2. These 256 bits are interleaved and passed through the diversity copier that copies each bit as many times as possible to fill in all the subcarriers enabled by the tone mask (the subcarriers of the frame control is always mapped to coherent QPSK, meaning each subcarrier carries 2 bits). This procedure increases the robustness of data transmission proportionally to the number of repetitions.

As an example, for the American standard tone mask, which uses 917 subcarriers, each encoded bit will be repeated 7.2 times. The 256 bits are mapped to the in-phase (I) and quadrature (Q) components with a bit index offset of 128, i.e., the first bit (index zero) is mapped to component I of the first subcarrier (index zero) and the 129th bit (index 128) is mapped to the Q component of the same subcarrier. This sequence continues to the 256th subcarrier (index 255), where the 256th and 128th bits of the frame control data are respectively mapped to I and Q components of this

Table 8.3   Bit ordering of the diversity copier of the FFT-OFDM PHY layer

Used subcarrier	Bit address mapped into I component	Bit address mapped into Q component
0	0	128
1	1	129
2	2	130
...	...	...
c	$c \bmod 256$	$(c+128) \bmod 256$
...	...	...
NumCarriers-1	(NumCarriers-1) $\bmod 256$	[(NumCarriers-1)$+128$] $\bmod 256$

Source: Reference [10].

subcarrier. Then, the same bit sequence is repeated until all 917 subcarriers are filled with redundant frame control information.

Table 8.3 shows the bit ordering of the diversity copier when a symbol is used to transmit frame control. The *NumCarriers* variable represents the number of active subcarriers of the tone mask, while c represents the subcarrier index.

ROBO interleaver

As explained in Section 8.1.1.1, the adaptive tone map function dynamically adjusts the FFT-OFDM PHY layer parameters to a communication link according to the channel conditions. However, for initial communication between two nodes and for broadcast and multicast communication to multiple nodes, the channel conditions are unknown. Regarding this situation, three ROBO modes with more reliable communication, and consequently lower throughput, are supported by the FFT-OFDM PHY layer. The ROBO modes are used for beacon, control and data broadcast, multicast communication, session setup and initial unicast communication for channel estimation. The three ROBO modes use coherent QPSK modulation and TC coding with code rate of 1/2.

When one of the ROBO modes is enabled, the payload bits (encoded and interleaved) are sent to the ROBO interleaver, which introduces redundancy to the encoded bits by means of a repetition encoding and spreads the copies of these new encoded bits in both frequency and time. Table 8.4 shows the parameters of the three available ROBO modes, including the data rate provided at the FFT-OFDM PHY layer when the default broadcast tone mask is used.

The Standard ROBO (STD-ROBO) mode uses a repetition code of factor four with PHY blocks of 520 bytes, whereas the high-speed ROBO (HS-ROBO) mode, for the same PHY block size, uses a repetition code of factor two, providing a twice higher data rate (used when the channel conditions, estimated by the receiver, allow to achieve reliable communication with the HS-ROBO mode). The mini-ROBO (MINI-ROBO) mode has the highest number of copies and works with PHY blocks of 136 bytes, being normally used when small payloads need a highly reliable communication

Table 8.4   ROBO mode parameters of the FFT-OFDM PHY layer

ROBO mode	Number of copies	PHY data rate (Mbps)	PHY block size (bytes)
STD-ROBO	4	4.9226	520
HS-ROBO	2	9.8452	520
MINI-ROBO	5	3.7716	136

Source: Reference [10].

(the beacon frame uses the MINI-ROBO mode). Details about the ROBO interleaver can be found in [10].

8.1.3.2   W-OFDM PHY layer—forward error correction coding and interleaving schemes

Similar to FFT-OFDM PHY layer, the FEC schemes of the W-OFDM PHY layer also process the header and payload fields differently. The frame control passes through a concatenated FEC encoder composed by an outer (50, 34) RS encoder and an inner CONV encoder with code rate 1/2 and generator polynomials $(171, 133)_8$. The TMI and FL fields of the header are encoded similarly, except that they do not pass through the RS encoder and the bit interleaver (only by the CONV encoder).

The frame body, that carries the payload, passes through a scrambler, a concatenated FEC scheme composed by a (255, 239) RS encoder and a CONV encoder, a puncturing block and a bit interleaver. There is also a selectable RS mode in which the scrambled data passes only through the RS encoder (puncturing and interleaving blocks are not used). Additionally, for the frame body, there is also an optional low density parity check convolutional (LDPC-CONV) encoder (details about these encoder can be found in [10]) in the place of the concatenated encoder.

Figure 8.13 shows the block diagram of the W-OFDM PHY layer FEC system. The header (frame control) and payload fields are encoded by the FEC scheme, interleaved and then sent to the OFDM generation block, where the data is mapped into subcarriers and converted to OFDM symbols by the IDWT operation.

The frame control, TMI and FL fields of the W-OFDM PHY layer are always transmitted in diversity mode, in which data is repeated several times using different subcarriers within the same symbol (as redundant data are added, the diversity mode increases the robustness of the transmission). There is also a diversity mode for the frame body, used whenever there is no valid tone map for the communication link or when channel operates in poor conditions. The W-OFDM PHY layer has three different modes of diversity, one for frame control, one for TMI and FL and one for frame body. Details about these diversity modes can be found in [10].

Reed–Solomon encoder
The RS coding is applied to the frame control of the header and also to the frame body field (when RS only mode is enabled, the frame body is only codified by the

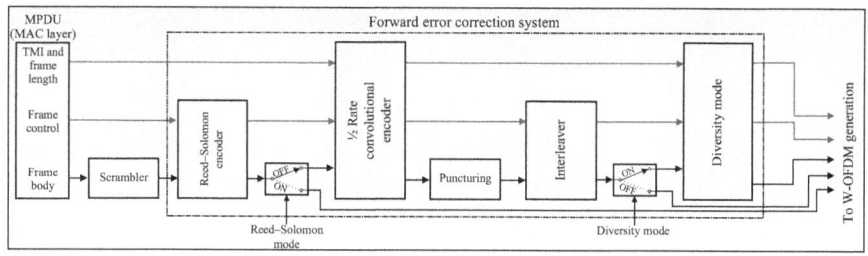

Figure 8.13 Block diagram of the W-OFDM PHY layer FEC system

RS encoder). The input data bits must be divided into blocks with size equal to the message length parameter of the RS encoder. The blocks are encoded separately and in sequence (the MSB is the first bit to the input/output of the encoder). The RS code has the following parameters:

- Codeword length (maximum): $n = 2^s - 1 = 255$, with $s = 8$;
- Message length (maximum): $k = n - 2t = 239$, with $t = 8$;
- RS coding for the frame body: $(255, 239)$ and $(56, 40)$ (diversity mode);
- RS coding for the frame control: $(50, 34)$.

Where n represents the codeword length in bytes (data bits plus 8 bytes of parity bits), k the message length in bytes (input data bits must be organized in blocks with this length), s is the symbol length in bits and t is the number of symbols in which the encoder is able to correct (i.e., errors in up to 8 bytes of any part of the codeword can be corrected).

The RS codeword is obtained by the generator polynomial of (8.3) with $(t = 8)$. The RS parity symbols are given by (8.4), corresponding to a Galois Field $GF(2^8)$.

$$g(x) = \prod_{i=0}^{2t}(x - \alpha^i) \tag{8.3}$$

$$p(x) = x^8 + x^4 + x^3 + x^2 + 1 \tag{8.4}$$

Convolutional encoder

The CONV encoder is applied to all fields of the PPDU, except when RS only mode is enabled. Data from the RS encoder (frame control and frame body) or directly from the MAC layer (TMI and FL) are inserted into the CONV encoder input (MSB first).

The CONV encoder has a code rate of $1/2$ (for each input bit, there will be two output code bits), constraint length $K = 7$ (seven memory elements) and generator polynomials $(171, 133)_8$. The generator polynomials correspond to the tap connections of the LFSRs that integrate the encoder, represented in binary by the polynomials $(1111001)_2$ for the X_{output} and $(1011011)_2$ for the Y_{output}.

Figure 8.14 shows the structure of the CONV encoder. In the beginning and at the end of each frame, the encoder state is always set to zero (the error probability in the CONV decoder is lower when the initial and final states are known) by means

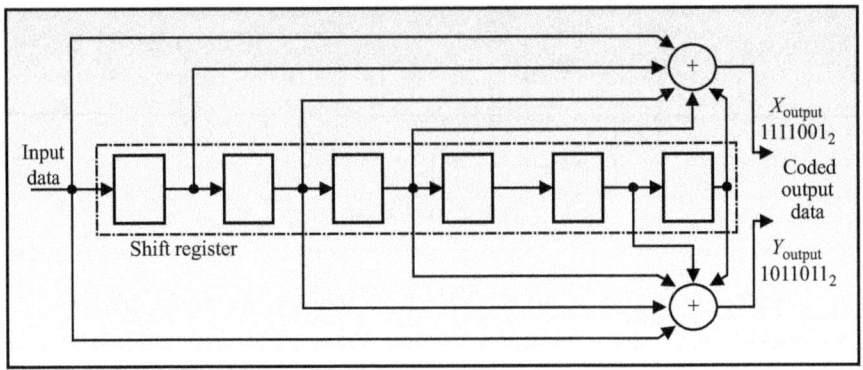

Figure 8.14   Convolutional encoder structure of the W-OFDM PHY layer

Table 8.5   Puncture patterns of W-OFDM PHY layer

Coding rate	Puncture pattern (X)	Puncture pattern (Y)
2/3	10	11
3/4	101	110
4/5	1000	1111
5/6	10101	11010
6/7	100101	111010
7/8	1000101	1111010

Source: Reference [10].

of the flushing zeros appended to the end of each field of the PPDU (tail bits) as described in Section 8.1.1.2.

Puncturing

The code rate of the CONV encoder is chosen based on the channel conditions estimated by the RCE/REC frames. TMI, frame control, FL and frame body (in diversity mode) always use a code rate of 1/2; however, the frame body can use code rates of 1/2 to 7/8 (the better the channel conditions, the higher the code rate and the effective data rate). Since the CONV encoder has a fixed code rate of 1/2, the puncturing block has the function of increasing the code rate by removing some parity bits shortly after the CONV encoding process. Table 8.5 shows the code rates according to the puncturing patterns.

Interleaver

The encoded data (CONV or LDPC-CONV coding) is sent to an interleaver to increase the system robustness to burst errors that can be caused by the PLC channel. TMI and FL fields of the header and frame body in RS only mode are not interleaved. For W-OFDM standard, interleaving can be performed for each W-OFDM symbol, individually, or for multiple consecutive OFDM symbols (optional).

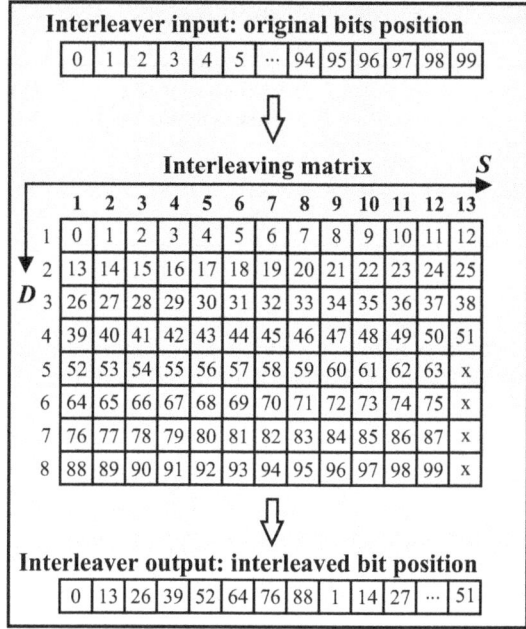

Figure 8.15   Interleaving example for one W-OFDM PHY layer symbol

When interleaving is performed for one W-OFDM symbol, the bit stream is organized in a matrix in which the number of rows is equal to the interleaver depth parameter D. Consequently, the number of columns, denoted as the horizontal size parameter S, is obtained dividing the number of bits N_{bits} in a W-OFDM symbol by the depth parameter (if this operation do not return an integer number, it is rounded up). The bit stream sequence is organized in a way that the first bit (MSB) is inserted in the first column followed by the other bits in column direction, as shown in Figure 8.15 with the bits identified in numerical order (example with $N_{bits} = 100$ and $D = 8$). The interleaved output consists of the bits of the matrix read in row direction, starting from the first column, as also shown in Figure 8.15. The D and N_{bits} parameters for the different parts of the frame are shown in Table 8.6.

Details about the optional interleaver for multiple W-OFDM symbols can be found in [10].

8.1.4   OFDM generation

FFT-OFDM PHY layer uses a special form of multicarrier (MC) modulation called OFDM, where the subcarriers are orthogonal in the frequency domain by means of an IFFT operation, which is an efficient computational implementation of the inverse discrete Fourier transform, at the transmit side. This feature allows subcarriers to be closely spaced without interfering with each other providing higher spectrum efficiency when compared to other MC modulation techniques.

Table 8.6   Interleaver parameters of the W-OFDM PHY layer

Frame portion	Number of bits per symbol (N_{bits})	Interleaver depth (D)
Frame control (header)	102	8
Frame body in diversity mode (payload)	84	8
Frame body	Variable	16

Source: Reference [10].

In comparison to SC modulation systems, OFDM systems present more robustness against interference and frequency selective fading (the spectrum is divided in multiple narrowband subchannels that are individually affected by frequency flat fading instead of frequency selective fading). Then, parts of the spectrum with higher level of noise or fading can be avoided by not transmitting data on the subcarriers inside them. Moreover, in certain conditions, the FEC schemes can recover corrupted data on the affected subcarriers.

The W-OFDM PHY layer employs the W-OFDM technique, a different type of OFDM based on the IDWT operation which exhibits even greater spectrum efficiency than the conventional FFT-OFDM technique, requiring no GIs in the frequency domain, and nor cyclic prefix (CP) in the time domain to mitigate intercarrier interference. Because it has very low side lobes, it provides good noise rejection and limited spectral leakage.

For both PHY layers (FFT-OFDM and W-OFDM), the encoded data coming from the FEC schemes are mapped into subcarriers and then converted to OFDM symbols.

8.1.4.1   FFT-OFDM PHY layer—OFDM generation

After passing through the encoder and interleaver, the resulting encoded bits stream is sent to the mapper to be transformed into a modulated symbols stream. Then, the resulting symbols are applied to the OFDM generator. In this step, they are grouped into blocks (each symbol corresponding to a subcarrier) and applied to the IFFT operation to form the OFDM symbols. Each OFDM symbol is cyclically extended by adding a CP block before the beginning of the symbol. Each CP block is a reproduction of an end portion of the corresponding OFDM symbol. In the windowing and overlay processes, a time domain window is applied to each OFDM symbol with CP, causing adjacent symbols to be partially overlapped.

Finally, the preamble previously generated by the IFFT block is appended in order to form the PPDU (preamble, header and payload). The resulting symbols are sent to the AFE interface to be coupled to the power line and transmitted by the PLC channel. Figure 8.16 shows the OFDM generation block diagram of the FFT-OFDM PHY layer.

Subcarrier mapper

The bit stream coming from the interleaver (or from ROBO interleaver when one of the ROBO modes is enabled) is mapped into subcarriers with coherent PSK/QAM

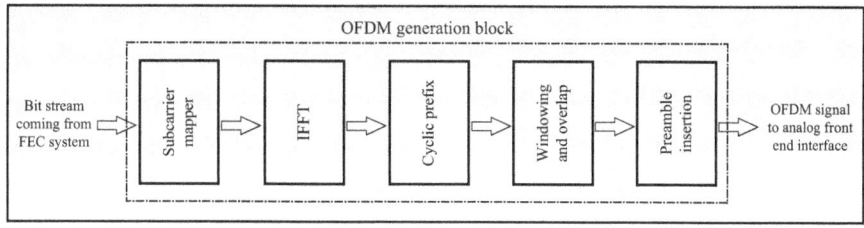

Figure 8.16 OFDM generation in the FFT-OFDM PHY layer

Table 8.7 Subcarrier modulation characteristics of the FFT-OFDM PHY layer

Frame field	Bits per carrier	Modulation type
Frame control (header)	2	Coherent QPSK
Payload (STD-ROBO, HS-ROBO, MINI-ROBO)	2	Coherent QPSK
Payload	1	Coherent BPSK
	2	Coherent QPSK
	3	Coherent 8-QAM
	4	Coherent 16-QAM
	6	Coherent 64-QAM
	8	Coherent 256-QAM
	10	Coherent 1,024-QAM
	12	Coherent 4,096-QAM

Source: Reference [10].

modulation by means of the subcarrier modulator block. The frame control field of the PPDU is always QPSK modulated, whereas the payload can be mapped into different modulations (BPSK, QPSK, 8-QAM, 16-QAM, 64-QAM, 256-QAM, 1,024-QAM or optionally 4,096-QAM) according to the TMI information.

Table 8.7 shows the different configurations of subcarrier modulation and their respective bit load. Different modulations can be used for different subcarriers inside the same symbol (except for ROBO mode). The subcarriers not carrying data, according to the tone map, carry dummy bits (generated by means of a PN sequence) with coherent BPSK modulation.

IFFT, cyclic prefix, windowing and overlap

OFDM symbols are generated by the IFFT operation. The mapped data are grouped and allocated into the inputs of a 4,096-point IFFT block (according to the predefined tone mask), resulting in an OFDM symbol consisting of 4,096 time samples (IFFT range). Then, a CP is created by copying a pre-set number of samples from the end of the OFDM symbol and inserting it before the beginning of the symbol. After insertion of the CP, a time domain window is applied to the extended OFDM symbol (details can be found in [7]) causing an overlap with the adjacent symbols. Figure 8.17 shows the structure of the resulting OFDM symbol after the windowing process, and Table 8.8 presents its specifications considering a sampling rate of 100 MHz.

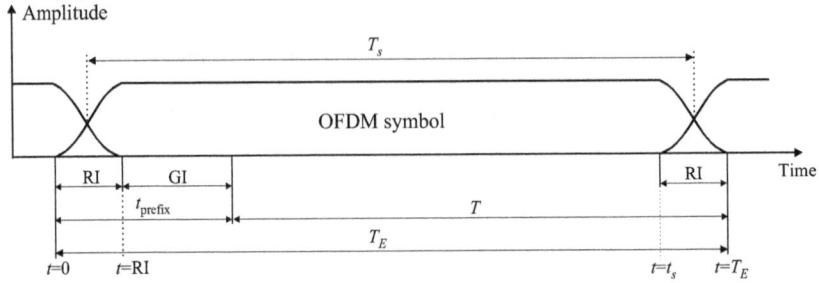

Figure 8.17   OFDM symbol structure of the FFT-OFDM PHY layer

Table 8.8   Specifications of the OFDM generation block of FFT-OFDM PHY layer

Symbol	Description	Time samples	Time (μs)
T	IFFT interval (mandatory)	4,096	40.96
	IFFT interval (optional for access stations)	4,096	81.92, 163.84
t_{prefix}	CP interval	RI+GI	4.96+GI
T_E	Extended symbol interval	$T + t_{prefix}$	45.92+GI
RI	Roll-off interval	496	4.96
T_S	Symbol period	4,096+GI	40.96+GI
GI_{FC}	Frame control guard interval	1,832	18.32
GI	Payload symbol guard intervals (mandatory)	556, 756, 4,712	5.56, 7.56, 47.12
	Extended smaller guard intervals (optional)	160, 392, 208, 256	1.60, 3.92, 2.08, 2.56
	Extended larger guard intervals (mandatory)	956, 1,156, 1,556, 1,956	9.56, 11.56, 15.56, 19.56
GI_{SR}	STD-ROBO payload symbol guard interval	556	5.56
GI_{HR}	HS-ROBO payload symbol guard interval	556	5.56
GI_{MR}	MINI-ROBO payload symbol guard interval	756	7.56
F	Frequency range	1.8–30 MHz	
	Frequency range (optional)	1.8–50 MHz	

Source: Reference [10].

The OFDM symbols may contain up to 1,974 subcarriers (spaced by approximately 24.414 kHz), ranging from 1.8 to 50 MHz. In the particular case of the IEEE 1901-2010 FFT-OFDM standard, the FFT-OFDM PHY layer specifies the frequency band of 1.8 to 30 MHz as default and leaves the range above 30 MHz as optional.

There are two types of tone masks used in the FFT-OFDM PHY layer for subcarrier selection, the broadcast tone mask, used for frame control data symbols and ROBO, and the extended tone mask, used for preambles, data, and priority resolution

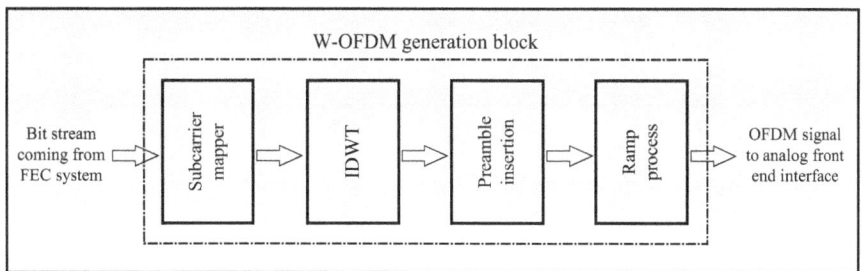

Figure 8.18   OFDM generation in the W-OFDM PHY layer

symbols (below 30 MHz, 917 subcarriers can be used by both tone masks). Details on the tone masks can be found in [10].

The preamble construction is described in Section 8.1.1.1.

8.1.4.2 W-OFDM PHY layer—OFDM generation
In the OFDM generation process of the W-OFDM PHY layer, the encoded bit stream coming from the interleaver is sent to the mapper to be transformed into a PAM symbols stream. Then, the resulting symbols are applied to the OFDM generator, where these are grouped into blocks and applied to the IDWT operation to form the W-OFDM symbols. After that, the preamble previously generated by the IDWT block is appended to form the PPDU (preamble, header and payload) and ramping process is performed on the composite waveform. Lastly, the resulting symbols are sent to the AFE interface to be coupled to the power line and transmitted by the PLC channel. Figure 8.18 shows the W-OFDM generation block diagram of the W-OFDM PHY layer.

Subcarrier mapper
The encoded bits stream from the interleaver is mapped to a PAM symbols stream. The resulting symbols are grouped in blocks, and each symbol of a given block is associated with a corresponding subcarrier.

The frame control, TMI and FL fields of the PPDU are always mapped to 2-PAM, while the frame body can be mapped in different ways (2-PAM, 4-PAM, 8-PAM, 16-PAM or optionally 32-PAM). The voltage levels of the PAM schemes can be found in [10]. In diversity modes, the frame body can use 2-PAM or differential 2-PAM (D2-PAM) [10]. Table 8.9 shows the different subcarrier mapping configurations and their respective bit loads.

IDWT operation, preamble and ramp process
The frame of the PHY W-OFDM layer can be transmitted directly in baseband (for residential and access applications, the baseband is mandatory) or optionally in bandpass. It should be noted that baseband transmissions employ a cosine modulated filter bank (CMFB). On the other hand, in bandpass transmissions, it is necessary to include a sine modulated filter bank (SMFB) for the generation of the quadrature signal, in order to preserve the spectral efficiency. Thus, bandpass transmissions employ both CMFB and SMFB to generate in-phase and quadrature signals [10].

Table 8.9 Subcarrier modulation characteristics of W-OFDM PHY layer

Frame field	Bits per carrier	Modulation type	Mode
TMI and frame length	1	2-PAM	Diversity
Frame control	1	2-PAM	
Frame body	1	2-PAM	
	1	D2-PAM	High speed
	1	2-PAM	
	2	4-PAM	
	3	8-PAM	
	4	16-PAM	
	5	D32-PAM	

Source: Reference [10].

For baseband transmission, each group of PAM symbols is associated with 512 equally spaced subcarriers of 61.03515625 kHz (the spacing of the subcarrier varies according to the size of the IDWT block), occupying a bandwidth from direct current up to 31.25 MHz and allocated in the fixed frequency range of 1.8–28 MHz with notches used to avoid amateur radio frequencies. The upper limit of the frequency range may vary below or above 28 MHz according to country regulations. The PAM symbols are applied to the inputs of a 512-point IDWT block to obtain an OFDM symbol composed of 512 time samples.

For the optional bandpass transmission, each PAM symbol group is associated with 1,024 uniformly spaced subcarriers, ranging from 1.8 to 50 MHz. The PAM symbols are applied to the inputs of a 1,024-point IDWT block, resulting in a symbol OFDM composed of 1,024 time samples. Table 8.10 shows the specifications of the W-OFDM generation block of W-OFDM PHY layer. More details about the IDWT block, subcarrier frequencies and baseband and bandpass transmissions can be found in [10].

The preamble construction and ramp process is described in Section 8.1.1.2 and in more detail in [10].

8.1.5 Tone mapping

The channel adaptation process is initiated by estimating some important channel characteristics (e.g., SNR) at the receiver and determining the best tone maps. Tone maps are lists of subcarriers to be used with their corresponding FEC encoding and modulation mapping configurations and the length of the GI to be employed on a specific unicast communication link between two devices to optimize throughput.

8.1.5.1 FFT-OFDM PHY/MAC layers tone mapping

The frame control field contains the TMI information employed to properly demodulate the PPDU payload. The TMI is a 5-bits field that indicates which tone map was

Table 8.10   *Specifications of W-OFDM generation block of W-OFDM PHY layer*

Parameter	Value
IDWT block size	512, 1,024 (optional) and 2,048 (optional)
Subcarrier spacing	61.03515625, 30.517578125 (optional) and 15.2587890625 kHz (optional)
Symbol duration	8.192, 16.384 (optional) and 32.768 μs (optional)
Frequency range	1.8–28 and 1.8–50 MHz (optional)
Maximum physical data rate	220 and 420 Mbps (optional)

Source: Reference [10].

used to generate the PPDU payload. The tone map is only valid for the payload (robust modes do not support tone map), and it contains information on which subcarriers are being used to carry data, the modulation method and coding rate of each subcarrier and the OFDM symbol GI.

As the FFT-OFDM PHY/MAC layers operate with multiple tone maps, the receiver stores a list of tone maps that are individually selected according to the TMI field information. The tone map list is obtained by means of channel estimation procedure handled by the MAC layer in conjunction with the PHY. It consists in obtaining the characteristics of the power line channel for a specific communication link (such as SNR of each subcarrier and channel delay) in order to choose the best PHY configuration (indicated in the tone map) and consequently to optimize the network throughput. The subcarriers with higher SNR will use a higher modulation order and code rate, whereas the ones with lower SNR a lower modulation order and code rate (or even will not carry data).

The tone map is constantly updated as the channel conditions are dynamic. One important characteristic of the power line channel is that the SNR and noise characteristics vary according to the alternating current (AC) line cycle (typically, the noise level is lower during zero crossing and higher during peaks). For this reason, the 50/60 Hz cycle is divided in regions with different tone maps (multiple tone maps operation). Then, the PPDUs are transmitted using the specific tone map for the region of the AC cycle that is being crossed during the transmission. Such mechanism provides performance improvements to the point the throughput in better conditions regions are often 50 percent higher than in the worse ones [9].

The channel estimation procedure is divided in two processes, the initial channel estimation (ICE) and the dynamic channel adaptation (DCA). When the transmitter needs to send data to a specific destination in which there is no tone map available, it performs the ICE. For that, the transmitter sends one or more sound MPDUs to the destination. The receiver processes these MPDUs, estimates the channel conditions in the PHY layer, creates and stores the default tone map and sends it to the transmitter that will store and start using the tone map immediately after receiving it. Then, the receiver may also generate one or more AC line cycle adapted tone maps, providing a tone map list for the communication link.

On the other hand, the DCA is performed by the receiver subsequently to the ICE requested by the transmitter. For that, the receiver continuously monitors the channel conditions based on the received MPDUs (either sound or data) and provides dynamic updates to the tone map list, adapting the communication link to the variable conditions of power line channel.

8.1.5.2   W-OFDM PHY/MAC layers tone mapping

Similar to the FFT-OFDM PHY/MAC layers, W-OFDM PHY/MAC layers also support the tone map function that selects the most appropriate PPDU configuration (payload part) for power line channel conditions. For the channel estimation procedure, the transmitter sends a RCE PPDU to the destination. This PPDU has an evaluation sequence that is used by the PHY of the receiver to estimate the channel conditions. Based on this estimation, the new tone map is created, stored in the receiver and sent to the transmitter together with its identification number (TMI) by means of a CER PPDU. Immediately after processing the CER PPDU, the transmitter will store the new tone map and begin using it for the following transmissions of this communication link (the PPDU header will take the TMI to the receiver so that it will have the information needed to properly demodulate and decode the payload).

The tone map indicates the modulation of each subcarrier and the FEC system type of the payload field. It allows masking the subcarriers that present the worse noise conditions so that they will not be used for transmission. Different from FFT-OFDM, W-OFDM tone map function does not support multiple tone map operation.

Channel estimation is always performed when there is no valid tone map for the communication link. In order to keep the tone map constantly updated, it will have a lifetime defined for the MAC layer (maximum of 30 s) so that the devices will have to perform the channel estimation procedure every time it expires. Channel estimation is also performed when there is insufficient transmission capacity between devices or it was detected an improvement or deterioration of channel conditions.

8.2   Simulation of FFT-OFDM and W-OFDM PHY layers

As explained in Section 8.1.3 and in previous chapters, PLC channels present multipath fading and high levels of noise. In order to evaluate the performance of IEEE 1901-2010 PHY layers regarding the data information transmission (payload) over some of these typical characteristics of the PLC channel, some simulations were performed according to the specifications presented in Section 8.1 and analyzed for four distinct scenarios:

- AWGN channel
- Multipath fading channel
- AWGN channel with periodic impulsive noise
- Multipath fading channel with periodic impulsive noise

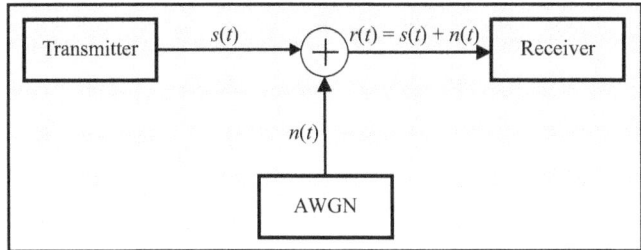

Figure 8.19 Simulation model for an AWGN channel

The FFT-OFDM PHY layer was implemented using a 4,096 samples OFDM symbol with duration of 40.96 μs and a fixed GI (CP) of 556 samples (5.56 μs). The FEC system consists of a $(1, 13/15)_8$ TC coding scheme, the code rate of which was set to $1/3$ rather than $1/2$ and the data block interleaving was set randomly, for simplicity of implementation. The subcarrier mapping considered in the simulations were BPSK, QPSK, 16-QAM and 64-QAM (higher mapping order specified in the standard were omitted for simplicity).

The W-OFDM PHY layer was implemented using a 512-point IDWT block for generating the W-OFDM symbols, each with a duration of 8.192 μs. The FEC system consists of the concatenation of $(255, 239)$ RS and $(171, 133)_8$ CONV coding schemes (all simulations were performed without puncturing) with random data block interleaving. The subcarrier mapping considered in the simulations were 2-PAM, 4-PAM, 8-PAM and 16-PAM (32-PAM was not analyzed since it is optional).

For each simulation point, it was transmitted 10^4 frames, each one was composed by $20,480$ digital coded symbols (equivalent to 10 OFDM symbols for FFT-OFDM and 40 OFDM symbols for W-OFDM). Depending on the subcarrier mapping, bit-filling can be employed to match the FL.

8.2.1 AWGN channel

In order to validate the simulation model and to verify the system performance in the simplest noise scenario, some analysis in an AWGN channel (considering only thermal background noise) were performed [16].

In the adopted simulation model, described in Figure 8.19, the AWGN $n(t)$ is simply added to the transmitted OFDM signal $s(t)$ and the resulting received signal $r(t)$ is processed at the receiver.

The simulation results for both FFT-OFDM and W-OFDM in the AWGN scenario are shown, respectively, in Figures 8.20 and 8.21. The curves express the system performance in terms of BER as a function of the energy per bit to the noise power spectral density ratio (E_b/N_0).

For FFT-OFDM, it can be verified in Figure 8.20 that for reaching a BER of $1 \cdot 10^{-5}$, BPSK, QPSK, 16-QAM and 64-QAM mapping require, respectively, an E_b/N_0 of approximately 0.85, 0.85, 2.9 and 5.0 dB.

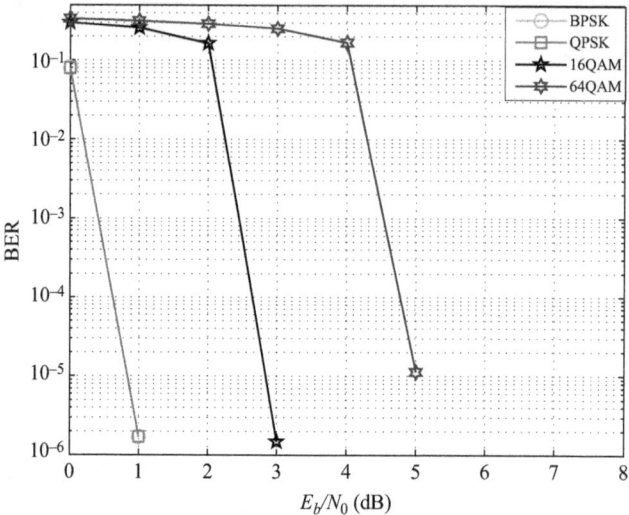

Figure 8.20   BER performance of FFT-OFDM PHY layer over an AWGN channel

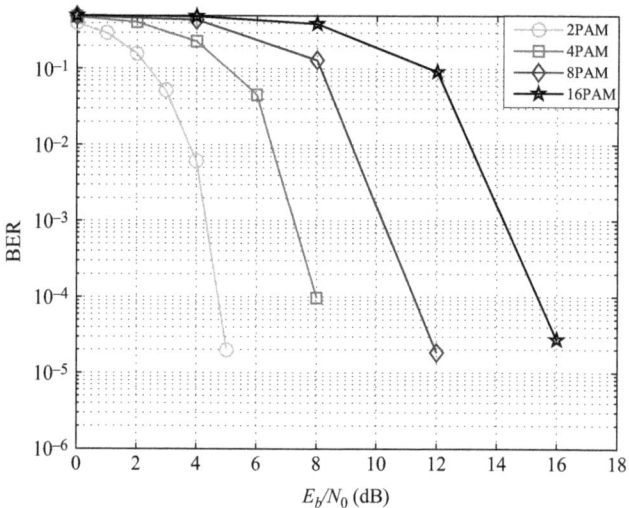

Figure 8.21   BER performance of W-OFDM PHY layer over an AWGN channel

For W-OFDM, it can be noted in Figure 8.21 that for reaching a BER of $1 \cdot 10^{-5}$, 2-PAM, 4-PAM, 8-PAM and 16-PAM mapping require, respectively, an E_b/N_0 of approximately 5, 8.5, 12.2 and 16.5 dB.

Due to TC coding and employed modulation mapping techniques, FFT-OFDM clearly presented a better performance in comparison to W-OFDM (that employs

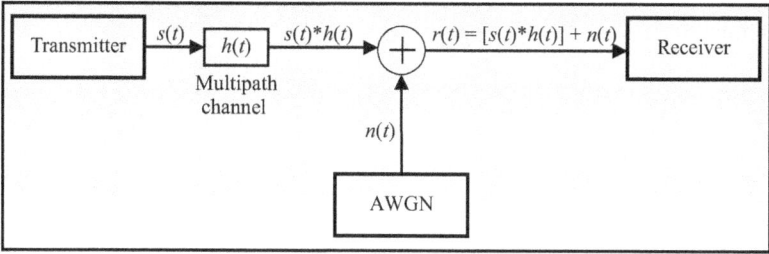

Figure 8.22 Simulation model for multipath fading channel with AWGN

RS/CONV coding and M-PAM) in AWGN channels for equivalent modulation mapping schemes (carrying same number of bits per digital symbol).

8.2.2 Multipath fading channel

The multipath fading propagation effects, usually presented in PLC channels, can be depicted by the tapped delay line (TDL) model (representing the channel impulse response). This model has been often used in the literature for modeling different bandlimited communication channels, including PLC channels [17–22], and it represents the multipath fading propagation effects (e.g., reflections, scattering) caused by different resolvable paths (corresponding to distinct symbol time intervals) by varying the complex coefficients of a TDL filter.

In this work, it was considered a 5-taps digital coded symbol-spaced TDL model (five resolvable paths) whose coefficients have Rayleigh distributed magnitudes and uniform distributed phases. An exponential decay profile with factor 2 between coefficients was also employed.

In the adopted simulation model, described in Figure 8.22, the resulting signal $r(t)$ at the receiver is obtained by the convolution of the transmitted OFDM signal $s(t)$ and the 5-tap multipath channel $h(t)$ and by the addition of AWGN $n(t)$.

In both FFT-OFDM and W-OFDM, the received OFDM symbols were equalized using the minimum mean square error (MMSE) frequency domain equation (FDE) technique. The only difference between them is that the MMSE-FDE is processed before the DWT operation on the W-OFDM receiver (requiring FFT/IFFT operations as in single carrier systems), while it is conventionally applied after the FFT operation on the FFT-OFDM receiver.

Figures 8.23 and 8.24 present, respectively, the performance results in the multipath fading scenario for both FFT-OFDM and W-OFDM in terms of BER as a function of E_b/N_0.

For FFT-OFDM, it can be verified in Figure 8.23 that for reaching a BER of $1 \cdot 10^{-4}$, BPSK, QPSK and 16-QAM mapping require, respectively, an E_b/N_0 of approximately 4.5, 4.5, and 11.5 dB. On the other hand, 64-QAM requires approximately 14 dB for reaching a BER of $2 \cdot 10^{-3}$ (did not reach $1 \cdot 10^{-4}$ in the investigated E_b/N_0 interval).

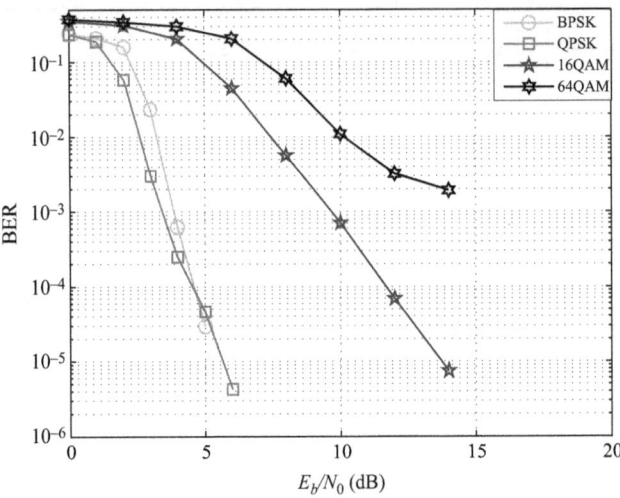

Figure 8.23   BER performance of FFT-OFDM PHY layer over multipath fading with AWGN

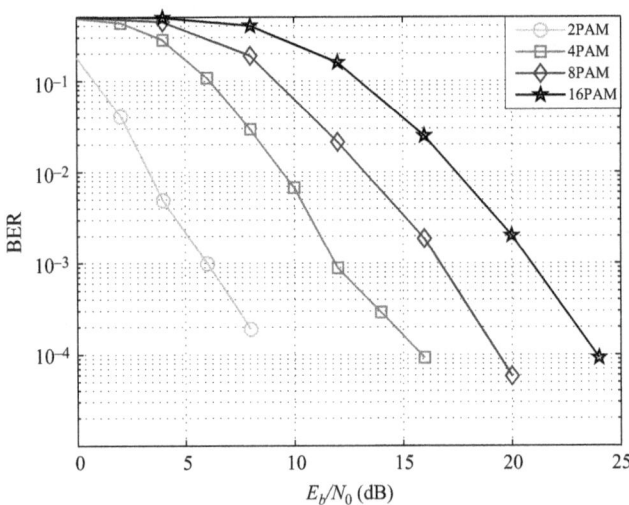

Figure 8.24   BER performance of W-OFDM PHY layer over multipath fading with AWGN

For W-OFDM, it can be noted in Figure 8.24 that for reaching a BER of $1 \cdot 10^{-4}$, 2-PAM, 4-PAM, 8-PAM and 16-PAM mapping require, respectively, an E_b/N_0 of approximately 8.5, 16, 19 and 23.5 dB.

Although the multipath channel causes a larger degradation compared to the AWGN channel, the FFT-OFDM also performed better than the W-OFDM in this scenario for equivalent modulation mapping schemes.

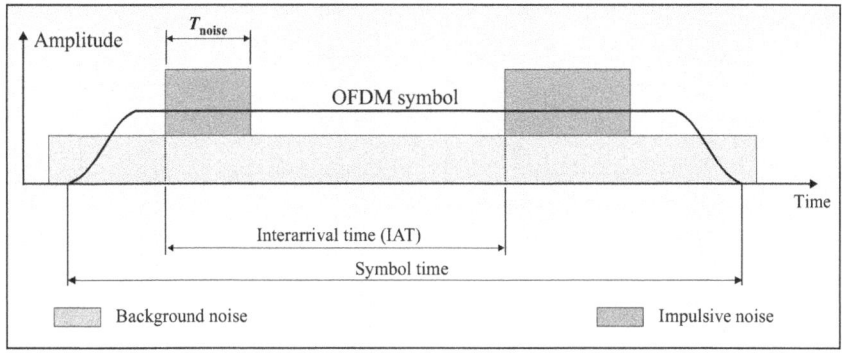

Figure 8.25   Representation of AWGN and impulsive noise over an OFDM symbol

Table 8.11   Periodic impulsive noise parameters

Impulsive noise scenario	IAT (s)	T_{noise} (ms)
Heavily disturbed	0.0196	0.0641
Medium disturbed	0.9600	0.0607
Weakly disturbed	8.1967	0.1107

Source: Reference [24].

8.2.3   AWGN channel with periodic impulsive noise

Regarding the noise scenario, it is not possible to analyze the performance of PLC systems considering only AWGN, as in general, the noise in power lines consists of background noise and impulsive noise (very destructive). The impulsive noise occurs generally in bursts, as represented in Figure 8.25, and it can be classified as periodic (occurs at regular time intervals, synchronous or asynchronous to the mains frequency) or aperiodic (random occurrence) [13].

In [23], periodic impulsive noise parameters were evaluated by means of a measuring campaign that established three different scenarios: heavily disturbed, medium disturbed and weakly disturbed. Table 8.11 shows the impulsive noise parameters for these three scenarios, where *IAT* is the interarrival time of the impulsive noise and T_{noise} is its average duration.

For evaluating the performance of FFT-OFDM and W-OFDM PHY layers, the periodic impulsive noise $n_{imp}(t)$, generated according to [24] for the heavily disturbed scenario of Table 8.11, was added to the background noise $n(t)$ (represented by the AWGN), as shown in Figure 8.26.

The simulations were performed considering two levels of impulsive noise, characterized by the impulsive noise to AWGN power ratio (F_{imp}), one setting $F_{imp} = 10$ and the other $F_{imp} = 100$. Although $F_{imp} = 100$ represents an impulsive noise level much higher than the one considered in [23,24] ($F_{imp} = 10$), it was also used in

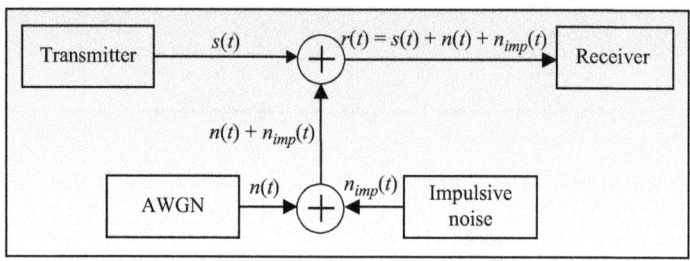

Figure 8.26   Simulation model for AWGN channel with periodic impulsive noise

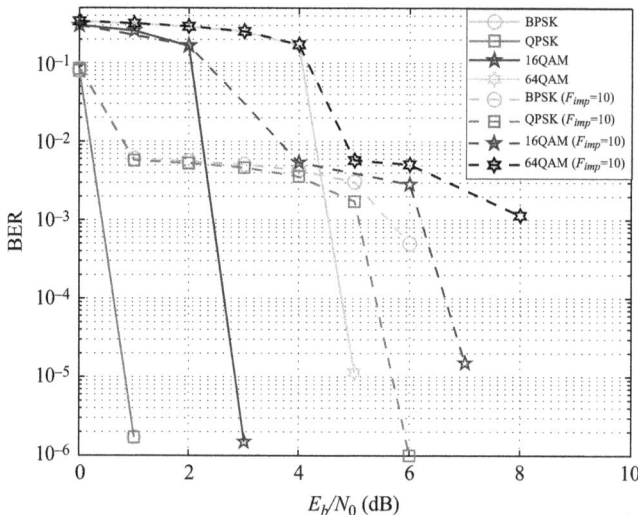

Figure 8.27   BER performance of FFT-OFDM PHY layer over AWGN channel with periodic impulsive noise ($F_{imp} = 10$)

this work only for investigation purposes (verify when impulsive noise levels start to become more critical).

Figures 8.27 and 8.28 present the performance results for FFT-OFDM in a AWGN channel with impulsive noise in terms of BER as a function of E_b/N_0, respectively, for $F_{imp} = 10$ and $F_{imp} = 100$. For comparison purpose, the curves without impulsive noise (only AWGN) were also inserted to the figures.

It can be verified in Figure 8.27 that for reaching a BER of $1 \cdot 10^{-4}$, the E_b/N_0 has to be increased by around 4–6 dB (when compared to the no impulsive noise scenario), depending on the modulation mapping schemes, due to the presence of the periodic impulsive noise, highlighting the degradation effect of the impulsive noise in FFT-OFDM PHY layer even using TC coding, interleaving and OFDM.

It can be noted in Figure 8.28 that with $F_{imp} = 100$, the higher impulsive noise level degraded significantly the performance of FFT-OFDM. It created a BER floor

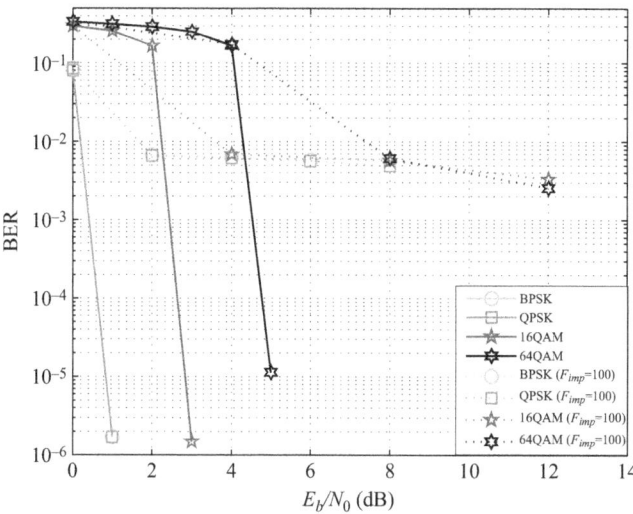

Figure 8.28 *BER performance of FFT-OFDM PHY layer over AWGN channel with periodic impulsive noise ($F_{imp} = 100$)*

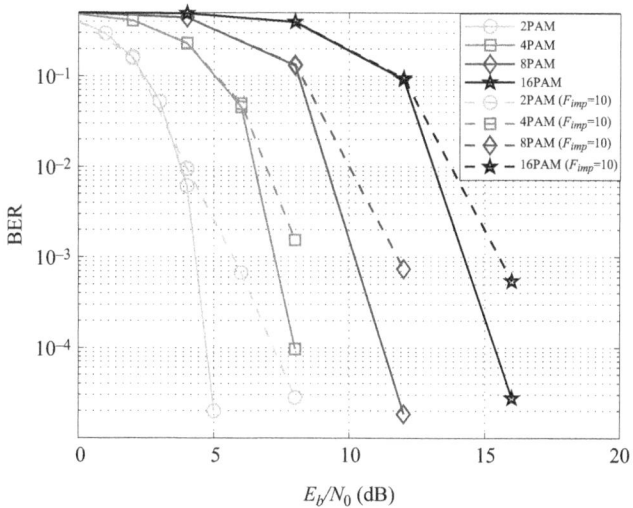

Figure 8.29 *BER performance of W-OFDM PHY layer over AWGN channel with periodic impulsive noise ($F_{imp} = 10$)*

at around $5 \cdot 10^{-3}$ that goes at least to E_b/N_0 of 8 dB to BPSK and QPSK and to 12 dB for 16-QAM and 64-QAM.

Figures 8.29 and 8.30 show the performance results for W-OFDM in a AWGN channel with impulsive noise, respectively, for $F_{imp} = 10$ and $F_{imp} = 100$.

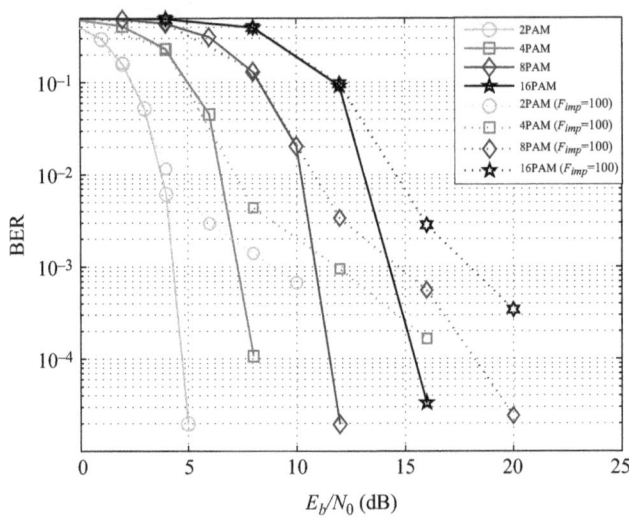

Figure 8.30   BER performance of W-OFDM PHY layer over AWGN channel with periodic impulsive noise ($F_{imp} = 100$)

From Figure 8.29, it can be noted that the presence of impulsive noise with $F_{imp} = 10$ requires an increase in E_b/N_0 (when compared to no impulsive noise scenario) by around 1.5–2 dB for a BER of $1 \cdot 10^{-3}$, 2-PAM, 4-PAM, 8-PAM and 16-PAM mapping. Although some of the mapping schemes studied did not reach a BER of $1 \cdot 10^{-4}$ during the simulations (BER drops very rapidly with the increase of E_b/N_0), it can be easily noted that all schemes require an increase in E_b/N_0 to achieve the same level of performance obtained without the presence of impulsive noise.

Analyzing the results of Figure 8.30, it can be noted that the presence of impulsive noise with $F_{imp} = 100$ starts to cause a similar BER floor as presented in FFT-OFDM.

8.2.4   Multipath fading channel with periodic impulsive noise

In this scenario, all PLC channel characteristics previously discussed are combined, as shown in Figure 8.31. The transmitted OFDM signal $s(t)$ pass through the multipath channel $h(t)$ with five resolvable paths, and then AWGN $n(t)$ and periodic impulsive noise $n_{imp}(t)$ are added to the resulting signal at the receiver. The multipath channel was generated according to the parameters shown in Section 8.2.2 and the periodic impulsive noise was configured according to the heavily disturbed scenario presented in Table 8.11, for $F_{imp} = 10$ and $F_{imp} = 100$. The performance for $F_{imp} = 100$ was also analyzed only for investigation purposes.

Figures 8.32 and 8.33 present the performance results for FFT-OFDM in a multipath channel with AWGN and impulsive noise in terms of BER as a function of E_b/N_0, respectively, for $F_{imp} = 10$ and $F_{imp} = 100$. For comparison purpose, the curves without impulsive noise (only multipath fading) were also inserted into the figures.

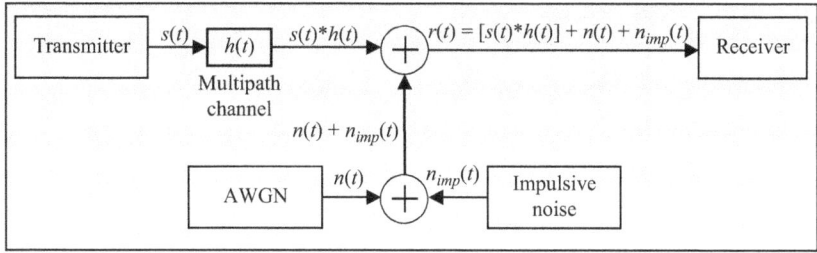

Figure 8.31 Simulation model for multipath channel with AWGN and periodic impulsive noise

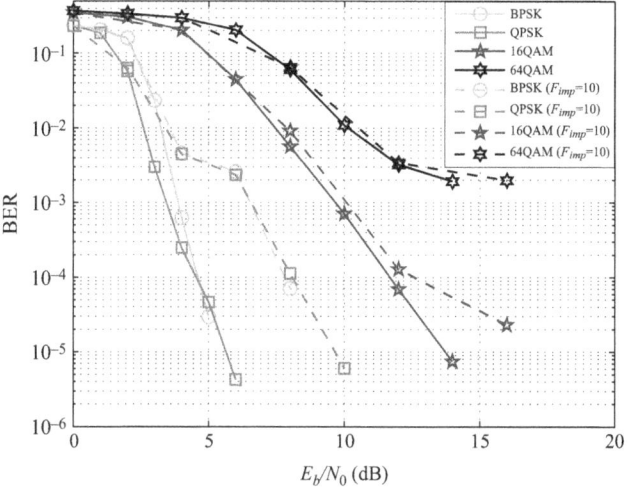

Figure 8.32 BER performance of FFT-OFDM PHY layer over multipath fading channel with AWGN and periodic impulsive noise ($F_{imp} = 10$)

It can be noted in Figure 8.32 that with $F_{imp} = 10$, there are a BER floor at different levels (e.g., around $4 \cdot 10^{-3}$ to BPSK from 4 to 6 dB), independent to the modulation mapping scheme. Due to the multipath effect, the contribution of the impulsive noise in the performance degradation is reduced, although it is still significant (because the multipath degradation by itself is high). For BPSK and QPSK, the degradation due to the multipath effect when compared to AWGN is around 5 dB to a BER of $1 \cdot 10^{-5}$ and the additional degradation caused by the impulsive noise is around 3.5 dB at the same BER.

It is possible to verify in Figure 8.33 that the increase in the impulsive noise caused by $F_{imp} = 100$, generate a performance floor of around $6 \cdot 10^{-3}$ from 5 to 12 dB for BPSK and QPSK, around $8 \cdot 10^{-3}$ that goes from 8 to 12 dB for 16-QAM

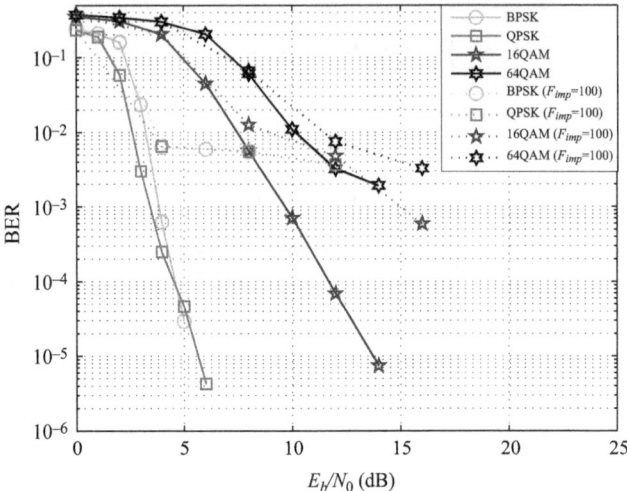

Figure 8.33   BER performance of FFT-OFDM PHY layer over multipath fading channel with AWGN and periodic impulsive noise ($F_{imp} = 100$)

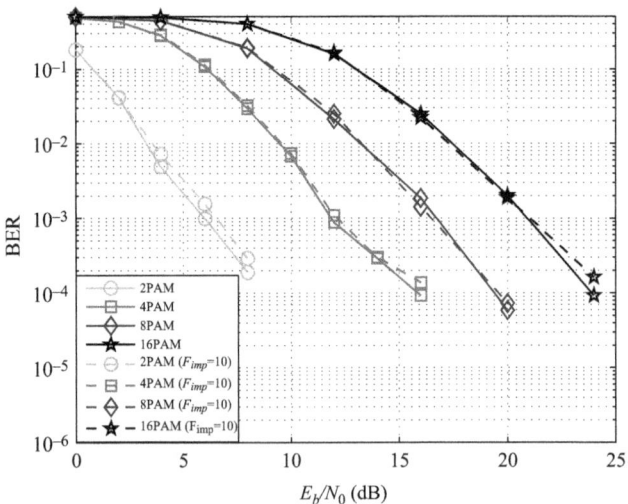

Figure 8.34   BER performance of W-OFDM PHY layer over multipath fading with AWGN and periodic impulsive noise ($F_{imp} = 10$)

and around $4 \cdot 10^{-3}$ that goes from 12 to 16 dB for 64-QAM (partially caused by the multipath effect).

Figures 8.34 and 8.35 show the performance results for W-OFDM in a multipath channel with AWGN and impulsive noise, respectively, for $F_{imp} = 10$ and $F_{imp} = 100$.

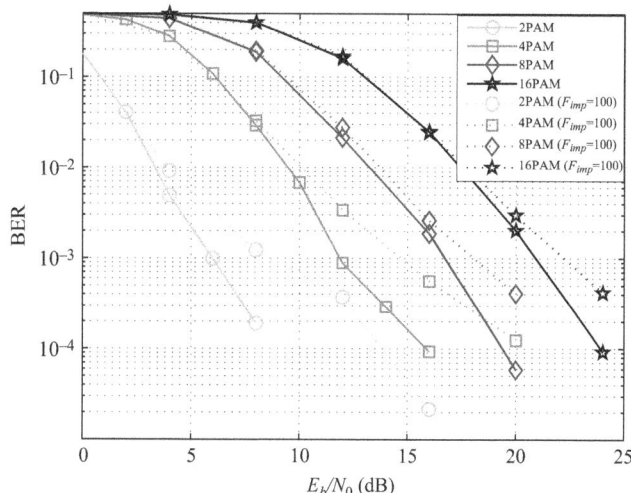

Figure 8.35 *BER performance of W-OFDM PHY layer over multipath fading with AWGN and periodic impulsive noise ($F_{imp} = 100$)*

Analyzing Figure 8.34, it can be noted that for a $F_{imp} = 10$, the impulsive noise did not affect W-OFDM performance significantly up to a BER of $1 \cdot 10^{-4}$. The worst impulsive noise effect appears in 2-PAM, which presented a performance degradation of around 0.5 dB at a BER of $1 \cdot 10^{-4}$.

On the other hand, Figure 8.35 for $F_{imp} = 100$ shows that the increase in the impulsive noise caused a performance degradation of up to 2 dB at $1 \cdot 10^{-3}$ for the considered modulation mapping. However, the performance degradation can increase significantly to lower BERs.

8.3 Conclusion remarks

In the analysis conducted in this chapter, it was possible to conclude that the FFT-OFDM PHY layer presents better performance than the W-OFDM PHY layer in AWGN (between 4 and 11 dB at $1 \cdot 10^{-5}$) and multipath (between 4.5 and 12.5 dB at $1 \cdot 10^{-4}$) PLC channels for equivalent modulation order (i.e., bits per digital symbol).

On the other hand, it seems that the FFT-OFDM PHY layer is less robust than the W-OFDM PHY layer to periodic impulse noise. This becomes clearer when the results of $F_{imp} = 100$ are analyzed, and it is possible to verify the presence of a more pronounced BER floor for FFT-OFDM than for W-OFDM.

Despite this, FFT-OFDM PHY layer still performed better than W-OFDM PHY layer for AWGN (e.g., QPSK required 5.5 dB at $1 \cdot 10^{-5}$ and 4-PAM required 9 dB) and multipath (e.g. QPSK required 8.0 dB at $1 \cdot 10^{-5}$ and 4-PAM required 17.0 dB) channels corrupted by impulsive noise with $F_{imp} = 10$.

It is important to mention that the degradation caused by the periodic impulsive noise can be reduced by increasing the length of the transmitted frame. As the number of OFDM symbols increases, the effectiveness of coding (mainly to TC coding) and interleaving schemes can be improved considerably (however, the Doppler effect might be a concern as FL increases).

References

[1]   Galli S, Scaglione A, Wang Z. For the Grid and Through the Grid: The Role of Power Line Communications in the Smart Grid. Proceedings of the IEEE. 2011;99(6):998–1027.

[2]   Tsuchiya F, Misawa H, Nakajo T, *et al.* Interference Measurements in HF and UHF Bands caused by Extension of Power Line Communication Bandwidth for Astronomical Purpose. In: 2003 International Symposium on Power Line Communications and Its Applications (ISPLC); 2003 March 26–28; Kyoto, Japan. IEEE; 2003. p. 265–269.

[3]   Wheen A. The last mile. In: Webb S, editor. Dot-Dash to Dot.Com: How Modern Telecommunications Evolved from the Telegraph to the Internet. New York, NY: Springer; 2011. p. 103–111.

[4]   Meng A, Ponzelar S, Koch M. The ITU-T G.9960 Broadband PLC Communication Concept for Smart Grid Applications. In: 2017 IEEE International Conference on Smart Grid Communications (SmartGridComm); 2017 October 23–27; Dresden, Germany. IEEE; 2017. p. 492–496.

[5]   Graf N, Tsokalo I, Lehnert R. Validating Broadband PLC for Smart Grid Applications with Field Trials. In: 2017 IEEE International Conference on Smart Grid Communications (SmartGridComm); 2017 October 23–27; Dresden, Germany. IEEE; 2017. p. 497–502.

[6]   IEEE. P1901.1 – IEEE Draft Standard for Medium Frequency (less than 15 MHz) Power Line Communications for Smart Grid Applications. Institute of Electrical and Electronics Engineers; 2018. 978-1-5044-4784-3.

[7]   Yonge L, Abad J, Afkhamie K, *et al.* An Overview of the HomePlug AV2 Technology. Journal of Electrical and Computer Engineering. 2013;(892628):1–20.

[8]   Simply Connect with HomePlug networking. HomePlug Powerline Alliance; [cited 2017 Sep 25]. Available from: http://www.homeplug.org/media/filer_public/6b/05/6b05d6a2-0e29-4b35-bb06-d9610a896711/homeplug_brochure.pdf.

[9]   Latchman HA, Katar S, Yonge L, *et al.* Introduction. In: Homeplug AV and IEEE 1901: A Handbook for PLC Designers and Users. Piscataway (NJ): Wiley-IEEE Press; 2013. p. 1–9.

[10] IEEE. IEEE Standard for Broadband over Power Line Networks: Medium Access Control and Physical Layer Specifications. Institute of Electrical and Electronics Engineers; 2010. 978-0-7381-6472-4.

[11] Clark D. Powerline Communications: Finally Ready for Prime Time? IEEE Internet Computing. 1998;2(1):10–11.

[12] Lin J, Nassar M, Evans BL. Impulsive Noise Mitigation in Powerline Communications Using Sparse Bayesian Learning. IEEE Journal on Selected Areas in Communications. 2013;31(7):1172–1183.

[13] Gotz M, Rapp M, Dostert K. Power Line Channel Characteristics and Their Effect on Communication System Design. IEEE Communications Magazine. 2004;42(4):78–86.

[14] Razazian K, Umari M, Kamalizad A, *et al.* G3-PLC Specification for Powerline Communication: Overview, System Simulation and Field Trial Results. In: 2010 IEEE International Symposium on Power Line Communications and Its Applications (ISPLC); 2010 March 28–31; Rio de Janeiro, Brazil. IEEE; 2010. p. 313–318.

[15] Alliance H. HomePlug 1.0.1 Specification. HomePlug Powerline Alliance; 2001.

[16] Liu W, Li C, Dostert K. Emulation of AWGN for Noise Margin Test of Powerline Communication Systems. In: 2011 IEEE International Symposium on Power Line Communications and Its Applications (ISPLC); 2011 April 3–6; Udine, Italy. IEEE; 2011. p. 225–230.

[17] Hoque KMR, Debiasi L, Natale FGB. Performance Analysis of MC-CDMA Power Line Communication System. In: IFIP International Conference on Wireless and Optical Communications Networks (WOCN); 2007 July 2–4; Singapore, Singapore. IEEE; 2007. p. 1–5.

[18] Liu L, Cheng T, Yanan L. Analysis and Modeling of Multipath for Indoor Power Line Channel. In: 10th International Conference on Advanced Communication Technology (ICACT); 2008 February 17–20; Gangwon-Do, South Korea. IEEE; 2008. p. 1966–1969.

[19] Tan B, Thompson JS. Powerline Communications Channel Modelling Methodology Based on Statistical Features. CoRR – Computing Research Repository. 2012;1203.3879:1–9.

[20] Lampe L, Vinck AJ. Cooperative Multihop Power Line Communications. In: 2012 IEEE International Symposium on Power Line Communications and Its Applications (ISPLC); 2012 March 27–30; Beijing, China. IEEE; 2012. p. 1–6.

[21] Mathur A, Bhatnagar MR, Panigrahi BK. PLC Performance Analysis Over Rayleigh Fading Channel Under Nakagami-m Additive Noise. IEEE Communications Letters. 2014;18(12):2101–2104.

[22] Rjeily CA. Power Line Communications under Rayleigh Fading and Nakagami Noise: Novel Insights on the MIMO and Multi-hop Techniques. IET Communications. 2018;12(2):184–191.

[23] Zimmermann M, Dostert K. Analysis and Modeling of Impulsive Noise in Broad-Band Powerline Communications. IEEE Transactions on Electromagnetic Compatibility. 2002;44(1):249–258.

[24] Ma YH, So PL, Gunawan E. Performance Analysis of OFDM Systems for Broadband Power Line Communications Under Impulsive Noise and Multipath Effect. IEEE Transactions on Power Delivery. 2005;20(2):674–682.

Chapter 9

Power line communications for smart grids applications

Ivan R. S. Casella[1]

Smart grids (SGs) can be considered as an evolution of the current energy model to optimally manage the balance between power supply and demand and meet the energy needs of the modern world.

One of the key elements of the SGs to achieve this goal is a bidirectional information and communication technology (ICT) infrastructure, with real-time monitoring, control and self-healing capability.

Among potential telecommunications technologies, power line communications (PLCs) appear as a strong candidate to integrate this ICT infrastructure by having some interesting features for applications in SGs. For example, they can exploit the existing electric grid infrastructure to reduce deployment costs, provide a low-cost alternative to complement existing technologies in the search of ubiquitous coverage and establish high data rate communication through obstacles that typically degrade wireless communications.

In order to understand the role of PLCs in SGs applications, this chapter will introduce a brief overview of how power grids are structured, then some fundamental characteristics of SGs will be shown and, finally, some possible applications of PLC technologies in SGs will be presented. Some of the key topics discussed here will be dealt with more deeply in next chapters.

9.1 Conventional power grids

Conventional power grids are one of the most remarkable conquests of the nineteenth and twentieth centuries. They are large complex systems responsible for delivering energy to an enormous number of consumers (residential, commercial and industrial) spread across a vast region. Historically, this process was built on the principle of unidirectional energy flow to provide electricity to passive consumers. The energy is

[1]Center for Engineering, Modeling and Applied Social Sciences (CECS), Federal University of ABC (UFABC), Brazil

generated in some few large power sources and pervasively transported over long-distance transmission and distribution networks, until reaching the end consumers [1,2]. This fact, associated with regulatory, technical, commercial and economic issues, reflected in the monopoly of the power facilities in the early years of electricity [3].

Thus, it can be noted that the architecture of conventional power grids is highly centralized and hierarchical. As shown in Figure 9.1, it consists of the stages of generation, step-up conversion (inside step-up substations), transmission, step-down conversion (inside step-down substations), distribution and consumption [4,5].

Although the main specifications of each stage may vary from country to country, it is possible to describe, in a general way, the whole process. The generation corresponds to the conversion of different sources of energy, such as hydraulic, wind, solar, tide, biomass, fossil fuel and nuclear, into electricity. In general, due to the technical and economic characteristics of the generators, three-phase medium voltages (MVs) are produced between 1 and 35 kV,[1]. in AC (alternating current) mode at 50 or 60 Hz.

As generation is often located far from large consumption centers for technical, legal and safety reasons, the voltage levels are elevated within step-up substations in order to reduce energy transmission losses over long distances. The transmission networks connect the generation plants to the regions of consumption. They usually cover hundreds of kilometers and are basically composed of high voltage (HV) [between 35 and 230 kV[1]] or extra HV (EHV) [above 230 kV[1]] transmission lines, in AC or DC (direct current) mode.

Lastly, the distribution networks play the role of safely and reliably delivering electricity from the transmission networks to the final customers within the consumption regions. They encompass step-down substations (to reduce the HV and EHV of the transmission lines to MV), three-phase MV lines, MV/LV transformers and three- and single-phase low voltage (LV) lines, generally below 1 kV[1] (e.g., 400/230, 380/220, 240/120 and 230/115 V), both in AC mode. In the distribution networks, the distances are smaller than in the transmission networks; thus, the voltages can be reduced to privilege the safety of the consumers instead of providing lower losses [6].

A step-down substation can supply power at different voltages for different areas and is usually composed by transformers, relays and circuit breakers, which can disconnect the substation from different distribution lines or from the power grid, if necessary. For large consumers, such as large industries, hospitals and universities, three-phase MV lines can be used directly. For small and medium consumers, a MV/LV transformer is employed to reduce the voltage at the consumer side. For residential applications, for example, two LV phases of the MV/LV transformer are connected to the power meters. For other applications that require more power, such

---

[1]This voltage classification is in accordance with IEC 60038 (International Electrotechnical Commission 60038) [7]

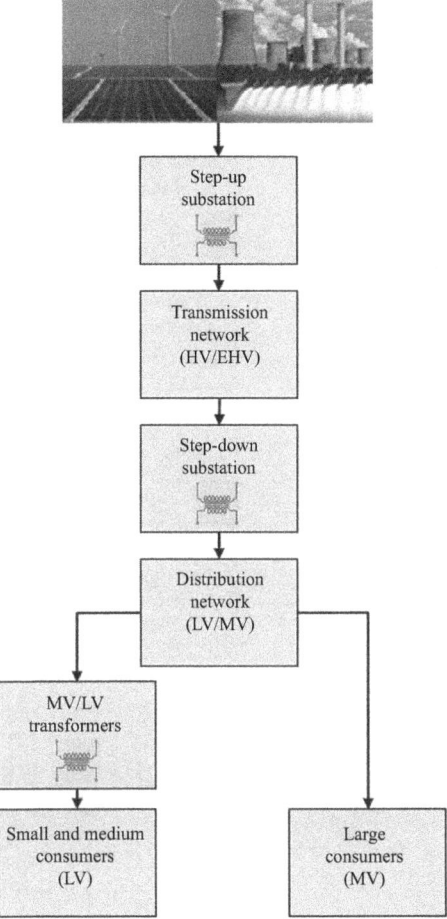

Figure 9.1   Architecture of the conventional power grids

as residential and commercial buildings and small industries, three-phase LV can be used.

Despite its importance, this centralized and hierarchical model, generally managed in part by the old cybernetic supervision and data acquisition control (SCADA) system, is aging and is becoming inadequate to meet the new electricity needs of the twenty-first century (e.g., accurate integration of renewable sources, more efficient mechanisms of generation, transmission and consumption of energy, high capacity of monitoring and control). This fact has motivated the development of a new power grid concept, the so-called SGs [8–10].

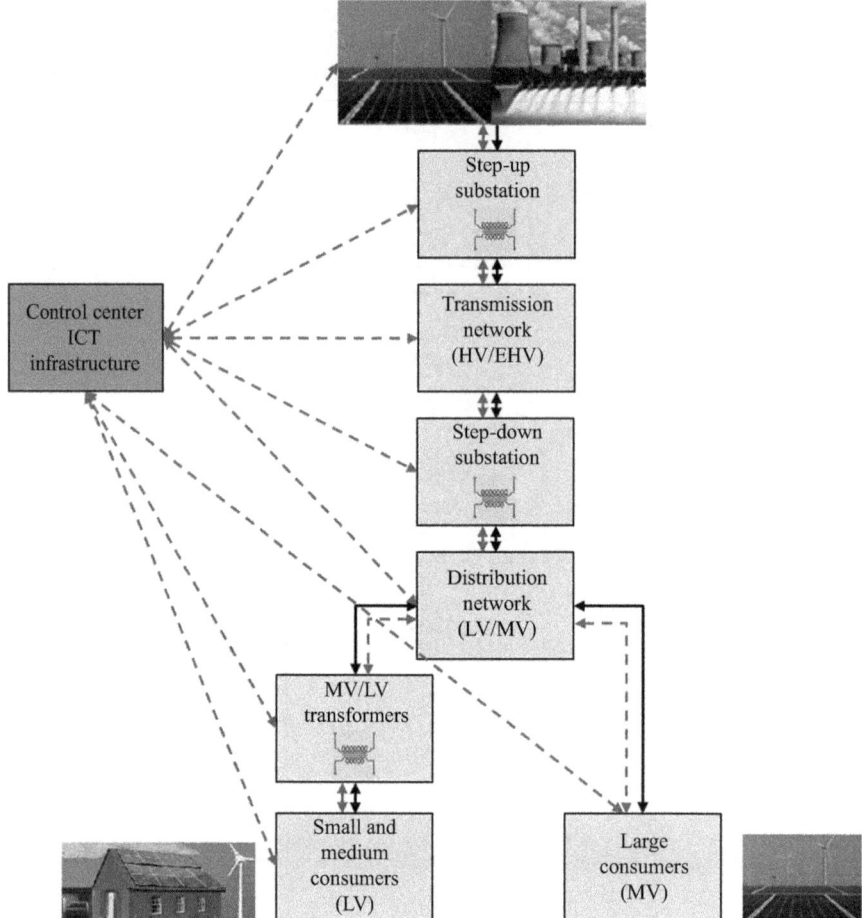

Figure 9.2   Architecture of the SGs

9.2   Smart grids

SGs can be considered as an evolution of conventional power grids. They employ modern ICT to build a complete interconnected bidirectional infrastructure composed of communication networks, data processing systems and real-time monitoring and control applications to optimally manage power generation, transmission, distribution and consumption and increase the efficiency, reliability and safety of energy use [1,2,11,12]. A description of the architecture of the SGs is presented in Figure 9.2.

SGs have the potential to achieve several interesting objectives for both power utilities and consumers [11,13,14], such as the following:

- Power quality suitable for twenty-first-century necessities
- Reduction of greenhouse gas emissions (GHGEs)

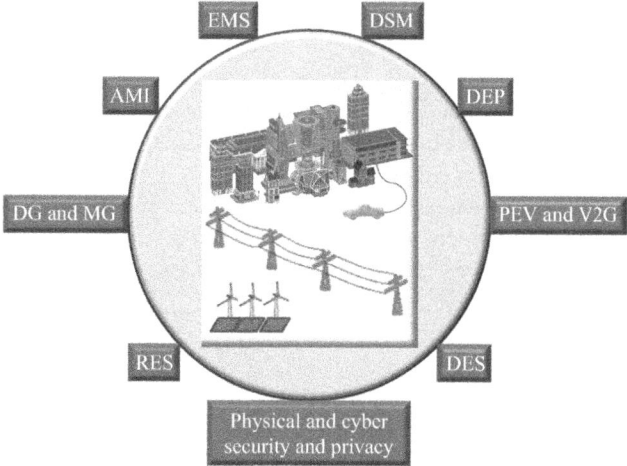

Figure 9.3   Main techniques used in the SGs

- Increased use of digital technologies to improve reliability, security and efficiency of the power grid
- Fast self-healing capability to minimize the impact of power outages on consumers
- Deployment and integration of distributed generation, including renewable resources
- Dynamic resources management to optimize the supply of energy and minimize operation and maintenance costs
- Real-time monitoring and control of the power delivered to consumers
- Active participation of the consumers in the power grid operations
- Resiliency against physical and cyberattacks;
- Development of new smart products, services and markets;
- Development of standards for communication and interoperability of appliances and equipment connected to the power grid.

As shown in Figure 9.3, SGs can offer these benefits through the use of the following techniques:

- Smart metering (SM) and advanced metering infrastructure (AMI) to enable energy monitoring and control in the consumer domain;
- Optimization of energy resources use and integration of different renewable energy sources (e.g., solar and wind) to existing power grids to increase energy supply in a sustainable, reliable, safe and low-cost manner, supported by the energy management system (EMS);
- Distributed generation (DG) and microgrids (MGs) in order to increase grid reliability (e.g., more than one supply of energy), simplify grid maintenance and reduce energy losses and costs (e.g., locating power generators close to the consumers);

- Decentralized energy storage (DES) to compensate time varying nature of the renewable energy sources (RESs);
- Plug-in electric vehicles (PEVs) and vehicle-to-grid (V2G) in order to coordinate the balance of energy generation and consumption along the day by discharging energy into the SGs, such that electrical vehicles (EVs) can act as a mobile energy source to further stabilize the grid;
- Demand side management (DSM) and dynamic energy pricing (DEP) programs, allowing active consumers participation to full coordination between the generated and consumed energy and to reduce energy losses, peaks demand and energy costs;
- Physical and cybernetic security and privacy.

Traditionally, energy companies have invested in increasing the power grid infrastructure to meet high peak power demand to avoid blackouts; however, this strategy is expensive and inefficient [15]. Within the SGs concept, power utilities can use, for example, DSM programs to modify the energy use patterns of the consumers and distribute energy usage more uniformly throughout the day. DEP programs can also be employed to encourage off-peak energy consumption by charging for electricity differently throughout the day.

9.2.1 Advanced metering infrastructure

Smart meters are devices initially designed to monitor energy use of the consumers in a much more accurately way than conventional power meters and to automatically delivery collected data to the power utility in regular intervals [16].

One of the first large initiatives for the deployment of SM are the automatic meter reading (AMR) systems. These are basically composed by smart meters and a unidirectional communication infrastructure with a capability to automatically collect data from power meters and periodically send them to the power utility for billing and troubleshooting analysis [17,18].

The natural evolution of AMR systems toward the SGs are the so-called AMIs. These are designed for the monitoring and control of devices and equipment within the consumer domain and are essentially comprised of smart meters, smart plugs (sensors and actuators remotely assisted), a bidirectional communication network, and a home EMS (HEMS), which is responsible for monitoring, analyzing and controlling the energy use pattern of each consumer in the power grid [19–21].

In this way, AMIs can provide power utilities and each consumer with much more information on the individual consumption pattern, allowing real-time energy management in each consumer domain, especially during peak demand times (e.g., DSM). With the participation of consumers, power utilities can reduce grid operating and expansion costs and share their earnings with consumers (e.g., DEP) [9,18,22]. Thus, AMIs are essential for the success of the DSM and DEP programs expected for the SGs and, consequently, for the GHGEs reduction.

AMR and AMI systems are not exclusively dedicated to consumer electricity services, they can also be employed for water and gas services.

9.2.2 Optimization of energy resources use and integration of renewable energy sources

The growing demand for energy from the modern world and recent environmental concerns make the study of RESs extremely important, especially due to the arising of the SGs concept [23]. The use of RESs can increase the power generation capacity of the grid and reduce dependence on fossil fuels, which are among the main responsible for GHGEs.

Most RESs currently installed around the world are wind or solar. They can be used both in the deployment of large power systems such as wind and solar farms with capacities from hundreds of megawatts to tens of gigawatts, as well as in small- and medium-sized systems for homes, buildings and industries applications with capacities of tens of kilowatts to dozens of megawatts, providing the dissemination of the DGs and MGs techniques [24].

However, the intermittency and unpredictability of RESs and the generalized use of DG and MGs bring several additional challenges in optimizing the operation and planning of the power grid, as they can cause an imbalance between the supply and demand of energy, requiring new flexible strategies and a different use of energy reserves (employing conventional energy generation and/or new energy storage systems) to supply this sui generis instantaneous power demand in the grid.

In this sense, some of unique features of the SGs, such as advanced EMS, DSM, DES and an ICT infrastructure with real-time monitoring and control capabilities, turn possible an accurate integration of these energy sources for a clean and sustainable energy generation.

9.2.3 Distributed generation and microgrids

Conventional power grids have currently faced some problems, such as constant growth of power demand, high complexity and costs of centralized power plants, gradual depletion of fossil fuel reserves and GHGEs. These issues have led to the emergence of a new trend of power generation in the vicinity of the regions of consumption (reducing power transmission losses) and employing different possible sources of energy (especially renewable sources, such as wind and solar) [25,26].

This strategy, called DG, can then be defined as the generation of energy from the connection of renewable or nonrenewable energy sources to the distribution network or directly to the consumer. Thus, larger power generation systems can also be considered part of a DG system if they are connected to the distribution network instead of the transmission network. The term DG was designed to distinguish this generation concept from conventional centralized generation and, in this case, the energy sources can be called distributed energy resources.

DG systems can be connected to or disconnected from the grid and allow consumers to produce some or all of the energy they need. The main advantage of considering grid-connected DG systems is the guarantee of receiving power from the utility when they are not producing as much energy as necessary. This is crucial for systems that use intermittent RESs (e.g., solar and wind) and useful for power supply in the event of failure of the DG systems or during their maintenance period [27,28].

The use of grid-connected DG systems can make the distribution network active. This means that energy can flow in two directions; in the direct direction, from the distribution network to the consumption region, and in the reverse direction, from the consumption region to the distribution network. For example, in the reverse direction, the surplus energy generated by the solar panels in the house of a prosumer (i.e., producer and consumer) can be delivered to the distribution network and sold to the power utility.

In turn, MGs are a wider way to deploy the concept of DG. They can be considered as a cluster of consumers, DG and DES systems (interconnected with each other), operating as small-scale power grids around a predefined consumption region and connected to the distribution network of the power utility, but capable of operating as self-sufficient islands [25,29,30].

An interesting feature of MGs operating in island mode is that they can supply power to remote regions without electrification at a much lower cost than necessary for the expansion of the utility's power grid up to these regions (e.g., construction of new long-distance transmission lines).

EMS is an essential component for the deployment of MGs. It is responsible for the decision-making related to the balance between generation and consumption, local and global operating conditions (inside and outside the MGs) and market information. One of the main differences between the EMS of MGs and the conventional EMS is the control of the energy exchange between the MGs and the utility's power grid, including islanding operations [31,32]. EMS can also monitor intermittent RESs to control DG and DES systems and optimize energy flow within the context of SGs [33].

9.2.4   Decentralized energy storage

Energy storage systems have been used naturally for a long time in power systems such as fossil fuel tanks in thermoelectric plants and water reservoirs in hydroelectric plants. However, the recent deployment of DG and MGs techniques based on RESs has introduced a new approach for energy storage.

DG systems and MGs based on intermittent RESs present significant power fluctuations over time, especially during islanding operations. The use of DES systems in these grids can contribute to a better balance between energy generation and consumption, allowing the storage of the excess energy when intermittent sources are available for later use by means of batteries or other resources (e.g., fossil fuel and hot water tanks). This mechanism can be employed in conjunction with conventional power generation methods to increase generation diversity and ensure uninterrupted power supply [23,33].

EMS can play an important role in identifying periods of excess or lack of energy generated by RESs and in controlling the charging and discharging cycles of DES in order to maximize the use of renewable energy and reduce the use of energy from the burning of fossil fuels [23,31,32].

From the perspective of the prosumer, the DES systems installed in the DG systems and MGs can allow that EMS execute different energy management strategies, according to the electricity pricing policy (purchase and sale), the availability of RESs and the desired demand.

9.2.5 Plug-in electric vehicles and vehicle-to-grid

Electrification of the transport systems is one of the most promising trends in reducing dependence on fossil fuels in both urban and rural regions. The deployment of hybrid electric vehicles (HEVs), plug-in HEVs (PHEVs) and PEVs has recently gained popularity because of their ability to reduce GHGEs and costs per distance traveled or per task performed [34].

HEVs are powered by two complementary drive systems, one based on a conventional internal combustion engine (e.g., gasoline or ethanol) and another based on batteries-coupled electric motors. The batteries are charged by regenerative braking and by a converter connected to the combustion engine. Unlike the PEVs and PHEVs, the HEVs cannot be connected to the mains to charge the batteries. PHEVs also employ combustion and electric motors. However, batteries can be additionally charged by connecting them to the mains or to an appropriate external power source. Unlike the HEVs, the PHEVs may operate in electric mode, but for a specific distance, usually of the order of tens of kilometers. On the other hand, the PEVs are only driven by electric motors and the batteries can be partially charged by regenerative braking and completely through the connection to external power points. Unlike the HEVs and PHEVs, which can generate a reduced amount of GHGEs, the PEVs do not generate any GHGEs for not using fossil fuels [35].

The introduction of a large number of PHEVs and/or PEVs can result in some important technical issues that can affect the entire power grid, particularly, at low voltage levels. PHEVs and PEVs can significantly increase energy consumption. The International Energy Agency estimates that together these can increase the transport share to about 10 percent of the total electricity consumption by 2050 [36].

Also, the increase in the amount of PHEVs and PEVs can cause high peaks of consumption in the power grid if the charging of the vehicles is not coordinated properly. In this sense, SGs with their real-time ICT infrastructures can assist in the management of smart PHEVs and PEVs charging operations at different locations (e.g., at home, work and recharging stations), so that they can be performed during off-peak hours (e.g., during night shift) and uniformly distributed according to DR and/or DEP programs [37].

Once the PHEVs and the PEVs are integrated into the SGs infrastructure, these can work as DES systems, supplying the stored energy to the distribution network when necessary. This concept associated with a bidirectional power flow is known as V2G. Thus, V2G systems can supply energy to the grid under conditions of power failure, increased consumption or RESs intermittence, in order to mitigate power variations in the distribution network and keep the power supply and demand balance [38–41].

9.2.6 Demand side management and demand response

DSM systems can be considered as one of the most important features of SGs by allowing the power grid to operate more efficiently, reliably and economically. These can take full advantage of the generation capacity available to meet the increasing power demand without the need to expand the power grid by optimally allocating the energy generated and controlling energy use patterns of consumers (i.e., instead of increasing the power generation infrastructure to meet the new demand needs, the demand itself can become more flexible to meet the power supply capacity of the power grid). With the increasing use of RESs, DSM techniques are especially important because they can adjust demand to accommodate the intermittent generation patterns of RESs [22].

Although DSM techniques can be applied in conventional power grids, only in the SGs, they can provide significant results by integrating all levels of the grid, from energy generation to consumption (including inside consumer points), through a bidirectional ICT infrastructure and a sophisticated EMS.

Thus, DSM can be defined as the set of techniques and policies used during the operation and planning of the SGs to adequately change the energy use patterns of consumers (individual control of the use of each smart device), in order to optimize the balance between energy generation and consumption [22,42–45].

DSM can be decomposed into the following elements [44,46]:

• Load management (LM);
• Demand response (DR) programs;
• Energy efficiency (EE) programs.

LM is an automated technique used (by the power utility or consumers) to control the timing and intensity of the electricity used by the consumers. The most common LM methods applied in DSM and DEP strategies are as follows:

• Peak clipping, which can reduce demand when the predefined demand limits are exceeded by turning off noncritical loads through. This scheme is commonly used by power utilities that do not have enough generation capacity to meet peak demand.
• Load shifting, which basically aims to reduce the energy consumption during the peak hours by shifting the use of loads (e.g., smart devices) to off-peak periods.
• Valley filling, which can be used for economic advantages by building loads during off-peak hours (when the electricity price is lower). This strategy may be desirable when the costs of building loads in off-peak hours can reduce the long-term average price (e.g., thermal heat storage and PEVs).
• Strategic conservation, which is the reduction of consumption over a long period due to programs promoted by the power utility. It can encompass increase of energy cost, changes in consumer behavior or use of more efficient devices (e.g., motivated by the utility).
• Strategic load growth, which represents an increase in energy consumption over a long period (beyond the valley filling method) motivated by utility programs to use surplus energy (e.g., due to technological advances or the installation of new plants).
• Flexible load shape, which involves the periodic energy use control of each consumer throughout the day to improve the balance between generation and consumption of the power grid.

DR programs are based on dynamic price DSM techniques in which consumers can automatically react to price changes over time through a flexible real-time LM strategy in order to improve grid reliability and reduce electricity costs. Hence, these encourage consumers to make short-term reductions in power demand in response to price changes, the values of which are usually significantly higher during peak periods.

The deployment of DR programs can become feasible and efficient due to the high granularity of communication between consumers and the power utility and the advances of the EMS guaranteed by the SGs.

EE programs allow the reduction of energy consumption without changing the quality of the power offered. These programs are based on the use of devices with technological advances that allow a more efficient energy consumption. As a result, peak demand can be permanently reduced by EE programs, directly impacting on the overall energy consumption.

9.2.7 Dynamic energy pricing

The ongoing balance between power supply and demand is a considerable challenge for future SGs, mainly due to the constant growth of demand and the uncertainties and high power fluctuations of RESs-based DG systems. DEP is a strategy to reach this balance by adjusting the price of energy consumed over time [47]. The fact that the costs of increasing the energy supply do not increase linearly with demand makes this technique even more relevant [24].

The success of DR programs relies on the participation of consumers. Usually, DR programs can be classified into incentive based and price based (also known as time based) [46,48–50].

In incentive-based programs, consumers are encouraged to reduce energy consumption through contractual arrangements and direct-load intervention. Agreements generally specify participation payments to consumers that reduce electricity use below some predefined limits and can impose restrictions or penalties for noncompliances. These programs often have some disadvantages related to consumer privacy and system scalability. They are generally adopted by industries and, according to Federal Energy Regulatory Commission (FERC), nearly 80 percent of the total load reduction potential in the United States of America (USA) may come from incentive-based DR programs [51].

Incentive programs can be associated to direct load control (DLC), in which the power utility can remotely disconnect equipment or devices from a consumer (e.g., air conditioner or water heater) for a period of time to meet the balance between generation and consumption or maintain consumption below predefined limits.

In price-based programs, the power demand profiles of consumers are modified (manually or automatically) according to the price of electricity over time. Following the pricing policy of the power utility, each consumer is encouraged to manage their loads in order to reduce energy consumption at peak times (e.g., peak clipping method), transfer part of the consumption to less congested periods (e.g., peak shifting method) or increase consumption in less congested periods (e.g., valley filling method), improving power balance [49,51,52]. Different pricing policies have been considered to motivate consumers to participate in DR programs. Among the most used are

- Time-of-use (TOU)
- Day ahead pricing (DAP)
- Real-time pricing (RTP)
- Critical peak pricing (CPP)

In TOU, the price of electricity depends on the period of day (e.g., two different periods: low and high peak demand periods or three different periods: low, mid and high peak demand periods) and year (e.g., monthly or seasonal) and is known in advance. In this way, consumers can schedule the use of their devices throughout the day to reduce electricity costs. In practice, electricity prices may vary according to supply and demand, which is the essence of DR programs. Therefore, prices should be adjusted often to correct variations in energy use behavior of consumers. DAP is similar to TOU; however, prices can vary from one day to another. RTP policy is also similar to TOU; however, energy prices vary on a higher granularity, on an hourly or subhourly basis (e.g., 15 min). Thus, consumers should respond periodically to price changes to reduce electricity costs. On the other hand, in the CPP policy, the price of electricity increases significantly during the occurrence of some critical events of high demand (e.g., hot summers). Thus, if consumers do not respect these predetermined periods, they will be heavily penalized. CPP is simple to implement and is often implemented in conjunction with other pricing policies, such as TOU or RTP [52].

HEMS (on the consumer side) can automatically adjust the consumption of smart devices (consumer loads) according to the pricing policy established by the power utility to minimize energy costs and contribute to the balance between the dynamic power demand profile of the consumer and the generation capacity of the power grid.

9.2.8   Physical and cyber security and privacy

SGs rely on the use of a complex bidirectional ICT infrastructure to significantly improve the quality of power generation, transmission, distribution and consumption. However, using this large heterogeneous infrastructure to interconnect millions of new smart devices and equipment along the power grid can increase problems related to data integrity and the security and privacy of consumers and the power utility, making them more vulnerable to physical and cybernetic attacks such as power theft, improper charging of energy, unwanted device control, power supply and ICT infrastructure failures, human risks, damage to smart devices and equipment and blackouts [53–55].

Thus, the study and development of physical and cybernetic surveillance systems have become crucial in improving the monitoring of network access (physical and logical), protection of data integrity of consumers and the power grid and efficient processing of millions of data that travels through the SGs, making possible the analysis of the network status, verification of occurrence of vulnerability failures, prevention of attacks and application of countermeasures [56].

Any system has vulnerabilities and the SGs are no exception. Several new issues can emerge with the integration of ICT and power infrastructures, along with other relevant factors such as human behavior, business interests and regulatory policies. Some issues are similar to those of conventional networks but involving more complex interfaces [54,57].

Some of the most important aspects in protecting the physical and cybernetic systems that compose the SGs are integrity, security and privacy. Integrity can be defined as the confidence that a given consumer or the power utility can access at a certain instant of time the data created in a specific smart device in the expected location and at the appropriate time, sent through the correct communication protocol and maintained unchanged throughout the process [57].

Security is the capacity of protect data from nonauthorized access. SG applications will handle very sensitive private data. Thus, the intrusion of hackers into the ICT infrastructure can lead to serious economic and reliability issues. The ICT infrastructure of the SGs can use several new communication protocols to interconnect all the smart devices within the whole power grid and to connect them to the Internet (e.g., for remote access, monitoring and control). It is quite different than old communication systems involving devices that were physically inside a restrict area (e.g., within fences or enclosed buildings). Consequently, the new ICT systems require different methods of cyber security to protect the SGs, as advanced user login access management, permissions management, surveillance systems, data classification and anomaly behavior analysis [55,57–59].

Privacy is another important deployment aspect of SGs. Authentication, access and admission of smart devices in the ICT infrastructure must be strictly controlled to avoid undue access. As the SGs incorporate some applications such as AMI and EMS, privacy of the consumers is increasingly becoming an issue. Electricity use patterns can disclose not only how much energy was consumed but also if the consumer is at home, work or traveling. It may also be possible to deduce the specific activities that are being performed in each environment (e.g., when at home, whether consumer is sleeping, showering or watching TV) and find out what types of devices are present. Consumer privacy violation can lead to various identity issues, such as increase in energy consumption, change in business models and hiring of undue services. Consequently, such privacy violation can support criminal activities or provide business information to competitors [55,57,58].

Data cryptography can improve the privacy of the SGs. However, current smart devices for monitoring and control may not be able to adequately support the high volume and speed required to perform all necessary encryption operations (although they usually can support some sort of protection). In addition, cryptographic solutions in this context should include a key management system to periodically update or at least revoke them, and this may cause concerns because of the increase in both time and processing required [57].

9.3 Information and communication technologies for smart grids

The ICT infrastructure of conventional power grids is basically composed of a wide area networks (WANs) with dedicated communication links and a SCADA system associated with an EMS [8].

SCADA system is responsible for remote monitoring and control of power grids. It collects data measured from remote terminal units (RTUs) installed in important elements of the power grid and sends them to the master terminal unit (MTU) located in the control center (CC) through a WAN. It may also send control data from the MTU to RTUs for optimization and protection actions of the grid. Data are usually transmitted through unencrypted communication links, making this critical infrastructure vulnerable to cyberattacks [53–55].

EMS is a specific application software framework that assists the SCADA system to reliably perform the management, optimization and protection operations of the power grid [53,60–63].

At the generation and transmission levels, the ICT infrastructure of conventional power grids is permeated locally in some important parts of the power system facilities, such as power generation plants, transmission networks and step-up and step-down substations, but does not provide complete interconnection between all parties and with the CC. On the other hand, at the distribution level, it is very limited in terms of functionality and availability and, at the consumption level, it is virtually nonexistent [10,64].

Over the last years, energy distribution networks have undergone constant changes to meet the growing use of DG (mainly based on RESs) in order to increase power generation capacity and reduce GHGEs. With the use of DG, the distribution networks become an active system, allowing bidirectional energy flows. Since most of the clean sources considered for DG applications are intermittent, managing and controlling the energy flow become very difficult, requiring a more complex ICT infrastructure. In addition, these new energy needs of the modern world require a precise balance between energy generation and consumption and a more participative behavior of consumers. However, conventional distribution networks do not have the appropriate technologies (e.g., communication, monitoring and control) to provide this balance with the necessary grid stability [9,10,12,65].

To overcome these new energy challenges, conventional power grids need to evolve into the so-called SGs. As discussed earlier, the great differential of the SGs is the presence of a complex real-time multi-level ICT (generation, transmission, distribution and consumption) of real time. This infrastructure can connect the various smart devices (e.g., smart plugs and smart meters), positioned at different points (nodes) and at different levels of the power system (to be monitored, controlled and/or protected), to each other and to the CC of the SG. The huge amount of information exchanged between these smart devices, upon reaching the CC, is processed and used by the management system to optimally perform all necessary operations of the power grid [2,9,12,65].

Cloud computing, data mining and big data techniques can help to efficiently organize, store and access all information generated in the power grid to improve decision making and facilitate the execution of the different management, optimization and protection services expected for the SGs [56,66–68].

As the SCADA/EMS presents in general a high latency and low data transmission capacity, associated with an asynchronous data processing and transmission architecture, it is also necessary to develop alternative systems that are distributed by nature such as the flexible AC transmission system, distributed web-based SCADA system or distributed EMS [8–10].

Another important system that is usually associated with the SCADA system and WANs is the wide area measurement system (WAMS). WAMS is responsible for the monitoring, control and some protection mechanisms of the transmission network. It uses phasor measuring units (PMUs), capable of measuring the voltage and current phasors at points of the transmission network (e.g., transmission substations), synchronized through satellites, to estimate the state of the network and to avoid failures and blackouts. According to the WAMS hierarchical architecture, the collected data are transmitted to the phasor data concentrator and further to the CC, where they can be used in the SCADA system for monitoring and control of the power grid. However, depending on the amount of PMU data, more sophisticated systems and high-rate communication networks may be required to ensure higher reliability and lower latency for the required actions [63,69].

In addition, the ICT infrastructure of the SGs should rely on the Internet Protocol (IP) to provide a high degree of interoperability between different applications and, depending on the level of criticality of the application (e.g., grid protection), also on the transport control protocol to ensure end-to-end data reliability [70].

The ICT infrastructure of SGs presents some specific requirements from both technical and economic point of view [5,6,10,19,26,71,72], such as:

- QoS (quality of service). The ICT infrastructure should provide the desired level of quality of each of the applications expected for the SGs. The main parameters usually considered to quantify the QoS are as follows:
  - Latency. It can be defined as the end-to-end delay to transmit data.
  - Throughput. It can be defined as the aggregated data rate associated to a given application. It can depend on the number of devices, size of the data packets and data traffic pattern.
  - Transmission error rate. It can be defined as the ratio between the total number of data received with error over the total number of data transmitted for a given signal-to-noise ratio. It is usually expressed by the bit error rate (BER), frame error rate or packet error rate.
- Reliability. It can be measured by the percentage of time over a year that the ICT infrastructure works properly.
- Interoperability. The ICT infrastructure should allow hardware and software from different manufacturers to interact with each other in a perfect and imperceptible way. Standardization is essential to effectively achieve this objective.
- Scalability. The ICT infrastructure should allow for a high degree of scalability from both technical and economic standpoints. Due to the large number of smart devices expected in the SGs, communication technologies should have low deployment, maintenance and operation costs. In addition, it should be able to add new devices and services.
- Security and privacy. SGs applications will handle very sensitive private data. Thus, the intrusion of hackers into the ICT infrastructure can lead to serious economic and reliability problems. Physical and cybernetic security and privacy

Table 9.1   Requirements of latency, data rate and transmission error rate
            of some SGs applications

Application	Latency	Throughput	BER
AMI	2–15 s	10–100 kbps (500 kbps for backhaul)	$< 10^{-6}$
DSM	500 ms	14–100 kbps (per node or device)	$< 10^{-6}$
DES	20 ms to 15 s	9.6–56 kbps	$< 10^{-6}$
Distribution grid management	100 ms to 2 s	9.6–100 kbps	$< 10^{-8}$
PHEVs/V2G	2 s to 5 min	9.6–56 kbps	$< 10^{-6}$
Wide area situational awareness	20–200 ms	600–1,500 kbps	$< 10^{-8}$
Substation automation	15–200 ms	9.6–56 kbps	$< 10^{-8}$
Overhead transmission line monitoring	15–200 ms	9.6–56 kbps	$< 10^{-6}$

are important deployment aspects and, therefore, authentication, access and admission of devices in the ICT infrastructure must be strictly controlled to avoid undue access.

With these requirements in mind, a simple summary of some SGs applications and their respective QoS requirements of latency, data rate and transmission error rate are presented in Table 9.1 [6,13,73–76].

The ICT infrastructure of the SGs is much more complex than the one of conventional networks. In order to effectively meet the requirements presented above and achieve several functionalities of the SGs, it probably will be composed of several different technologies [65,72,77,78]. Some standards, such as the IEEE 2030-2011 of the Institute of Electric and Electronic Engineering (IEEE), have provided some guidelines for defining an interconnected and interoperable systemic ICT model to provide end-to-end communication in the SGs [79,80].

With the view to facilitate the study, deployment and optimization of the ICT infrastructure, regarding the SG requirements and types of applications, it can be represented through a core-edge architecture composed of the following hierarchical levels, defined according to the coverage area and functionalities [72,77]:

• Home area network (HAN), building/business area network (BAN) and industrial area network (IANs), which correspond to the infrastructures for interconnecting smart devices (e.g., for energy monitoring and control of homes, building or industries), such as smart plugs or intelligent electronic devices (IEDs), inside each consumer domain. They can be coordinated, respectively, by HAN/BAN/IAN gateways (e.g., smart meters).

- Neighborhood area network (NAN), which corresponds to the infrastructure that serves a cluster of HANs/BANs/IANs. It is usually coordinated by a NAN gateway (e.g., data collector or aggregator).
- Field area network (FAN), which corresponds to the infrastructure that serves a cluster of NANs and/or some important IEDs of the power distribution network. It is usually managed by a FAN gateway (e.g., data router).
- WAN, which corresponds to the backbone infrastructure that connects all NANs and FANs to the CC of the power utility.
- Utility control area network (UAN), which corresponds to the communication infrastructure inside substations and the CC of the power utility.

In addition, it is still possible to define another type of network specifically to provide maintenance services and routine operations for workforce/utility employees, called mobile workforce network.

A HAN is a private network that interconnects smart devices and appliances in the consumer domain. It can be used in conjunction to a HEMS and DSM programs to control the use cycle of household appliances (e.g., lighting, refrigerators, heaters, washers/dryers, air conditioners), the charge and discharge of PHEVs, the operations of photovoltaic panels and other common types of DG, in order to reduce the energy consumption during peak hours and to exchange consumption pattern information with the CC of the power utility through the HAN gateway, that usually is the smart meter.

Thus, the HEMS enables the consumers to customize their energy use pattern and take into account DSM and DEP programs in order to minimize their electricity costs. Also, it also plays an important role for AMI applications by allowing the consumption data of all household appliances to be sent out of the HAN through the smart meter. Typically, a HAN covers areas of up to few hundreds of square meters and needs data rates in the order of tens of kilobits per second [65,72,77].

A BAN is also responsible for interconnecting smart devices in the consumer domain but, in this case, the consumer domain comprises an entire building with several apartments and/or business offices. A BAN can be considered a collection of HANs connected to a building smart meter (BAN gateway). Due to the high number of smart devices and the energy management applications, a BAN usually covers areas around few thousand square meters (vertically distributed) and requires data rates in the order of hundreds of kilobits per second [65,72,77].

An IAN plays the same role as the two previous networks, the only difference being the presence of a more powerful EMS with more sophisticated applications and smart devices, such as the PMU, specific for industrial environments. An IAN is also coordinated by a smart meter (IAN gateway) and may generally cover areas around tens of thousands square meters and demand data rates around hundreds of kilobits per second [65,72,77].

Regardless of the differences between HANs, BANs and IANs, they share many characteristics in common. They are primarily deployed in indoor environments and need to support short-range communications between smart devices for applications of monitoring and control. Also, they use a smart meter as a gateway to connect them with other networks and the utility. Thus, from the system point of view, they usually can be analyzed as one.

On the other hand, a NAN interconnects a cluster of HANs/BANs/IANs, in the distribution domain, to a NAN gateway called data aggregator unit (DAU). It is responsible for delivering the energy monitoring data from each smart meter to the DAU and sending real-time control information from the CC of the utility through the DAU to each smart meter. A NAN can support multiple SGs applications such as AMI, DSM, meter data and management system (MDMS) [72,77].

Depending on the communication technology, coverage area and number of smart meters of the HANs/BANs/IANs, it can be used an additional element called data collection unit (DCU) between the smart meters and the DAU. The smart meters can be divided in groups, each group can be connected to a DCU and a group of DCU can be connected to the DAU of the NAN. For example, the smart meters from the consumers on the same street could first connect to a DCU installed on the pole of the MV/LV transformer, and then all DCU of all the streets covered by the NAN could connect to the DAU. The great advantage of this strategy is that data traffic can be better organized locally, reducing network congestion and contention problems [81].

The number of smart meters in a NAN may range from a few hundred to a few thousand, depending on the country, SGs topology and communication technology. Thus, a NAN can cover areas of up to a few square kilometers and require data rates of up to a few tens of megabits per second [65,72,77].

In a complementary way, a FAN can provide connectivity between field IEDs such as feeders equipment, power line monitors, switches, circuit breaker controllers, transformers, public lighting, data collectors (including DAU of NAN) and some power substations, in the distribution domain, and the FAN gateway called field area router (FAR). It can support multiple SGs applications such as DG, rapid anomalies and failures detection (RAFD), power quality monitoring (PQM), distribution automation (DA) and substation automation (SA) systems to improve reliability and quality of distribution energy services.

A FAN can contain thousands of field IEDs, according to the SGs topology and communication technology. Thus, a FAN can cover areas of hundreds of square kilometers and require data rates of tens of megabits per second [65,72,77].

In some applications such as AMI, depending on the proposed communication technology, adopted topology and service coverage area, the NAN and FAN can be treated as a single network [18,82].

A WAN forms the communication backbone of the ICT infrastructure of the SGs. It provides a connection between several IEDs spread across transmission lines, power substations and power generation plants, in the generation and transmission domains, various IEDs within NAN, FAN, power substations and DG plants, in the distribution domain, and a WAN gateway called wide area router (WAR), in order to enable monitoring, control, protection (e.g., self-healing capability), security and automation operations in a large service coverage area. A WAN may cover thousands of square kilometers and require data rates of hundreds of megabits per second.

Lastly, a UAN locally connects all communication equipment inside power substations or the CC of the power utility to the monitoring, control, security, management

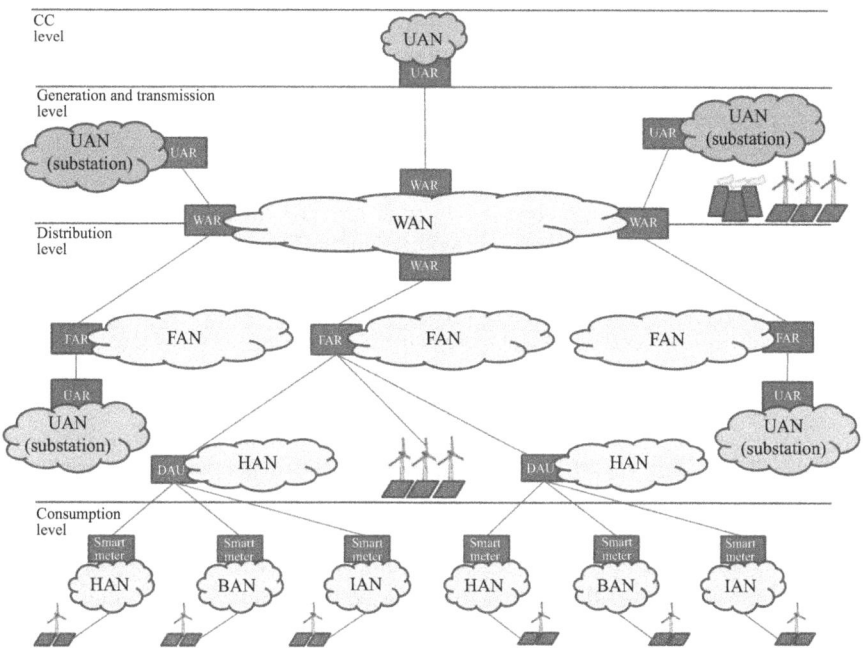

Figure 9.4   Hierarchical ICT architecture of the SGs

and automation of all levels of the grid, in order to meet the energy services offered by the SGs. The utility area router is responsible for coordinating the communication of the UAN with other levels of the ICT infrastructure over a WAN connection through a WAR.

In Figure 9.4, a diagram with the main communication networks that compose the hierarchical ICT architecture of the SGs is presented.

All these networks can use wireless and wired communication technologies in a complementary and, sometimes, competitive manner to meet all of these requirements [8,9,12]. However, this issue is not trivial since it involves the convergence of different areas of knowledge, design aspects and levels of interoperability.

Wireless communication technologies such as Zigbee, Bluetooth, BLE (Bluetooth low energy), LoRa (long range), Wi-Fi (wireless fidelity), WiMAX (world interoperability for microwave access), GSM (global system mobile), GPRS (general packet radio service), EDGE (enhanced data rate for GSM evolution) and LTE (long-term evolution), appear as an interesting solution as they can provide many benefits to the SGs, such as low deployment cost, ease of expansion, ability to use the technologies currently applied in mobile phone systems, flexibility of use and distributed management. However, they also present some limitations, as for example, the degrading characteristics of wireless propagation channels (e.g., attenuation, noise, multipath fading, Doppler propagation) that can reduce the quality of the communication with smart devices (placed for monitoring, controlling and/or protecting specific nodes of the grid), especially those located inside buildings and tunnels (e.g., difficulty in controlling lighting and charging consumers) or below street level (e.g., difficulty in supervising underground cables), low coverage in remote electrified areas (e.g., rural areas, wind farms) and the cost, reliability and safety constraints of using the wireless infrastructure of mobile service providers (if this strategy of wireless communication is adopted).

Wired communication technologies, such as SDH (synchronous digital hierarchy), DSL (digital subscriber line), MPLS (multiprotocol label switching), GbE (gigabit ethernet), WDM (wavelength division multiplexing) and PLCs, also appear viable because they offer high data rates, high reliability and high security. However, most of them have several disadvantages, as for example, high deployment costs, high maintenance costs, low flexibility and difficulty in expanding and accessing remote areas.

PLC technologies can be considered an exception to all this. As they present some of the good advantages of wireless communications and do not present the mentioned disadvantages of wired communications, they emerge themselves as solid candidates to integrate the ICT infrastructure of the SGs for technical and economic reasons.

For simplicity, next sections will explore only the use of PLC technologies in the ICT infrastructure of the SGs. A study on the integration of PLCs and wireless technologies for SGs applications will be presented in another chapter.

9.4   Power line communication technologies for smart grids

The success of the SGs deployment requires the use of bidirectional communication for interconnecting the most important nodes of the grid. PLC technologies can contribute with this new power model since they can exploit the existing electric grid infrastructure to establish pervasive communication across all nodes of the grid, potentially reducing costs of investments (the only wired technology that has deployment costs comparable to wireless solutions). Also, they can provide a low-cost alternative to complement existing technologies when aiming for ubiquitous coverage, establish high data rate communication through obstacles that commonly degrade wireless communications, and use PLC technologies currently applied in other sectors (e.g., Internet access). Moreover, chip manufacturers of PLCs devices for in-home and for SGs applications reported that they are selling millions of devices each year and expect the numbers continue to grow in the future [83].

As mentioned before in the book, one important aspect to be considered in the development of PLC technologies is that the power grid was not designed to be a data transmission medium. Thus, its use for communications usually addresses a number of issues that depend on the type of application, network topology and operating frequency band, as for instance [84]:

- High levels of background noise
- High interference coupled into the lines
- Attenuation and multipath fading (usually time varying and frequency selective)

- Low line impedance
- Electromagnetic radiation effects

Another important issue is the interoperability between existing PLCs standards. Several noninteroperable PLCs standards have been developed over the past few years, and PLCs devices can interfere with one another as they can share the same power lines for signal transmission and can cause significant performance degradation or even service disruption.

To overcome these problems, PLCs can use some modern techniques of digital communication, such as orthogonal frequency division multiplexing, forward error correction coding, channel equalization, transmission and reception diversity, that are already used in wireless communication systems [85,86] and some coexistence mechanisms to limit the interference caused by neighboring devices as specified, for instance, in the Priority Action Plan 15 of the National Institute of Standards and Technology, in the standard ITU-T G.9972 developed by the Institute of Telecommunications Union, and in the standards IEEE 1901 adopted by the IEEE.

A promising indication attesting to the technological maturity of PLCs, even with the degraded characteristics of power grid for data transmission, is their widespread use today in Europe for AMI applications [87]. In fact, the use of PLC technologies to exploit energy infrastructure as a communication medium emerged more than a century ago [9,88,89]. PLCs were initially used by power utilities for monitoring applications (status and alarms information) over MV and HV networks between power plants and substations and, since then, their areas of application have expanded significantly [20,90].

Currently, PLC technologies can provide communication over LV, MV, HV and EHV networks to drive several different applications with specific requirements of throughput (data rate), latency and transmission error rate [6]. The amount of traffic transported by the PLCs networks will depend on the corresponding power grid level. Although each single connection may have low data rate requirements, the overall traffic can be high, especially when fast real-time services are offered [84].

For example, for some applications such as DSM, DG and DEP, a FAR connected to a step-down substation at the MV level and to several DAUs at the LV level (e.g., small and medium consumers) may have to deal with a data traffic of the order of hundreds of Mbps and maintain the overall latency below 100 ms. On the other hand, some applications at MV, HV or EHV levels, such as fault detection, circuit breaker control and islanding control, may generate a total data traffic of tens of kbps, but require extremely high reliability and a latency of less than 15 ms.

In order to analyze in more detail the deployment of PLC technologies in SGs, some specific aspects of PLC technologies for HANs/BANs/IANs, NANs, FANs and WANs applications will be shown in the next sections.

9.4.1 PLCs applications in HAN/BAN/IAN

A HAN/BAN/IAN corresponds to the communication infrastructure for connecting smart devices (e.g., for energy monitoring and control) inside each consumption point (e.g., a home, building, business, or industry) to the smart meter, for a better energy use management. As it covers the restricted area within a consumption point, the use of PLC technologies for this type of application encompasses basically an indoor LV network (single- or three-phase, depending on the type of consumer point).

In general, this indoor LV network has a high capillarity, heterogeneity and variability due to the reasonable number of connected devices and equipment with different characteristics and energy consumption patterns (e.g., devices with constant or intermittent consumption, random variation of the amount of devices and equipment connected to the grid over time). This leads to different types of noise (e.g., Gaussian, impulsive, asynchronous, synchronous) and variations in the communication channel characteristics over the time (due to changes in the impedance of the medium over time), which can degrade the quality of the PLC transmissions.

Considering these types of networks, some of the possible applications supported by the PLC technologies would be

- SM and AMI
- HEMS
- DSM and DR
- DEP
- DG (e.g., local power generation using renewable energy sources, such as solar and wind)
- PEVs and V2G
- Multimedia monitoring

Due to the distances involved and amount of data and devices, both multicarrier (MC) NB-PLC (e.g., IEEE 1901.2) and BB-PLC (e.g., IEEE 1901-2010) technologies can be considered for HANs/BANs/IANs deployments.

9.4.2   PLCs applications in NAN

A NAN corresponds to the communication infrastructure that serves a cluster of HANs/BANs/IANs. As it covers the area around several consumption points, the use of PLC technologies may have to encompass LV networks or both LV and MV networks at the distribution level (although can be more suitable encompass just the LV network). A DAU may be located in the LV or MV network, depending on the PLC technology considered (bypass circuits or couplers may be necessaries).

The pervasive deployment of PLC technologies is expected to mainly occur in this level of the power grid. In this case, the NAN and FAN will have the important mission and to support a voluminous amount of data generated, respectively, from the HANs/BANs/IANs and IEDs (e.g., active loads, sensors, actuators, transformers) spread across the distribution network [84].

The outdoor LV and MV networks at the distribution level may be classified as overhead or underground, depending on the positioning of cables; this classification is not only important in terms of cost and installation requirements but also relevant for the applicability and performance of PLC technologies. The power lines impedance, cables access and network architecture are different for each case and PLC signal propagation is generally favored in overhead networks. Typically, overhead networks use a bus topology with mechanical switches in parts of the grid, while underground networks have a point-to-point topology that is very attractive for communication applications [20].

There are some factors that need to be highlighted specifically for the NAN based on LV and MV networks. Regarding the LV networks, distances, number of consumption points and its density are very important parameters. If the distance between smart devices and equipment in communication is too large (in terms of attenuation) for a specific implementation of a PLC technology, or if the density of consumers is low enough to consider repetition of the PLC signals, or if the number of consumers per cluster (i.e., subnetwork) is not high enough to allow for a significant economic return of investments, PLCs might not be an attractive technology. Regarding the MV networks, the voltage level is an additional important parameter, critical in the design of couplers responsible for connecting PLC signals to the MV networks [20].

For this type of network, some possible applications supported by the PLC technologies would be:

• AMI backbone network
• EMS (a variety of applications of monitoring and control for small consumers as homes, buildings, businesses and industries)
• DSM and DR
• DEP
• Public lighting monitoring and control

Depending on the application requirements and power grid level, MC NB-PLC technologies, MC BB-PLC technologies or a combination of both can be employed for a NAN deployment [84]. However, despite offering a lower data rate than BB-PLC, the use of NB-PLC allows data to pass easily through MV/LV transformers, making possible a significant reduction in deployment costs and operation management. NB-PLC technologies have millions of devices deployed around the world [91]. In many circumstances, NB-PLC technologies do not require the installation of bypass circuits or couplers at MV/LV transformers, allowing the connection of a large number of smart meters to the DAU on the MV network side. Contrariwise, BB-PLC technologies in general require the installation of bypass circuits or couplers and, even with the installation of these devices, there are higher signal attenuations than those observed in NB-PLC technologies [92].

These technologies have also proven to be able to prevent network congestions when cooperative techniques are employed, making possible to predict more accurate message deliveries, and reduce the transmission power consumption when compared to wireless solutions [87].

9.4.3   PLCs applications in FAN

A FAN corresponds to the communication infrastructure to interconnect some distribution substations, several field IEDs along the distribution network and also to serve a cluster of NAN. As it covers a large area around a segment of the distribution network, the use of PLC technologies should be delimited to MV networks. A FAR located in the MV network can collect data from all elements of the FAN and send them to the CC via a WAN and receive the specific control information to each element sent by the CC through the WAN. A FAR at the MV level can connect to a specific DAU of a NAN at the LV level through bypass circuits or couplers (in general causing signal attenuations).

Considering this type of network, PLC technologies could provide the following applications:

- AMI backbone network
- EMS (a variety of applications of monitoring and control inside large consumers as industries)
- DG and MGs
- RAFD in the distribution network (for rapid fault detection, isolation and self-healing)
- PQM
- Islanding detection in the distribution network
- Substations connectivity (sharing state of equipment and power flow status)
- DA and SA (to improve reliability and quality of distribution network services)

The deployment of a FAN can be based on both NB-PLC and BB-PLC MC technologies [84]. Some low data rate access points could use NB-PLC technologies, but high data rate access points resulting from voluminous aggregate traffic (e.g., several NAN connecting to a FAR) would require the use of BB-PLC technologies. The feasibility of using BB-PLC (IEEE 1901-2010) in an MV network was demonstrated, for example, in [93]. The authors performed a field test covering 1.6 km through 7 BB-PLC modems, achieving data rates of approximately 27 Mbps.

9.4.4 PLCs applications in WAN

The WAN corresponds to the communication backbone of the ICT infrastructure of SGs. It may be used to connect several power substations, FANs and IEDs for monitoring, control, protection, management and automation of long-distance distribution and transmission networks to the CC. As it covers a very large region, the use of PLC technologies for WAN applications encompasses mainly MV, HV and EHV networks of the power grid. As mentioned before, the outdoor MV networks can be classified as overhead or underground. Due to the variations of the power lines impedance, cables access and network topology, they can present a complex medium for PLCs signal propagation.

On the other hand, the HV and EHV networks are usually deployed by large overhead structures connecting step-up substations to the step-down substations. Although, historically, the communication network at this level was based on low data rate single carrier PLC technologies (e.g., ripple carrier signaling), optical communications technologies came to dominate this segment due to its high data rate capability, its high electromagnetic immunity at HVs and electrical surges and the simplicity of the topology of these networks.

New MC BB-PLC technologies appears a promising alternative at HV and EHV networks since they can offer a reduction in the costs of deploying new ICT infrastructures (e.g., grid expansion), when compared with optical technologies, and act as a secondary communication network (e.g., redundancy) to existing ICT infrastructures. Also, as HV and EHV have few connections and transformers and better impedance when compared to LV and MV networks, they work as a good transmission medium for PLCs. Coupling BB-PLC devices to the electrical network may be critical and expensive due to the wide frequency bandwidth required and the HVs of this part of the power grid.

Another promising technique at these levels of the power grid are multiple-input and multiple-output PLCs (MIMO-PLCs). MIMO-PLC technologies can form parallel communication channels between the coupled pairs of wires of a multiphase power grid to offer significantly higher data rates and increased communication reliability when compared with single-input–single-output technologies, without any additional cost regarding transmitter power and bandwidth. However, coupling of MIMO-PLC signals can be even more complex and expensive than coupling BB-PLC signals [94].

Considering this type of network, BB-PLC technologies can provide the following WAMS services:

• Remote fault detection in the distribution and transmission networks (broken insulator detection, insulator short circuit, cable rupture, circuit breaker activation and deactivation)
• Remote station surveillance or state estimation
• Sag monitoring in the transmission network (determination of the average height change above ground of horizontal HV overhead conductors)

The use of PLC technologies for high data rate WAN applications still requires attention and deep research.

9.5 Conclusion remarks

In this chapter, the main characteristics and problems of conventional power grids were presented and some strategies for them to evolve into the SGs were discussed. It was shown that, among existing telecommunication technologies, PLC technologies are particularly interesting for integrating the ICTs of the SGs and to provide essential real-time monitoring, control and self-recovery capabilities by using the utility's own electrical infrastructure as means of communication, significantly reducing deployment costs. Also, it was shown that they can also be associated with other existing technologies, as wireless technologies, in the search for ubiquitous coverage and to establish high data rate communication, where other technologies would have difficult to work reliably alone.

References

[1] Simoes MG, Roche R, Kyriakides E, *et al.* A Comparison of Smart Grid Technologies and Progresses in Europe and the U.S. IEEE Transactions on Industry Applications. 2012;48(4):1154–1162.

[2] Baimel D, Tapuchi S, Baimel N. Smart Grid Communication Technologies. Journal of Power and Energy Engineering. 2016;04(08):1–8.

[3] Pagani GA. From the Grid to the Smart Grid, Topologically; Groningen: NetzoDruk, 2014.

[4] Nghia Le T, Chin WL, Chen HH. Standardization and Security for Smart Grid Communications Based on Cognitive Radio Technologies: A Comprehensive Survey. IEEE Communications Surveys & Tutorials. 2017;19(1):423–445.

[5] Zhou S, Brown MA. Smart Meter Deployment in Europe: A Comparative Case Study on the Impacts of National Policy Schemes. Journal of Cleaner Production. 2017;144:22–32.

[6] Gungor VC, Sahin D, Kocak T, *et al.* A Survey on Smart Grid Potential Applications and Communication Requirements. IEEE Transactions on Industrial Informatics. 2013;9(1):28–42.

[7] IEC 60038. *IEC standard voltages*. Published By: International Electrotechnical Commission (IEC), 7th Edition, 2009.

[8] Luka MK, Pallam SW, Thuku IT, *et al.* Narrowband Power Line Communication for Smart Grid. International Journal of Scientific & Engineering Research. 2015;6(7):1244–1252.

[9] Galli S, Scaglione A, Wang Z. For the Grid and Through the Grid: The Role of Power Line Communications in the Smart Grid. Proceedings of the IEEE. 2011;99(6):998–1027.

[10] He Y. Smart Metering Infrastructure for Distribution Network Operation; Cardiff: Cardiff University; 2016.

[11] Fang X, Misra S, Xue G, *et al.* Smart Grid The New and Improved Power Grid: A Survey. IEEE Communications Surveys & Tutorials. 2012;14(4):944–980.

[12] Kabalci Y. A Survey on Smart Metering and Smart Grid Communication. Renewable and Sustainable Energy Reviews. 2016;57:302–318.

[13] US Department of Energy. Communications Requirements of Smart Grid Technologies; 2010.

[14] Hammoudeh MA. Comparative Analysis of Communication Architectures and Technologies for Smart Grid Distribution Network. Denver, CO: University of Colorado; 2012.

[15] Albadi MH, El-Saadany EF. A Summary of Demand Response in Electricity Markets. Electric Power Systems Research. 2008;78(11):1989–1996.

[16] Rashed Mohassel R, Fung A, Mohammadi F, *et al.* A Survey on Advanced Metering Infrastructure. International Journal of Electrical Power & Energy Systems. 2014;63:473–484.

[17] Lo CH, Ansari N. The Progressive Smart Grid System from Both Power and Communications Aspects. IEEE Communications Surveys & Tutorials. 2011;14(3):799–821.

[18] Sharma K, Mohan Saini L. Performance Analysis of Smart Metering for Smart Grid: An Overview. Renewable and Sustainable Energy Reviews. 2015;49:720–735.

[19] Yan Y, Qian Y, Sharif H, *et al.* A Survey on Smart Grid Communication Infrastructures: Motivations, Requirements and Challenges. IEEE Communications Surveys & Tutorials. 2013;15(1):5–20.

[20] Sendin A, Pea I, Angueira P. Strategies for Power Line Communications Smart Metering Network Deployment. Energies. 2014;7(4):2377–2420.

[21] Lopez G, Moreno JI, Amars H, *et al.* Paving the Road Toward Smart Grids Through Large-Scale Advanced Metering Infrastructures. Electric Power Systems Research. 2015;120:194–205.

[22] Deng R, Yang Z, Chow MY, *et al.* A Survey on Demand Response in Smart Grids: Mathematical Models and Approaches. IEEE Transactions on Industrial Informatics. 2015;11(3):570–582.

[23] Wang Y, Lin X, Pedram M. Adaptive Control for Energy Storage Systems in Households With Photovoltaic Modules. IEEE Transactions on Smart Grid. 2014;5(2):992–1001.

[24] Georgievski I, Degeler V, Pagani GA, *et al.* Optimizing Energy Costs for Offices Connected to the Smart Grid. IEEE Transactions on Smart Grid. 2012;3(4):2273–2285.

[25] Lasseter RH. Microgrids and Distributed Generation. Journal of Energy Engineering. 2007;133(3):144–149.

[26] Jiang J, Qian Y. Distributed Communication Architecture for Smart Grid Applications. IEEE Communications Magazine. 2016;54(12):60–67.

[27] Jiang H, Wang K, Wang Y, *et al.* Energy Big Data: A Survey. IEEE Access. 2016;4:3844–3861.

[28] Kakran S, Chanana S. Smart Operations of Smart Grids Integrated with Distributed Generation: A Review. Renewable and Sustainable Energy Reviews. 2018;81:524–535.

[29] Jiayi H, Chuanwen J, Rong X. A Review on Distributed Energy Resources and MicroGrid. Renewable and Sustainable Energy Reviews. 2008;12(9):2472–2483.

[30] Ton DT, Smith MA. The U.S. Department of Energy's Microgrid Initiative. The Electricity Journal. 2012;25(8):84–94.

[31] Zaheeruddin, Manas M. Renewable Energy Management through Microgrid Central Controller Design: An Approach to Integrate Solar, Wind and Biomass with Battery. Energy Reports. 2015;1:156–163.

[32] Pourbabak H, Chen T, Zhang B, *et al.* Control and energy management system in microgrids. In: Obara S, Morel J, editors. Clean Energy Microgrids. Institution of Engineering and Technology; Submitted on 29 May 2017. p. 109–133. DOI: 10.1049/PBPO090E_ch3.

[33] Bando S, Sasaki, Y, Asano, H. Balancing control method of a microgrid with intermittent renewable energy generators and small battery storage. In: IEEE Power and Energy Society General Meeting – Conversion and Delivery of Electrical Energy in the 21st Century; 2008. p. 1–6.

[34]   Liu R, Dow L, Liu, E. A survey of PEV impacts on electric utilities. In: IEEE Innovative Smart Grid Technologies; ISGT 2011; Anaheim, CA. IEEE; p. 1–8.

[35]   Tan KM, Ramachandaramurthy VK, Yong JY. Integration of Electric Vehicles in Smart Grid: A Review on Vehicle to Grid Technologies and Optimization Techniques. Renewable and Sustainable Energy Reviews. 2016;53:720–732.

[36]   International Energy Agency. Technology Roadmaps: Smart Grids. International Energy Agency; 2011.

[37]   Lopez G, Custodio V, J Herrera F, *et al.* Machine-to-Machine Communications Infrastructure for Smart Electric Vehicle Charging in Private Parking Lots. International Journal of Communication Systems. 2014;27:643–660.

[38]   Guille C, Gross G. A conceptual framework for the vehicle-to-grid (V2G) implementation. Energy Policy. 2009;37(11):4379–4390.

[39]   Pang C, Dutta P, Kezunovic M. BEVs/PHEVs as Dispersed Energy Storage for V2B Uses in the Smart Grid. IEEE Transactions on Smart Grid. 2012;3(1): 473–482.

[40]   Goebel C, Callaway DS. Using ICT-Controlled Plug-in Electric Vehicles to Supply Grid Regulation in California at Different Renewable Integration Levels. IEEE Transactions on Smart Grid. 2013;4(2):729–740.

[41]   Han W, Xiao Y. Privacy Preservation for V2G Networks in Smart Grid: A Survey. Computer Communications. 2016;91–92:17–28.

[42]   Gellings CW. The Concept of Demand-Side Management for Electric Utilities. Proceedings of the IEEE. 1985;73(10):1468–1470.

[43]   Gellings CW, Smith WM. Integrating Demand-Side Management into Utility Planning. Proceedings of the IEEE. 1989;77(6):908–918.

[44]   Logenthiran T, Srinivasan D, Shun TZ. Demand Side Management in Smart Grid Using Heuristic Optimization. IEEE Transactions on Smart Grid. 2012;3(3):1244–1252.

[45]   Gelazanskas L, Gamage KAA. Demand Side Management in Smart Grid: A Review and Proposals for Future Direction. Sustainable Cities and Society. 2014;11:22–30.

[46]   Vardakas JS, Zorba N, Verikoukis CV. A Survey on Demand Response Programs in Smart Grids: Pricing Methods and Optimization Algorithms. IEEE Communications Surveys & Tutorials. 2015;17(1):152–178.

[47]   Alagoz BB, Kaygusuz A. Dynamic Energy Pricing by Closed-Loop Fractional-Order PI Control System and Energy Balancing in Smart Grid Energy Markets. Transactions of the Institute of Measurement and Control. 2016;38(5):565–578.

[48]   Bu S, Yu FR, Liu PX. Dynamic pricing for demand-side management in the smart grid. In: IEEE Online Conference on Green Communications; 2011. p. 47–51.

[49]   Haider HT, See OH, Elmenreich W. A Review of Residential Demand Response of Smart Grid. Renewable and Sustainable Energy Reviews. 2016;59:166–178.

[50] Muratori M, Rizzoni G. Residential Demand Response: Dynamic Energy Management and Time-Varying Electricity Pricing. IEEE Transactions on Power Systems. 2016;31(2):1108–1117.

[51] Amer M, Naaman A, M'Sirdi NK, *et al.* Smart home energy management systems survey. In: Renewable Energy for Developing Countries. IEEE; 2014. p. 167–173.

[52] Bergaentzl C, Clastres C, Khalfallah H. Demand-Side Management and European Environmental and Energy Goals: An Optimal Complementary Approach. Energy Policy. 2014;67:858–869.

[53] Teixeira A, Dn G, Sandberg H, *et al.* A Cyber Security Study of a SCADA Energy Management System: Stealthy Deception Attacks on the State Estimator. IFAC Proceedings Volumes. 2011;44(1):11271–11277.

[54] Yu X, Xue Y. Smart Grids: A Cyber Physical Systems Perspective. Proceedings of the IEEE. 2016;104(5):1058–1070.

[55] Cintuglu MH, Mohammed OA, Akkaya K, *et al.* A Survey on Smart Grid Cyber-Physical System Testbeds. IEEE Communications Surveys & Tutorials. 2017;19(1):446–464.

[56] Hu J, Vasilakos AV. Energy Big Data Analytics and Security: Challenges and Opportunities. IEEE Transactions on Smart Grid. 2016;7(5):2423–2436.

[57] McDaniel P, McLaughlin S. Security and Privacy Challenges in the Smart Grid. IEEE Security & Privacy Magazine. 2009;7(3):75–77.

[58] Liu J, Xiao Y, Li S, *et al.* Cyber Security and Privacy Issues in Smart Grids. IEEE Communications Surveys & Tutorials. 2012;14(4):981–997.

[59] Shashanka M, Shen MY, Wang J. User and entity behavior analytics for enterprise security. In: IEEE International Conference on Big Data. IEEE; 2016. p. 1867–1874.

[60] Kokai Y, Masuda F, Horiike S, *et al.* Recent Development in Open Systems for EMS/SCADA. Electrical Power & Energy Systems. 1998;20(2):111–123.

[61] Marihart DJ. Communications Technology Guidelines for EMS/SCADA Systems. IEEE Transactions on Power Delivery. 2001;16(2):181–188.

[62] Amanullah MTO, Kalam A, Zayegh A. Network security vulnerabilities in SCADA and EMS. In: IEEE/PES Transmission and Distribution Conference & Exhibition: Asia and Pacific. IEEE; 2005. p. 1–6.

[63] Arghira N, Hossu D, Fagarasan I, *et al.* Modern SCADA Philosophy in Power System Operation: A Survey. University Politehnica of Bucharest Scientific Bulletin, Series C: Electrical Engineering. 2011;73(2):153–166.

[64] Elyengui S, Bouhouchi R, Ezzedine T. The Enhancement of Communication Technologies and Networks for Smart Grid Applications. International Journal of Emerging Trends & Technology in Computer Science. 2013;2(6):107–115.

[65] Gao J, Xiao Y, Liu J, *et al.* A Survey of Communication/Networking in Smart Grids. Future Generation Computer Systems. 2012;28(2):391–404.

[66] Kezunovic M, Xie L, Grijalva S. The role of big data in improving power system operation and protection. In: IREP Symposium-Bulk Power System Dynamics and Control; 2013. p. 1–9.

[67]   Baek J, Vu QH, Liu JK, *et al.* A Secure Cloud Computing Based Framework for Big Data Information Management of Smart Grid. IEEE Transactions on Cloud Computing. 2015;3(2):233–244.

[68]   Simmhan Y, Aman S, Kumbhare A, *et al.* Cloud-Based Software Platform for Big Data Analytics in Smart Grids. Computing in Science & Engineering. 2013;15(4):38–47.

[69]   Suljanovic N, Borovina D, Zajc M, *et al.* Requirements for communication infrastructure in smart grids. In: IEEE International Energy Conference. IEEE; 2014. p. 1492–1499.

[70]   Uribe-Prez N, Angulo I, de la Vega D, *et al.* Smart Grid Applications for a Practical Implementation of IP over Narrowband Power Line Communications. Energies. 2017;10(11):1–24.

[71]   Kuzlu M, Pipattanasomporn M, Rahman S. Communication Network Requirements for Major Smart Grid Applications in HAN, NAN and WAN. Computer Networks. 2014;67:74–88.

[72]   Dehalwar V, Kalam A, Kolhe ML, *et al.* Compliance of IEEE 802.22 WRAN for field area network in smart grid. In: IEEE International Conference on Power System Technology; 2016. p. 1–6.

[73]   IEC. Communication networks and systems for power utility automation part 90-1: Use of IEC 61850 for the communication between substations. IEC; 2010. IEC/TR 61850-90-1.

[74]   Wang J, Ou Q, Shen H. An overview of smart grid routing algorithms. In: AIP Conference Proceedings; 2017. p. 1–6.

[75]   Gao Y. Performance and Applicability of Candidate Routing Protocols for Smart Grid's Wireless Mesh Neighbor Area Networks. Montreal: McGill University; 2014.

[76]   Perez-Guzman RE, Salgueiro-Sicilia Y, Rivera M. Communications in smart grids. In: Conference on Electrical, Electronics Engineering, Information and Communication Technologies; 2017. p. 1–7.

[77]   Tsado Y, Lund D, Gamage KAA. Resilient Communication for Smart Grid Ubiquitous Sensor Network: State of the Art and Prospects for Next Generation. Computer Communications. 2015;71:34–49.

[78]   Saputro N, Akkaya K, Uludag S. A Survey of Routing Protocols for Smart Grid Communications. Computer Networks. 2012;56(11):2742–2771.

[79]   Annaswamy, AM. IEEE vision for smart grid controls: 2030 and beyond. New York, NY: Institute of Electrical and Electronics Engineers; 2013. OCLC: 861074253.

[80]   IEEE Standards Committee, IEEE Standards Coordinating Committee 21 on Fuel Cells P Dispersed Generation and Energy Storage, Institute of Electrical and Electronics Engineers, *et al.* IEEE Guide for Smart Grid Interoperability of Energy Technology and Information Technology Operation with the Electric Power System (EPS), End-Use Applications and Loads. New York, NY: Institute of Electrical and Electronics Engineers; 2011. OCLC: 756647294.

[81]   Xu S, Qian Y, Hu RQ. On Reliability of Smart Grid Neighborhood Area Networks. IEEE Access. 2015;3:2352–2365.

[82]　Ikpehai A, Adebisi B, Rabie K. Broadband PLC for Clustered Advanced Metering Infrastructure (AMI) Architecture. Energies. 2016;9(7):1–19.

[83]　Cano C, Pittolo A, Malone D, *et al.* State of the Art in Power Line Communications: From the Applications to the Medium. IEEE Journal on Selected Areas in Communications. 2016;34(7):1935–1952.

[84]　Tonello AM, Pittolo A. Considerations on narrowband and broadband power line communication for smart grids. In: IEEE International Conference on Smart Grid Communications. IEEE; 2015. p. 13–18.

[85]　Goldsmith A. Wireless Communications. New York, NY: Cambridge University Press; 2005.

[86]　Proakis JG, Salehi M. Digital Communication. 5th ed. New York, NY: McGraw-Hill; 2007.

[87]　Galli S, Scaglione A, Wang Z. Power line communications and the smart grid. In: IEEE International Conference on Smart Grid Communications; 2010. p. 303–308.

[88]　Schwartz M. Carrier-Wave Telephony Over Power Lines-Early History. IEEE Communications Magazine. 2009;47(1):14–18.

[89]　Dostert K. Telecommunications over the power distribution grid – possibilities and limitations. In: International Symposium on Power-Line Communications and its Applications; 1997. p. 1–9.

[90]　Peck M, Alvarez G, Coleman B, *et al.* Modeling and analysis of power line communications for application in smart grid. In: International Multi-Conference for Engineering, Education, and Technology Conference; 2017. p. 1–6.

[91]　Andreadou N, Guardiola M, Fulli G. Telecommunication Technologies for Smart Grid Projects with Focus on Smart Metering Applications. Energies. 2016;9(5):1–35.

[92]　Papadopoulos TA, Kaloudas CG, Chrysochos AI, *et al.* Application of Narrowband Power-Line Communication in Medium-Voltage Smart Distribution Grids. IEEE Transactions on Power Delivery. 2013;28(2):981–988.

[93]　Castor LRM, Natale R, Favero JP, *et al.* The Smart Grid Concept in Oil & Gas Industries by a Field Trial of Data Communication in MV Power Lines. Journal of Microwaves, Optoelectronics and Electromagnetic Applications. 2016;15(2):81–92.

[94]　Berger LT, Schwager A, Pagani P, *et al.* MIMO Power Line Communications. IEEE Communications Surveys & Tutorials. 2015;17(1):106–124.

Chapter 10

An overview of quad-generation system for smart grid using PLC

Muhammad Kashif[1], Muhammad Naeem[1], Muhammad Iqbal[1], Waleed Ejaz[2], and Alagan Anpalagan[3]

Electricity produced through conventional power systems is both expensive and inefficient. A portion of useful fuel is wasted as heat and greenhouse gas emissions (GHGEs). Therefore, conventional power generation systems are not only inefficient but also causing environmental pollution. In conventional power systems, different energy requirements, namely, daily electricity consumption, chiller consumption, and heating are usually supported by purchasing electricity from the utility. Contrary to the conventional power systems, in combined heating and power (CHP) systems, different energy requirements are satisfied from a generation system coupled with a heat recovery system. After the integration of few thermally activated technologies, namely, absorption and adsorption chillers into the CHP, the system is transformed into combined cooling, heating, and power (CCHP) system which is more efficient as compared to CHP system. A CCHP system can be further integrated with the carbon dioxide extractor to transform the CCHP system into a new generation of power system called quad-generation system. Complex integration of various components requires careful selection of their rated capacity and the number of equipment giving rise to the various optimization formulations. In this chapter, we review different optimization formulations being used to optimize the design, operation, and planning of the CHP, CCHP, and quad-generation systems. Mathematical formulations of commonly used objective functions, namely, cost minimization, efficiency maximization, and GHGEs minimization have been elaborated. We also present various optimization algorithms and the simulation tools being used to solve the optimization formulations. The chapter can serve as a foundation stone for the beginners in this research area. It can also serve as a guide for the practitioners to optimally design, deploy, and operate the multigeneration power systems.

[1]Department of Electrical Engineering, COMSATS Institute of Information Technology-Wah, Pakistan
[2]Department of Electrical and Computer Engineering, Sultan Qaboos University, Oman
[3]Department of Electrical and Computer Engineering, Ryerson University, Canada

10.1 Introduction

The concept of smart grid opens many corridors in terms of energy transfer. This energy can be in the form of electrical energy, heat energy, etc. One can get the optimal benefit of the smart grid if we use features of smart grid for CCHP systems. The concept of smart grid-enabled decentralized energy system has facilitated the use of CCHP systems. The CCHP systems are not only more energy efficient but also help to reduce GHGEs [1–5]. The idea of CCHP can be traced from the concept of combined heating and power (CHP) systems being used in large-scale centralized power plants and other commercial buildings [6]. CCHP is now becoming a key sector to the development of smart grid. The CCHP systems are more energy efficient as compared to conventional power systems [7]. Figure 10.1 shows an illustration of conventional and CCHP systems. It is assumed that any type of fuel can be used for the conventional and CCHP systems. In conventional power systems, different energy requirements, namely, daily electricity consumption, chiller consumption, and heating, are usually met by purchasing electricity from the utility. Conventional electricity production systems are inefficient in utilization of fossil fuels [8]. Much of this costly fuel is being wasted in the form of heat, and only a portion of fuel energy is converted into electricity. Contrary to the conventional power systems, in CHP systems, different energy requirements are satisfied from a generation system coupled with a heat recovery system [9]. Unmet demands can be satisfied through the electricity purchased from the utility or the nearby local grid. After the addition of few thermally activated technologies, namely, absorption and adsorption chillers into the CHP, the system is transformed into CCHP system [10] which is also known as a trigeneration system. A typical CCHP system can achieve 50% higher efficiency as compared to a CHP plant of the same capacity [11]. The CCHP system can be integrated with carbon dioxide extractor which transforms it into a new generation of power system known as quad-generation system.[1] This includes not only the combination of heating, cooling, and electricity but it also extracts carbon dioxide from exhaust gases [12]. Figure 10.2 shows a typical CCHP system that generates electricity, in which wasted heat is extracted to fulfill the heating and cooling demands by using absorption chillers. CCHP and trigeneration systems consist of various subsystems including power generation equipment, absorption chillers/heat pump, and heat storage subsystem [13]. Therefore, these systems are the integration of many thermal processes, energy supply subsystems, and energy/mass conversion subsystems. It is, therefore, imperative that the optimal utilization of fuel/energy requires optimization of power production, cost, reliability, and sizing of the CCHP systems.

One issue in the data transfer of CCHP smart grid system is the selection of communication technology. One option is the power line communication (PLC). PLC can play a key role in the trading of electricity between the smart buildings [14–18]. PLC has a decade of history and tradition to be used in electric companies with a number of different applications and implementations. PLC systems can work as an

---

[1] http://www.clarke-energy.com/gas-engines/quadgeneration/.

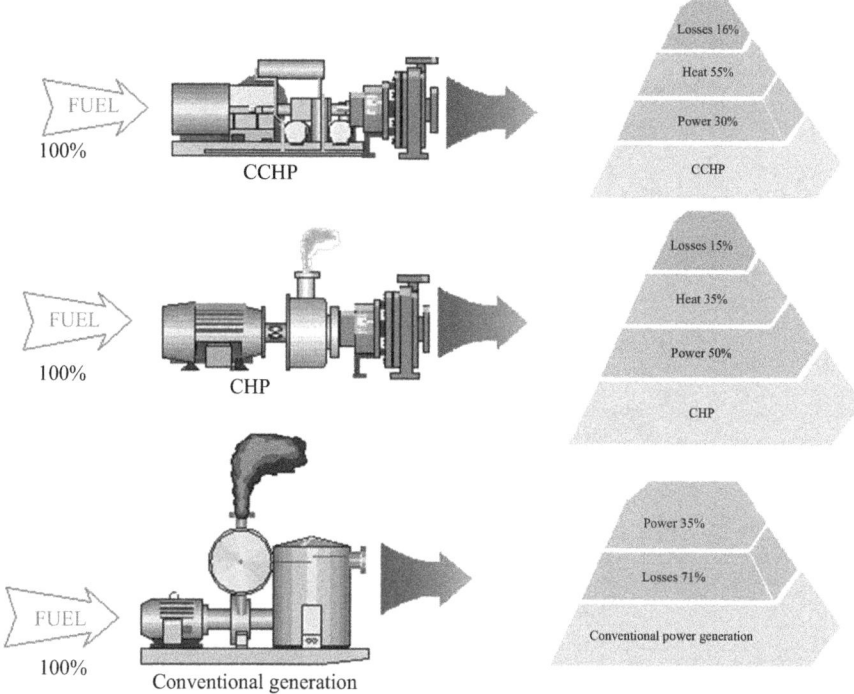

Figure 10.1   Conventional single generations versus poly-generation techniques

ultra-narrow-band system with the range of 0.3–3 kHz or narrow-band system from 3 to 500 kHz [18]. The data rate of PLCs can be from few kbps to hundreds of kbps. Another advantage of PLC is that the issue of congested spectrum was overcome, which helps the information to flow along the same path as power, hence simplify the deployment of smart grid based CCHP system.

Various optimization strategies have been proposed for the optimization of poly-generation systems in order to come up with optimal selection of prime movers, thermally activated technologies, system configuration, system management, and sizing. For example, an optimization strategy has been suggested in [19] to optimize the size of the poly-generation system powered by natural gas, solar energy, and gasi-fied biomass. The authors used mixed integer nonlinear programming (MINLP) to solve the optimization formulation for finding the trade-off between primary energy savings, GHGEs, and economic viability. A trigeneration system has been modeled as a MINLP problem in [20] for optimal heating of several buildings. Long-term optimization problem of a CCHP system in a hospital environment has been modeled as a MINLP formulation in [21] to obtain optimal size and operational conditions. The authors argued that the optimized sizing of heat pumps could result in better economic, energetic, and environmental aspects of the system. Thermoeconomical

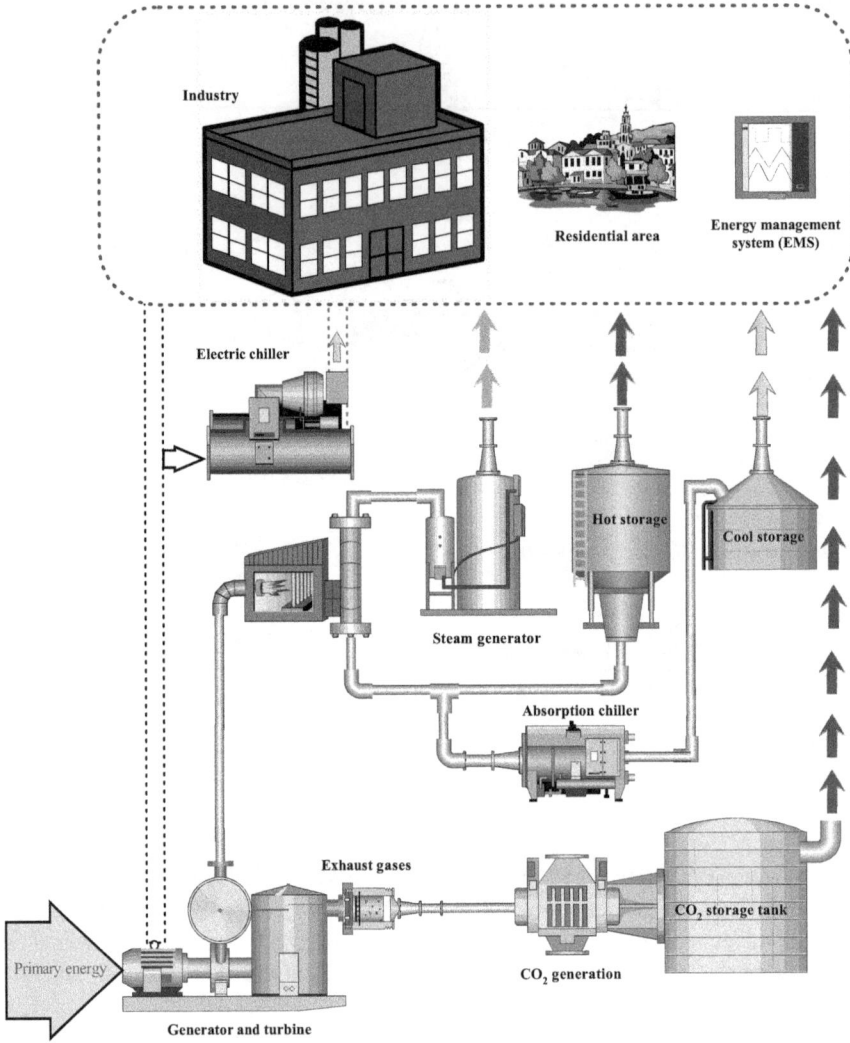

Figure 10.2   Quad-generation system showing integration of CCHP and carbon dioxide extractor

optimization of a distributed trigeneration system has been investigated in [22] by considering thermodynamic, economic, and GHGE aspects of the system. The optimization problem focused on system configuration and operation strategy with the objective function of net present value using a MINLP formulation. In [23], the authors formulated a trigeneration system using a fuzzy multiobjective optimization strategy to find optimal configuration of the system.

Literature is brimming with numerous types of optimization formulations being used to model different optimization problems relating to CCHP systems. This chapter presents an updated review of the optimization techniques being used to optimize CCHP systems for the objective of finding the optimal type/number of prime movers, thermally activated technologies, system configuration, system management, and sizing. The chapter provides an insight into varying degree of preferences for different conflicting objectives. Various real-life optimization formulations relating to CCHP systems are reviewed, and the corresponding solution approaches are compared to conclude their applicability and scalability. Therefore, it can serve as a guideline to configure CCHP systems for different trade-offs between various performance parameters depending upon the application environment of the systems.

A snapshot of the previous reviews/surveys on the topic is depicted in Table 10.1. It can be inferred that the existing surveys do not encompass the subject completely. For example, [24–30] have not considered the optimization aspect of the poly-generation systems rather presented the review with respect to different energy generation technologies and application scenarios. Feasibility of micro-CHP systems has been considered in [31] to meet the energy demand of household consumers. The authors reviewed numerous energy generation types including micro gas turbines, micro-Rankine cycles, Stirling engines, and thermophotovoltaic generators for CHP requirements. A comprehensive review of CCHP systems is presented in [32], where the authors have also depicted operational strategies to optimize the system performance and improve its overall efficiency.

Despite there exists abundant literature on optimization of renewable energy in smart grid, the study and optimization of fossil fuel resources for power and heat in the smart grid context are also important and so on. Therefore, this chapter reviews the

Table 10.1 Existing surveys related to the subject

Reference	Review type			Energy type		
	Optimization	TS	AS	Power	Heating	Cooling
[24]			✓			✓
[25]						✓
[26]		✓	✓	✓		✓
[27]					✓	✓
[28]				✓	✓	
[29]		✓	✓		✓	✓
[30]		✓		✓	✓	
[31]	✓			✓	✓	
[33]				✓	✓	
[34]					✓	✓
[35]				✓	✓	✓
[32]	✓			✓	✓	✓
[36]				✓		✓
[37]	✓			✓		✓

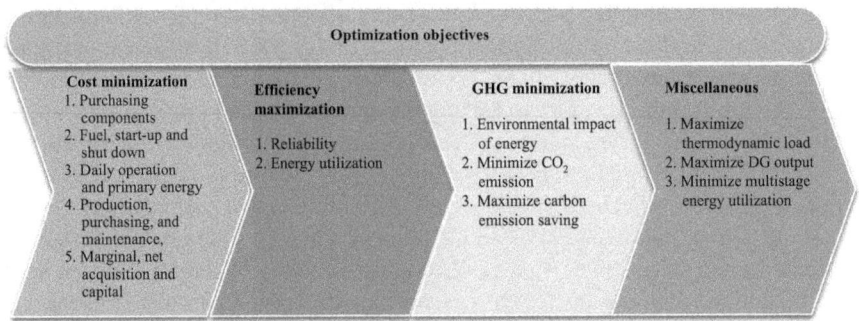

Figure 10.3   Objective functions of CHP and CCHP

recent work related to the application of optimization techniques in the area of CCHP. The existing literature in this research area has been classified with respect to different objective functions, different optimization algorithms, different solution types, and related tools. We also compare different solution algorithms and relevant tools to facilitate the readers/practitioners to decide which optimization technique to be used in a specific scenario.

The rest of the chapter is organized as follows: Section 10.2 presents commonly used objective functions for the optimization of CHP and CCHP. Types of optimization techniques are elaborated in Section 10.3, and solution approaches along with simulation tools are depicted in Section 10.4. Finally, Section 10.5 concludes the paper by reflecting some future research directions.

10.2   Objective functions being used for the optimization of CHP and CCHP

CHP and CCHP systems are an intricate integration of various components and subsystems. Therefore, these require complex mathematical modeling and optimization techniques to efficiently design, plan, deploy, and operate them. These systems are optimized for various objective functions including but not limited to the cost minimization, efficiency maximization, and GHGEs minimization as shown in Figure 10.3. Detailed formulations and description of objective functions are given in the following subsections.

10.2.1   Cost minimization and economic analysis

Minimization of capital and operational costs are critical objectives while optimizing the design of any project/system. Economic feasibility of any project against the capital cost can be measured by using various criteria. For example, net present worth (NPW) of the system and payback period are two commonly used criteria to measure the feasibility of the project [38]. In [39], the authors have investigated the optimization of a CHP system based on steam turbine by using the criteria of NPW which has been modeled as follows:

$$NPW = \sum_{j=1}^{K} \left\{ \frac{SV_j}{(1+i)^{LT}} - CC_j \right\} - \sum_{y=1}^{LT} \left(\frac{1}{(1+i)^y} \right) CP$$

$$+ \sum_{m=1}^{N} \left[\sum_{j=1}^{k} (OM_j + COF_j) + Pwt_b \times C_{el,b} - Pwt_{CHP,\tau} \right.$$

$$\left. \times C_{el,b} - H_{CHP,\tau} \times C_h - Pwt_{CHP,s} \times C_{el,s} \right]_m \times \tau_m, \tag{10.1}$$

where LT is the lifetime in years, SV is the salvage value in dollars, T is the temperature, PWB is the present worth of benefits in dollars, CC is the capital cost, $C_{el,b}$ is the buying/selling electricity, COF is the cost of fuel in dollars, $C_{el,s}$ is the selling electricity, Cel is the cost of generated electricity, Ch is the cost of generated heat, K is the number of equipment, τ_m is the time interval of demand profile in hours, H is the heart, and P is the electricity in watts. Further, NPW is a cost indicator calculating the difference between the dollar value of the present worth of benefits and the present worth of costs. The higher value of *NPW* represents higher feasibility of the project. A negative *NPW* indicates that the project is uneconomical. On the other hand, N depicts the amount of time required to recover the capital costs of the system.

Minimization of the operation cost of a typical CCHP system has been addressed in [40], where the authors have formulated the minimization of the operational cost as under:

$$COST = (F_{pgu} + F_{boiler})(C_f + \mu_f C_c) + E_{grid} C_e + \frac{R}{365} \sum_{j=1}^{m} N_j C_j, \tag{10.2}$$

where F_{boiler} is the fuel consumption of auxiliary boiler (kW h), F_{pgu} is the fuel consumption of PGU (kW h), C_f is the unit prices of natural gas, E_{grid} is the electricity provided by power grid (kW h), μ_f is the emission factor of natural gas (g/kW h), C_c is the unit carbon tax (dollar/g), C_e is the unit price of grid electricity (dollar/kW h), C_j is the initial capital cost of jth equipment (dollar/kW h), and N_j is the installed capacity of jth equipment (kW h). Here, *COST* is an operational cost criterion.

Economic, energy, and environmental analysis has been performed in [41] to find the optimal number and power of prime movers. The authors suggested an optimization strategy by defining an objective function called the relative annual benefit (RAB) which is given as follows:

$$RAB \left(\frac{\$}{year} \right) = TAC_{trad} - TAC_{CCHP}, \tag{10.3}$$

where TAC_{CCHP} is the total annual cost of CCHP system and TAC_{trad} represents the cost for production of cooling, heating, and power energies in traditional system.

A grid connected trigeneration plant has been investigated in [42] to minimize the total energy cost and plant maintenance cost. The minimization of cost has been modeled as under:

$$G(K) = \sum_{t=1}^{N} C(t, K),$$
(10.4)

where $C(t, K)$ collectively represents the fuel cost, maintenance cost, and the system on–off cost, and N is the number of time steps in the whole simulation period.

An optimization formulation has been proposed in [43] to find the optimal capacities for the CHP and boiler while satisfying the electrical and thermal demands with high cost efficiencies. The proposed formulation is given in the following equation:

$$G(K) = \max \left(\sum_{j=1}^{N} \frac{d^j}{1+p} \left[\sum_{i=1}^{8,760} c_{3ij} e_{ij} + \sum_{i=1}^{8,760} h_{ij} - f_2 \right] - f_1 - f_3 \right),$$
(10.5)

where c_{3ij} is the cost of electricity imported from EG (dollar/kW h), e_{ij} is the electrical demand in time ith and year jth (kW h), f_1 is the operation strategy cost in dollars, f_2 is the O&M cost in dollars, f_3 initial investment cost in dollars, and N is the investment lifetime in years. In the above equation, the upper limit of the summation, 8,760, is the number of hours in 1-year period.

CHCP system design optimization has been proposed in [44] to minimize the total annual cost. The formulated objective function for the total annual cost minimization is given as under:

$$Ctot = C_{fix} + C_{var} = \sum_{i} I_i + \sum_{j} c_j X_j,$$
(10.6)

where C_{fix} is the equipment amortization, maintenance cost, C_{var} is the annual energy cost, I_i is the purchase and installation cost of the ith equipment, c_j is the cost in dollars per kW h, and X_j is the energy flow. The authors demonstrated the proposed model by a case study of a set of buildings constituted of 5,000 apartments located in Spain. The authors argued that the significant reduction in the annual energy cost with a payback period less than 4 years in relation to the conventional energy system was possible.

Optimization formulation for the cost minimization of a trigeneration system has been proposed in [45]. The hourly trigeneration planning model for the simultaneous minimization of production and purchase costs as well as carbon dioxide emissions costs is given as follows:

$$\sum_{j \in J} c_j x_j + c^{p-} x^{p-} + c^{p+} x^{p+} + c^{q-} x^{q-} + c^{q+} x^{q+} c^{r-} x^{r-} + c^{r+} x^{r+}$$
(10.7)

where x^{p+}, x^{q+}, and x^{r+} are surplus variable used to absorb any excess of the three products, namely, cooling, heating, and power. Similarly slack variables x^{p-}, x^{q-}, and x^{r-} are used to compensate for the lack of any of the three commodities and c^{p-}, c^{q-}, c^{r-}, c^{p+}, c^{q+}, and c^{r+} are present in the objective function to represent the penalties for the slack or surplus in the energy balance.

A trigeneration system has been investigated in [46] to find the optimal configuration of the system for maximizing the annual savings. The authors have introduced a terminology called "gross operational margin (*GOM*)" to represent the annual saving.

$$\langle maxGOM \rangle = \sum_{i=1}^{3} \sum_{j=1}^{2} \sum_{k=1}^{12} (1 - I)\{R_{ij(k)} - C_{ij(k)}\}\Delta td_{ij}, \tag{10.8}$$

$$\langle maxNPV \rangle = PVF \sum_{y=1}^{n} \{GOM_{(y)} + DEPR_{(y)}I\} - INV, \tag{10.9}$$

where *PVF* represents the present value factor, *DEPR* depicts the economic depreciation, *INV* represents the investment costs, and *R* represents the returns. Indices *i*, *j*, and *k* represent division of solar months into winter, intermediate, and summer as working or nonworking, respectively.

10.2.2 Energy efficiency maximization

Contrary to the conventional approach of producing heat and power in separate plants, CHP or CCHP systems are potentially more fuel economical [47]. Various studies have been conducted to investigate different optimization formulations of CHP systems for the maximization of energy efficiency. For example, in [48], the authors have set up a natural gas fueled CHP system based on internal combustion engine (ICE) at the building energy research center in Beijing, China. The authors showed that the new CHP system could increase the heat utilization efficiency by 10% as compared to conventional systems. A trigeneration system has been analyzed in [49] to compare the performance of the system in terms of primary energy saving and energy conversion efficiency. The authors have formulated the primary energy saving of a trigeneration system as under:

$$TPES = \frac{F^{SP} - F^{CCHP}}{F^{SP}}, \tag{10.10}$$

where F^{SP} and F^{CCHP} are the total fuel energy input to the trigeneration system and the total fuel energy input required for the separate production of the same energy vectors, respectively.

A CHP plant based on coal has been investigated for energetic and exergetic efficiencies in [50]. The authors compared the efficiency of the CHP plant with that of the separate heat and power plant. Exergetic efficiency ε^{tm} and energetic efficiency Ψ are modeled as under:

$$\varepsilon^{tm} = h - h_0 - T(s - s_0), \tag{10.11}$$

$$\Psi = \frac{E_{Wchp} + E_{Qchp}}{E_F}, \tag{10.12}$$

where *h* and *s* are the specific enthalpy in (J/kg) and entropy of the stream, respectively. E_F is the rate of fuel energy in [LHV (W)], E_{Qchp} is the rate of heat (W), *T* is the temperature, and E_{Wchp} is the rate of work or electric energy in (W). In the exergy analysis, the exergetic efficiency Ψ is an indicator of the performance of a system or a component.

The efficiency of a plant is measured for legal purposes by the power energy saving (PES) index, which quantifies the *PES* of a combined, heat, cooling, and power (CHCP) plant with respect to the conventional separate generation of electricity and heat. In [51], an optimization formulation has been proposed for the optimal design of a CHCP plant to maximize the PES.

$$PES = 1 - \frac{1}{(\eta_E/Ref_E) + (\eta_V/Ref_V)},$$ (10.13)

where Ref_E is the reference for electricity efficiency, Ref_V is the reference for thermal efficiency, η_E is the electricity efficiency of a CHCP system, and η_V is the total useful thermal energy efficiency.

In practical scenarios, the load conditions vary with time, which make the efficiency of the plant dependent on the operation strategy. Therefore, it is imperative to optimize the operation strategies of CCHP for varying load conditions. In [52], the authors have formulated a multiobjective optimization strategy to optimize the operation of a CCHP for the objective function of the energy saving ratio (ESR) and the cost saving ratio (CSR).

$$ESR_{opt} = \max \frac{Q_{f,conv} - Q_{f,cchp}}{Q_{f,conv}},$$ (10.14)

$$CSR_{opt} = \max \frac{COST_{conv} - COST_{cchp}}{COST_{conv}},$$ (10.15)

where $COST_{cchp}$ represents the operation cost of CCHP and $COST_{conv}$ depicts cost of the conventional separate generation system.

A general model of a complex CCHP system has been considered in [53] which is based on theory of exergy cost and structural coefficients of internal links. The electric efficiency of the system is modeled as under:

$$\eta_{ele} = \frac{W_{ICE} - W_C}{Q_{inLHV}} \times 100\%,$$ (10.16)

where W_{ICE} is the power output generated by the ICE, W_C is the system power consumption, including the fans and pumps of the system, and Q_{inLHV} is the natural gas input.

A micro-CCHP system is proposed and investigated in [54] to optimize the performance of the system. Primary energy ratio (PER) has been used to measure the efficiency of the micro-CCHP system as follows:

$$PER = \frac{P_{el} + Q_h + Q_r}{Q_f},$$ (10.17)

where P_{el}, Q_h, Q_r, and Q_f represent electric power output, heating output of the system, cooling output of the system, and fuel input of the system, respectively.

Load forecasting is critical for the optimization strategies for large energy systems on dynamic basis. Various load forecasting models have been evaluated in [55] for their ability to accurately forecast hourly loads for a district energy system up to 24 h in advance using weather conditions as input. Autoregressive model with exogenous inputs (ARX) is one of the models evaluated. ARX autoregressive model with exogenous input model uses past load information to predict future loads, making it recursive in nature. Exogenous inputs such as weather or time can be added to improve model accuracy by representing the dependence of the load on these additional inputs. The model is given as follows:

$$L_k = \sum_{i=1}^{N} a_i L_k + \sum_{i=0}^{N} b_i \theta_{k-i} + c_{h,k} + c_{d,k} + c_{m,k} + d, \tag{10.18}$$

where the parameters corresponding to different hours, days, or months. The linear load-forecasting model takes on the form shown in (10.20), where a, b, and c are the model fitting coefficients, L_k is the load at time k, N is the model order (i.e., the number of time steps back to retrieve inputs), and q represents the values of weather variables. The subscripts h, d, and m refer to the hour of the day, day of the week, and month of the year, respectively. Therefore, there are different additive constants ($c_{h,k}$, $c_{d,k}$, and $c_{m,k}$) for each time period.

Parametric optimization of a CCHP system is conducted in [56] to maximize the efficiency of the system. The system was composed off a Rankine cycle and an ejector refrigeration cycle to produce cooling output, heating output, and power output simultaneously. The authors evaluated the overall performance of the system by the thermal efficiency and exergy efficiency which are given as under:

$$\eta_{thm} = \frac{W_{NET} + Q_E + Q_H}{Q_U}, \tag{10.19}$$

where W_{NET} is the power output from the turbine, reduced by the power input to the pump, Q_H is the heat output, Q_E is the cooling output, and Q_U is the total heat added to the system from the solar collectors. Similarly, exergy efficiency can be defined as the exergy output divided by the exergy input to the overall system, which can be modeled as follows:

$$\eta_{exg} = \frac{W_{NET} + E_E + E_H}{E_U}, \tag{10.20}$$

where E_U is the exergy input to the overall system from the solar collector, E_E is the exergy associated with the refrigeration output, which is calculated as the working fluid exergy difference across the evaporator and E_H is the heat exergy output.

10.2.3 GHGEs minimization

Buildings usually use large amount of resources namely, land, energy, water, and air. In the United States, 40% of the energy is consumed in buildings to satisfy the cooling, heating, and power requirements. Similarly, 40% of the carbon dioxide emissions are contributed by the buildings [57]. Therefore, the environmental efficiencies of building cooling, heating, and power (BCHP) systems require optimal design and operation strategies. For example, in [58], the authors have analyzed the energy consumptions of a traditional separation production (SP) system and BCHP system, respectively, and formulate their corresponding environmental impact model. The authors formulated the optimization problem relating to BCHP for the maximization of the primary ESR (PESR) and CO_2-equivalent emission reduction ratio (EERR), i.e., CO_2-EERR as under:

$$PESR = \frac{F^{SP} - F^{BCHP}}{F^{SP}} \times 100\%, \qquad (10.21)$$

$$CO_2 - EERR = \frac{\alpha^{SP} - \alpha^{BCHP}}{\alpha^{SP}} \times 100\%, \qquad (10.22)$$

where F^{SP} and F^{BCHP} are the life cycle energy consumptions for SP and BCHP, respectively. α^{SP} and α^{BCHP} are the equivalent CO_2 production because of SP and BCHP, respectively.

Environmental impact of BCHP system for a commercial building in China has been analyzed in [59] with respect to global warming, acid precipitation, and ozone layer depletion. The authors formulated the environmental impact of the SP system and the BCHP system and then compared the emission reduction potential of the BCHP system over SP system in different weather conditions. Emissions from BCHP system are modeled as under:

$$M_{total}^{BCHP} = X_p + Xpgu + X_b + Xch + \mu_c \cdot F_c + \mu_g \cdot F_{pgu}$$
$$+ \mu_g \cdot F_b + X_{ch} + \mu_c \cdot F_c + \mu_g \cdot F_g + X_{ch}, \qquad (10.23)$$

where X_p, X_{pgu}, X_b, and X_{ch} are the emission vectors from the central power plant, the recovered heat system, the gas boiler, and the absorption chiller, respectively. μ is the emission factor. F_c and F_g are the energy consumptions of coal and gas, respectively. F_{pgu} and F_b are the gas consumptions of the PGU and the boiler, respectively.

The authors were of the view that the BCHP system is more energy efficient in the colder weather or cold zone as compared to SP system, whereas in the hot summer and warm winter zones, BCHP system consumes more energy as compared to the SP system.

A multiobjective design and operation strategy has been proposed in [60] to optimize the capacity of BCHP system based on gas engine for the objective functions of global warming, acid precipitation, and respiratory effects. The optimization formulation for the objective function of primary energy saving ratio (PESR) is given as follows:

$$PESR = \frac{F_{IC}^{SP} - F_{IC}^{BCHP}}{F_{IC}^{SP}}, \qquad (10.24)$$

where F_{IC}^{SP} and F_{IC}^{BCHP} are the life cycle energy consumptions of the SP system and the BCHP system, respectively. Positive PESR reflects that the BCHP system is more energy efficient as compared to the SP system. Contrarily, the negative PESR will represent that the BCHP system is less energy efficient as compared to the SP system.

Similarly, global warming potential reduction ratio (GWP-RR) of BCHP system as compared to the SP system is given as follows:

$$GWP - RR = \frac{CO_2 - equiv_{IC}^{SP} - CO_2 - equiv_{IC}^{BCHP}}{CO_2 - equiv_{IC}^{SP}}, \quad (10.25)$$

when GWP-RR is positive, the global warming impacts caused by BCHP systems are relatively less severe than those of the SP system. Conversely, when GWP-RR is negative, BCHP system is unable to reduce the CO_2-equivalent emissions as compared to the SP system.

A multiobjective optimization formulation has been proposed in [61] to optimize a trigeneration system for the objective functions of economic and environmental impacts. The proposed formulations are given as follows:

$$\min C_{tot} = C_{fix} + C_{ope}, \quad (10.26)$$

$$\min CO_{2tot} = CO_{2fix} + CO_{2ope}, \quad (10.27)$$

$$\min SS_{tot} = SS_{fix} + SS_{ope}, \quad (10.28)$$

where SS_{fix} is the annual fixed impact of the equipment, SS_{tot} is the total annual impact, SS_{ope} is the annual operation impact, CO_{2tot} is the total annual emissions, CO_{2fix} represents annual fixed emissions of the equipment, CO_{2ope} depicts annual operation emissions, and C_{tot} represents the total annual cost.

10.3 Optimization types used in CCHP

Aforementioned optimization formulations relating to CCHP can be solved using various optimization techniques. Figure 10.4 depicts various types of optimization techniques being used to solve the optimization problems relating to CCHP. The preceding subsections discuss the relevant work related to each specific technique.

10.3.1 Linear programming

A @@linear programming (LP) problem can be defined as the problem of maximizing or minimizing a linear function subject to linear constraints [62]. These constraints may be equalities or inequalities. LP uses a mathematical model to describe the problem of concern [63]. In LP, all the mathematical functions are required to be linear. LP involves the planning of activities to obtain an optimal result [64].

Optimization algorithms used to solve real-world problems relating to CCHP systems that are usually divided into LP and nonlinear programming. For example, in [65], an optimization model based on hourly load forecasts has been proposed for the optimal planning of a trigeneration system. The authors model the hourly trigeneration system as a linear programming model with a joint characteristic of three energy components to simultaneously minimize the production and purchase costs of the system along with CO_2 emissions costs. The problem of optimal energy management of cogeneration system consisting of combined cooling, heating, and

Figure 10.4   Types of optimization used in CCHP system

power production has been addressed in [3] to minimize the overall cost of energy for the CCHP system. The proposed CCHP system consists of a gas turbine, an absorption chiller, and a heat recovery boiler. The authors were of the view that the optimal system operation depends upon the load conditions being satisfied. An economic unit dispatch problem of a cogeneration system consisting of CHP, compression heat pump, and cold storage using flue gas heat has been modeled as linear programming problem in [66] to maximize the electricity sale price and minimize the energy purchase price.

10.3.2   NLP and MINLP

Nonlinear programming (NLP) refers to that class of optimization problems in which decision variables are continuous. On the other hand, mixed integer nonlinear programming (MINLP) refers to the optimization problems with continuous and discrete variables and nonlinear functions in the objective function and/or the constraints [67]. Many real-world problems relating to optimization of CCHP systems can be formulated as MINLP problems. For example, a MINLP formulation has been proposed in [40] to find the optimal operation strategy and the corresponding implemented decision-making process. The authors were of the view that the proposed formulation could result in reduction of primary energy consumption, carbon dioxide emissions, and operational costs.

A natural gas CHP system based on an ICE is described in [48], which has been set up at the building energy research center in Beijing, China. The system is composed of an ICE, a flue gas heat exchanger, a jacket water heat exchanger, and other auxiliary facilities. The author used MINLP approach to improve the heat utilization efficiency. As the practical load conditions vary with time, the performance of micro-CCHP system depends a lot on the operation strategy, with the equipment and energy balance constraints. A MINLP formulation has been used in [52] to find the optimal operation strategy to obtain the energy saving optimization and cost saving optimization. A mixed integer linear programming (MILP) model to determine the optimal capacity and operation of seven CCHP systems in the heating and cooling network of a residential district in east Tehran has been suggested in [68]. The authors argued that the economic and environmental results achieved from the case study could revealed saving in costs and reduction in CO_2 emissions in the optimal cogeneration system as compared to the traditional systems.

10.3.3   BIP, DP, and MILP

Binary integer linear programming also known as binary integer programming (BIP) refers to that class of optimization problem in which decision variables are binary integers. Backward dynamic programming (DP) is used in [42] to determine the economically optimal plant state, i.e., the set of the components loads to minimize the total cost. On the other hand, MILP theory and practice has been appreciably flourished and is now a paramount tool for modeling various engineering problems [69,70]. Due to peculiar characteristics of MILP, many real-world optimization problems relating to CCHP can be modeled as MILP problems. For example, in [71], the optimization of a CCHP-based waste heat recovery system has been modeled as MILP problem for the maximization of profit, minimization of the primary fuel cost, and maximization of production. In [72], MILP formulation is used to optimize the CCHP system for maximization of energy saving, reduction of annual fuel cost, and to minimize the capital and maintenance cost of the system.

The optimization of synthesis, design, and operation of trigeneration systems for building applications is a quite complex task, due to the high number of decision variables, the presence of irregular heat, cooling, and electric load profiles, and the variable electricity price. The related optimization problem has been formulated as MILP in [73] to address the aforementioned issues relating to CCHP system. The optimal operation strategy has been proposed in [74] using MILP while considering the cost allocation to all cogeneration products of a CCHP system developed for a hospital with 500 beds located in Spain.

In [75], the authors have proposed a MILP model for developing an accurate simulation model applicable to building clusters for energy sharing and competition. A distributed energy system has been investigated in [76] to obtain optimal investment and unit sizing by using MILP formulations.

10.4   Solution approaches and tools used to solve optimization problems related to CCHP

In this section, we will discuss solution approaches and tools used to solve optimization problems related to CCHP.

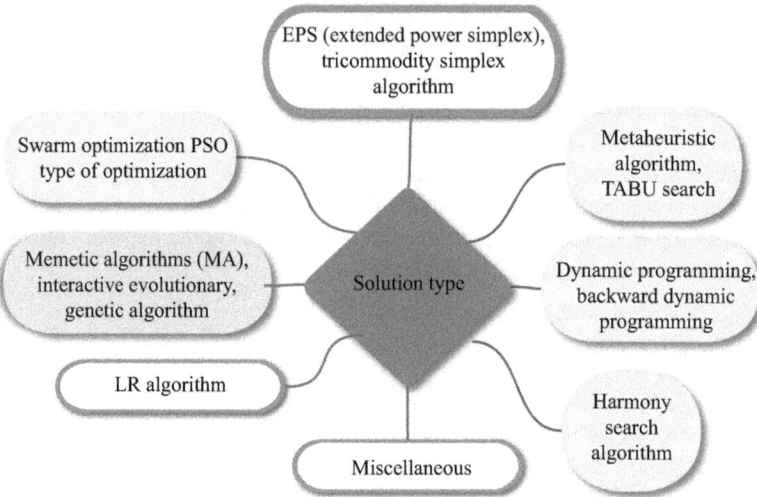

Figure 10.5　Solution type used in CHP and CCHP

10.4.1　Solution approaches

Literature is brimming with numerous solution approaches including but not limited to tricommodity simplex (TCS), memetic algorithms, ant colony optimization, genetic algorithm (GA), harmony search (HS), backward DP, and evolutionary algorithm. Figure 10.5 presents various types of solution approaches used to solve optimization problems relating to CHP and CCHP. For example, in [77], the authors used HS algorithm to solve a multiobjective optimization model to obtain a trade-off between various desirable objectives, namely, the low emissions and energy efficiency. TCS algorithm has been used to solve the problem of optimal planning of trigeneration to minimize the energy production and purchase costs, as well as CO_2 emission cost during the planning horizon [78]. TABU search technique is suggested in [79], which is a form of metaheuristic search utilizing local search methods to obtain the optimality condition.

Optimization of capacity and operation of a CCHP system has been addressed by using GA in [80]. A modified dichotomic search algorithm for a general biobjective LP problem considering a fuel mix setting has been proposed in [81] to find different trade-off solutions between various conflicting objectives.

10.4.2　Tools used to solve CCHP

Numerous tools are being used to solve the optimization problems relating to CHP and CCHP. For example, in [55], the authors have used MATLAB® to solve the optimization formulation for the optimal forecasting of the loads on hourly basis, whereas

CPLEX[2] is used to solve optimization problems involving integer and continuous variables. CPLEX is also highly recommended to obtain convex and nonconvex quadratic objectives. Basically, CPLEX is termed as multipurpose commercial software that goes a long way in solving huge mathematical programming problems. In [78], the authors have used CPLEX to solve the optimization formulation for the planning of a trigeneration system. A CCHP system consisting of gas turbine combined cycle unit has been analyzed in [82] by using an application specific heat balance engineering software called GT PRO.[3] The energy price and carbon balance scenarios (the ENPAC tool) has been used in [83] to maximize the efficiency of the heat recovery of a CCHP system.

10.5   Conclusion and future work

Electricity produced through conventional power systems is not only expensive but also inefficient. A portion of useful fuel goes wasted as heat and GHGEs. Therefore, conventional power generation systems are not only inefficient but also causing environmental pollution. A heat recovery system is used in the CCHP system which has better efficiency as compared to the conventional separate systems. After the integration of few thermally activated technologies into the CHP, the system is transformed into CCHP system which is more efficient as compared to CHP system. A CCHP system can be converted into a quad-generation system by integrating carbon dioxide extractor. Complex integration of various components requires careful selection of capacity and the number of equipment giving rise to the various optimization formulations. In this chapter, we reviewed different optimization formulations being used to optimize the design, operation, and planning of the CHP, CCHP, and quad-generation systems. Mathematical formulations of commonly used objective functions, namely, cost minimization, efficiency maximization, and GHGEs minimization have been elaborated. We also presented various optimization algorithms and the simulation tools being used to solve the optimization formulations. The chapter can serve as a foundation stone for the beginners in this research area. It can also serve as guide for the practitioners to optimally design, deploy, and operate the multigeneration power systems. This chapter considered the optimization classification, algorithms, and tools related to multigeneration system in general. The future study may be more specific to the application of optimization related to real-world problem arising by the use of CCHP in various buildings in different time zones. The problems may be further classified as single-objective and multiobjective optimization problems which may prove more fruitful for the optimization of multigeneration systems under various conditions and scenarios.

---

[2]http://ampl.com/products/solvers/solvers-we-sell/cplex.
[3]http://www.thermoflow.com/products appspecific.html.

References

[1]   Huangfu Y, Wu J, Wang R, *et al*. Evaluation and analysis of novel micro-scale combined cooling, heating and power (MCCHP) system. Energy Conversion and Management. 2007;48(5):1703–1709.

[2]   Bilgen E. Exergetic and engineering analyses of gas turbine based cogeneration systems. Energy. 2000;25(12):1215–1229.

[3]   Kong X, Wang R, Huang X. Energy optimization model for a CCHP system with available gas turbines. Applied Thermal Engineering. 2005;25(2): 377–391.

[4]   Khan K, Rasul M, Khan MMK. Energy conservation in buildings: cogeneration and cogeneration coupled with thermal energy storage. Applied Energy. 2004;77(1):15–34.

[5]   Havelský V. Energetic efficiency of cogeneration systems for combined heat, cold and power production. International Journal of Refrigeration. 1999;22(6):479–485.

[6]   Wu D, Wang R. Combined cooling, heating and power: a review. progress in energy and combustion science. 2006;32(5):459–495.

[7]   Wang JJ, Jing YY, Zhang CF, *et al*. Performance comparison of combined cooling heating and power system in different operation modes. Applied Energy. 2011;88(12):4621–4631.

[8]   Serra LM, Lozano MA, Ramos J, *et al*. Polygeneration and efficient use of natural resources. Energy. 2009;34(5):575–586.

[9]   Liu M, Shi Y, Fang F. Combined cooling, heating and power systems: a survey. Renewable and Sustainable Energy Reviews. 2014;35:1–22.

[10] Xu J, Sui J, Li B, *et al*. Research, development and the prospect of combined cooling, heating, and power systems. Energy. 2010;35(11):4361–4367.

[11] Hernández-Santoyo J, Sánchez-Cifuentes A. Trigeneration: an alternative for energy savings. Applied Energy. 2003;76(1):219–227.

[12] Liu P, Gerogiorgis DI, Pistikopoulos EN. Modeling and optimization of polygeneration energy systems. Catalysis Today. 2007;127(1):347–359.

[13] Weber C, Maréchal F, Favrat D, *et al*. Optimization of an SOFC-based decentralized polygeneration system for providing energy services in an office-building in Tōkyō. Applied Thermal Engineering. 2006;26(13): 1409–1419.

[14] van Rensburg PAJ, Sibanda MP, Ferreira HC. Integrated impedance-matching coupler for smart building and other power-line communications applications. IEEE Transactions on Power Delivery. 2015;30(2):949–956.

[15] Lin YJ, Latchman HA, Lee M, *et al*. A power line communication network infrastructure for the smart home. IEEE wireless communications. 2002;9(6):104–111.

[16] Mainardi E, Bonfè M. Powerline communication in home-building automation systems. In: Robotics and Automation in Construction. InTech, London; 2008.

[17] Berger LT, Schwager A, Escudero-Garzás JJ. Power line communications for smart grid applications. Journal of Electrical and Computer Engineering. 2013;2013:3.

[18] Sendin A, Peña I, Angueira P. Strategies for power line communications smart metering network deployment. Energies. 2014;7(4):2377–2420.

[19] Rubio-Maya C, Uche-Marcuello J, Martínez-Gracia A, *et al.* Design optimization of a polygeneration plant fuelled by natural gas and renewable energy sources. Applied Energy. 2011;88(2):449–457.

[20] Buoro D, Casisi M, Pinamonti P, *et al.* Optimization of distributed trigeneration systems integrated with heating and cooling micro-grids. Distributed Generation & Alternative Energy Journal. 2011;26(2):7–34.

[21] Arcuri P, Florio G, Fragiacomo P. A mixed integer programming model for optimal design of trigeneration in a hospital complex. Energy. 2007;32(8): 1430–1447.

[22] Li H, Nalim R, Haldi PA. Thermal-economic optimization of a distributed multi-generation energy system A case study of Beijing. Applied Thermal Engineering. 2006;26(7):709–719.

[23] Wang JJ, Jing YY, Zhang CF, *et al.* A fuzzy multi-criteria decision-making model for trigeneration system. Energy Policy. 2008;36(10):3823–3832.

[24] Yu T, Heiselberg P, Lei B, *et al.* A novel system solution for cooling and ventilation in office buildings: a review of applied technologies and a case study. Energy and Buildings. 2015;90:142–155.

[25] Baldini L, Kim MK, Leibundgut H. Decentralized cooling and dehumidification with a 3 stage LowEx heat exchanger for free reheating. Energy and Buildings. 2014;76:270–277.

[26] Liu Z, Zhang L, Gong G, *et al.* Review of solar thermoelectric cooling technologies for use in zero energy buildings. Energy and Buildings. 2015;102:207–216.

[27] Hu R, Niu J. A review of the application of radiant cooling & heating systems in Mainland China. Energy and Buildings. 2012;52:11–19.

[28] Zeghici RM, Damian A, Frunzulică R, *et al.* Energy performance assessment of a complex district heating system which uses gas-driven combined heat and power, heat pumps and high temperature aquifer thermal energy storage. Energy and Buildings. 2014;84:142–151.

[29] Sarbu I, Sebarchievici C. General review of ground-source heat pump systems for heating and cooling of buildings. Energy and Buildings. 2014;70:441–454.

[30] Antonanzas J, Alia-Martinez M, Martinez-de Pison F, *et al.* Towards the hybridization of gas-fired power plants: a case study of Algeria. Renewable and Sustainable Energy Reviews. 2015;51:116–124.

[31] Maghanki MM, Ghobadian B, Najafi G, *et al.* Micro combined heat and power (MCHP) technologies and applications. Renewable and Sustainable Energy Reviews. 2013;28:510–524.

[32] Jradi M, Riffat S. Tri-generation systems: energy policies, prime movers, cooling technologies, configurations and operation strategies. Renewable and Sustainable Energy Reviews. 2014;32:396–415.

[33] Ersayin E, Ozgener L. Performance analysis of combined cycle power plants: a case study. Renewable and Sustainable Energy Reviews. 2015;43:832–842.

[34] Chan HY, Riffat SB, Zhu J. Review of passive solar heating and cooling technologies. Renewable and Sustainable Energy Reviews. 2010;14(2):781–789.

[35]   Rosiek S, Batlles FJ. Renewable energy solutions for building cooling, heating and power system installed in an institutional building: case study in southern Spain. Renewable and Sustainable Energy Reviews. 2013;26: 147–168.

[36]   Sahoo U, Kumar R, Pant P, *et al.* Scope and sustainability of hybrid solar–biomass power plant with cooling, desalination in polygeneration process in India. Renewable and Sustainable Energy Reviews. 2015;51:304–316.

[37]   Ayou DS, Bruno JC, Saravanan R, *et al.* An overview of combined absorption power and cooling cycles. Renewable and Sustainable Energy Reviews. 2013;21:728–748.

[38]   Biezma M, San Cristobal J. Investment criteria for the selection of cogeneration plants—a state of the art review. Applied Thermal Engineering. 2006;26(5):583–588.

[39]   Gibson CA, Meybodi MA, Behnia M. Optimisation and selection of a steam turbine for a large scale industrial CHP (combined heat and power) system under Australia's carbon price. Energy. 2013;61(0):291–307.

[40]   Fang F, Wei L, Liu J, *et al.* Complementary configuration and operation of a CCHP-ORC system. Energy. 2012;46(1):211–220.

[41]   Sanaye S, Khakpaay N. Simultaneous use of MRM (maximum rectangle method) and optimization methods in determining nominal capacity of gas engines in CCHP (combined cooling, heating and power) systems. Energy. 2014;72:145–158.

[42]   Facci AL, Andreassi L, Ubertini S. Optimization of CHCP (combined heat power and cooling) systems operation strategy using dynamic programming. Energy. 2014;66:387–400.

[43]   Moradi MH, Hajinazari M, Jamasb S, *et al.* An energy management system (EMS) strategy for combined heat and power (CHP) systems based on a hybrid optimization method employing fuzzy programming. Energy;49:86–101.

[44]   Lozano MA, Ramos JC, Serra LM. Cost optimization of the design of CHCP (combined heat, cooling and power) systems under legal constraints. Energy. 2010;35(2):794–805.

[45]   Rong A, Lahdelma R. An efficient linear programming model and optimization algorithm for trigeneration. Applied Energy. 2005;82(1):40–63.

[46]   Arcuri P, Florio G, Fragiacomo P. A mixed integer programming model for optimal design of trigeneration in a hospital complex. Energy. 2007;32(8): 1430–1447.

[47]   Fragaki A, Andersen AN, Toke D. Exploration of economical sizing of gas engine and thermal store for combined heat and power plants in the UK. Energy. 2008;33(11):1659–1670.

[48]   Zhao XL, Fu L, Zhang SG, *et al.* Performance improvement of a 70 kWe natural gas combined heat and power (CHP) system. Energy. 2010;35(4): 1848–1853.

[49]   Jannelli E, Minutillo M, Cozzolino R, *et al.* Thermodynamic performance assessment of a small size CCHP (combined cooling heating and power) system with numerical models. Energy. 2014;65:240–249.

[50]   Liao C, Ertesvåg IS, Zhao J. Energetic and exergetic efficiencies of coal-fired CHP (combined heat and power) plants used in district heating systems of China. Energy. 2013;57(0):671–681.

[51]   Martınez-Lera S, Ballester J. A novel method for the design of CHCP (combined heat, cooling and power) systems for buildings. Energy. 2010;35(7):2972–2984.

[52]   Wu J-y, Wang J-l, Li S. Multi-objective optimal operation strategy study of micro-CCHP system. Energy. 2012;48(1):472–483.

[53]   Dıaz PR, Benito YR, Parise JAR. Thermoeconomic assessment of a multi-engine, multi-heat-pump CCHP (combined cooling, heating and power generation) system: a case study. Energy. 2010;35(9):3540–3550.

[54]   Wu JY, Wang JL, Li S, *et al.* Experimental and simulative investigation of a micro-CCHP (micro combined cooling, heating and power) system with thermal management controller. Energy. 2014;68:444–453.

[55]   Powell KM, Sriprasad A, Cole WJ, *et al.* Heating, cooling, and electrical load forecasting for a large-scale district energy system. Energy. 2014;74:877–885.

[56]   Wang J, Dai Y, Gao L, *et al.* A new combined cooling, heating and power system driven by solar energy. Renewable Energy. 2009;34(12):2780–2788.

[57]   Agency USEP. EPA green building strategy; 2008. [Online; accessed 19-January-2016]. http://www.epa.gov/greenbuilding/pubs/about.htm.

[58]   Wang J, Zhai ZJ, Jing Y, *et al.* Optimization design of BCHP system to maximize to save energy and reduce environmental impact. Energy. 2010;35(8): 3388–3398.

[59]   Wang J, Zhai ZJ, Zhang C, *et al.* Environmental impact analysis of BCHP system in different climate zones in China. Energy. 2010;35(10):4208–4216.

[60]   Jing YY, Bai H, Wang JJ. Multi-objective optimization design and operation strategy analysis of BCHP system based on life cycle assessment. Energy. 2012;37(1):405–416.

[61]   Carvalho M, Lozano MA, Serra LM. Multicriteria synthesis of trigeneration systems considering economic and environmental aspects. Applied Energy. 2012;91(1):245–254.

[62]   Bertsimas D, Tsitsiklis JN. Introduction to Linear Optimization. vol. 6. Athena Scientific, Belmont, MA; 1997.

[63]   Roos C, Terlaky T, Vial JP. Theory and algorithms for linear optimization: an interior point approach. Wiley, Chichester; 1997.

[64]   Zimmermann HJ. Fuzzy programming and linear programming with several objective functions. Fuzzy Sets and Systems. 1978;1(1):45–55.

[65]   Rong A, Lahdelma R. An efficient linear programming model and optimization algorithm for trigeneration. Applied Energy. 2005;82(1):40–63.

[66]   Blarke MB, Dotzauer E. Intermittency-friendly and high-efficiency cogeneration: operational optimisation of cogeneration with compression heat pump, flue gas heat recovery, and intermediate cold storage. Energy. 2011;36(12): 6867–6878.

[67]   Bussieck MR, Pruessner A. Mixed-integer nonlinear programming. SIAG/OPT Newsletter: Views & News. 2003;14(1):19–22.

[68]   Ameri M, Besharati Z. Optimal design and operation of district heating and cooling networks with CCHP systems in a residential complex. Energy and Buildings. 2016;110:135–148.

[69]   Artigues C, Koné O, Lopez P, *et al.* Mixed-integer linear programming formulations. In: Handbook on Project Management and Scheduling Vol. 1. Springer, New York, NY; 2015. p. 17–41.

[70]   Benichou M, Gauthier JM, Girodet P, *et al.* Experiments in mixed-integer linear programming. Mathematical Programming. 1971;1(1):76–94.

[71]   Mitra S, Sun L, Grossmann IE. Optimal scheduling of industrial combined heat and power plants under time-sensitive electricity prices. Energy. 2013;54: 194–211.

[72]   Li CZ, Shi YM, Huang XH. Sensitivity analysis of energy demands on performance of CCHP system. Energy Conversion and Management. 2008;49(12):3491–3497.

[73]   Piacentino A, Cardona F. EABOT Energetic analysis as a basis for robust optimization of trigeneration systems by linear programming. Energy Conversion and Management. 2008; 49(11):3006–3016.

[74]   Lozano MA, Carvalho M, Serra LM. Allocation of economic costs in trigeneration systems at variable load conditions. Energy and Buildings. 2011;43(10):2869–2881.

[75]   Hu M, Weir JD, Wu T. Decentralized operation strategies for an integrated building energy system using a memetic algorithm. European Journal of Operational Research. 2012;217(1):185–197.

[76]   Akbari K, Nasiri MM, Jolai F, *et al.* Optimal investment and unit sizing of distributed energy systems under uncertainty: a robust optimization approach. Energy and Buildings. 2014;85:275–286.

[77]   Fazlollahi S, Becker G, Maréchal F. Multi-objectives multi-period optimization of district energy systems. Computers & Chemical Engineering. 2014;66: 82–97.

[78]   Rong A, Lahdelma R, Luh PB. Lagrangian relaxation based algorithm for trigeneration planning with storages. European Journal of Operational Research. 2008;188(1):240–257.

[79]   Jannelli E, Minutillo M, Cozzolino R, *et al.* Thermodynamic performance assessment of a small size CCHP (combined cooling heating and power) system with numerical models. Energy. 2014;65(0):240–249.

[80]   Wang JJ, Jing YY, Zhang CF. Optimization of capacity and operation for CCHP system by genetic algorithm. Applied Energy. 2010;87(4):1325–1335.

[81]   Rong A, Figueira JR, Lahdelma R. An efficient algorithm for bi-objective combined heat and power production planning under the emission trading scheme. Energy Conversion and Management. 2014;88:525–534.

[82]   Wu B, Wang L. Comparable analysis methodology of CCHP based on distributed energy system. Energy Conversion and Management. 2014;88: 863–871.

[83]   Viklund SB, Johansson MT. Technologies for utilization of industrial excess heat: potentials for energy recovery and CO_2 emission reduction. Energy Conversion and Management. 2014;77:369–379.

Chapter 11
Demand side management through PLC: concepts and challenges
Nikolaos G. Paterakis[1], Samina Subhani[1],
and Muhammad Babar[1]

11.1 Introduction

The demand for electrical energy is increasing worldwide due to the intensified electrification of the society in the developed industrialized countries, in addition to the economic growth and the improvement of living standards in emerging economies. In the same time, the energy sector of almost every country faces significant challenges such as inefficient and polluting energy generation and its adverse effects on the climate change. Moreover, the dependence on fossil fuels imported from politically unstable regions increases the vulnerability to fluctuating fuel prices and raises concerns related to a stable and sustainable growth of the electrical energy production. More specifically, both from a technical and institutional viewpoint, the electricity sector should be able to accommodate the following challenges [1]:

- The ever-increasing demand for electrical energy. The International Energy Agency estimates an increase by up to 50% by 2040 in comparison with today [2].
- The limited investments in conventional generation and power transfer capacity of the electrical network.
- The need for efficient (i.e., transport loss reduction, increased efficiency of generators) and reliable operation (i.e., adequate reserve levels in order to face contingencies, uncertainty management) of the power system. This holds especially true in the light of increasing penetration levels of renewable energy sources (RES), which require innovative monitoring and control technologies for efficient and reliable operation and incorporation into the existing power system.
- Enabling emerging business opportunities such as innovative products and services for the end-users, manufacturing of smart-grid enabling technologies [electric vehicles (EVs), new types of electrical apparatus, etc.].
- Greenhouse gas emissions related to the electrical energy sector account for a significant percentage of the global emissions.

[1]Department of Electrical Engineering, Eindhoven University of Technology, Netherlands

Transitioning from the current power system structure to a smart grid is the projected solution to the challenges that are emerging, which can be defined as a series of technical and institutional innovations, aiming to render the power system more efficiently and environment friendly. Although there is not a commonly accepted definition, it might be argued that the Smart Grid should serve as a hub that interfaces all the components of the power system (interoperability) using advanced metering, communication, and control infrastructure. The ultimate goal is to establish communication between the various stakeholders (system operators, energy providers, and consumers) with the aim to enable efficient management of electricity production and consumption. In the context of energy management, two types can be discerned:

- Supply side management (SSM)
- Demand side management (DSM)

SSM refers to actions taken by system operators to ensure that generation, transmission, and distribution of electrical energy are efficient. In the past, SSM mainly concerned the generation of electricity using almost exclusively fossil fuels, but nowadays, it is a term also related to actions, planning, and development of procedures and services concerning the supply of electrical power by RES such as solar and wind generation. Effective SSM enables the usage of energy resources in such a way that economic efficiency is improved and the environmental impacts of the power system operation are minimized while considering reliability and quality of electrical power supply.

In order to meet the increasing demand and confront the variability of RES generation, optimal control and operational SSM strategies for the full exploitation of the existing infrastructure are sought. Nevertheless, additional costly investments in information and communications technology (ICT) infrastructure might be needed to achieve these targets. These challenges have been the main motivation for enabling the active participation of resources at the demand side in the planning and operation of power systems. The term DSM refers to the analysis and control of energy consumption. The goal of DSM is to find areas of high usage and electricity waste and to determine services and/or systems that will reduce ensuring the balance within the electricity supply–demand chain. One of the most widely accepted definitions of DSM is the following [3]:

> Demand-side management is the planning, implementation, and monitoring of those utility activities designed to influence customer use of electricity in ways that will produce desired changes in the utility's load shape, that is, changes in the time pattern and magnitude of a utility's load. Utility programs falling under the umbrella of demand-side management include: load management, new uses, strategic conservation, electrification, customer generation, and adjustments in market share.

This definition renders evident that DSM is a broad term. Nonetheless, it is important to notice that DSM is often used in order to refer to any of its individual components or to characterize actions that result in permanent load reductions (i.e., energy efficiency). Moreover, it should be highlighted that the different DSM programs are not mutually exclusive.

Among the different components of DSM, energy efficiency refers to policies or measures taken by a community to use appliances that improve the efficiency of the energy infrastructure and to reduce energy wasting. In this way, a reliable, affordable, and sustainable energy system for the future can be ensured. Moreover, the concept of customer generation (so-called prosumers) or net-zero buildings/regions refers to the strategic development of an individual consumer's premises or a whole region such that the total energy consumption within the area of interest is zero over a particular period. Arguably, one of the most important aspects of DSM is load management, or more commonly referred to as demand response (DR). DR provides business opportunities to change wholesale and retail electricity prices with the purpose of inducing temporary effects on the energy consumption of the end-users, for instance in order to facilitate energy service companies or network operators.

DR usually refers to short-term effects on the load pattern of utilities and thereby the implementation of DR can potentially defer investment costs related to the construction of copper plate solutions (i.e., to enhance the transfer capacity of the electrical network). One distinct characteristic of DR is that it depends on the establishment of ICT solutions. This is required in order to physically implement DR programs and allow all the involved parties to harness the underlying benefits. There are many different definitions of DR but a commonly cited one comes from the US Department of Energy [4]:

> Changes in electric usage by end-use customers from their normal consumption patterns in response to changes in the price of electricity over time, or to incentive payments designed to induce lower electricity use at times of high wholesale market prices or when system reliability is jeopardized.

Over the last few years, it has been found that DR has many direct and indirect benefits [5]. First and foremost, DR provides benefits to electricity consumers by directly paying incentives for the shifting of their demand flexibility. Second, it encourages market competition, by allowing the participation of third-party service providers (i.e., aggregators). Apart from direct benefits, DR is capable of increasing the efficiency of power systems and electricity market by optimizing the consumption during peak demand periods. Besides, it substantially lowers the energy cost by reducing the need for investments in generation and transmission capacity to accommodate the ever-increasing peak demand.

This chapter focuses on the DR component of DSM and provides an overview of key concepts pertaining DR, its benefits as well as challenges, and implementation requirements. DR is enabled through a communication and control infrastructure that is embedded within the electronics (e.g., grid-tied inverters) by which the DR-capable appliances are connected to the electrical network. For this reason, a link to power electronics embedded in the smart grid will be made and the challenges that thereby arise related to power line communication (PLC) will be highlighted.

11.2   Overview of demand response

11.2.1   Types of demand response programs

DR delivers its benefits to participating customers, i.e., residential, industrial, and commercial customers, via direct financial incentives in exchange for their DR flexibility or dynamic pricing. Hence, DR may be broken down into two major categories: (1) explicit (incentive-based DR) and (2) implicit (price-based DR).

Explicit DR schemes generally enforce frequency control and direct-load control of large consumers, i.e., industrial and large commercial customers. Furthermore, in explicit DR, the demand flexibility is traded through centralized electricity market structures. Customers can profit by trading their demand flexibility either directly or by using an aggregator as a proxy agent who acts on their behalf. The customer generally commits to long-term fixed contracts, and thereby the options for further incentivization for DR services are limited. Explicit DR is predictable and therefore suitable to be scheduled in advance. However, due to the lack of customer satisfaction and privacy, explicit DR schemes currently face the highest degree of customer renunciations.

Implicit DR refers to all kinds of price-based DR programs which are aimed to facilitate voluntary participation from large communities of customers to lower effectively the peaking generation. The demand flexibility allows the demand to shift away from critical periods for the system and thereby defer pending investment costs to reinforce the network. In implicit DR, a customer is exposed to time-varying electricity prices, time-varying network tariffs with a number of price levels over a particular time-period or season, or a combination of both. Customers react to changes in rates depending on their possibilities and individual constraints. ICT infrastructure must be present to monitor the consumption and the alterations of price levels because the customer dispatches demand on a voluntary basis.

Both types of DR are distinct with respect to their purposes and timing; however, it is important to underline that the design and implementation of these schemes do not necessarily render them mutually exclusive. For instance, a customer may participate in explicit DR through an aggregator and, at the same time, respond to dynamic tariffs and, therefore, participate in implicit DR. In this way, a customer has greater flexibility in achieving higher economic remuneration on the basis of optimally deploying its response. From the market operators' point of view, explicit DR provides competitive and sustainable means of maintaining the balance between the supply and the demand and to control the exercise of market power by larger market players. Moreover, explicit DR constitutes a reliable and directly accessible resource for network operators that can exploit the flexibility of the demand side with the purpose of resolving operational issues such as congestion. Furthermore, although implicit DR is provided on a voluntary basis by the involved consumers, it can prove an effective way of deferring investments as it may lower the need for peaking generation by causing demand to shift away from critical periods for the system.

The implementation of both types of DR in a particular market context is influenced by regional market design and primordial network issues. For instance, in liberalized electricity markets, it is easier to deploy explicit DR schemes than in regulated markets. Moreover, customer responsiveness to market change depends highly on demographics, e.g., income and house ownership. Thus, the existing DR programs could be further segregated within the spectrum of these two broader classes. As a result, deriving a unique classification is not straightforward.

Below a widely accepted taxonomy is given:

- **Explicit demand response**
  - **Direct load control**
    A DR service provider (e.g., an aggregator) engages a large number of relatively small consumers and has direct control on the consumption of their appliances. End-users are typically compensated both for their enrollment and their actual participation in a DR event.
  - **Interruptible/Curtailable load**
    These are DR programs similar to direct load control with the difference that they are addressed to larger consumers (e.g., commercial and industrial) and that the demand reduction might be the responsibility of the consumer in response to an instruction by a procuring entity such as a market operator.
  - **Demand side bidding**
    Demand flexibility is traded in a short-term forward market such as the day-ahead and intraday markets or the spot balancing market. This can be accomplished directly by larger consumers or via an aggregator.
  - **Ancillary services**
    Flexibility offering in terms of demand side bidding can be extended to the provision of ancillary services in various time scales (e.g., regulation, reserves).
  - **Emergency DR**
    In response to emergency signals (e.g., due to major system contingency), participating demand side resources (individual customers or aggregators) can respond on a voluntary basis.
  - **Capacity market programs**
    These programs are usually designed in order to procure reliable commitments from the demand side with the purpose of avoiding system reinforcement.
- **Implicit demand response**
  - **Time-of-use tariffs**
    Consumers are exposed to static tariffs with respect to season, week, or day of the week. They usually reflect average electricity supply costs and consist of a peak tariff, an off-peak, and potentially a shoulder-peak tariff.
  - **Critical peak pricing**
    In combination with a time-of-use tariff scheme or a flat rate, consumers are exposed to an extra tariff component that is triggered by critical system conditions and, therefore, capture shorter term costs of electricity supply.

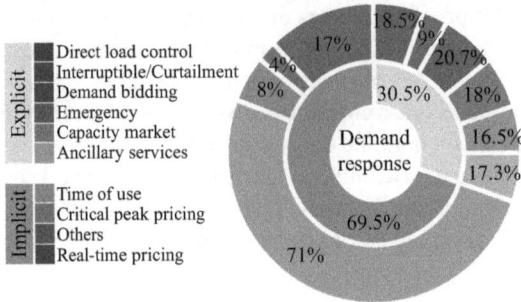

Figure 11.1   Quantitative classification of research related to DR program types

- **Real-time pricing**
  The actual wholesale market prices are passed on to consumers enrolled in real-time pricing programs.
- **Dynamic pricing**
  Dynamic pricing is a term that is used in case of other programs reflecting more sophisticated tariff structures.

In the recent years, a significant amount of research has been devoted to exploring the use and improving the different DR programs. In fact, considering only the IEEE journal and conference publications during the last 5 years (2011–16), it has been estimated that almost a thousand papers have focused on DR. Figure 11.1 provides an overview of this quantitative analysis. Based on the findings, it can be inferred that most of the relevant research has focused on implicit DR and, especially time-of-use tariff dynamic structures, which directly or indirectly is the subject of 71% of the respective literature.

11.2.2   Types of customer response

The DR programs discussed in Section 11.2.1 aim at engaging different types of consumers and to control their load demand via various mechanisms in exchange for payments and incentives. Traditionally, three types of consumers are discerned, namely, industrial, commercial, and residential.

Industrial customers are responsible for a major portion of energy demand, and therefore, they have historically been at the epicenter of attention of DSM and in particular DR programs [6]. The industrial sector presents significant potential in terms of response volumes, yet the adoption of DR programs is not straightforward for industrial customers. For other types of consumers, DR might entail temporary loss of comfort due to load reduction; however, it is possible for industrial customers that participating in DR programs might compromise their economic objectives, and as a result, a number of constraints (e.g., criticality of processes, inventory restrictions) have to be considered. Thus, it might not be sufficient to apply DSM solely in terms of DR but also by relying upon on-site generation and energy storage.

Commercial customers can contribute to load reductions and ancillary services, especially via subjecting thermostatically controllable loads to DR. Such loads include for example individual air-conditioning units in hotels and heating-ventilation and air-conditioning systems in large commercial premises. Such systems typically employ variable frequency drivers, which allow high-speed and continuous adjustment of their power demand [7]. More recently, the idea of energy intelligent buildings that can interact with the power system has been introduced, encompassing premises such as office buildings equipped with energy monitoring and control systems in order to take advantage of incentives, e.g., by participating in implicit DR programs [8].

Residential customers typically own appliances with significant power ratings such as electric water heaters, air-conditioning units, and washing machines. Implicit as well as explicit load control DR programs can be designed in order to exploit this potential.

Finally, it is worth mentioning that there is an emergence of two distinct new types of load, namely, EVs and data centers, that exhibit a more distinct behavior and response. These loads present both a challenge for the power system and a significant potential of being involved in DR activities.

The electrification of the transportation sector might lead to significant stress of the power system due to additional high power requirements of EVs [9]. Two measures that have been proposed in order to address the adverse impact of the increasing penetration of EVs and facilitate their integration fall under the umbrella of DR: (1) controlled unidirectional charging and (2) vehicle-to-grid operational mode. By applying these two techniques on a fleet of EVs peak shaving, valley filling and more effective integration of RES can be achieved [10], while the vehicle owners can enjoy significant economic benefits.

Data centers, an emerging type of major consumer, are already equipped with power management mechanisms offering control flexibility, and in contrast with other industrial or commercial consumers, their tasks are tolerant of delays and performance degradation. This makes them particularly suitable for enrollment in both explicit and implicit DR programs. For this reason, the US Environmental Protection Agency has urged data centers to adopt DR programs in order to mitigate the strain on the power system [11].

11.3 Benefits of demand response

DR has the potential to present a diverse range of benefits depending on the design and the aim of the specific DR implementation. One of the most important advantages of harnessing the flexibility that the demand side can offer is related to easing concerns regarding the integration of RES in power systems. For example, it is desired to increase the customer load demand when photovoltaic (PV) or wind turbines are generating large quantities of power and thereby mitigate overvoltage and voltage congestion in the power system network. Moreover, enabling DR is expected to have favorable effects on the overall management of the power system, the underlying market mechanisms, and the society as a whole.

11.3.1   Integration of high amounts of renewable energy sources

Introducing significant amounts of RES in the power generation mix is central to various initiatives worldwide [12]. Especially, wind and solar energies are leading the booming of renewables, and their capacity is expected to increase dramatically in the next decades [13]. For instance, the wind energy that was produced in 2008 in the United States was 31 TW h and is anticipated to increase to 1,160 TW h by 2030, while the solar capacity is expected to reach 16 GW by 2020 [14]. Similar targets have been set by the EU countries as well. The most characteristic example is Ireland that aims to cover 40% of the electricity demand by wind resources by 2020 [15]. Despite the foreseen environmental benefits stemming from the increasing uptake of RES, their inherently stochastic nature of weather-dependent generators may jeopardize the security of the power system and thereby raise new challenges both from technical and economic perspectives. Wind and solar generation depends highly on weather conditions and therefore is spatially and temporarily variable. In addition to that, the power output of these resources cannot be controlled in the same way with the output of conventional generators with the purpose of economically matching the system demand. At any rate, DR can be used in order to address these challenges.

The uncertainty linked to the generation of RES will augment the need for scheduling and deploying reserves in order to balance generation and demand. Traditionally, reserves have been procured from conventional generators in order to secure the system against load fluctuations and contingencies. Nevertheless, generators that are committed to providing reserves incur severe costs due to efficiency loss and increased emissions due to power output ramping, increased wear and tear, and opportunity costs in the energy market because of operating partly loaded [16]. In other words, on top of the technical drawbacks, energy that could have otherwise been sold in the energy market has to be withheld as a reserve, and therefore, the attainable revenue for the generator owners is decreased. On the contrary, certain types of loads such as air conditioners and electric space heaters are not characterized by such drawbacks and have the ability to adjust their power to changes in demand instantaneously [17]. Moreover, the increasing penetration of RES might increase the need of relying on interconnections with neighboring national power systems in order to stay balanced. Deploying DR resources may facilitate the economical use of interconnections [18].

DR can be also exploited in order to address another challenge related to the limited controllability of RES and the absence of correlation between renewable production and load demand. Namely, this is the problem of RES "over-generation" during off-peak periods [19]. As an alternative to unnecessarily (i.e., inefficiently) dispatching conventional generators or to spilling the excessive RES production, DR solutions can be employed. A variety of loads can be controlled in such a way that allows the otherwise curtailed RES energy to be used by the system, including water pumping, irrigation, municipal treatment facilities, thermal storage in large buildings, industrial electrolysis, aluminum smelting, EV charging, etc. [20].

11.3.2   System-wide benefits

The potential of DR to induce system-wide benefits has also been recognized. For this reason, the regulatory framework obliges utilities in the United States and in the EU to consider the peak shaving effect of DR in their resource planning, as opposed to the classical view to system upgrading that considers the projected gradual increase in demand. Hence, DR mechanisms can be used to defer investments in generation and transmission system capacity, while making sure that existing capacity is not underutilized. Similar benefits are foreseen for the operation of distribution systems. Problems related to voltage magnitude, increased losses, and congestions may be confronted by designing appropriate DR programs that take advantage of the abundance and dispersed nature of potentially active demand side resources. Apart from facilitating the integration of higher amounts of RES in the power system, DR can help reduce the carbon footprint of the power system by promoting energy efficiency overall and by peak shaving. As a matter of fact, in California, the carbon intensity is 33% higher in peak times in comparison with off-peak times [21].

It is also argued that DR could improve the efficiency of electricity markets by reducing and stabilizing electricity prices, controlling the market power of large generators and benefiting consumers. In general, when the demand reaches the capacity of the system, generation side bids tend to increase exponentially. Thus, a slight decrease in demand as a response to high prices can significantly reduce the market prices. Interestingly, electricity market crises have been linked to the absence of DR programs [22]. In addition to that, price-responsive demand may serve as an alternative to market monitoring and price caps that can reduce the overall efficiency of the electricity market. Finally, in the long run, DR may benefit participant consumers in terms of monetary gain but also consumers that do not participate in any DR program to the permanent effects of DR on the magnitude and the stability of prices.

11.3.3   Societal benefits

In the context of DR, societal benefits are benefits generated primarily as a result of actions undertaken by consumers. A significant amount of social benefits associated with DR have been reported, out of which a few distinct benefits are highlighted herein.

The most important societal benefit of consumers participating in relevant DR programs is the reduction of energy consumption and energy savings, and thereby improving ecological, political, and economical prospects. For instance, the Dutch Energy Savings Monitor Project estimated 3% of savings with indirect DR (using bimonthly home energy reports) and 6.4% with direct DR (using an in-home Display) [23]. Another project analyzed energy saving results from customers either with real-time display or energy savings advice based on historic consumption. An average 3% of energy savings across customers with real-time display and 5% of savings for the provision of energy savings advice were reported [24]. Another societal aspect of DR is that it inherently increases the awareness of energy savings among consumers, and this promotes the purchase of more energy-efficient appliances and more efficient use of the electricity. For instance, the E-Energy Projects in Germany have tested various tariff systems, and it was found that tariffs do not only increase the efficiency of energy use but also improve the efficiency of network operation [25]. Less energy consumption and higher energy efficiency will also help in protecting the environment and might play a role in achieving emission reduction targets.

Innovative services, such as smart energy management systems (EMS), will grant the consumer with the possibility of choosing among a broader range of service providers as well as of enrolling in different pricing mechanisms. In this way, DR will create significant new opportunities: service providers to offer competitive services, product companies to provide smart appliances and technology companies to develop intelligent solutions for effective communication and control. Hence, DR may contribute to the overall economic development of the society. Last but not the least, DR will have a considerable impact on the retail electricity market because of increased consumer awareness regarding electricity usage. Finally, the provision of reliable and accurate information on retail electricity prices is expected to allow for quick switching between energy suppliers.

11.4   Demand response implementation requirements

The primordial role of ICT in the establishment and evolution of DSM and DR programs is undoubtable. In this section, the metering, control and communication technologies that are required in order to physically implement DSM and DR programs and related challenges are reviewed. Standardization approaches in order to promote a consensus as regards the objectives and deployment of smart grids and in particular DR are also briefly discussed.

11.4.1   Metering, control, and communication infrastructure

The implementation of DR strategies relies heavily on being able to monitor and control the energy usage of the participating consumers. A smart meter is an electronic meter that is more advanced in comparison with conventional meters in the sense that it can identify power consumption with higher time granularity and communicate the collected information back to the entity that is serving the particular customer. The capabilities of smart meters extend beyond automatic meter reading (AMR), which serves for monitoring (e.g., power quality, antitheft protection) and billing purposes, in the sense that real-time bidirectional communication can be established between the consumer and the utility in such a way that smart meters can serve as the gateway for receiving explicit or implicit DR control signals, namely, pricing information or commands for load reduction, respectively [26].

To enable more effective participation in DR programs decision support systems can be employed in order to fully or partially automate the response of a consumer to explicit or implicit signals. Such systems are typically referred to as EMS and are installed behind-the-meter, that is, in the domain of consumer's responsibility. Depending on the type of the end-user, they are responsible to customize and monitor their own DR parameters such as analyzing information regarding the state of appliances or electricity-enabled processes, their power consumption and their criticality. In addition to that, the end-user can receive information about distributed generation or storage units the consumer might own or external factors such as temperature. Having received all the required information, these systems can decide an optimal operating schedule for all the assets that are involved in such a way that one or more objectives are optimized while satisfying a set of constraints. Objectives may refer to maximizing financial benefits that individual end-users are receiving in compensation for their participation or global objectives such as active power losses reduction of the local distribution system. Constraints that are being taken into account depend on the particular type of the end-user. For example, constraints encompass limits on the violation of the comfort for residential end-users or technical and economic constraints of commercial and industrial customers [27]. The market penetration of EMS is demonstrating significant asymmetry among different regions. Major industrial players have developed commercially available EMS solutions, while industrial consumers are investing in customized EMS with the purpose of participating in DR. The United States is currently leading in the adoption of EMS, especially in the case of residential customers, while in Europe, several initiatives have been set in place [5].

Smart metering and automated decision support systems are individual technology prerequisites for effective automated DR applications. However, advanced metering infrastructure (AMI) is the key component that characterizes the transition from the traditional power system to a smart grid by enabling the interconnection of the different participants. AMI does not refer to a single metering or communication technology; rather, it describes the integration of various metering, data management, and communication technologies that are configured in order to network end-users, grid operators, and service providers [28]. The AMI infrastructure encompasses numerous components, e.g., a network that connects smart appliances, sensors, actuators, and the EMS.

Depending on the type of end-user, the network can be characterized as follows:

• The home area network (HAN) or building area network (BAN) or industrial area network (IAN), depending on the type of end-user it refers to, is a domain that consists of a network of appliances, sensors, and actuators, the EMS, as well as their energy and communication interactions.

Other common AMI are as follows:

• The neighborhood area network (NAN) that extends the range of the communication area that is covered by a number of HAN, BAN, or IAN and incorporates information that is forwarded by the smart meter gateways. The objective of NAN is to enable energy management and DR at a larger area.
• The wide area network (WAN) is a further extension to NAN in the sense that it transmits metering and response information to further separated decentralized control centers. The target is to increase the awareness of system operators and service providers for the energy consumption of end-users and facilitate actions to enhance the overall reliability of the power system.

The different AMI domains have different communication requirements due to the spatial range, the traffic to be served and the data rate that have to be achieved in order to fulfill their targets, i.e., to support a particular DR implementation. Thus, different technologies might be suitable for each domain. The specific communication technologies to be used in order to implement the infrastructure that will support DSM and DR programs must be selected on the basis of numerous criteria [29,30]:

- Quality of service (QoS)
- Interoperability
- Scalability
- Security

Guaranteed QoS is essential for an effective implementation of DR programs. QoS encompasses latency, bandwidth, and communication reliability qualifications that should characterize the infrastructure in the path between the involved parties, for example, between the system operator and a particular controllable load. Latency is the time delay between the emission of a control signal and its reception by the responding party. Depending on the time restrictions that are imposed by the particular DR application, very low-latency may need to be achieved. For instance, direct reaction to emergency signals requires a latency of a few milliseconds. Moreover, the number of end-users that are involved in DR activities must not affect the overall performance of the communication infrastructure. Thus, depending on the frequency of the communication between the end-user and the service providers, sufficient bandwidth must be provided in order to fulfill the specification of a particular DR program. Finally, the operation of the communication and control infrastructure must be dependable and, therefore, present minimal failures.

The infrastructure that supports bidirectional communication capabilities between the providers and users of DR services comprises a number of different components and technologies. The effective establishment of the benefits of DR depends highly on the ability of all the elements of the communication chain and different actors to be interfaced cooperate with each other. In other words, interoperability has to be provided. In this respect, the role of standardization is primordial.

The smart grid vision projects the participation of a large number of DR capable resources in the provision of system services. The reliability and affectivity of the DR is highly correlated with the number of participating loads. Thus, the communication infrastructure must be designed on the basis of covering current needs but also considering scalability in anticipation of the future growth in the number of participating end-users.

Lastly, security of communications is essential in implementing DR applications, since relevant concerns might hinder both the involvement of end-users but also the trust of other DR beneficiaries in the provided services. A detailed treatment of the security dimension of smart grid communications can be found in [31].

11.4.2 Communication technologies

The latency and bandwidth requirements generally become more stringent as the geographical extent of the communication domain increases from a few tens of meters in the case of a HAN to tens of kilometers that is covered by a WAN. Various wireline and wireless communication technologies may satisfy the technical requirements that were discussed in Section 11.4.1 in different data communication domains. The main advantage of wireless communication technologies is the flexibility in areas where access to the end-points is problematic and the fact that additional wiring costs can be avoided. A manifold of wireless communication technologies could be exploited, including for instance ZigBee, Z-Wave, Wi-Fi, Wi-MAX, cognitive radio and cellular technologies. Nonetheless, wireless communication technologies are more vulnerable to signal loss and security breaches, and therefore, stronger security mechanisms are necessary in order to prohibit unauthorized access.

Among the different wired technologies, such as fiber-optics and Ethernet, this section focuses on one particular technology, PLC, the carrier of which is inherent to the power system. Section 11.4.2.1 provides an overview of the main challenges regarding simultaneous usage of PLC and the supraharmonics produced by increasingly larger power electronic devices, and thereby potentially reduce the QoS of DR programs. In Section 11.4.2.2, an introduction to supraharmonics is given, and the potential negative effect supraharmonics can have on PLC are described. The challenging compatibility of PLC and the low impedance paths in the network induced by capacitive electromagnetic interference (EMI) filters, i.e., filters that are built into power electronic devices to mitigate high frequency (HF) disturbances such as supraharmonics, are outlined in Section 11.4.2.3.

11.4.2.1   Electromagnetic compatibility of PLC in the smart grid

Over the last few years, a renewed interest in PLC has been sparked by the impending energy transition [32,33]. The energy transition entails a significant increase of distributed energy resources (DER), such as renewable generators, electrical energy storage units, and large consumer loads such as EVs and heat pumps [34]. DER units are prominent controllable loads that can participate in DR programs. These DER units are connected to the distribution network via grid-tied power electronics. These power electronics regulate the power conversion (e.g., DC-to-AC conversion) and power output (e.g., active power factor correction or total delivered power) of the DER unit, and thereby control the demand flexibility available for DR. As of now, power electronics are already present in most household appliances. The proliferation of power electronics is expected to continue [35], and due to the energy transition, the size of the power electronics tied to the distribution grid is expected to increase as well. The trend of increasing amounts of RES and the increasingly larger individual DER units facilitates both the need as well as the feasibility of DR schemes.

In the future grid, electricity will be increasingly generated via DER units that are dispersed throughout the distribution grid. To incorporate these DER units into the day-to-day operation of electrical power systems and to enable participation in DR programs, it is required to connect the power electronics of the DER units to an extensive AMI in order to coordinate and control the many individual DER units into coherent actions that enable aggregation, DR market participation, and support grid operation. Hereby, the so-called smart grid emerges. The basic incentives to use PLC technology in the smart grid ICT infrastructure are that the deployment costs are considered comparable to wireless since the lines are readily available and that the power grid provides an infrastructure that is in principle more extensive and pervasive than any other wired or wireless alternative [32].

PLC has been used extensively by European utilities for decades, e.g., to control day/night tariff switching via low data rate ripple carrier signaling (~100 bps, 0.1–2 kHz). However, the emergence of high data rate narrowband (NB) PLC (<500 kb/s, 3,148.5 kHz) opens up the possibility for more advanced power system applications to support grid operation in the future smart grid [32]. Sustainable generators produce energy according to stochastic weather conditions. Hence, the electrical energy production will generally not correlate with the consumer energy demands. Consequently, the energy transition requires a paradigm shift in which the electric energy demand follows the generation of electrical energy. A principle focus of DR is to enable this paradigm shift. In order to support the implementation of DR, modern NB PLC can be deployed for, e.g., accurate real-time one-way communication AMR and two-way communication AMI [36]. European utilities in France, Sweden, and Spain have deployed PLC for such metering applications. Other potential smart grid PLC applications will entail comprehensive control schemes for DER units to adjust demand flexibility, e.g., control of power output of microinverters PV installations, EV charging, i.e., vehicle-to-grid communications for DSM, or in-home EMS. Furthermore, future PLC smart grid applications will be geared toward extending the Supervisory Control and Data Acquisition model, by providing e.g., overvoltage alarms, synchronous measurement data deliverance, power quality monitoring, phase detection, multiagent communication, fault detection online diagnostics, or fraud detection.

Modern power electronics contain switching circuits that produce HF disturbances. These HF disturbances may produce EMI that causes household equipment, DER units, and utility assets to malfunction [37]. Standards applicable to the 2–150 kHz range are limited. Thus, emission limits at lower frequencies (<2 kHz) have been taken into account in the design of power electronic devices, but emissions in the 2–150-kHz range have increased [38]. The term supraharmonics refers to these HF voltage and current disturbances in the range of 2–150 kHz [39]. The EMI induced by these supraharmonics have been reported to disturb PLC, and thereby reduce feasibility of PLC as the main communication channel in DR applications [37,40]. In order to mitigate supraharmonic EMI, capacitive filters are built into the power electronic devices. However, the added capacitance in the power electronic devices may also reduce the efficacy of PLC. The number of power electronic devices is projected to grow significantly in the future grid [35], and thereby the applicability of PLC as a smart grid communication channel can be affected.

11.4.2.2　Similarities between PLC signals and supraharmonics

A brief introduction to supraharmonics

DER units are power electronic devices (e.g., distributed generators, energy storage devices, smart household appliances) that form a fundamental group of controllable loads in DR programs. The original components used in switching circuits of power electronic devices were diodes and thyristors. These switches were line-commutated and thereby in principle can only produce lower order harmonics (Hz–kHz).

The development of self-commutating switches, such as transistors, enabled significantly faster switching, and subsequently, such switching circuits generate residual electromagnetic ripples with significantly higher frequencies (kHz–MHz) that can propagate from device to device [41]. Commonly used applications of power electronics are, e.g., active power factor correction [i.e., the reduction of lower order emission (<2 kHz) for which the emission limits are regulated more stringently], pulse width modulation (i.e., the regulation of the magnitude and/or conversion of the power output), or switch mode power supplies (i.e., an application aimed to increase the overall energy efficiency of a device) [38]. Such applications enable demand flexibility and thereby DR and several other aspects of DSM.

HF disturbances that fall into the range of 2–150 kHz are often referred to as supraharmonics. These HF disturbances occur both as NB emission (bandwidth <5 kHz) and as broadband emission and, in this regard, are similar to PLC. The amplitude of supraharmonic emission in general is small, the supraharmonic current disturbance has an amplitude in the range of milliamperes. For example, in [42], the supraharmonic current disturbances of an LED lamp (8 W) were measured and a disturbance of a mere 0.75 mA was found. For larger devices such as PV installations (20 kW), disturbances of 120 mA and voltage distortions up to 1 V have been measured [43–46]. The amplitude of the current and voltage distortions depends on different factors such as network impedance, network topology, or number of HF emitting inverters connected locally to the grid. Under certain grid conditions, the HF voltage and current levels may increase, and the generated EMI has been reported to disturb electronic equipment, such as digital clocks, street lights, traffic lights, and touch dimmers. More specifically related to upholding the QoS of DR programs, these HF disturbances affect AMI components such as smart household appliances (e.g., EV charging stations, or PV inverters), produce erroneous smart meters readings, and compromise PLC.

The phase angles of traditional harmonic distortion (i.e., <2 kHz) among devices connected within an installation have a high degree of coherence [47]. Therefore, the amplitude of the aggregated lower order harmonic emissions at the point of connection (PoC) generally has a positive correlation to the number of power electronic devices, i.e., a larger number of devices produces a larger total harmonic amplitude. This is in contrast to supraharmonics, where a negative correlation is observed between the total emission from an installation at the PoC and the number of supraharmonic emitting devices [43]. However, at the terminals of each individual device, the measured current increases. This can be attributed to phase angle homogenization and due to so-called primary and secondary emission [46]: supraharmonic emission at the terminal of a device consists of two accumulative current disturbances. The current that is produced by the internal switching circuits within the power electronic device itself is referred to as primary emission. Primary emission has been reported to be influenced by three factors, i.e., the topology of the switching circuits within the electronic device, the local network impedance at the terminals of the device, and whether resonances are present in the network. Secondary supraharmonic emission of a device is believed to be the remnants of primary emission of other devices connected elsewhere in the installation or grid and that has propagated to the terminals of the device under test

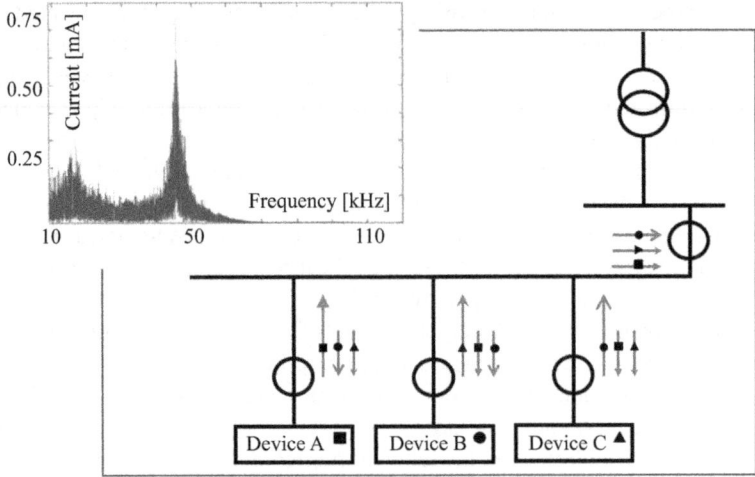

Figure 11.2   A simple schematic depiction of the propagation of supraharmonic
primary (large arrows) and secondary (small arrows) emission [46].
The inset (top left) illustrates the emission spectra of an LED lamp
(8 W), a common source of supraharmonics

(DUT). Secondary emission is affected by the impedance values of the grid as well
as the impedance values at the terminals of the DUT. Figure 11.2 depicts a simple
schematic representation of the propagation of supraharmonic primary and secondary
emission. As mentioned above, the total supraharmonic emission measured at the
terminals of the DUT is the accumulated effect of primary and secondary emission.
Furthermore, Figure 11.2 depicts that supraharmonic emission propagates between
the electronic devices connected within an installation or in a local area of the grid.
Hence, supraharmonics can potentially disrupt smooth operation of in-home EMS
and controllable DR loads. Another part of the emission migrates toward the PoC
and thereby further into the grid. However, grid impedance is frequency dependent,
and due to the inductive nature of the grid, the grid impedance is relatively high
for high-frequency distortions such as supraharmonics. Hence, it is expected that
the fraction of the supraharmonic emission that propagates into the public network
will be small. Moreover, due to the presence of capacitive EMC filters within the
electronic devices connected to the installation network will induce a relatively low
impedance at the supply side of the installation's PoC. At the grid side of the PoC
for traditional operating frequencies, the impedance is determined for the greatest
part by the inductance values of the transformer and the cables. However, a lot of
unknowns and uncertainty exists regarding the impedance values of the grid with
respect to HF voltages and currents. Therefore, ongoing research is required to give
insight into the effects of grid conditions on the propagation of supraharmonics, and
their undermining effect on the proper function of power electronic devices such as
DER units and thereby the QoS of DR programs [48].

Standardization in the supraharmonic and NB PLC frequency ranges

Standards for the traditional operating frequency range (<2 kHz) are sufficiently covered to insure reliable operation of power systems, as the amplitude of lower order harmonics is larger and can more readily disrupt normal operation. Thus, the presence and the efforts to mitigate such low-order harmonics have been more prominent [44]. Furthermore, nonconducted electromagnetic radiation can start to become significant for frequencies above 150 kHz. Thereby, broadcast telecommunication technologies such as radio can be disrupted. Consequently, strict emission limits for signals and disturbances with frequencies above 150 kHz have been set in place. Hence, for the 2–150-kHz range, a void had existed regarding standardization. Consequently, power electronic engineers designed devices that comply with existing emission limits (<2 or >150 kHz) and utilized the supraharmonic frequency band of 2–150 kHz to discard the switching circuit emissions [35]. However, the number of EMI related complaints are growing as the size and number of the power electronic devices grow, such as controllable DR loads like PV inverters and EV chargers. As a result, efforts have been directed by issuing bodies regarding the development of feasible standards regarding compatibility, emission, and immunity levels for coexistence between different electronic devices and communication technology such as PLC and radio, which will promote among others the operability of devices participating in DR programs.

NB PLC is achieved by ejecting intentional emissions within the frequency range of 3–148.5 kHz in Europe, with the upper frequency extending up to 500 kHz in the United States and Japan. Hence, the European NB PLC is encompassed by the supraharmonic frequency range. Supraharmonics are a form of nonintentional emission that can appear both as broadband and NB emission. Standards regarding the emission of NB signals can be found in EN 50065-1 and IEC 61000-3-8. The emission limits for the NB signaling range of 3–148.5 kHz coincide for the two standards [see Figure 11.3 (solid line, triangular marker)]. Limits for intentional signaling emission that include, i.e., limits for the emission of conducted disturbances are given in EN 50160 [49] [see Figure 11.3 (solid line, square marker)]. In general, the PLC signal power is dynamically reduced for frequencies higher than 9 kHz with the aim to minimize the interference with radio signals. Thus, PLC standards aim to allow side-by-side operation with radio communication technology which has intrinsic merit with respect to interoperability of the diverse devices that can participate in DR programs.

Nonintentional broadband emission limits (i.e., noncommunication signals) are defined in two standards: the CISPR 11 and 15 [50]. CISPR 11 aims to protect undisturbed functioning of high performance scientific, medical, or industrial equipment and sets mandatory emission and immunity limits for these devices. The standard CISPR 15 is aimed specifically at lighting equipment, a well-known source of supraharmonic emission [44]. The emission and immunity limits sets in CISPR 15 are equivalent to the limits set in CISPR 11, with the distinction that the limits in CISPR 15 are provisional, not mandatory. The technical specification IEC TS 62578 describes recommended maximum (nonintentional) emission limits typical of active infeed converters for 2–150 kHz. These converters are often used for, e.g., AC-to-DC conversion to store electrical energy and can provide high demand flexibility for DR schemes.

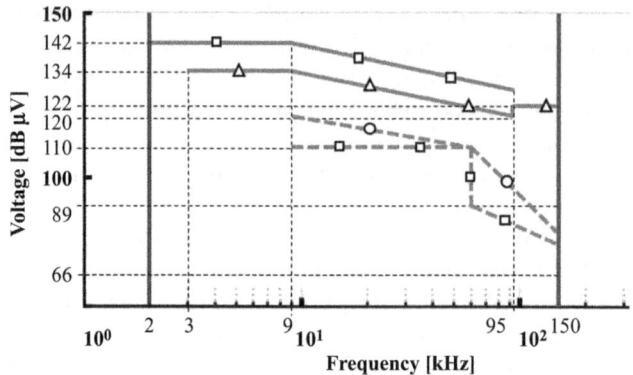

Figure 11.3   Narrowband emission limits according to EN 50160 (solid line, square marker) IEC 61000-3-8 (solid line, triangle marker). Broadband emission IEC TS 62578 (dashed line, circle marker) and CISPR 15 (dashed line, square marker)

These emission limits are shown in Figure 11.3 (dashed lines, circular and square marker).

Note that all broadband emission limits are lower than the NB limit to protect the integrity of communication technology such as PLC, and thereby protect the integrity of the AMI for DR. The International Electrotechnical Commission founded the technical committee TC 77. This committee aims to develop EMC standards [51]. TC 77 proposed for PLC that the frequency range of 2–30 kHz is the designated frequency band that can be used to discard emissions such as switching circuit remnants, and the frequency band of 30–148.5 kHz should be preserved for PLC.

False-positive PLC signals and degradation of signal-to-noise ratio
The previous section provides an outline of the standardization progress that has been made to support the integrity of PLC by curbing supraharmonic voltage variations produced by, e.g., lighting, and recommendations regarding converter emission limits have been initiated, which is an important work in progress as the number of converters will increase in the smart grid as the number of DER units in the distribution grid is expected to grow and to enable DR. The emission limits are given in maximum voltage deviations (dBV) which is a common practice in power system standards. However, there is a lack of knowledge on grid impedance values at HF. Thus, it is ambiguous what level of current deviations are produced by supraharmonics. When the grid impedance is high (e.g., due to low power demand), the current can increase significantly [52]. Moreover, the impedance values of the grid will be time-varying as the impedance is correlated to the to the highly varying load demand of feeders in the distribution grid. Hence, it is unclear what EMI levels and degrading effects on the AMI of DR are produced by supraharmonics solely based on voltage emission limits alone. Moreover, the research into supraharmonics is still relatively novel and

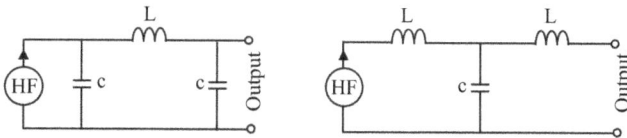

Figure 11.4   *Common low-pass filters used to mitigate high frequency (HF)*
                emission. (Left) A CLC-filter in which the capacitors provides a low
                impedance paths to drain high frequency emissions. (Right) A
                LCL-filter in which the inductors provides a high impedance path to
                obstruct HF emission

(publication of) extensive measurement studies are still lacking. Therefore, the possibility of false-positive PLC signals induced by suraharmonic emission at the PLC receiver cannot be excluded. Moreover, to successfully establish PLC, the signal-to-noise ratio (SNR) must be sufficiently high when the signal is detected by the receiver in order for the message carried in the signal to be decoded correctly. However, with an adequate selection of design criteria for signal modulation and coding, the receiver will be able to detect and decode the information even in environments with a relatively high degree of noise [53]. Still, it can be insightful to chart the suraharmonic emission levels in the future grid in order to operate and maintain an effective SNR, the effects on the AMI, and thereby the QoS of DR.

11.4.2.3   Attenuation of PLC signals due to capacitive shunting

EMC filters of power electronics

The switching circuits within power electronic devices such as DR controllable loads generate suraharmonics. To mitigate suraharmonics from entering the electrical network, an EMC filter is positioned between the device and the grid. The EMC filter is commonly an LCL and CLC low-pass filters (see Figure 11.4) [45,54]. For HF disturbances, capacitors induce a low-impedance pathway. Thus, suraharmonic emission will tend to flow between neighboring power electronic devices [46]. The HF current produced by a large electronic device can induce relatively high secondary emission that may flow toward the terminals of a small device connected to the same installation or local area network and damage the device itself or the filter. In practice, however, larger devices, such as DER units that participate in DR programs, will have an accordingly larger capacitive filter, and therefore, these larger devices are more prone to receive secondary emission from other devices connected to the installation or local area network [55]. As larger power electronic devices, such as DER units, i.e., prominent components of the DR AMI, are expected to proliferate in the future smart grid, adequate protection of these devices requires that the expected secondary emission within an environment have to be considered when designing the EMC filters of these devices. However, this is currently not the case although it does require further examination by manufacturers and task forces of standardization bodies.

PLC signal attenuation due to nondesignated shunting

Measurements taken from a laboratory setting as well as taken from field experiments have shown that PLC may interact with power electronic devices and controllable DR loads such as PV inverters and electrical storage devices [53]. The capacitive filters within the power electronic devices create a low impedance path. For example, PV inverters have shown to absorb supraharmonic emission from the grid as well as HF signals originating from other sources like PLC signals, i.e., a potential DR AMI communication channel [41]. Consequently, only a fraction of the injected PLC signal will reach the PLC receiver. The partially shunted PLC signal will inherently lead to a reduced SNR at the receiver end and thereby degrade the quality of the PLC, and thereby jeopardize the QoS of the DR program. Based on measurements taken from a lab setting and in the field, it has been concluded that that PLC signal attenuating due to nondesignated shunting by low impedance paths will most likely be the foremost reason for PLC signal degradation [53]. Furthermore, similar to supraharmonic disturbances, the shunted PLC signal can disturb normal operation of unintended recipient devices. Moreover, it has to be taken into consideration that the PLC current ripple can damage the (EMC filter of) unintended recipient devices as the emission levels of PLC are generally much higher than the design requirements of the electronic device generally allow. Hence, the reliability and compatibility of PLC as the main communication channel in a DR AMI with a large number of DER units, which induce inherent nondesignated low impedance paths for PLC signals, requires careful further study.

11.4.3   Standardization regarding demand response

The asymmetric development of DR programs in different electricity markets around the world [5] in combination with the manifold of enabling technologies that are available suggests that DR is highly fragmented from the real-world application and business point of view. Thus, developing standards in order to establish consistent protocols and promote compatibility and interoperability is essential in order to facilitate the design and increase the adoption rate of DR programs. Standards are also going to play an important role for end-users willing to enroll by allowing them to easily compare different DR program specifications and, therefore, minimize the discomfort or costs that are related to their participation and increase the received benefits. As a result, there are many efforts to standardize operational and business aspects of DR.

The US National Institute of Standards and Technology (NIST) is maintaining efforts to develop smart grid interoperability standards [56] in coordination with the government, industrial, and academic stakeholders. Standards related to smart grid operation, including a significant number of standards strongly related to DR, focusing particularly on communications have also been published by IEEE [57]. Moreover, the OpenADR Alliance that was formed in 2010 and is currently encompassing more than 130 industrial partners is aiming at standardizing and simplifying DR implementations [58]. At a national level, in Australia and New Zealand a common standard exists, namely, the AS/NZS 4755 entitled "DR capabilities and supporting technologies for electrical products" [59], while evaluation of standardization approaches is taking place in European countries [60].

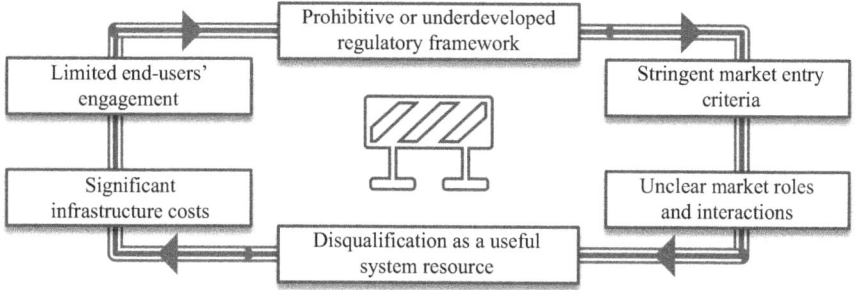

Figure 11.5 Barriers to demand response

11.5 Challenges and barriers to the development of demand response

Driven by the foreseen benefits, many countries around the world have initiated the development of DR programs. However, across different regions, the progress and the degree of adoption of DR programs is highly varying. This is the consequence of a series of challenges and barriers that need to be effectively addressed in order to enable active participation of DR in power system operations. In this section, a compilation of challenges and barriers to the development of DR are discussed. The main challenges are portrayed in Figure 11.5.

The first major obstacle that prevents the exploitation of DR is typically the regulatory framework that governs the operation of electricity markets and power systems. Such sets of rules have historically been developed in favor of large centralized generators. As a result, on many occasions, the regulatory framework in place is not in the position to reflect the technical diversity and dispersed nature of DR resources and, therefore, treat them equally with large generators in power system operation and planning. Subsequently, the feasible market participation of available demand flexibility is undermined. Typically, market rules specify the following criteria that the resources must fulfill in order to participate:

- The minimum amount of power a participant may bid and the bid direction: in many countries, e.g., in Europe [60], the minimum bid size remains large and, therefore, constitutes a direct barrier toward exploiting the flexibility that can be offered by relatively small consumers. In addition to that, many market structures are based on symmetric bids, implying that every participant must be able to change its market position in both upward and downward direction. In the case of DR, loads should be equally capable of decreasing and increasing their consumption, and thereby, effectively limiting the types of loads that can technically participate in the market.
- Possibility and extent of aggregating small consumers: in several markets aggregation is either not allowed or is geographically constrained. This might prevent

aggregators from complying with market entry criteria and might restrict DR to large industrial consumers [61].

• Number and characteristics of DR events: DR deployment on the basis of existing service definitions typically implies the unnecessarily prolonged interruption of service for the consumers that are involved which might deter utilities and/or consumers from engaging in DR.

The dominant liberalized environment in electricity markets has given way to multiple entities with different roles, responsibilities, and objectives. Although such a context might at first be deemed to be a fertile ground for DR applications, the absence of clear market roles and interactions between the stakeholders might in fact prove to be the second significant impedance to the widespread uptake of DR. Most DR resources are connected at the distribution system, i.e., at the jurisdiction area of distribution system operators (DSO). DSO would view DR as a valuable resource in order to ease local congestions and reduce losses in their system [62]. At the same time, the transmission system operators (TSO) would require DR to be used in such a way that facilitates the overall balance of the system. Such contrasting views and absence of coordination between DSO and TSO can potentially complicate the design of DR programs.

In the forefront of the challenges that need to be solved in order to enable the uptake of DR is the business model under which DR resources could participate in the market. The main business models that may be identified are direct contracts with TSO, aggregation, and response to real-time prices. Intense criticism is addressed to all the three. The first participation scheme, i.e., direct contracts with TSO, is characterized by stringent technical requirements for the participating resources, and therefore, it is potentially only suitable for a few large industrial customers. The second scheme, i.e., aggregation, might be suitable in order to involve smaller consumers. However, aggregations might compromise the key benefits of dynamic pricing both from the system and individual participant perspectives. The reason for this is that the aggregator that is participating in the market on behalf of the consumers might alter prices that are offered to the latter in order to induce demand behavior that will fulfill its obligations as an entity. As a result, these prices might no longer reflect the actual condition of the market or the power system [63]. Finally, real-time pricing based DR raises concerns regarding volatility both in demand and prices, as a result of the asymmetry between the communication of the price and the actual response of the load due to latency both in the communication network and human decision making [64].

Another category of barriers is related to the effects that significant deployment of DR might have on the established modus operandi of the power system and the market structures. Such concerns might underestimate the benefits of DR and pose it as a challenging and complicating factor instead. It has already been mentioned that potentially the most appealing application of DR is to facilitate the integration of RES in terms of responding in real time in order to compensate for their stochastic and intermittent nature. Nevertheless, the response of the load could limit the capacity factors and, thereby, reduce the revenues of conventional peaking and intermediate generators that have been traditionally providing regulation and reserve services. This can evidently lead stakeholders that own generating capacity to oppose to the introduction of DR in order to secure their investments or would relinquish conventional units and thereby reduce the generating capacity of the system. A strong argument in support of this controversy is that DR resources cannot provide critical services such as voltage support and do not have black start capability. Moreover, from the operational viewpoint, there are three main concerns related to the significant penetration of DR. First, the value of DR might not be significant in systems that already comprise flexible generating resources (e.g., fast natural gas fired turbines), as opposed to systems with high penetration of RES production and a relatively inflexible generator mix [65]. In addition to that, economically compensating DR is not as straightforward as valuing the contribution of generators [66]. The third complication is related to the lack of suitable and transparent tools for the measurement and verification of the effectivity of load response, since reducing consumption is not generally equivalent to increasing production [67]. This set of challenges must be carefully addressed in order to qualify DR as a useful system resource.

The fundamental technologies required for the development of DR applications are generally available. Nevertheless, the adoption levels of control, metering, and communication technologies in the power system should be significantly increased in order to enable the widespread uptake of DR [65]. Evidently, this constitutes a financial challenge as regards the investments that need to take place. Investment costs as a function of the technical requirements imposed by the existing regulatory framework have also been a barrier for industrial and commercial customers. The metering equipment that needs to be installed has to be in compliance with the telemetry requirements which have often been characterized unreasonable and prohibitive [68]. Residential DR might also be underexploited to an extent due to the increased cost of residential EMS in conjunction with social parameters [69].

Last but not the least, developing successful DR programs requires consumer engagement. This is perhaps one of the greatest challenges since lack of consumer interest is a definite undermining factor [1]. Consumer engagement is hindered mainly due to three reasons. First, electricity end-users, especially residential and small commercial consumers, do not necessarily behave rationally from an economic perspective. The perception that consuming energy on demand is more valuable than the compensation that is offered discourages consumers enrolling in DR programs. The second challenge that needs to be addressed in this domain is that retail contracts are typically perceived as complicated and unclear [70]. Finally, the absence of a clear regulatory framework that guarantees the preservation of privacy, as well as cyber security concerns, is generally considered as challenges that need to be addressed prior enabling the widespread roll-out of DR programs.

11.6 Conclusions

This chapter provided an overview of DSM and especially, DR. In the past years, a wide range of DR programs have been developed in different power systems across the world. These programs aim at engaging all the types of consumers: from large industrial customers to residential end-users with relatively small electricity consumption, considering also special types of consumers such as the EV and data centers. The primary motive for developing DR programs is that by enabling the participation of the demand side in electricity markets significant benefits are anticipated: more efficient and sustainable system planning, enhancement of the operation of the distribution system, lower and more stable electricity prices in the long run, mitigation of the market power of several participants and promotion of competition, economic benefits for the consumers, and increased operational flexibility. Increased operational flexibility is directly linked to accommodating the handicaps of the trend that indicates that significant amount of variable RES generation will be introduced in power systems in the future. The current advancement in metering, information, communication, and control infrastructure (ICT) allows for the development of DR programs aiming to engage different types of customers by giving them appropriate incentives. It is interesting to notice that despite the absence of homogeneity, there are efforts to develop DR programs at an international level, clearly indicating that utilities have started considering DR as a useful addition to their system rather than a complicating factor. The required infrastructure to implement DR programs is nowadays generally available, and therefore, the barriers that should be addressed in order to further promote the participation of the demand are mainly regulatory and economic.

The practical implementation of DR programs highly depends on the availability of ICT technology and in particular the communication means that are used in order to monitor and control loads. Among the different wired and wireless options, PLC is a technology that has attracted particular interest due to the fact that its carrier is the power system itself. The basic incentives to use PLC technology in the smart grid ICT infrastructure and for the DR AMI are that (1) the deployment costs are considered comparable to wireless since the lines are readily available and (2) the power grid provides an infrastructure that is in principle more extensive and pervasive than any other wired or wireless alternative. The emergence of high data rate NB PLC (<500 kb/s, 148.5 kHz) opens up the possibility for more advanced power system applications to support DSM and grid operation in the future smart grid.

Modern power electronic devices, such as controllable DR loads, contain switching circuits that produce HF disturbances in the frequency range of 2–150 kHz, the so-called supraharmonics. The EMI induced by these supraharmonics have been reported to disturb PLC. Standardization efforts are ongoing to support the integrity of PLC by curbing supraharmonic voltage variations. Emission limits of lighting and active infeed inverters, i.e., well-known sources of supraharmonics, are given in maximum voltage deviations (dBV) which is common practice in power system standards. However, due to the lack of knowledge of grid impedance values for high frequency phenomena, the emission limits are ambiguous regarding the current deviations that are produced. Hence, it is not clear what EMI levels are produced solely based on voltage emission limits alone. Therefore, the possibility of false-positive PLC signals induced by supraharmonics at the PLC receiver cannot be excluded and undermine the reliability of DR programs. Moreover, to successfully establish PLC, the SNR should be large enough for the receiver to correctly interpret the data. It is necessary to collect more measurement data to get insight into the supraharmonic emission levels in the future grid in order to operate and maintain an effective SNR and QoS of DR programs.

An EMC filter is placed at the grid side of the power electronic switching device such as DR controllable loads in order to mitigate the supraharmonic ripple that is injected into the network. The capacitive filters create a low impedance path within an installation and in the grid along LV feeders. For example, PV inverters have shown to absorb supraharmonic emission from the grid as well as HF signals originating from other sources like PLC signals. The partially shunted PLC signal will inherently lead to a reduced SNR at the receiver's end and thereby degrade the quality of the PLC. Based on measurements taken from a lab setting and in the field, it is believed that attenuating by shunting is, and will be, the most common reason for PLC degradation. Likewise, to supraharmonics EMI disturbances, the shunted PLC signal can disturb normal operation of unintended recipient devices. Moreover, it has to be taken into consideration that the high emission levels of PLC can damage the unintended recipient devices that absorb the HF PLC current ripple. Hence, the reliability and compatibility of PLC as the main communication channel in a DR AMI with a large number of power electronic controllable loads, which induce nondesignated low impedance paths for PLC signals, require careful further examination by both manufacturers as well as task forces of standardization bodies.

References

[1]   Luthra S, Kumar S, Kharb R, *et al.* Adoption of smart grid technologies: an analysis of interactions among barriers. Renewable and Sustainable Energy Reviews. 2014;33:554–565.

[2]   World Energy Outlook 2016. Part B: special focus on renewable energy. International Energy Agency; 2016.

[3]   Gellings C, Parmenter K. Chapter 15 – Demand-Side Management. In: Kreith F, Goswami DY, editors. Energy Efficiency and Renewable Energy Handbook, Second Edition. Boca Raton, FL: CRC Press; 2016. p. 289–310.

[4]   Benefits of Demand Response in Electricity Markets and Recommendations for Achieving Them. A report to the United States Congress Pursuant to Section 1252 of the Energy Policy Act of 2005. U.S. Department of Energy; February 2006.

[5]   Paterakis NG, Erdinç O, Catalão JPS. An overview of demand response: key-elements and international experience. Renewable and Sustainable Energy Reviews. 2017;69(Supplement C):871–891.

[6]   Henriksson E, Sderholm P, Wrell L. Industrial electricity demand and energy efficiency policy: the role of price changes and private R&D in the Swedish pulp and paper industry. Energy Policy. 2012;47(Supplement C):437–446.

[7]   Hao H, Lin Y, Kowli AS, *et al.* Ancillary service to the grid through control of fans in commercial building HVAC systems. IEEE Transactions on Smart Grid. 2014;5(4):2066–2074.

[8]   Georgievski I, Degeler V, Pagani GA, *et al.* Optimizing energy costs for offices connected to the smart grid. IEEE Transactions on Smart Grid. 2012;3(4):2273–2285.

[9]   Sundstrom O, Binding C. Flexible charging optimization for electric vehicles considering distribution grid constraints. IEEE Transactions on Smart Grid. 2012;3(1):26–37.

[10]   Richardson DB. Electric vehicles and the electric grid: a review of modeling approaches, impacts, and renewable energy integration. Renewable and Sustainable Energy Reviews. 2013;19(Supplement C):247–254.

[11]   Report to Congress on Server and Data Center Energy Efficiency. U.S. Environmental Protection Agency; August 2007.

[12]   Parvania M, Fotuhi-Firuzabad M. Integrating load reduction into wholesale energy market with application to wind power integration. IEEE Systems Journal. 2012;6(1):35–45.

[13]   Yousefi A, Iu HHC, Fernando T, *et al.* An approach for wind power integration using demand side resources. IEEE Transactions on Sustainable Energy. 2013;4(4):917–924.

[14]   Moslehi K, Kumar R. A reliability perspective of the smart grid. IEEE Transactions on Smart Grid. 2010;1(1):57–64.

[15]   Finn P, Fitzpatrick C, Connolly D, *et al.* Facilitation of renewable electricity using price based appliance control in Ireland's electricity market. Energy. 2011;36(5):2952–2960.

[16]   Kirby B, Milligan M. Capacity Requirements to Support Interbalancing Area Wind Delivery. Golden, CO: National Renewable Energy Laboratory; 2002.

[17]   Stadler I. Power grid balancing of energy systems with high renewable energy penetration by demand response. Utilities Policy. 2008;16(2):90–98. Sustainable Energy and Transportation Systems.

[18]   O'Connell N, Pinson P, Madsen H, *et al.* Benefits and challenges of electrical demand response: a critical review. Renewable and Sustainable Energy Reviews. 2014;39(Supplement C):686–699.

[19]   Hanser P, Madjarov K, Katzenstein W, *et al.* Chapter 10 – Riding the Wave: Using Demand Response for Integrating Intermittent Resources. In: Sioshansi FP, editor. Smart Grid. Boston, MA: Academic Press; 2012. p. 235–256.

[20]   Milligan M, Kirby B. Utilizing Load Response for Wind and Solar Integration and Power System Reliability. National Renewable Energy Laboratory; 2010 [cited 2017 Oct 10]. Available from: http://www.nrel.gov/docs/fy10osti/48247.pdf.

[21]   Dahlke S, McFarlane D. The environmental benefits of demand response; [cited 2017 Oct 10]. Available from: http://www.betterenergy.org/blog/environmental-benefits-demand-response.

[22]   Albadi MH, El-Saadany EF. A summary of demand response in electricity markets. Electric Power Systems Research. 2008;78(11):1989–1996.

[23]   Elburg HV. Dutch Energy Savings Monitor for the Smart Meter; March 2014 [cited 2017 Oct 10]. Available from: http://english.rvo.nl/

sites/default/files/2014/06/Dutch%20Smart%20Meter%20Energy%20savings
%20Monitor%20final%20version.pdf.

[24] OFGEM. UK Energy Demand Research Project (EDRP); January 2011 [cited 2017 Oct 10]. Available from: https://www.ofgem.gov.uk/gas/retail-market/metering/transition-smart-meters/energy-demand-research-project.

[25] E-Energy. Development of Digital Technologies; November 2017 [cited 2017 Oct 10]. Available from: http://www.digitale-technologien.de/DT/Redaktion/EN/Downloads/Publikation/entwicklung-digitaler-technologien-en .pdf?__blob=publicationFile&v=3.

[26] Khalifa T, Naik K, Nayak A. A survey of communication protocols for automatic meter reading applications. IEEE Communications Surveys Tutorials. 2011;13(2):168–182.

[27] Bayram IS, Ustun TS. A survey on behind the meter energy management systems in smart grid. Renewable and Sustainable Energy Reviews. 2017;72(Supplement C):1208–1232.

[28] Sharma K, Saini LM. Performance analysis of smart metering for smart grid: an overview. Renewable and Sustainable Energy Reviews. 2015;49(Supplement C):720–735.

[29] Yan Y, Qian Y, Sharif H, *et al.* A survey on smart grid communication infrastructures: motivations, requirements and challenges. IEEE Communications Surveys Tutorials. 2013;15(1):5–20.

[30] Vardakas JS, Zorba N, Verikoukis CV. A survey on demand response programs in smart grids: pricing methods and optimization algorithms. IEEE Communications Surveys Tutorials. 2015;17(1):152–178.

[31] Yan Y, Qian Y, Sharif H, *et al.* A survey on cyber security for smart grid communications. IEEE Communications Surveys Tutorials. 2012;14(4): 998–1010.

[32] Galli S, Scaglione A, Wang Z. For the grid and through the grid: the role of power line communications in the smart grid. Proceedings of the IEEE. 2011;99(6):998–1027.

[33] Bartak GF, Abart A. EMI of emissions in the frequency range 2 kHz–150 kHz. In: 22nd International Conference and Exhibition on Electricity Distribution (CIRED 2013); 2013.

[34] Subhani S, Gibescu M, Kling WL. Autonomous control of distributed energy resources via wireless machine-to-machine communication; a survey of big data challenges. In: 2015 IEEE 15th International Conference on Environment and Electrical Engineering (EEEIC); 2015. p. 1437–1442.

[35] Bollen M, Olofsson M. Consumer electronics and the power grid: what are they doing to each other? IEEE Consumer Electronics Magazine. 2015;4(1):50–57.

[36] Huh JH, Otgonchimeg S, Seo K. Advanced metering infrastructure design and test bed experiment using intelligent agents: focusing on the PLC network base technology for Smart Grid system. The Journal of Supercomputing. 2016;72(5):1862–1877.

322 *Power line communication systems for smart grids*

[37]   205A CS. Study Report on Electromagnetic Interference between Electrical Equipment/Systems in the Frequency Range below 150 kHz. CENELEC; 2010. SC205A/Sec0260/R.

[38]   Bollen M, Olofsson M, Larsson A, *et al.* Standards for supraharmonics (2 to 150 kHz). IEEE Electromagnetic Compatibility Magazine. 2014;3(1):114–119.

[39]   Rönnberg SK, d Castro AG, Bollen MHJ, *et al.* Supraharmonics from power electronics converters. In: 2015 9th International Conference on Compatibility and Power Electronics (CPE); 2015. p. 539–544.

[40]   Rönnberg SK, Bollen MHJ, Wahlberg M. Interaction between narrowband power-line communication and end-user equipment. IEEE Transactions on Power Delivery. 2011;26(3):2034–2039.

[41]   Rönnberg SK, Bollen MHJ, Larsson A. Emission from small scale PV-installations on the low voltage grid. Renewable Energy and Power Quality Journal. 2014. ISSN 2172-038 X.

[42]   Rönnberg SK, Bollen MHJ. Emission from four types of LED lamps at frequencies up to 150 kHz. In: 2012 IEEE 15th International Conference on Harmonics and Quality of Power; 2012. p. 451–456.

[43]   Larsson EOA, Bollen MHJ. Measurement result from 1 to 48 fluorescent lamps in the frequency range 2 to 150 kHz. In: Proceedings of 14th International Conference on Harmonics and Quality of Power – ICHQP 2010; 2010. p. 1–8.

[44]   Larsson EOA, Bollen MHJ. Emission and immunity of equipment in the frequency range 2 to 150 kHz. In: 2009 IEEE Bucharest PowerTech; 2009. p. 1–5.

[45]   Klatt M, Meyer J, Schegner P, *et al.* Emission levels above 2 kHz – Laboratory results and survey measurements in public low voltage grids. In: 22nd International Conference and Exhibition on Electricity Distribution (CIRED 2013); 2013. p. 1–4.

[46]   de Castro AG, Rönnberg SK, Bollen MHJ. A study about harmonic interaction between devices. In: 2014 16th International Conference on Harmonics and Quality of Power (ICHQP); 2014. p. 728–732.

[47]   Mansoor A, Grady WM, Chowdhury AH, *et al.* An investigation of harmonics attenuation and diversity among distributed single-phase power electronic loads. IEEE Transactions on Power Delivery. 1995;10(1):467–473.

[48]   Schöttke S, Rademacher S, Meyer J, *et al.* Transfer characteristic of a MV/LV transformer in the frequency range between 2 kHz and 150 kHz. In: 2015 IEEE International Symposium on Electromagnetic Compatibility (EMC); 2015. p. 114–119.

[49]   European Standard for Engineering. Voltage Characteristics of Public Distribution Systems; 2010. Standard EN 50160.

[50]   European Standard for Engineering. Signaling on Low Voltage Electrical Installations in the Frequency Range 3 kHz to 148.5 kHz; 1991–2002. Standard EN 50160.

[51]   Moreno-Munoz A, de Castro AG, Rönnberg S, *et al.* Ongoing work in CIGRE working groups on supraharmonics from power-electronic converters. In: 23rd International Conference on Electricity Distribution; 2015.

[52]   Verhelst B, Ryckeghem JV, Desmet J. Influence of grid-connected inverters on the power quality of the distribution grid. In: 2014 International Symposium on Power Electronics, Electrical Drives, Automation and Motion; 2014. p. 670–675.

[53]   Rönnberg SK, Bollen MHJ, Wahlberg M. Interaction between narrowband power-line communication and end-user equipment. IEEE Transactions on Power Delivery. 2011;26(3):2034–2039.

[54]   Bollen M, Shyar HH, Rönnberg S. Spread of high frequency current emission. In: 22nd International Conference and Exhibition on Electricity Distribution (CIRED 2013); 2013. p. 1–4.

[55]   de Castro AG, Rnnberg SK, Bollen MHJ. Harmonic interaction between an electric vehicle and different domestic equipment. In: 2014 International Symposium on Electromagnetic Compatibility; 2014. p. 991–996.

[56]   NIST. Smart Grid Interoperability Standards; [cited 2017 Oct 10]. Available from: http://www.nist.gov/smartgrid/.

[57]   IEEE Smart Grid Standards; [cited 2017 Oct 10]. Available from: http://smartgrid.ieee.org/resources/standards/.

[58]   OpenADR Standard; [cited 2017 Oct 10]. Available from: http://www.openadr.org.

[59]   Australia/New Zealand AS/NZS 4755 Standard – Demand Response Capabilities Supporting Technologies for Electrical Products; 2014.

[60]   Mapping Demand Response in Europe Today. Brussels, BE: Smart Energy Demand Coalition (SEDC); 2015.

[61]   STOR Market Information Report. National Grid; 2013.

[62]   ENTSO-E. Demand Side Response Policy Paper; September 2014 [cited 2017 Oct 10]. Available from: http://www.entsoe.eu.

[63]   Oh H, Thomas RJ. Demand-Side Bidding Agents: Modeling and Simulation. IEEE Transactions on Power Systems. 2008;23(3):1050–1056.

[64]   Nyeng P, Kok K, Pineda S, *et al.* Enabling demand response by extending the European electricity markets with a real-time market. In: IEEE PES ISGT Europe 2013; 2013. p. 1–5.

[65]   Strbac G. Demand side management: benefits and challenges. Energy Policy. 2008;36(12):4419–4426.

[66]   Cutter E, Woo CK, Kahrl F, *et al.* Maximizing the value of responsive load. The Electricity Journal. 2012;25(7):6–16.

[67]   Hu Z, Kim J-h, Wang J, *et al.* Review of dynamic pricing programs in the U.S. and Europe: status quo and policy recommendations. Renewable and Sustainable Energy Reviews. 2015;42:743–751.

[68]   Zimmerman M. The industry demands better demand response. In: 2012 IEEE PES Innovative Smart Grid Technologies (ISGT); 2012. p. 1–3.

[69]   Asare-Bediako B, Kling WL, Ribeiro PF. Home energy management systems: evolution, trends and frameworks. In: 2012 47th International Universities Power Engineering Conference (UPEC); 2012. p. 1–5.

[70]   He X, Keyaerts N, Azevedo I, *et al.* How to engage consumers in demand response: a contract perspective. Utilities Policy. 2013;27:108–122.

Chapter 12

PLC for monitoring and control of distributed generators in smart grids

Anton Poluektov[1], Antti Pinomaa[1], Antti Kosonen[1],
Aleksei Romanenko[1], and Jero Ahola[1]

12.1 Introduction

12.1.1 Grid faults and islanding

Recently, smart grids employing distributed generation (DG) technologies have evolved rapidly and become widespread. By definition, a smart grid has implemented functionalities such as load management, demand response, and grid protection. These functionalities aim at providing a higher efficiency, a lower environmental impact, and a higher safety compared with conventional/traditional power grids. Nevertheless, new safety-related issues may arise. One of these challenges is an unintentional islanding condition, which is a particular case of a loss-of-mains (LoM) fault.

In a conventional power-distributed grid, a LoM in the grid branch causes a power outage in the supplied grid segment if no auxiliary power distribution line feeder is available. In a smart grid with DG units, a LoM can result in islanding. Islanding describes a condition where a customer(s) becomes disconnected from the main power utility grid being still supplied by DG units. An islanding condition may interfere with the power balance and produce overcurrents and overvoltages, which may harm electrical loads. Moreover, the grid personnel may not be aware that a part of the grid is energized, which may lead to personal injuries. Obviously, an islanding condition should be detected in a shortest period of time in order to apply safety mechanisms and prevent an aftermath.

12.1.2 Standardization and legislation

Introduction of DG units and their functionalities (including fault detection) is a subject of international and local standardization and legislation. Global standards institutions, such as IEEE and IEC, provide requirements and testing conditions for

[1]LUT School of Energy Systems, Lappeenranta University of Technology (LUT), Finland

islanding detection [1]. The IEC 61727 standard addresses photovoltaic (PV) generation systems and power quality issues, such as the voltage and frequency range in the
normal operating mode. Moreover, this standard investigates safety-related matters.
On islanding, the standard states that a PV unit should disconnect from the utility
network within 2 s after the fault has been detected. The IEC 62116 standard defines
the test conditions and the test procedure for islanding prevention measures for PV
systems.

The IEEE 1547 standard defines the introduction of DG units into a power grid.
This standard provides a time period of 2 s for fault detection and interruption of power
production in the DG unit [2]. The IEEE 929 standard addresses PV systems and
provides procedures to examine the islanding-detection capability of the system [3].
The UL 1741 standard is often applied by PV inverter manufacturers. In issues related
to islanding detection and operating ranges, the standard is based on the IEEE 1547
standard [2,4].

12.1.3   Islanding-detection methods

A number of islanding-detection solutions are available nowadays. For convenience,
they can be divided into three groups by their operating principles. These groups
cover passive, active, and communication-based islanding-detection methods. Anti-
islanding solutions can vary by complexity, cost, fault detection accuracy, and speed.

Passive methods apply monitoring of sudden deviations in the characteristics of
the supplied power [1]. Such passive systems typically consider voltage, frequency,
phase angle, particular harmonics, or total harmonic distortion. When a monitored
parameter exceeds a set threshold, it is interpreted as a fault, and protective measures
are triggered, disconnecting the DG power converter from the grid or switching the
grid to a controlled island mode [1,5]. It is pointed out that passive methods may be
ineffective when the total power consumption of an island nearly matches the power
production of the DG units in the island [6,7]. In such a scenario, disconnection from
the main power utility may not lead to severe fluctuation in the monitored parameters,
and the protective system may not detect a fault.

When the accuracy of a fault detection system is discussed, the term of nondetection zone (NDZ) is commonly used. This term describes the scope of scenarios
under which the islanding condition is not detected by the fault-detection system.

In contrast to passive methods, active solutions aim at minimizing the NDZ
in a case when the island power consumption nearly matches the power supplied
by the DG units. Active anti-islanding solutions employ artificial disturbance injection and grid feedback monitoring [1,5]. The first subgroup of solutions considers
disturbances in grid variables that are injected into the DG control unit. These disturbances should be corrected by the voltage and frequency control of the grid under
normal operating conditions [5]. The second subgroup of solutions addresses channel impedance. This approach was adopted in [8,9]. Transient active methods apply
an impedance measuring device, which injects a short disturbing signal and ana-
lyzes the channel feedback (e.g., voltage is injected and current is analyzed) [5].

In steady-state solutions, a harmonic component is injected and the channel impedance is then analyzed [5]. A drawback of this particular approach is that an islanding condition can be detected only at the injection point. Moreover, a major drawback of active solutions is that artificial disturbances decrease the quality of the supplied power. Therefore, the disturbances to be used are designed to be as small as possible. When a high number of DG units using active anti-islanding are present in the grid, an issue of mutual interference arises and the fault-detection accuracy decreases.

The last group of islanding-detection methods covers communication-based solutions, which employ the communication between the power utility and the DG unit. Several groups of communication-based methods can be distinguished, which differ in the grid components employed in communication. The first group of solutions is based on supervisory control and data acquisition (SCADA). Voltage sensors are installed at DG units, and they are connected to a SCADA system. When islanding is detected, sensor(s) are triggered, and they transmit a warning signal over the SCADA network. The second group of methods applies communication with a utility recloser. When the recloser opens, a warning signal is generated in the signaling device installed at the recloser, and it is transmitted to other DG units using a designated communication channel [1,5]. The last group of methods applies power line communication (PLC). The main power supply utility is communicating with the inverters installed at DG units on the power line. When the signal is lost, an islanding condition is detected.

Communication-based methods, in general, are characterized by a higher cost and complexity compared with active and passive methods. Moreover, interruption in data transmission for instance as a result of a communication medium fault makes the system inactive [1]. On the other hand, communication-based solutions have a smaller negative effect on the power quality than active methods, and they have the smallest NDZ compared with active and passive methods [1]. Moreover, these methods provide a wider functionality, which includes data transmission over the grid [1].

A PLC-based approach for fault detection can be distinguished. Such a system shares the typical benefits of the communication-based methods compared with active and passive systems. A PLC system, unlike other communication-based systems, uses a communication channel, which is already there in the power grid. Thus, a separate communication network is not needed, which provides a beneficial economic effect. In order to define and design a PLC-based islanding-detection system, the application field, that is, power distribution grids, has to be investigated.

12.2 Application field

The application field is investigated by considering noise scenarios, channel attenuation characteristics, grid topology, power-distribution transformers, and signaling standards. These together have an effect on the PLC system characteristics and performance.

12.2.1   Noise scenario

The first aspect under discussion is channel noise scenarios. According to [10], five types of channel noise can be described:

1. Colored background noise,
2. Narrowband noise,
3. Periodic impulsive noise synchronous to the mains frequency,
4. Periodic impulsive noise asynchronous to the mains frequency, and
5. Asynchronous impulsive noise.

Colored background noise can be characterized by a relatively low power spectral density (PSD). Nevertheless, PSD is inversely related to the frequency, and it increases in the low-frequency range. This type of noise is caused by several noise sources with different noise amplitudes, for instance by common household appliances. Noise can cause disturbances in the frequency range up to 30 MHz, and its PSD may vary over time [10,11].

Narrowband noise typically consists of modulated sinusoids. The source of this noise is broadcast radio stations. This type of noise can be active in the range of 1–22 MHz [10,11].

Periodic impulsive noise asynchronous to the mains has a repetition rate of 50–200 kHz, and it is caused by switching power supplies [10]. Similar to the two previous types of noise, this noise is generally stationary over periods of seconds, minutes, and in some cases, hours [10]. Thus, these three types of noise can be interpreted as background noise [10].

Periodic impulsive noise synchronous to the mains frequency has a repetition rate of multiples of the mains frequency. This noise is mainly caused by power supplies, for instance by switching of rectifiers in the DC supplies. The PSD of this noise decreases with an increase in frequency [10].

Asynchronous impulsive noise is caused by switching transients [10]. As the name indicates, this type of noise has a random occurrence [10]. Moreover, noise pulses may occur in a random frequency range, possibly up to 20 MHz [11]. The pulses may have a random duration, varying from several microseconds to milliseconds [10]. The PSD of the noise can exceed the background noise by 50 dB [10]. Because of the complexity of prediction, this type of noise was ignored in the PLC concept design for a long time [11]. Nevertheless, because of the high PSD, this noise may cause severe disturbances and affect the PLC performance [11].

Despite the fact that different noise components are active in different frequency bands, it can be concluded, in general, that the frequency range below 20 MHz is the most suitable one for the PLC considering the impact of noise components [11].

Noise power spectral measurements in the MV channel were made at the primary substation at Tommola, Finland (Figure 12.1). The frequency band below 1 MHz was studied. It can be seen that in the frequency range above 200 kHz, the PSD is nearly constant, being close to -105 dB m/Hz.

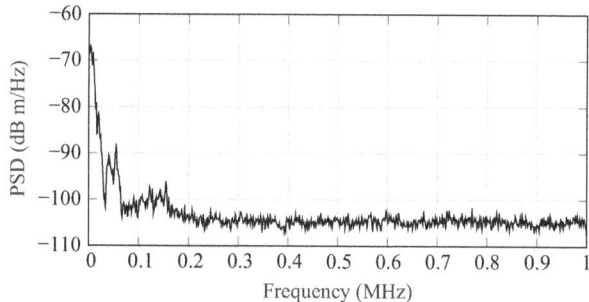

Figure 12.1 Noise PSD measurements in the MV channel at the primary substation

12.2.2 Channel attenuation

Channel attenuation is the most essential aspect to be investigated when a suitable frequency band for the LoM application is considered. The PLC channel attenuation can be categorized into the following classes: cable attenuation, attenuation caused by multiple power-dissipating loads, attenuation resulting from impedance mismatches, and attenuation caused by coupling interfaces [12].

Let us first consider cable attenuation, which is expressed by the channel attenuation coefficient (α). α is determined per cable kilometer, and it increases as a function of frequency (f) as

$$\alpha(f) = \alpha_R(f) + \alpha_G(f) = r \cdot \sqrt{f} + g \cdot f, \tag{12.1}$$

where $\alpha_R(f)$ represents attenuation generated by conductor losses (skin effect), $\alpha_G(f)$ represents attenuation generated by dielectric losses within the cable insulation, r denotes the frequency-dependent resistance, f denotes the signaling carrier frequency, and g denotes the conductance.

In this way, α is defined by the characteristics of the conductors and the insulation materials used in the power lines (underground cables, overhead lines). There are numerous technical differences between underground and overhead power delivery, which highly influence the attenuation characteristics of the power system. Underground cables have a higher conductor volume and insulation, which is not required for overhead lines. Typical insulation materials for low-voltage underground cables are polyvinyl chloride and polyethylene, which produce different dielectric losses and thereby different channel attenuation characteristics. In general, the structure of underground cables is more complex, including conductors, insulation materials, and cable shields, which raise the total cable cost. Nevertheless, damage to the cable insulation leads to a fault, which is more complicated to locate compared with an overhead power line damage. On the other hand, overhead power lines are more vulnerable to storms (weather influence) and radio interference.

Based on Shannon's theory, we may conclude that a frequency band with low attenuation and a low noise PSD is the most beneficial one for the PLC concept owing to the higher signal-to-noise ratio (SNR). The available bandwidth with the higher

SNR leads to a higher channel capacity, as demonstrated in the following equation, which is derived from Shannon's law:

$$Channel\ capacity = Bandwidth \cdot \log_2(1 + SNR) \tag{12.2}$$

Taking into account the channel noise scenarios, the channel attenuation, and the grid topology considerations given in the application field discussion, signaling in the frequency band below 1 MHz can be considered feasible for a PLC anti-islanding system.

12.2.3   Grid topology

Yet another aspect to be considered is the grid topology. A typical power distribution grid consists of two main parts—the medium-voltage (MV) and low-voltage (LV) grids (Figure 12.2). These grids have different power line lengths, line installation types, and branching.

An MV grid has power line lengths up to tens of kilometers, while LV lines are much shorter, typically from several hundred meters to a few kilometers.

Compared with an LV line, a typical MV line is less branched. Each line branch divides the signal power depending on the line impedance, thus limiting signal propagation through the grid. The power line installation type may vary between the MV and LV grids, meaning that they may contain overhead or underground lines. Moreover, the line installation type and the line length highly depend on the environment. Higher lengths and a low number of branches are typical for rural areas. Therefore, the grid topology, being a vital aspect of the PLC design, has to be considered in every particular case.

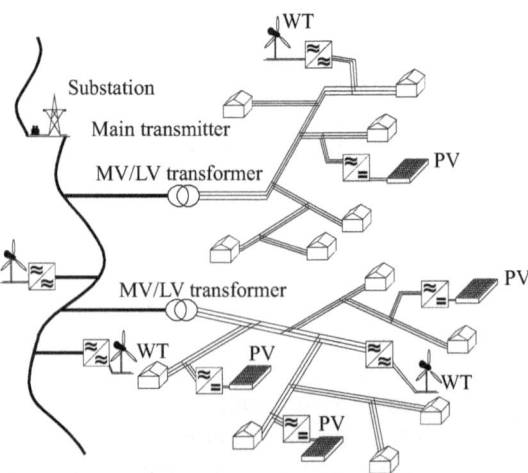

Figure 12.2   Typical MV/LV power distribution grid with DG units

12.2.4 Power distribution transformer

When considering the grid topology, the MV/LV power distribution transformer, being an essential grid component, cannot be neglected. The distribution transformer poses significant challenges, when PLC signaling is performed through the transformer, as it acts as a source of high channel attenuation (up to 30–40 dB) and disturbance [13]. Generally, there are two approaches to handle this aspect.

In the first approach, signaling is performed directly through the transformer [14]. In this case, channel analysis is carried out first in order to establish a frequency range with the highest gain [15]. Moreover, conventional signal carrier modulation techniques applied for signaling are often substituted by communication using wideband modulation and channel access technologies, such as direct sequence spread spectrum (DSSS) and orthogonal frequency division multiplexing (OFDM) [15–17].

The second approach is based on bypassing the transformer. Bypassing signaling concepts apply various schemes; signaling can be performed by using wireless communication or a bypassing communication circuit with additional coupling components [13,15].

In order to investigate the transformer channel characteristics, channel input impedance and gain-phase measurements were performed in the laboratory, examining a Dyn11 50 kVA, 50 Hz, 20,500/410 V oil-immersed distribution transformer (KONCAR, 5TBNp 50-24/ED). During the study in [18], an MV/LV transformer channel without coupling interfaces and channel loads was investigated. Based on the results, a frequency range above 200 kHz was chosen as the most beneficial one for PLC signaling, as it is characterized by a higher gain compared with lower frequencies. After that, channel gain measurements were performed using a laboratory setup, presented in Figure 12.5. On both sides of the transformer, an inductive coupling interface, a balun, and a signal filter were used. After the first test, a signal amplifier and a 15 dB attenuator were added at the MV side of the transformer. Channel measurements are presented in Figure 12.3. It can be seen that a frequency band

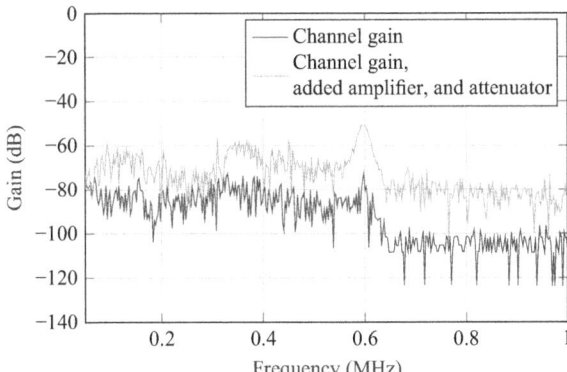

Figure 12.3   MV/LV distribution transformer channel gain measurements. In the second case, a signal amplifier and a 15 dB attenuator are added

of 200–600 kHz provides the highest gain and thus can be chosen for signaling over the transformer.

12.3   Design of a PLC solution

12.3.1   Signaling scheme

A typical PLC-based anti-islanding solution is based on signaling from the main power utility toward the DG units and the customers in the LV grid [1]. To this end, a signal transmitter is installed at the main substation, and the customers and the DG units are equipped with signal receivers.

Taking into consideration the application field characteristics discussed above, signal propagation through the grid is a matter of concern. Thus, taking into account the total signal attenuation and distortion, certain measures have to be introduced to provide communication over the grid. Signal repeaters installed on the power lines can improve signal propagation. These units operate as signal transceivers combining signal reception and transmission functionalities. At the same time, intermediate communication devices increase the communication latency. In the context of a PLC islanding-detection system, it means a decrease in the fault-detection speed.

12.3.2   Coupling interfaces

As discussed above, the total channel attenuation includes an attenuation component in the coupling interface. Therefore, the design of the coupling interface is a vital element of the system design. Two alternative coupling methods can be named; capacitive and inductive coupling.

A capacitive coupling interface is connected in parallel to the channel. In general, this solution can be considered a low-cost solution for low-voltage applications. Therefore, capacitive coupling is widely used in PLC applications using remote metering and in home applications [19]. The drawback of this type of coupling is the price of the capacitor, which is economically unfeasible for medium-voltage channels comparing to inductive coupling interfaces [19,20]. Moreover, a small load impedance resulting from a parallel connection is problematic for the capacitive coupling [19,20].

An inductive coupling interface, in turn, applies a series connection to the channel. Inductive coupling is not used in consumer PLC applications, even though it is widely applied to PLC in the electricity grids in the high-frequency range. Contrary to capacitive couplers, inductive interfaces are applied on MV lines, being less economically feasible for LV applications [19]. A drawback of this interface is the magnetization of the coupling ferrites, caused by a supply frequency line current in the couplers due to the serial connection to the power line, which leads to the core saturation. Therefore, inductive couplers are designed for a limited current to avoid such scenarios.

One of the most important aspects when considering coupling interfaces is the attenuation on the interface. Signal insertion and reception losses have a significant impact on the total attenuation. Measurements and simulation were performed comparing inductive and capacitive coupling interfaces in the band of 100 kHz–30 MHz for communication between an electrical motor and an inverter [19]. According to the results, an inductive coupling provides a higher attenuation when the output filter of the inverter is connected. The results are opposite when this filter is not applied.

Thus, we may conclude that an inductive coupling interface is a more feasible solution for a PLC application in the MV power grid. Comparing to capacitive solutions, an inductive interface provides a higher economical feasibility, an ease of installation, and a galvanic isolation from the power line.

12.3.3 Frequency band

An essential aspect of a PLC concept is the signaling bandwidth. Based on the investigation of the application field, we can consider a frequency band below 1 MHz to be a beneficial choice. Therefore, applicable standards and legislation have to be considered for the low-frequency and medium-frequency ranges.

Signaling in these ranges is regulated by the following standards: the European Standard CENELEC 50065-1 divides the 3–148.5 kHz band into four frequency bands; A, B, C, and D [21]. In the United States, again, the frequency band of 14–480 kHz is covered by a Federal Communications Commission (FCC) standard, which was developed by the FCC [22]. Any proposed PLC solution has to operate in compliance with these standards. A G3-PLC modem, for instance, is intended to operate in the CENELEC-A range, between 35.9 and 90.6 kHz, and in the FCC band, between 148 and 490 kHz [13].

The CENELEC EN 50065-1 standard distinguishes two frequency ranges, 3–95 and 95–148.5 kHz. In the first range, the maximum transmission level is 120–134 dBμV/Hz and the maximum disturbance level is 75.5–89 dBμV/Hz. In the second frequency range, the maximum transmission level is 116 dBμV/Hz and the maximum disturbance level is 65.97–75.5 dBμV/Hz [21].

12.3.4 Signaling modulation techniques

A vital component of a PLC system is the modulation technique employed for signaling. Besides conventional approaches, which include for instance single carrier modulation techniques, such as amplitude-shift keying, frequency-shift keying, phase-shift keying (PSK), quadrature amplitude modulation (QAM), and their modifications, there are spread spectrum modulation techniques, such as DSSS, frequency-hopping spread spectrum, chirp spread spectrum, and multicarrier modulation techniques, such as OFDM, which can all be employed for PLC. Wideband modulation techniques are gaining ground as they have distinct advantages over conventional modulation techniques, such as a greater robustness against channel attenuation and channel noise and a higher bandwidth efficiency [13,16,17,23].

12.3.5   *Concept evaluation, SDR platform*

One of the platforms that can be used for the design and laboratory evaluation of the PLC concept is the software-defined radio (SDR). An SDR is a radio system, in which components conventionally implemented by means of hardware are implemented by software. These components are for instance filters, amplifiers, (de)modulators, and mixers.

Signaling software, designed for instance in GNU Radio or LabVIEW, is operated on a personal computer (PC), which is connected to an SDR by an Ethernet cable. The signaling modulation settings and the signaling frequency band can be modified in software instead of hardware, which provides a higher flexibility of the test platform. The SDR is then connected to the coupling interface by a coaxial cable.

12.4   PLC concept implementation

As an example of a PLC-based anti-islanding system, a concept proposed in [24] can be considered. The concept evaluation covers fault-detection functionality, sensitivity to a variation of SNR, and a transformer bypassing scheme.

12.4.1   *Signaling concept*

The proposed PLC concept is based on continuous signaling from the main power utility in the MV grid toward DG units and customers in the LV grid. Islanding detection is carried out by processing of the received signal, which contains packets of known data. In the processing, bit error rate (BER), bit rate, and latency between subsequent data packets are considered.

The first signal transmitter is installed at the primary substation (Figure 12.4). Signal receivers are installed at the households on the LV grid and at the DG units. Intermediate signal transceivers installed on MV lines allow to analyze the grid state. Intermediate signal transmitters, which are installed on MV/LV transformers, are powered from the LV side. The signaling devices are coupled to the channel using inductive coupling interfaces. When a signal transceiver detects a fault, it is possible to define the faulty grid segment.

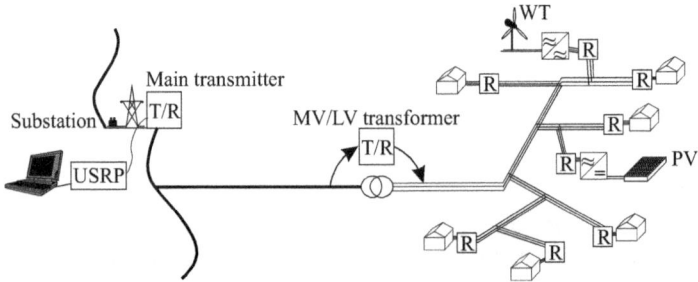

Figure 12.4   Intermediate signal repeater installed at the MV/LV transformer applying a bypassing scheme

Signaling is performed by applying a DSSS modulation technique with the carrier frequency of 200 kHz and the bandwidth of 200 kHz. The frequency band was chosen based on the MV/LV transformer channel analysis, performed in the laboratory environment and discussed above. The developed DSSS signaling solution is based on binary PSK (BPSK). A DSSS software interface has two adjustable parameters, such as a spreading code length (SC) and a synchronization key length (SK). The transmitted data packet has a constant length of 14 bits, where SK can be 8 or 12 bits. The SC has 6 available settings: Barker codes of 5, 7, 11, and 13 bits, and Walsh–Hadamard codes of 16 and 32 bits. First, a data packet is modulated in the transmitter applying BPSK with 2*SC samples per one symbol. Then the BPSK symbols are multiplied by the SC and continuously transmitted. The DSSS receiver, knowing the SC and the SK, performs the best correlation position search and extracts the BPSK symbols from the received signal. Then, SK bits and data bits are separated and BER and bit rate values are calculated. At the same time, a latency between subsequent data packets is measured.

The concept was evaluated with a laboratory setup featuring SDRs and a MV/LV distribution transformer presented in Figure 12.5.

12.4.2 Functionality

12.4.2.1 Fault detection

The idea behind the proposed fault detection algorithm is that under normal operating conditions, the monitored parameters fluctuate within certain limit values. Fluctuation can result from noise scenarios and disturbances in the channel. The limits can be determined when the concept is introduced to the grid. After the limits have been defined, thresholds for each parameter can be set. If a threshold is passed, a protection algorithm is triggered. The monitored parameters are evaluated in a derivative form. By adjusting the thresholds and the time during which parameters can leave the permitted value ranges, the concept sensitivity can be adjusted. Detection sensitivity contributes to the NDZ, and thus, insufficient sensitivity may lead to incorrect concept operation.

When a LoM occurs, signal transmission through the grid is interrupted. In this case, the BER and the latency between subsequent packets increase, whereas the bit rate decreases. After the fault condition is detected in the transceiver, the signal repeater modulates and transmits a warning signal with the identifier of the SDR downstream, toward the customers and the DG units in the LV grid. When the receiver detects a fault or receives a warning signal, the protection algorithm is triggered and the customer/DG units may either disconnect from the grid or stay connected while the controlled island mode is initiated.

The continuous parameter monitoring allows to evaluate the power line state. By investigating the parameter behavior, environmental impacts that do not always lead to LoM and islanding can be detected. In this case, signal parameters can fluctuate, yet being within the permitted range.

The concept evaluation performed in the laboratory considered a LoM scenario only when the signal transmission was interrupted.

12.4.2.2   Fault localization

The accuracy of the fault localization equals the distance between two signaling units. When a fault is detected, a warning signal is transmitted from the signaling device that detected the fault. The warning signal contains a unique device identifier. This identifier allows to determine the device and thereby the grid segment from which the warning was originally sent. The drawback of this method is the low fault location accuracy, even though it can be improved by introducing additional intermediate signaling devices in the grid. The outcomes will not only be the improved accuracy but also the increased communication latency between the main signaling device and the receivers at the DG units and the customers. Thus, the total number of installed devices is limited. The communication latency should be within certain limits so that the fault-detection speed will be in compliance with the standards.

12.5   Laboratory tests

12.5.1   Laboratory setup

The concept under study was evaluated with the laboratory setup presented in Figure 12.5. The main component of the setup is the MV/LV distribution 20/0.4 kV transformer.

On the MV side of the transformer, there is a signal transmitter, which is connected through an amplifier (TS200 by Accel Instruments), a band-pass filter (pass-band of 140–500 kHz), and a balun to the inductive couplers. The couplers are installed on phases A–B of a power cable. The cable has a section where a signal attenuator is embedded into phases A–B. The attenuator is used to adjust the SNR in the channel under study. The AXMK cable is then coupled to the transformer on the MV side. On the LV side of the transformer, the AXMK cable is connected to a load. The inductive couplers are connected to phases A–B. Further, the couplers are connected to the balun, then to the filter, the signal amplifier, and the signal receiver. Each SDR is connected to the respective PC, which is executing the DSSS software application in LabVIEW. Two USRP N200 SDRs by Ettus Research were employed in the tests.

12.5.2   Fault detection tests

To evaluate the signal analysis as a fault-detection method, a series of laboratory tests were conducted with the presented laboratory setup (Figure 12.5). The idea of the tests was to investigate whether it is possible for the designed solution to distinguish a fault and normal operating states based on the observed parameters. In the tests, the SNR had a constant value of 12 dB and constant signaling parameters of SC=5 SK=8.

The tests were performed in an automated manner, and they were cyclical. One test cycle consisted of two parts; 4 s of signaling and 1 s of an idle mode. The idle mode represented a loss of signal, which occurs in the case of islanding. To perform the signal analysis, three received signal parameters were considered; BER, bit rate,

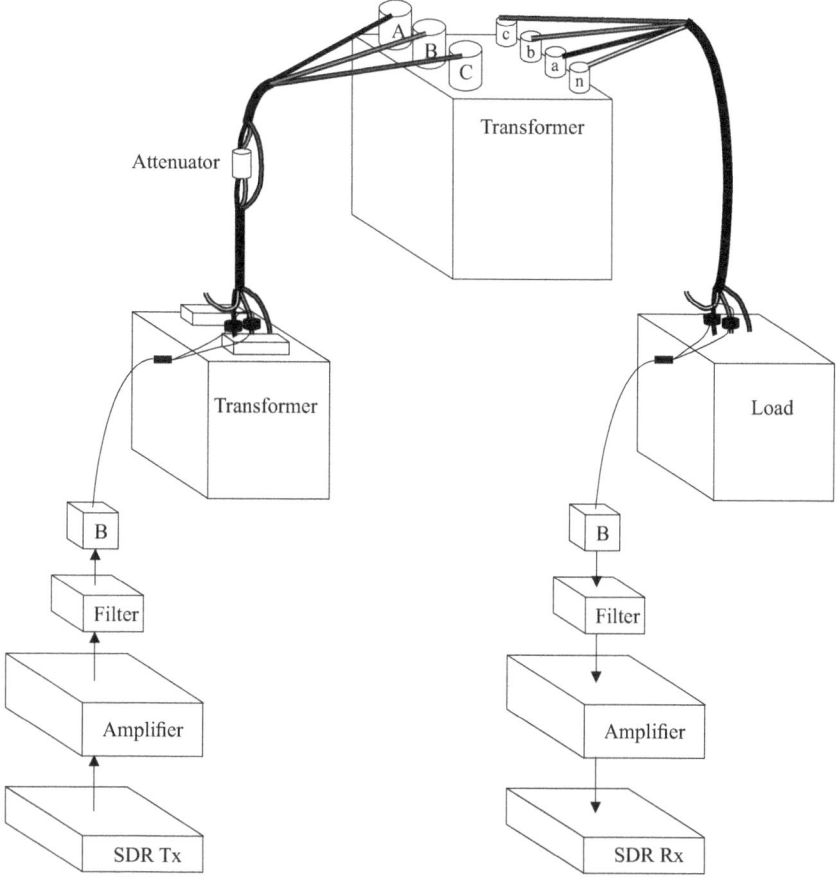

Figure 12.5   Laboratory setup employed in the tests. B: balun

and received packet processing latency. These parameters were calculated continuously in the receiver and written with a 1-ms resolution. To investigate whether the parameters fluctuate in the case of a fault (LoM), three time frames were considered: a time frame of 1 s before the fault, 1 s of the fault condition (LoM), and 1 s after the fault condition. The mean parameter values were calculated for each time frame for each test cycle. Based on the measurements, a scatter plot was made (Figure 12.6). In the case of a fault, the BER and latency parameters increase, whereas the bit rate decreases. The latency graph demonstrates more drastic changes in the observed parameter compared with the measurements of BER and bit rate. The BER and bit rate parameters should be evaluated in derivative forms in the final system. Though the SNR level was constant and equal to 12 dB during the tests, the BER values are relatively high. This can be explained by the fact, that signaling was performed through the transformer; therefore, the channel can be characterized by severe signal distortion. Moreover, application of different DSSS settings can decrease BER.

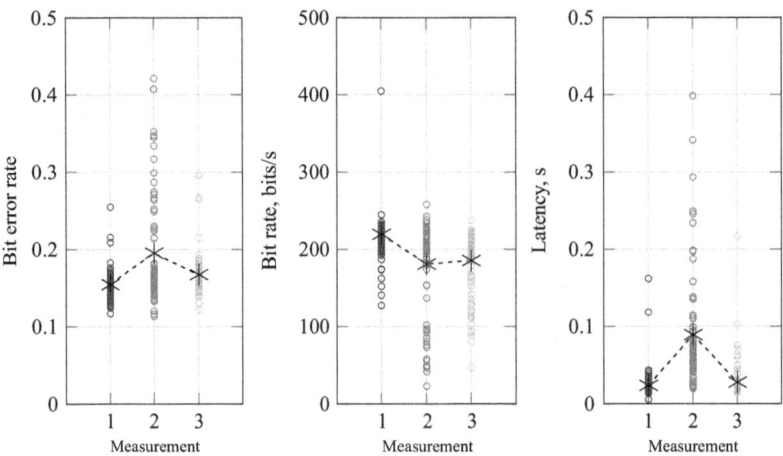

Figure 12.6   *Measurements of bit error rate, bit rate, and latency measurements during three periods: 1 – 1 s before a fault; 2 – during a 1 s fault; 3 – 1 s after the fault. The asterisks indicate the mean values during the period. Dotted line demonstrates a transition between the states*

The effect of the DSSS settings on the signaling parameters is investigated during the sensitivity analysis.

12.5.3   Sensitivity analysis

The DSSS solution has two main settings; SC and SK. The sensitivity analysis investigated the sensitivity of the DSSS solution to a variation of the SNR by observing signal parameters under different SNR values and DSSS settings. As a result, the SC and SK settings providing the highest islanding-detection speed and throughput were defined.

The test was performed in the laboratory setup (Figure 12.5) with SNRs in the range of −13–12 dB. First, signaling was performed with the same attenuator and a constant SNR, while the settings were changed periodically in an automated manner. Then, the attenuator was changed and the tests continued.

12.5.3.1   Bit error rate

The obtained measurement results were processed, and the concept performance for each setting was evaluated. The measurements for BER are depicted in Figure 12.7. For the sake of convenience, the results are presented in two graphs.

It can be observed in the figures for the BER results that a longer SC provides a lower BER. By comparing the figures, we may state that a longer SK decreases the BER.

At the same time, it can be seen that a setting of SC=32 and SK=12 bits has severe oscillations in the positive SNR range. During the tests it was observed that SC length has a significant effect on the processing speed of the application; longer

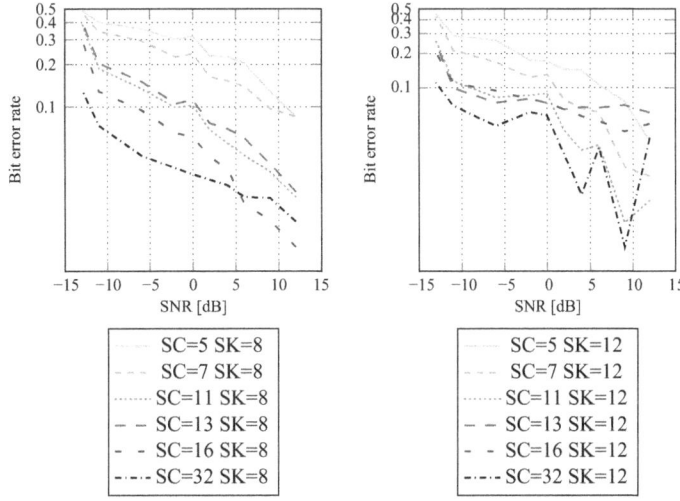

Figure 12.7   Comparing the measurement results for the bit error rate under different SNR values and different settings of SC, while the SK is 8 bits

SCs can increase processing latency and deteriorate the application performance. For this reason, measurement results for the Walsh–Hadamard codes of 64 and 128 bits were excluded from the analysis, and results for 32 bit code are provided to expose this issue.

12.5.3.2   Bit rate

The bit rate measurements with SK=8 and with SK=12 are presented in Figure 12.8. Observations from these figures differ. The first figure shows that a shorter SC provides a higher bit rate. In the second figure, the opposite can be observed: longer SCs have higher bit rates. At the same time, the settings from the second figure (with a longer SK) have a lower bit rate. It can be assumed that a receiver with a short SC interprets the channel noise as a valid transmitted signal. In this case, a higher bit rate can be achieved, while also the BER values will be higher. Moreover, it can be seen that the SK defines the number of data bits in the transmitted packet, and thus, a shorter SK can provide a higher bit rate.

12.5.3.3   Throughput

In order to combine the obtained results of the BER and bit rate measurements and evaluate the number of correct data bits that can be received per second, a throughput characteristic was introduced. Throughput was calculated by using the following equation:

$$Throughput = (1 - 2 \cdot BER) \cdot Bitrate \qquad (12.3)$$

The results are presented in Figure 12.9. As can be seen in the figures and as stated previously, a longer SC contributes to a lower BER, and a shorter SK to a higher bit rate.

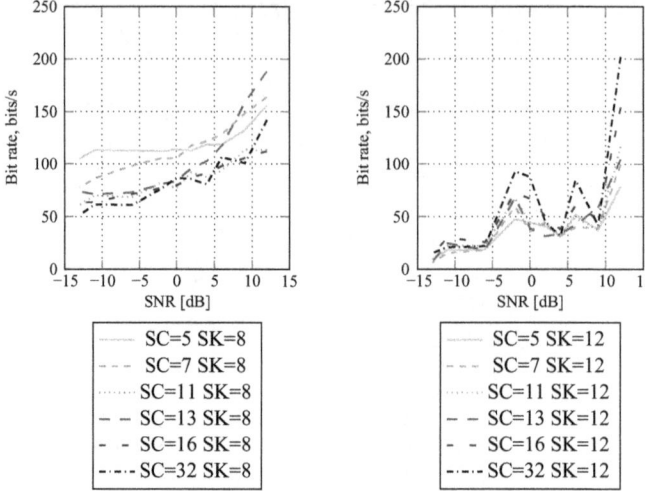

Figure 12.8   Comparing the measurement results for the bit rate under different SNR values and different settings of SC, while the SK is 8 bits

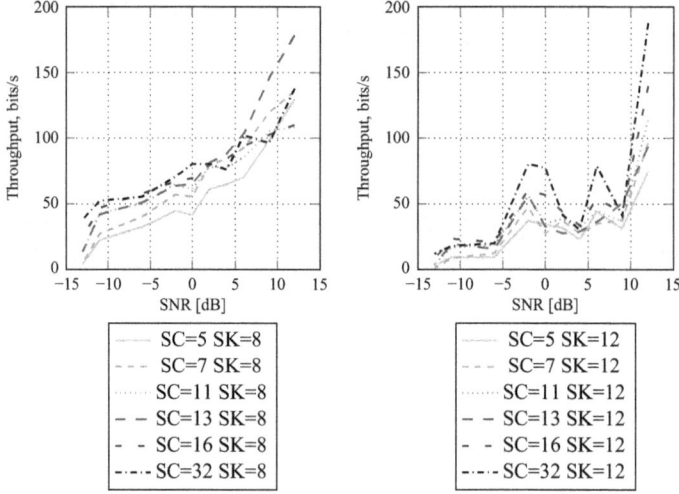

Figure 12.9   Comparing the measurement results for the throughput under different SNR values and different settings of SC, while the SK is 8 bits

12.5.3.4   Latency

The operational speed of a PLC-based islanding-detection solution depends highly on the processing latency in the receiver. The target is to minimize the latency between two subsequent received data packets. The measurement results are presented in Figure 12.10. The first figure shows that a longer SC contributes to a higher latency.

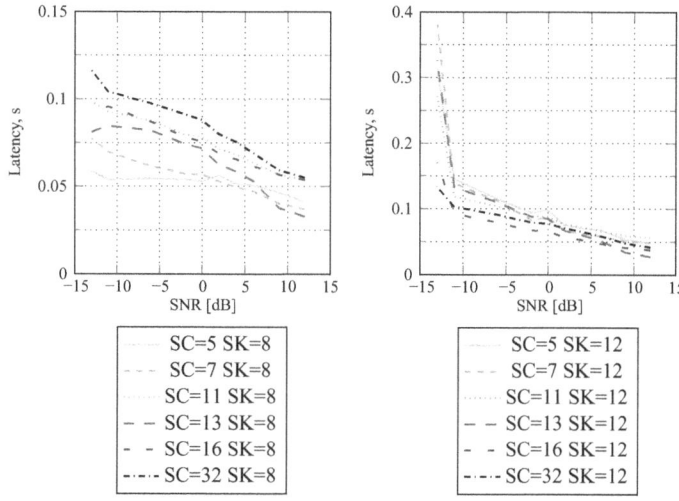

Figure 12.10   Comparing the measurement results for the processing latency under different SNR values and different settings of SC, while the SK is 8 bits

An opposite scenario is observed in the second figure. Comparing the two figures, we can state that a shorter SC provides a lower latency. A setting of SC=13 SK=8 was chosen for the bypassing scheme evaluation on the basis of the presented measurements. This setting was distinguished, as it provided a combination of a high bit rate, a low BER, and a low latency. The parameters were evaluated considering each parameter separately, and SC=13 SK=8 had the highest overall result.

12.5.4   Orthogonal frequency division multiplexing

An alternative PLC solution, applying OFDM modulation technique, was also evaluated during the laboratory tests. Signaling was performed through the MV/LV transformer using the setup previously employed in the DSSS tests (Figure 12.5).

An OFDM modulation scheme was implemented in LabVIEW. First, a data packet consisting of 20 data bits is coded using the 13 bit Barker code. Then data is mapped to QAM modulation symbols, divided into five subsets, to which reference symbols are added. Then zero padding is applied on the edges and on the DC. After that an inverse fast Fourier transform (FFT) is implemented, a cyclic prefix is added, and the signal is transmitted. The carrier frequency of 200 kHz and the bandwidth of 200 kHz were applied.

The signal receiver is applying the van de Beek algorithm on the obtained signal to detect the signal prefix. Then the prefix and the frequency offset are removed. An FFT is performed. After that data is separated from the reference bits and zero padding. Equalization coefficients are calculated, and equalization is applied to the data symbols. Using the symbol map data bits are then excluded and decoded using the Barker code. Majority voting is applied on data, obtained from the subchannels, and then BER and bit rate values are calculated. During a parallel process, latency between decoded data packets is measured.

As mentioned earlier, a DSSS-based solution can provide a benefit of signal reception in the negative SNR range, which cannot be achieved with OFDM. Therefore, laboratory measurements were performed in the SNR range of 0–12 dB with 1 dB steps. Obtained results are compared to the DSSS measurements with the setting of SC=13 SK=8. In the whole studied range BER of OFDM-based signaling is higher, comparing to DSSS-based signaling (Figure 12.11). At the same time, a higher bit rate can be achieved with an OFDM solution. With the SNR of 12 dB the bit rate of OFDM signaling is 1,815 bits/s, while for DSSS, it reaches only 188 bits/s. As observed earlier, a variation of the code length can affect BER and bit rate of signaling. This way, a shorter code length applied in OFDM can increase the bit rate and BER at the same time.

With the highest SNRs observed, DSSS and OFDM provide a major difference in bit rate, while difference in BER is relatively low. As a consequence, with SNRs above 5 dB, OFDM provides a higher throughput (Figure 12.12). With the maximum SNR of 12 dB, an OFDM solution can provide a throughput of 1,667 bits/s, while DSSS can provide 178 bits/s. Moreover, a major difference in the processing latency of DSSS and OFDM signaling can be seen. OFDM can provide a lower processing latency, being 7 ms with SNR of 12 dB, while a processing latency for DSSS is 33 ms.

To summarize the provided OFDM performance analysis, it may be concluded that in the range of high SNRs (above 5 dB), an OFDM solution is a more beneficial technique for PLC, comparing to DSSS. The distinct advantage of OFDM is a high signaling throughput and a low processing latency. However, the OFDM-based solution is less preferable than the DSSS with SNRs below 5 dB and cannot operate in the

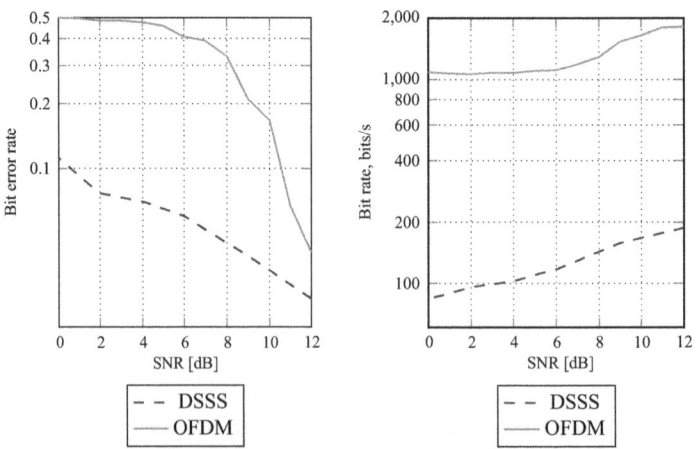

Figure 12.11   Comparing the measurement results for the bit error rate and bit rate under different SNR values for the OFDM solution and the DSSS solution (setting of SC=13 SK=8)

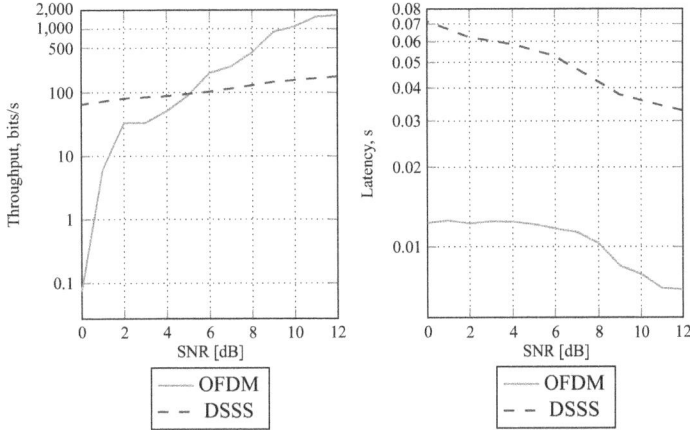

Figure 12.12 *Comparing the measurement results for the throughput and processing latency under different SNR values for the OFDM solution and the DSSS solution (setting of SC=13 SK=8)*

negative SNR range. In other words, an applicability range of OFDM in the proposed islanding-detection concept is quite limited. Therefore, the DSSS-based solution is preferred for that certain application field and environment.

12.5.5 Bypassing

The anti-islanding concept proposed here employs a bypassing scheme with PLC. The laboratory setup with a bypassing circuit is presented in Figure 12.13.

The signal from the coupling interface on the MV side of the transformer is amplified, filtered, and then processed in the PLC device, which operates as a signal repeater. The signal repeater performs the fault-detection algorithm and transmits the signal (regular signal or a warning). This signal is then amplified, filtered, and injected through the coupling interface on the LV side into the channel, that is, the LV grid power line.

To evaluate the bypassing concept, a series of tests were performed with a DSSS signaling system with a setting of SC=13 and SK=8. In these tests, the idea was to evaluate under which conditions in the original channel the introduction of bypassing can increase the throughput and the fault-detection speed. The SNR in the modified channel was not considered. For this reason, the obtained results (Figures 12.14 and 12.15) include SNR values only for the original channel.

When the SNR is above 0 dB, the mutual interference of the original signal and the bypassing signal increase the BER. Nevertheless, for SNRs below 10 dB, a bypassing circuit can still provide signaling with a lower BER, comparing with the original scheme. The signaling bit rate slightly increases with a decrease of SNR, being higher than bit rate of the original scheme in the SNR range below 10 dB. The same behavior can be seen, when we consider the throughput characteristic. At the same time, significant difference can be seen in latency characteristics with SNRs

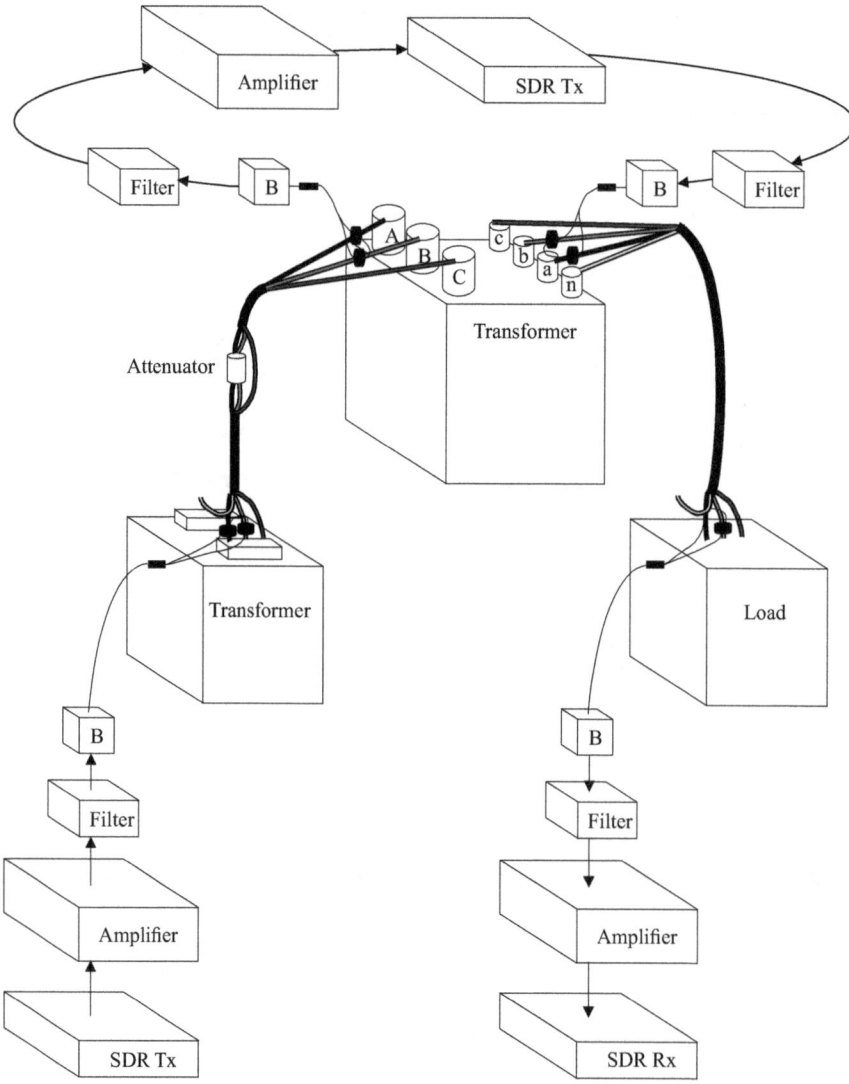

*Figure 12.13   Laboratory setup with a bypassing scheme applied to the tests.
            B: balun.*

above 5 dB. Application of the bypassing scheme is not beneficial with higher SNRs, as the processing latency increases.

Based on the tests, it can be concluded that introduction of a bypassing scheme is beneficial for the concept operation, when the SNR at the receiver at the MV side of the distribution transformer is below 5 dB. A condition with a low SNR can represent a case of a high channel noise amplitude and/or a high channel attenuation. A bypassing scheme can provide a lower BER, a higher bit rate, and, therefore, a higher

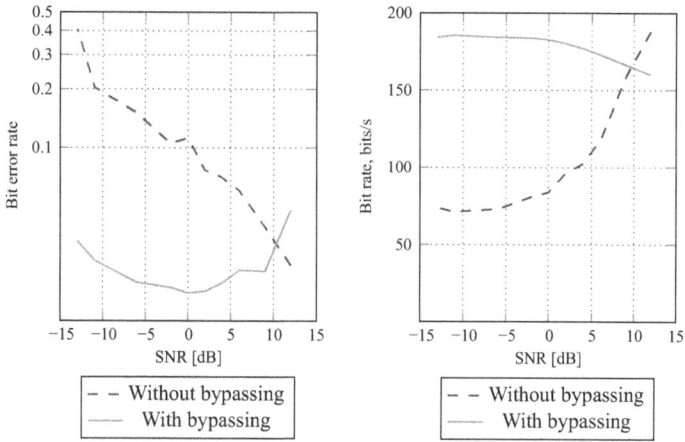

Figure 12.14   Comparing the measurement results for the BER and the bit rate under different SNR values for SC=13 SK=8 in the setup with and without a bypass circuit with an intermediate transmitter

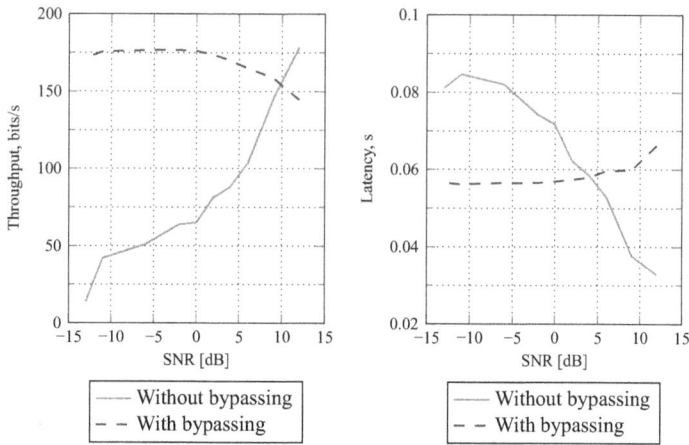

Figure 12.15   Comparing the measurement results for the throughput and the processing latency under different SNR values for SC=13 SK=8 in the setup with and without a bypass circuit with an intermediate transmitter

throughput comparing with the original signaling scheme. Moreover, a processing latency decreases.

References

[1]   Timbus A, Oudalov A, Ho CNM. Islanding detection in smart grids. In: 2010 IEEE Energy Conversion Congress and Exposition; 2010. p. 3631–3637.

[2]   IEEE. IEEE standard for interconnecting distributed resources with electric power systems. In: IEEE Std 1547-2003; 2003. p. 1–28.

[3]   IEEE. IEEE recommended practice for utility interface of photovoltaic (PV) systems. In: IEEE Std 929-2000; 2000.

[4]   UL. Inverters, converters, controllers and interconnection system equipment for use with distributed energy resources. In: UL Std UL 1741; 1999.

[5]   Reigosa D, Briz F, Charro CB, *et al.* Active islanding detection using high-frequency signal injection. IEEE Transactions on Industry Applications. 2012;48(5):1588–1597.

[6]   Ye Z, Kolwalkar A, Zhang Y, *et al.* Evaluation of anti-islanding schemes based on nondetection zone concept. IEEE Transactions on Power Electronics. 2004;19(5):1171–1176.

[7]   Chen X, Li Y. An islanding detection method for inverter-based distributed generators based on the reactive power disturbance. IEEE Transactions on Power Electronics. 2016;31(5):3559–3574.

[8]   Automatic disconnection device between a generator and the low-voltage grid. In: DIN-VDE Std 0126-1-1; 2005.

[9]   Photovoltaic semiconductor converters, Part 1: Utility interactive fail safe protective interface for PV-line commutated converters. In: Eur Std EN 50 330-1; 1999.

[10]   Zimmermann M, Dostert K. Analysis and modeling of impulsive noise in broad-band powerline communications. IEEE Transactions on Electromagnetic Compatibility. 2002;44(1):249–258.

[11]   Gotz M, Rapp M, Dostert K. Power line channel characteristics and their effect on communication system design. IEEE Communications Magazine. 2004;42(4):78–86.

[12]   Kosonen A, Ahola J, Pinomaa A. Analysis of channel characteristics for motor cable communication with inductive signal coupling. In: ISPLC2010; 2010. p. 72–77.

[13]   Kikkert CJ. MV to LV transformer PLC bypass coupling networks for a low cost smart grid rollout. In: 2011 IEEE PES Innovative Smart Grid Technologies; 2011. p. 1–6.

[14]   Galli S, Scaglione A, Wang Z. For the grid and through the grid: the role of power line communications in the smart grid. Proceedings of the IEEE. 2011;99(6):998–1027.

[15]   Black WC. Data transmission through distribution transformers without bypass components. In: ISPLC2010; 2010. p. 13–17.

[16]   Meng J, Ding X. A comparison study of three power-line communication techniques in low signal-to-noise ratio conditions. In: 2007 IEEE International Symposium on Power Line Communications and Its Applications; 2007. p. 407–412.

[17]   Meng J, Marble AE. Effective communication strategies for noise-limited power-line channels. IEEE Transactions on Power Delivery. 2007;22(2): 887–892.

[18]   Poluektov A, Pinomaa A, Ahola J, *et al.* Designing a power-line-communication-based LoM protection concept with application of software-defined radios. In: 2016 International Symposium on Power Line Communications and its Applications (ISPLC); 2016. p. 156–161.

[19]   Kosonen A, Ahola J. Comparison of signal coupling methods for power line communication between a motor and an inverter. IET Electric Power Applications. 2010;4(6):431–440.

[20]   Mlynek P, Koutny M, Misurec J, *et al.* Evaluation of load impedances, discontinuity and impedance mismatch in PLC networks by experimental measurements with coupling circuits. In: 2015 38th International Conference on Telecommunications and Signal Processing (TSP); 2015. p. 657–661.

[21]   IEEE. Signalling on low-voltage electrical installations in the frequency range 3 kHz to 148.5 kHz part 1 general requirements, frequency bands and electromagnetic disturbances. In: EN-50065-1; 2001.

[22]   FCC. Title 47 CFR part 15 subpart B. Federal Communications Commission Std. 1998.

[23]   Wang W, Kliber J, Zhang G, *et al.* A power line signaling based scheme for anti-islanding protection of distributed generators—Part II: Field test results. IEEE Transactions on Power Delivery. 2007;22(3):1767–1772.

[24]   Poluektov A, Romanenko A, Pinomaa A, *et al.* Sensitivity analysis of a PLC-based anti-islanding solution using DSSS. In: 2017 IEEE International Symposium on Power Line Communications and its Applications (ISPLC); 2017. p. 1–6.

Chapter 13

Performance evaluation of PRIME PLC modems over distribution transformers in Indian context

Konark Sharma[1] and Lalit Mohan Saini[2]

The past few years have witnessed a tremendous development in powerline intelligent metering evolution (PRIME) technology for high speed data communication across medium voltage (MV) and low voltage (LV) transmission/distribution networks based smart grid (SG) applications. PRIME PLC (PRIME power-line communication) technology also elucidates the importance of employing robust modulation schemes across distribution-transformers and motivates research in this direction. Indeed, the aim of the chapter is to investigate PRIME channel measurements through MV/LV distribution transformers by implementing experimental tests to analyze the signal-to-noise ratio (SNR), bit error rate (BER) and packet error rate (PER) performance of differential binary phase shift keying (DBPSK), differential quadrature phase shift keying (DQPSK) and eight-ary differential phase shift keying (D8PSK) modulation schemes over multipath PLC channels in Indian context.

13.1 Introduction

The uses of power-lines for long distance data communication solution over the last two decades have gained global interest. It is also well known that advanced control and operation systems for distribution grids and microgrids (μGs) [1], which are facilitated by the use of emerging measurement technologies such as smart metering [2] and phasor measurement units applications [3,4], which are providing more accurate and extensive measurement information to enable real-time data monitoring, state estimation and system analysis for demand side management (DSM) applications [5]. It has already proved with experimental results that latest communication technologies are contributing for voltage regulation as well as DSM applications for SGs [6]. Nowadays, most of the advanced distribution automation (ADA) systems are based on IEC 61850 standard, which is reliable for long-time SG-related analysis [7].

[1]Department of Electrical & Electronics Engineering, National Institute of Technology Delhi, India
[2]Department of Electrical Engineering, National Institute of Technology Kurukshetra, India

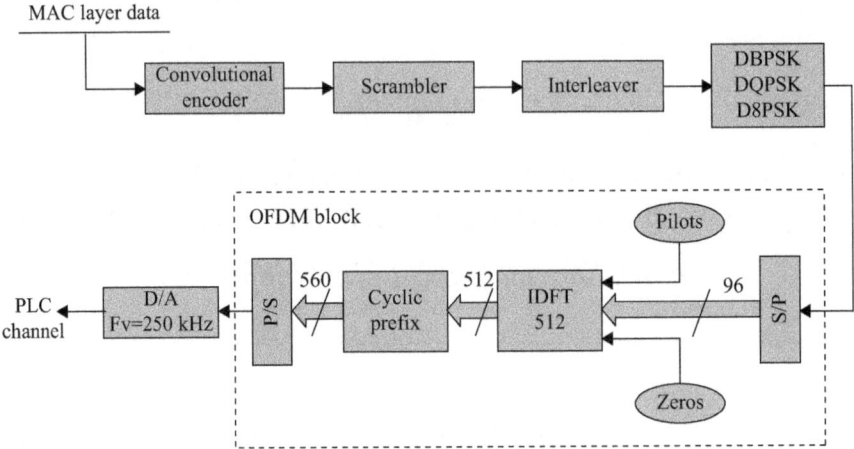

Figure 13.1   A typical block diagram of PRIME PLC transmitter (Source: Reference [15])

However, the ADA relies heavily on a secure and robust bidirectional communication system. Recently, the worldwide use of PLC technology for ADA systems has gained significant interest due to its low deployment costs [8]. The PLC reliability and varying distance of the LV systems/networks are mostly affected by the channel interference and the time varying load impedance, but alone in Europe, with more than 40 million PLC technology–based metering points installed [9]. Since 2010, worldwide increasing demand of data rates by smart metering companies led to rehabilitating the various next-generation narrow-band PLC (NB-PLC) standards like G3-PLC [10], PRIME [11], ITU G.9955/9956 standard [12] and IEEE 1901.2-2013 standard [13] has been introduced. Consequently, the PRIME PLC is widely deployed and mature PLC standard for Plug & Play, advanced metering infrastructure (AMI), grid controlling and asset monitoring applications to enable the SG in the last mile [14]. To date, the PRIME standard with an increased number of PRIME certified products/solutions interoperability among equipment and systems from different manufacturers has been achieved and is already being utilized/installed in over 15 countries worldwide. During 2007, the PRIMR PLC has been standardized by PRIME Alliance, with the help of Spanish distribution system operator, which was one of the founding members. The first PRIME PLC protocol was accepted during October 2012 and published in February 2013, available online in [15]. General block diagram of PRIME PLC transmitter scheme is shown in Figure 13.1, which supports IEC6134-4-32, IPv4 or IPv6 and modulation schemes; DBPSK, DQPSK and D8PSK at the network layers.

During the year 2009, PRIME PLC-based system-on-chip solution was first ever evolved and proposed for orthogonal frequency division multiplexing (OFDM) based NB-PLC solution to utilize the average transmission rate of PHY layer is around 70 kbps and the maximum is 120 kbps using from 42 to 88 kHz [16]. Present PRIME PLC solutions as per CENELEC A-band with OFDM scheme utilize the average transmission rate up to 61.4 kbps with forward error correction (FEC) and up to

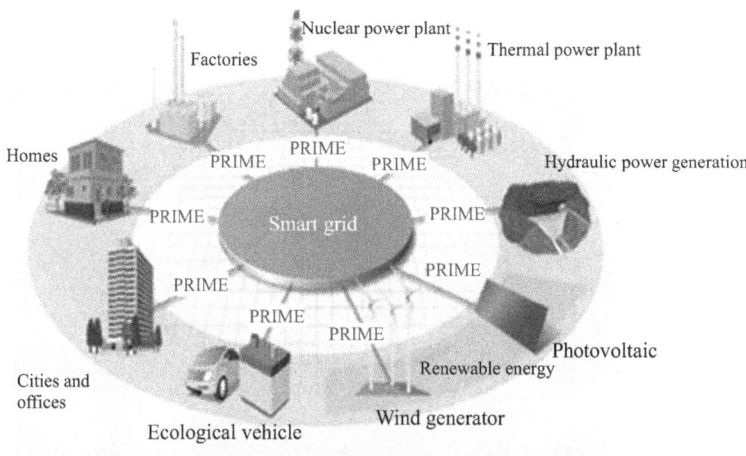

Figure 13.2   A typical PRIME based SG infrastructure

128 kbps without FEC [15]. During operation, the selection of constellation depends on the levels/types of noise present in the power-line channel. PRIME PLC (ITU-IT G.9904) standard is already being rolled out especially for regarding internet-of-things-based applications, smart homes/buildings energy management schemes and various SG applications, as are shown in Figure 13.2.

Modern OFDM technology based PLC modems (within 9–95 kHz CENELEC-A frequency band) can be directly connected to LV power-lines [17]. PRIME PLC technology with data rate up to 130 kbps at PHY layer is quickly gaining ground with latest Spain's smart metering standards and various SG applications [18–20]. The main strength of PRIME PLC is its interoperable structure as a multivendor alliance of various industry players like Device Language Message Specification/Companion Specification for Energy Metering (DLMS/COSEM). It has its origins in the car-manufacturing industry at the end of the 1970s [20]. Presently, DLMS (IEC 62056-53) with advanced encryption standard (AES)-128 encryption scheme has become the global choice among smart meter designers for interoperability among metering systems, including most energy types (electricity, gas, heat and water), multiple applications (residential, transmission and distribution), numerous communication media and secure data management schemes [21]. PRIME PLC technology also provides a reliable and secure connection between PLC-based devices, including crossing data from MV to LV distribution transformers (around 120 m per transformer in average distance) [22,23]. To the best of authors' knowledge, no PRIME PLC technology based testing to perform BER and PER analysis of two types of transmission (i.e., packet transfer and data file transfer) across two distribution-transformers in Indian context has been conducted yet. The remainder of this chapter is structured as follows: PRIME PLC technology–based proposed measurement system is described in Section 13.2. Test results are discussed in Section 13.3. And finally, the chapter is concluded in Section 13.4.

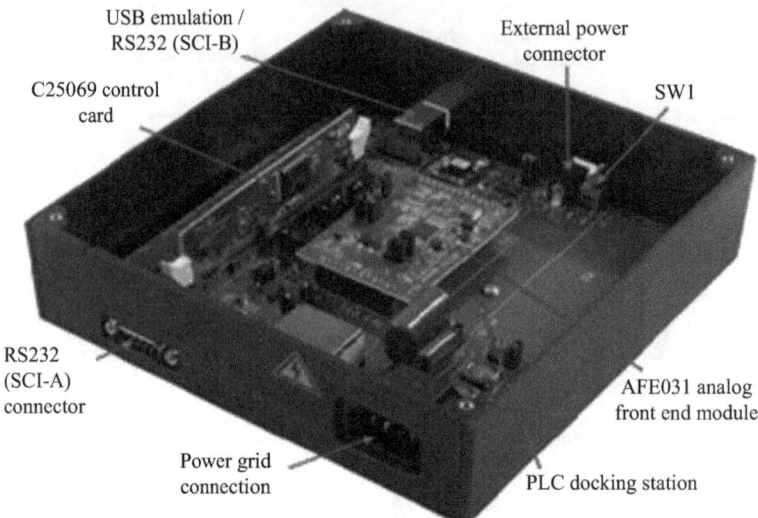

USB emulation /
RS232 (SCI-B)

External power
connector

C25069 control
card

SW1

RS232
(SCI-A)
connector

AFE031 analog
front end module

Power grid
connection

PLC docking station

Figure 13.3   Snapshot of Texas Instruments PLC development kit (Source: Reference [24])

13.2   Proposed algorithm

In this work, we used PLC development kit, as shown in Figure 13.3, containing a digital signal processor control card with F28069 series microcontroller [24]. It consists of integrated PLCs analog front-end AFE031, which is able to work uniquely in harsh conditions, where a negative SNR is also expected.

Development kit embodied with universal asynchronous receiver/transmitter (UART) and serial peripheral interface supports data rates up to 34.16 kbps (data carriers up to 36 tones per symbol) CENELEC A-band and up to 28 kbps (data carriers up to 72 tones per symbol) for FCC with differential phase modulation schemes (i.e., DBPSK/DQPSK/D8PSK/ROBO). In order to perform following tests (i.e., BER and PER analysis) with PRIME PLC modems in the industrial area of Panipat (Haryana), India, a transparent protocol using GUI software were implemented. A GUI software running on notebooks having Intel® CoreTM2 Duo CPUP8700 @2.53 GHz with 4 GB RAM configuration with Microsoft .net framework 2.0 on Windows XP environment was used to configure the PRIME PLC modems as RX-receiver and TX-transmitter to select suitable frequency band, selective gain, subcarrier modulation type and data packet size. During experimental investigation one PRIME PLC modem was termed as coordinator communication module (CCM), while the second PRIME PLC modem was termed as terminal communication module (TCM) that monitored the data rate estimation, number of transmitted/received and data packets loss between distribution-transformers. We implemented a transparent protocol between CCM and consumer device/application type1, as shown in Figure 13.4, where

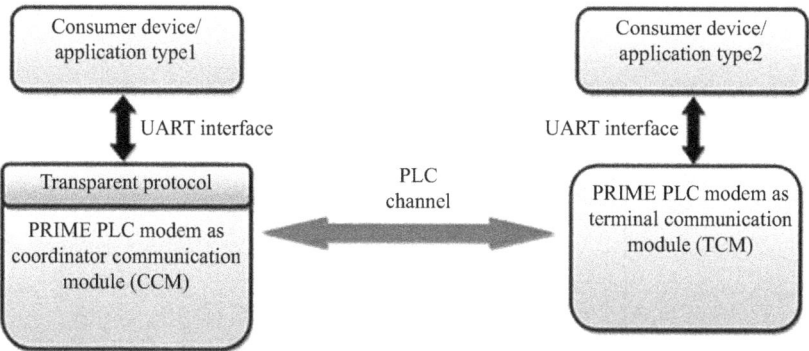

Figure 13.4   Block diagram of PRIME PLC based proposed algorithm

the TCM directly connects with consumer device/application type 2 through UART interface, and PRIME PLC modems can handle the network formation (i.e., connect, data transmission, etc.)

If we implement the application program in consumer device/application type1, then advanced protocol interface supports transparent protocol, which can communicate with consumer device/application type1 through PLC channel. During operation within the network, if more than one TCM nodes are required, then transparent protocol provides a peer-to-peer connectivity with other nodes. Flowchart of the proposed algorithm, as shown in Figure 13.5, is as follows:

Step 1: Start.
Step 2: Configure PRIME PLC modem A, PRIME PLC modem B and their host UARTs.
Step 3: Receive the sample data signal from via PRIME PLC modems host UARTs.
Step 4: Initialize PRIME PLC modems to take necessary sample data signal and display
Step 5: Check and wait for the sample data signal.
Step 6: Send the sample data signal via host UARTs
Step 7: Continue till all other PRIME PLC modems received their sample data signal
Step 8: Stop.

During experiments 1,000 packets (size: 256 B each packet) from one PRIME PLC modem to other PRIME PLC modem, and the GUI software were informed about the data transfer rate and the packets error rate (PER). Even afterward, the PER were not detected correctly, estimated through the following formula:

$$PER\% = \frac{\text{Number of packets loss}}{\text{Total number of packets sent}} \times 100 \qquad (13.1)$$

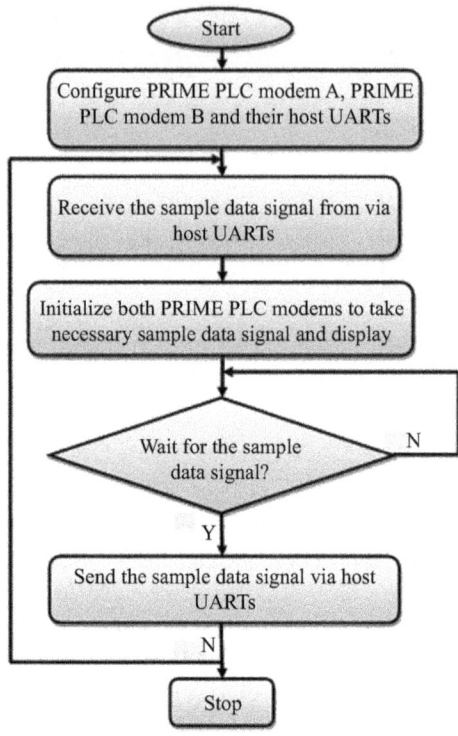

Figure 13.5   Flowchart of proposed algorithm

For the sake of reliability, the tests were repeated 10 times (after every 5 s interval) and the average PER of these tests were displayed.

13.3   Field trial results and analysis

Table 13.1 specifies the worldwide comparison of distribution transformers, which were used for various NBPLC based applications. PRIME PLC technology is robust and specially designed for LV (and low current) power lines and targets high data rates by employing high-order modulation schemes.

In power distribution substation, the voltage limit was set from 215 to 240 V, which can be normally observed during LV network μGs in India. G3-PLC and PRIME PLC standards normally support OFDM with DBPSK, DQPSK and D8PSK modulation schemes per carrier with inverse fast Fourier transform size of 256 and 512 B, respectively [30]. DBPSK modulation is an incoherent form of phase shift keying without restoring the coherent reference signal at the receiver, which reduces the complexity at receiver side. As per the operating principle, the input binary data should be in DBPSK mode, and the corresponding output does not change until input bit is 0;

Table 13.1 *Worldwide comparison of various distribution-transformers for narrow-band power-line communication based applications*

Distribution transformer	Modulation type/frequency range	Mode of operation		Country
		Real-time	Simulation	
400 kV A/10 kV–380 V	(–) 9–95 kHz	✓	✓	Netherlands [25]
(–)/20 kV/200 V	G3-PLC 35.9–90.6 kHz	✓	✓	France [26]
10–100 kV A/(–)	G3-PLC Up to 95 kHz	–	✓	Australia [27]
100 kV A/(–)	Frequency shift keying (FSK) 50–150 kHz	✓	✓	Italy [28]
11 kV/266.67 V	G3-PLC Up to 95 kHz	✓	✓	India [29]

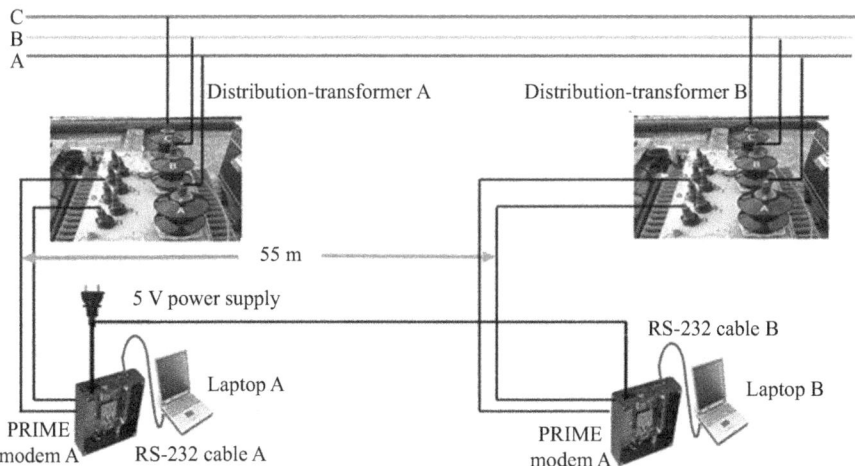

Figure 13.6 *Experimental set-up for PRIME PLC performance analysis in the industrial area, Panipat (Haryana), India*

otherwise, the output will turn over. Otherwise there is a simple mapping relationship: bit 1 is output 1 and bit 0 is output −1, which is termed as BPSK modulation.

During our experimental verification with PRIME PLC modems, two types of transmission (i.e., packet transfer and data file transfer) for the communication testing purpose were performed from both directions, as shown in Figure 13.6.

A transfer comparison of PLC signal using DBPSK, DQPSK and D8PSK modulation schemes connecting across secondary sides of two 11 kV–433/266.67 V

Table 13.2   11 kV distribution-transformer specifications

Distribution-transformer manufacturer	Century Infrapower (Pvt.) Ltd.
Rated voltage, HV	11 kV
Rated voltage, LV	433–266.67 V
Frequency	50 Hz ± 5%
Assistant channel	Channel 1
Connection HV	Delta
Connection LV	Star (neutral brought out)
Vector group	Dyn-11
Winding material	Aluminum
Cooling type	ONAN
Year make	2016

Table 13.3   Packets transmissions across secondary windings of two 11 kV distribution-transformers

Modulation type	Distribution-transformer A to distribution-transformer B			
	SNR 1 (dB)	BER 1	PER 1 (%)	Data transfer rate (kbps)
DBPSK	13	0.002	0	18.21
DQPSK	12	0.012	54.6	34.43
D8PSK	13	0.039	78	36.38
Modulation type	Distribution-transformer B to distribution-transformer A			
	SNR 2 (dB)	BER 2	PER 2 (%)	Data transfer rate (kbps)
DBPSK	12	0.012	0	18.21
DQPSK	12	0.017	52.3	34.43
D8PSK	13	0.035	73	36.38

distribution-transformers (L–MV–LV type) and nearly 55 m distance were covered. Typical Indian distribution-transformer specifications, which were used during experiment, are given in Table 13.2.

During a transfer block size of 256 B using DBPSK, DQPSK and D8PSK modulation schemes transmitted, as compared in Table 13.3 and shown in Figure 13.7, the channel performance was measured on notebooks through USB cables. During experimental verification, ROBUST mode was essential to establish reliable communication among distribution-transformers.

Packet transmissions from both directions were performed. In Table 13.4 and Figure 13.8, we compared the five data files (i.e., with different sizes in kB), which were successfully transferred using DBPSK and DQPSK modulation schemes.

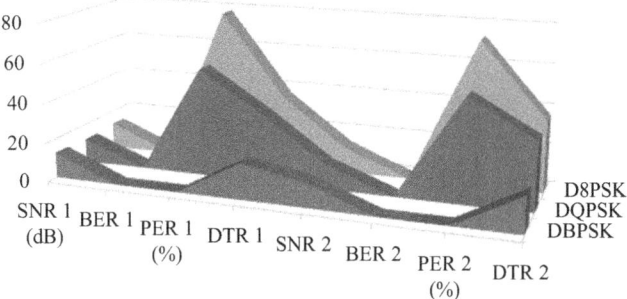

■ DBPSK ■ DQPSK ■ D8PSK

Figure 13.7 Packets transmissions across secondary windings of two 11 kV distribution-transformers

Table 13.4 Data file transmissions across secondary windings of two 11 kV transformers

Data file size (kB)	Distribution-transformer A to distribution-transformer B			Distribution-transformer B to distribution-transformer A		
	SNR 1 (dB)	PER 1 (%)	Data transfer rate 1 (kbps)	SNR 2 (dB)	PER (%)	Data transfer rate 2 (kbps)
25	13	0	6.6	13	0	6.4
50	13	0	6.1	13	0	5.9
75	13	0.7	6.5	13	0.6	6.2
100	13	1.1	6.3	13	1.2	6.1
125	13	1.2	6.4	13	1.5	6.2

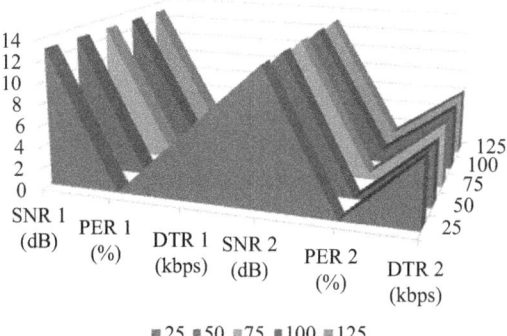

■25 ■50 ■75 ■100 ■125

Figure 13.8 Data file transmissions across secondary windings of two 11 kV distribution-transformers

During experimental verification, the best results for 25 and 50 kB size file transmission were achieved. It is found during high speed that the data files transmission decreased or packet lost and retransmissions of data files were performed. As presented in this chapter, the data transfer speed was slow, but the PRIME PLC technology through MV power-lines has proved the successful communication through two distribution-transformers, and it is observed that in the coming years, this technology will be useful and deployable for Indian SG applications.

13.4   Final summary

This chapter has aimed to measure PRIME PLC performance analysis practically over distribution transformers in the industrial area of Panipat (Haryana), India. In our test implementation, the PLC channel measurements were performed to achieve BER performance, and PRIME PLC using different modulation schemes (i.e., DBPSK, DQPSK and D8PSK modulation schemes) looks truly promising due to availability of Reed–Solomon codes to communicate very high volumes of data over distribution-transformers. With recent field measurement results, the following main discussion points are:

- First, to the best of author's knowledge, no PRIME PLC technology–based testing to perform BER and PER analysis of two types of transmission (i.e., packet transfer and data file transfer) across two distribution-transformers in Indian context has been conducted yet.
- Second, the PRIME PLC measurements in this study were realized only for specific industrial power-lines in India, the proposed algorithm in its present form is not generic to all other industrial power-lines. Therefore, further studies should be carried out to prove the accuracy of algorithm for other industry power-lines.
- And the last discussion is about this pilot is the real proof that PRIME PLC technology has become a communication foundation not only for global smart metering networks but soon in future it could also effectively help to implement the future Indian AMI systems/and SG applications such as demand response with a preferable rate and reliability, even in the noisy environments.

Acknowledgments

This research work was supported by Department of Electrical Engineering, National Institute of Technology Kurukshetra, India, and by the Ministry of Human Resource Development (MHRD), Government of India. The authors are also thankful to Narender Sharma, Junior Engineer-1 (Operation), Uttar Haryana Bijli Vitran Nigam (UHBVN), Panchkula, India, and Texas Instruments, India, for their continuous motivation, support and expert suggestions and direction for this work.

References

[1] Roy K., Mandal K.K., 'Hybrid optimization algorithm for modeling and management of microgrid connected system'. *Frontiers in Energy*. 2014;**8**(3): 305–314.

[2] Sharma K., Saini L.M., 'Performance analysis of smart metering for smart grid: an overview'. *Renewable and Sustainable Energy Reviews*. 2015;**49**: 720–735.

[3] Sharma K., Saini L.M., 'Smart grid energy management' in Dincer, I. (ed.). Comprehensive Energy Systems. Oxford: Elsevier; 2018. pp. 474–536.

[4] Gopakumar P., Reddy M.J., Mohanta D.K., 'Pragmatic multi-stage simulated annealing for optimal placement of synchrophasor measurement units in smart power grids'. *Frontiers in Energy*. 2015;**9**(2):148–161.

[5] Saravanan B., 'DSM in an area consisting of residential, commercial and industrial load in smart grid'. *Frontiers in Energy*. 2015;**9**(2):211–216.

[6] Stimoniaris, D., Kollatou, T., Tsiamitros, D., *et al.* 'Demand-side management by integrating bus communication technologies into smart grids'. *Electric Power Systems Research*, 2016;**136**:251–261.

[7] Hayati H., Ahadi A., Aval S.M., 'New concept and procedure for reliability assessment of an IEC 61850 based substation and distribution automation considering secondary device faults'. *Frontiers in Energy*. 2015;**9**(4):387–398.

[8] Sharma K., Saini L.M., 'Power-line communications for smart grid: progress, challenges, opportunities and status'. *Renewable and Sustainable Energy Reviews*. 2017;67:704–751.

[9] Landis+Gyr proves G3 PLC technology is right fit for smart metering. Successful field tests prove the technology's promise. Available from http://www. landisgyr.com/landisgyr-proves-g3-plc-technology-is-right-fit-forsmart-metering-2/ [Accessed 14 April 2018].

[10] Sharma K., Saini L.M., 'Reliability and survivability analysis of narrowband power-line communication links in low-voltage building network applications'. *The Journal of Central Power Research Institute*. 2015;**11**(3):505–516.

[11] Draft specification for PowerLine Intelligent Metering Evolution. Available from http://www.primealliance.org/wp-content/uploads/2013/04/PRIME-Spec_v1.3.6.pdf [Accessed 14 April 2018].

[12] G.9903: Narrowband orthogonal frequency division multiplexing power line communication transceivers for G3-PLC networks. Available from https://www.itu.int/rec/T-REC-G.9903-201508-I!Amd1/en [Accessed 14 April 2018].

[13] IEEE standard for low-frequency (less than 500 kHz) narrowband power line communications for smart grid applications. Standard of IEEE 1901.2-2013, pp. 1–269.

[14] PRIME Alliance. Interoperable standard for advanced meter management & smart grid. Available from http://www.prime-alliance.org/ [Accessed 14 April 2018].

[15]   Masood B., Baig S., 'Standardization and deployment scenario of next generation NB-PLC technologies'. *Renewable and Sustainable Energy Reviews*. 2016;**65**:1033–1047.

[16]   Sanz A., Nicolas J.I.G., Estopinan P., Miguel S., 'PRIME from the definition to a SoC solution'. Proceedings of Power Line Communications and its Applications, ISPLC2009. Dresden, Germany, April 2009, pp. 347–352.

[17]   Kikkert C.J., 'MV to LV transformer PLC bypass coupling networks for a low cost smart grid rollout'. Proceedings of Innovative Smart Grid Technologies Asia. ISGT2011. Perth, Australia, November 2011, pp. 1–6.

[18]   Sendin, A., Berganza, I., Arzuaga, A., Pulkkinen, A., Kim, I.H., 'Performance results from 100,000+ PRIME smart meters deployment in Spain'. Proceedings of Smart Grid Communications. SmartGridComm2012. Tainan, Taiwan, November 2012, pp. 145–150.

[19]   Gonzalez-Sotres, L., Mateo, C., Frías, P., Rodriguez-Morcillo, C., Matanza, J. 'Replicability analysis of PLC PRIME networks for smart metering applications'. *IEEE Transactions on Smart Grid*. 2018;**9**(2):827–835.

[20]   Lopez, G., Moreno, J.I., Amaris, H., Salazar, F., 'Paving the road toward smart grids through large-scale advanced metering infrastructures'. *Electric Power Systems Research*. 2015;**120**:194–205.

[21]   Alonso, E., Matanza, J., Rodriguez-Morcillo, C., Alexandres, S., 'Performance evaluation of AMR simultaneous polling strategies in a PRIME PLC network'. Proceedings of Power Line Communications and its Applications. ISPLC2015. Austin, TX, USA, April 2015, pp. 101–106.

[22]   Aruzuaga, A., Berganza, I., Sendin, A., Sharma, M., Varadarajan, B., 'PRIME interoperability tests and results from field'. Smart Grid Communications (SmartGridComm), 2010 First IEEE International Conference on, IEEE, October 2010, pp. 126–130.

[23]   Berganza, I., Sendin, A., Arzuaga, A., Sharma, M., Varadarajan, B. 'PRIME on-field deployment first summary of results and discussion'. Smart Grid Communications (SmartGridComm), 2011 IEEE International Conference on, IEEE, October 2011, pp. 297–302.

[24]   TI Designs. TI G3 power line communication developer's kit design guide. Available from http://www.ti.com/lit/ug/tidu237/tidu237.pdf [Accessed 14 April 2018].

[25]   Wouters, P.A., van der Wielen, P.C., Veen, J., Wagenaars, P., Steennis, E.F., 'Effect of cable load impedance on coupling schemes for MV power line communication'. *IEEE Transactions on Power Delivery*. 2005;**20**(2):638–645.

[26]   Razazian, K., Umari, M., Kamalizad, A., Loginov, V., Navid, M., 'G3-PLC specification for powerline communication: overview, system simulation and field trial results'. Proceedings of Power Line Communications and Its Applications. ISPLC2010. Rio de Janeiro, Brazil, March, 2010, pp. 313–318.

[27]   Kikkert C.J.,'Power transformer modelling and MV PLC coupling networks'. Proceedings of Innovative Smart Grid Technologies Asia. ISGT2011. Perth, Australia, November 2011, pp. 1–6.

[28] Cataliotti, A., Cosentino, V., Di Cara, D., Tine, G., 'Oil-filled MV/LV power-transformer behavior in narrow-band power-line communication systems'. *IEEE Transactions on Instrumentation and Measurement*. 2012;**61**(10): 2642–2652.

[29] Sharma K., Saini L.M., 'Performance evaluation of G3-PLC over distribution transformers in Indian context'. *The Journal of Central Power Research Institute*. 2016;**12**(3):463–472.

[30] Hoch M., 'Comparison of PLC G3 and PRIME'. Proceedings of Power Line Communications and Its Applications. ISPLC2011. Udine, Italy, April 2011, pp. 1–5.

Chapter 14
Analysis of hybrid communication for smart grids

Fabiano Salvadori[1], Camila S. Gehrke[1],
Fabrício B.S. de Carvalho[1], and Alexandre C. Oliveira[2]

The smart grid (SG) is a new and modern design for electric power systems (EPSs), leading to highly efficient, reliable, and safe electric power infrastructures. In addition, it provides a harmonious integration of renewable and alternative energy sources by means of modern communication technologies and automated control systems [1]. In the SG context, secure and real-time information becomes the key factor for the reliable energy supply from the generation units to the end users. Furthermore, the information can have decisive characteristics to provide self-healing abilities [2,3].

The SG also enables new power system-management strategies that provide effective grid integration for distributed generation (DG), demand side management (DSM), and energy storage (ES) [4]. There is a large variety of research results showing that more active market participation by the demand side could significantly benefit the whole market. In particular, Salvadori *et al.* [3] highlighted the following benefits:

1. Reduction in the energy cost for consumers who trade their demand during periods of high prices for periods of lower prices.
2. Reduction in the overall system generation cost, because changes on consumers behavior will eventually flatten the overall demand profile.
3. Even consumers who do not adjust their demand can make a profit if cost reduction translates into lower prices to avoid price spikes (i.e., very large price increase over short periods of time).
4. Reduction in the ability of generating companies to exert market power.

According to the International Energy Agency [5], "a Smart Grid is an electricity grid that uses digital technology to monitor and manage the transportation of electricity from all sources of generation, finding a variety of demands and users. These networks will be able to coordinate the needs and capabilities of all generators, operators, end users and stakeholders of the electricity market in order to optimize the

[1]Electrical Engineering Department, Federal University of Paraíba, Brazil
[2]Electrical Engineering Department, Federal University of Campina Grande, Brazil

use and operation of the assets in the process, minimizing environmental costs and impacts while maintaining reliability, resilience and stability of the system."

The SG design arose from the need for evolution and changes in the electricity sector, becoming an alternative solution to upgrade the existing power system, thus changing the paradigms of the last decades. Therefore, the SG cannot be considered just a single technology or a technology *per se*, but a set of technologies which may have different approaches for different contexts, moving toward an improvement of the electric power infrastructure.

Therefore, it is clear that the concept of SG is not unique, exclusive, but rather a set of definitions and technologies that leads to the use of an infrastructure composed by digital elements and communication mechanisms in the power system. Another reason driving the SG evolution results from an increasing energy demand and the need for the interconnection among systems, allowing more efficient energy usage [6,7].

SG technologies are often associated with an automated metering infrastructure (AMI), which involves not only measuring but also smart meters (SMs), protection and control equipment, communication systems, and a wide variety of sensing, monitoring, and controlling systems. So, providing a large amount of data, as means for allowing real-time pricing, demand response programs, and other customer facing applications. As more data becomes available across the power system, a greater level of automation can be introduced to the day-to-day grid management. Potential disturbances can be detected sooner, while being able to be addressed without human intervention. As more intelligence is built into the EPS (including low level devices), grid utilities are beginning to explore the potential for leveraging the vast amount of data these devices generate. The purpose is to gain a better understanding of the actual condition of grid assets, allowing operations and maintenance activities to be optimized along with the performance of the network as a whole.

Nevertheless, to add some intelligence to an EPS, independent processors are required for each component in the generators, substations, power plants, end users and stakeholders [8]. These processors must have a robust operating system and should be able to act as independent agents, which are able to communicate and cooperate with each other, forming a large distributed computing network. Each agent should be connected to sensors associated with its own component or substation. Therefore, it can assess its own operating conditions, reporting them to its neighboring agents through the communication paths.

Agents, such as intelligent electronic device (IED), SM, and digital relays, among others, are based on embedded systems which can be found everywhere in the daily life, from electrical commodities and appliances to nonlinear compensation mechanism, complex automation systems, and adaptive control systems. With embedded systems, it is possible to implement the integration systems to prevent unexpected failures, while improving system reliability and maintainability, and avoiding severe economic losses. There are several hardware and software solutions applied to SG, for the most varied scenarios, as presented in [3,9–11].

Integrated systems can consist of several devices and/or embedded systems connected to a computer through a communication network. In traditional power grids, automation usually employs wired communication [e.g., copper, optical fiber (OF)].

For the SG deployment, a very reliable and flexible communication network with wide area coverage and cost effective is required. Thus, the SG model takes into account both wired and wireless communication technologies, forming an hybrid network architecture (HNA).

The high number of existing communication technologies is an opportunity for SG applications, even though the debate on which technology is better suited for the SG requirements is still open. It results from the fact that different SG applications have different communication requirements, in terms of data payloads, sampling rates, latency, and reliability [12].

According to Shuaib *et al.* [13], an SG communication framework must address the following aspects:

1. Performance and reliability requirements in terms of data transfer for all applications under consideration.
2. Appropriate communication technologies for transferring the collected data to the energy service provider management/control node based on the location (e.g., home, substation, power plant), type of deployed sensor, and the type of collected data.
3. Seamless integration of the various network technologies and protocols employed for gathering and transferring the data.
4. Smart metering and smart interfaces/applications to be deployed with a variety of sensors for the smart relaying of information over a wireless/wired network or a heterogeneous network composed by several of such technologies.
5. Security measures for protecting consumers' privacy and to ensure the reliability and availability of the smart grid communication.

To the success of the emerging SG, a communication infrastructure is an essential part, and a scalable and pervasive communication infrastructure is crucial for deploying and operating an SG.

For the communication technologies, the main requirements, for defining the more suitable technology for each SG applications, are transfer time, time synchronization accuracy (evaluated in terms of maximum variation; that is, the jitter of the time offset among the nodes in the network), dependability (evaluated in terms of the packet-loss rate as defined in IEC 61850) [14,15].

Finally, defining the best communication approach relies on the proper selection of physical layer and protocols, based on the required characteristics: distance, speed, interference, latency, bandwidth, errors, security, and packet loss.

14.1 Wired communications for smart grid applications

SG data transmission is between intelligent electronic devices (IEDs) and electrical utilities, in which different communication technologies can be used in infrastructure based on wired or wireless [3]. In general, the wired technologies provide more capacity, reliability, and security deployment.

Furthermore, wired solutions do not face severe interference problems, and their functions are not battery dependent as wireless solutions often do. For example, OF can support data rate up to several Gbps with bit error rate (BER) as low as 10^{-15} [10]. The major limitation is its high renting and installation costs, and hence it should only be considered when high bandwidth and stringent performance guarantees are required.

In this section, the best wired mediums, i.e., electrical wiring, twisted-pair, and OF, and technologies implemented over them, i.e., power line communication (PLC), digital subscriber line (DSL), synchronous digital hierarchy (SDH), gigabit ethernet (GbE) and wavelength-division multiplexing (WDM), are evaluated as an alternative for data communication in SG.

14.1.1 Electrical wiring

Electrical grid is a highly interconnected network. For example, in Brazil, almost the entire country is connected by the electrical wiring. This high interconnection makes a great opportunity to communicate data for every place.

PLC is a well-known communication method implemented over the electrical wiring, and among wired communication technologies candidates for SG, PLC is frequently considered, particularly in AMI. The expression PLC, also known as power line telecommunications (PLT), is adopted to identify technologies, equipment, applications, and services that provide communication among energy lines. Data rates on power lines vary from a few hundred bps to Mbps [12].

PLC can guarantee reliable and fast data transmission services through electrical power system with reduction on cabling and integrating residences, buildings, and factories. The idea is to use the electrical distribution system to propagate communication data using the existed wired infrastructure [16]. It can be useful for substations (ground and underground), in rural areas (where alternative communication infrastructures are not deployed and/or very expensive), in residential, industrial and commercial automation, and another applications [9].

Different applications of PLC are envisioned in daily life, including several home automation applications (security, gaming, Wi-Fi access, among others) [17] and more complex scenarios as in SG systems [18]. Furthermore, PLC is the only wired technology that has deployment cost comparable to wireless technologies since the lines are already there [19].

However, the PLC technology also has some disadvantages: (i) the interruption of the communication channel among devices occurs when the interruption of electric service occurs; (ii) background noise due to the electrical devices connected to the EPS introduces a considerable error rate; (iii) the signal that is carried along the power cables may be intercepted, because of the lack of shielding of the electric wiring, representing a serious security problem; and (iv) the quality of the transmitted signal deteriorates with the load charging of the EPS, the higher the load the lower the quality [20,21].

Electrical wiring is a hostile medium for data transmission, since it was not originally designed for such purpose. Many problems such as variant impedance, considerable noise and high levels of attenuation hinder transmission in this medium and limit the use of this technology. Besides, the interference levels inside the transmission lines are elevated and limit the applications of PLC for larger distances [16].

The signal propagation does not occur in a single path between the transmitter and the receiver; consequently, the echo is also relevant and must be considered, due to the multiple paths in the channel.

Channel measurements show that, for high frequencies, channel attenuation increases. In this way, the channel can be described as being random and time variant, with a frequency-dependent signal-to-noise ratio (SNR) over the bandwidth of the transmission [22].

The noise problems could be minimized if a high amplitude modulation is used. However, the PLC communication is constrained by standards, which impose the maximum amplitude levels to the signal data. As the voltage amplitude depends on the grid impedance, for a good communication, a tuning should be performed for each scenario.

The PLC technology in low-voltage grids can be applied in narrow or broadband, depending on the application. For example, the PLC can broadcast the same data to several homes, industries and commerce, as the electric system represents a widespread infrastructure. However, when using broadband, the communication is limited by low–medium distances, as the broadband operates with high frequency (1.8–250 MHz) and high data rates (up to 100 Mbps). On the other hand, the narrowband operates at lower frequencies (3–500 kHz), lower data rates (up to 100 kbps), so it can reach longer range (up to several kilometers). However, broadband bandwidth leads to higher speed, reaching up to 20 Mbps. The narrowband is considered the most suitable for implementation in electric utilities [22]. The wide variation of frequencies for narrow or broadband communication depends on the different worldwide standards with different frequency bands allocated. Also, the distances and data speed can vary depending on the used modulation scheme.

The narrowband is regulated by commissions such as Federal Communications Commission (FCC), Association of Radio Industries and Businesses (ARIB), and European Committee for Electrotechnical Standardization (CENELEC), which determine, besides the frequency range, the maximum voltage output levels. As mentioned before, PLC is affected by the EPS interference in order to constrain the communication to the standard voltage limits.

The reduction of the attenuation and noise issues could be solved by raising the level of the signals containing data. But, as explained, the standards as CENELEC EN50065 [23] constrains the voltage level of the data signals injected to the electric power grids. The maximum level depends on the PLC operational frequency; for CENELEC A-band, the maximum allowed amplitude of the signal is 134 dB µV.

Furthermore, PLC communicates signals based on different modulations techniques, which can be frequency, amplitude and phase. Generally, the amplitude modulation is used. Although, multiple carrier can be also used, such as orthogonal frequency-division multiplexing (OFDM), implemented by powerline intelligent metering evolution (PRIME).

Other specifications for PLC have been proposed by HomePlug, Universal PowerLine, and HD-PLC Alliance. Since 2009, IEEE P1901 has been elaborated and the PLC has been incorporated to SG standards.

14.1.2 Twisted pair

Twisted pair is based on cooper and prevents crosstalk (noise generated by cable pairs when electric current passes through it). There are different categories of twisted pair cables which varies from 1 to 8. The difference between the categories basically depends on the type of shield. As the shield changes, the maximum rate and distance also vary. Two useful situations can be mentioned as an example of usage: locations where wired networks has already been installed; or places where wireless communication is hard to deploy (e.g., due to interferences and difficulties to install antennas) [3].

DSL also transmits data over a twisted telephone line. It can operate symmetrically (SDSL) or asymmetrically (ADSL). The SDSL operates with up/downstream with the same velocity, while in ADSL the upstream is more constrained. The bit rate ranges around 256 kbps to over 100 Mbps. The DSL technology is commonly used to provide internet for consumers.

Also some Ethernet networks implemented over the twisted pair can reach 100 Mbps up to 100 m; their variety is known as 1000BASE-T. IEEE 802.3ab is used as protocol for this technology.

Another serial bus that uses twisted pair is the controller area network (CAN), system originally conceived for passenger vehicle networks. Nowadays it is widely adopted in different transportation systems as well as several industrial and embedded electronics applications. CAN is another useful application deployed in twisted pairs which can be applied for SG monitoring applications [3].

14.1.3 Optical fiber

OF transmits data over long distances based on a total internal reflection of light pulses inside a cylindrical glass core. The glass core is surrounded by another glass layer named cladding. A buffer layer protects the cladding, while a plastic jacket layer completes the OF structure.

Today's transmitters and receivers are enabling a single wavelength to achieve rates around 100 Gbps over long distances (light traveling up to 200,000 km/s). The capacity and bandwidth of OF communication is many times greater than any other wired or wireless link. The OF communication provides excellent BER performance. As a typical fiber can accommodate up to 100 channels, the total capacity of a fiber can be extended to around 10 Tbps [24].

An OF can be made by a single or various multimode fibers. Single mode and multimode fibers are normally adopted in communication systems according to its requirements and particularities [25]. Also, depending on the necessary bandwidth, a different light modulation can be implemented.

Currently a great deployment of optical networks can be verified for terrestrial communication. The use of the already available fiber network can constitute an interesting alternative to serve as a backbone communication network in SGs. Due to the OF high speed, it is considered the best transmission medium for SG services, such as video, due to its high latency requirements [26,27].

The radio and electromagnetic immunity of optical signal makes OF a secure and reliable communication network. OF communication is the most cost-effective choice for long-distance communication than other available technologies, because few repeaters are required. The installation of OF network for remote control and monitoring is quite expensive. This disadvantage can be overcome by utilizing OF network backbones in the communication networks that are spare [28].

Indeed, communication over OF networks will be expected to increase in the future and will be one of the most important alternatives in the SG communication network.

GbE describes several technologies for transmitting Ethernet frames with 1 Gbps or higher bps over OF. Initially, GbE was deployed to be used in backbones, but a data communication was improved, for example, by SG applications, which now is widely used. Different standards define the Ethernet bases, i.e., IEEE 802.3z for 1000BASE-SX over multimode fiber and 1000BASE-LX over single-mode fiber. The bit per second reach 1,000–1,250 Mbps. The distances are up to 100 km.

SDH is a standard protocol for transmit multiple digital bit steam synchronously over OF. As it ensures all circuits to transmit simultaneously, it is more a transport protocol than a communication protocol. The protocol is multiplexed with a complex header interleaved with data, which allows a very low latency. Data rates reach up to 39.81 Gbps depending on the design, OC1 to OC278, STS-1 to STS-768 or STM-0 to STM-256.

WDM multiplexes a number of optical carrier signals over a single-mode fiber by using different wavelengths. This technique enables bidirectional communications and multiplication of capacity. As it can multiplex over 160 signals, 16 Tbps can be reached. WDM systems are divided into three different wavelength patterns, normal (WDM), coarse (CWDM), and dense (DWDM). Normal WDM, sometimes called BWDM, provides two wavelengths on one fiber, coarse WDM uses up to 16 channels across multiple transmissions and dense WDM uses the C-Band transmission, but with denser channel spacing.

When interoperability is required, those three technologies can work together. GbE is only compatible at low-level equipment, as SDH and GbE can be based on LAN PHY; however, it does not provide a bitstream level compatibility. When WDM, which includes both coarse and dense WDM systems, is considered, both communications are compatible.

14.2 Wireless communication in smart grid applications

Wireless communication expansion is enabling the rise of new applications as well as the consolidation of current services and techniques [29]. Wireless techniques are playing a vital role in creating a highly reliable SG, which rapidly responds to real-time events with appropriate actions. As a consequence, the implementation of the wireless technologies for SG communication has shown a dominating role in recent years.

However, developing an SG using wireless communication requires a multidisciplinary expertise such as [27,30]:

1.   A robust knowledge of the electrical power systems;
2.   Expertise on radio frequency (RF) design, propagation challenges, and RF interference problems in electrical power systems environments;
3.   Sensor technology to fully understand issues associated with sensor calibration, transducers, and clock-drift;
4.   Comprehension of the hierarchical network architectures to integrate different networks, which are required for SG to provide flexible and scalable architectures for hybrid applications.

To implement the envisioned SG in the electrical power system and, hence, take the advantages of the potential gains of wireless communication, effective communication protocols, which can address the unique challenges posed by such systems, are required. In this way, the wireless connections in SG, from user to grid utility, can be performed based on various technologies, as a number of different RF applications are potential, or have already been consolidated, to improve SG communication. Nowadays, the following wireless applications are highlighted for SG usage: cellular communication, Wi-Fi, Bluetooth, ZigBee, and more recently cognitive radio [29].

The wireless network technologies to SG infrastructure are usually considered for applications based on two possibilities: (i) to use the existing public network (such as cellular network) or (ii) to design a dedicated wireless communication network for the SG services.

14.2.1   Dedicated wireless networks

Different radio frequency technologies and standards have been developed to extend the access of wireless users to specific networks. Wi-Fi and Bluetooth are consolidated network standards that can be associated with other functionalities and lead to different applications.

Operating around the 2.4 GHz or in the 5.8 GHz industrial, scientific, and medical (ISM) bands, Wi-Fi (Wireless Fidelity) refers to different technologies based on the IEEE 802.11 standards with different data rates and coverage area. The goal is to enable different devices to be connected to the Internet via a wireless access point and a wireless local area network [31].

Bluetooth is another standard that operates around 2.4 GHz ISM band. It integrates devices located in the same geographical area. Devices can be connected via Bluetooth connection in a reduced area without the obligation of a wireless access point. Bluetooth is managed by the Bluetooth Special Interest Group [32].

Extending the concept of integrated devices in a geographic area, wireless sensor network (WSN) is a technology commonly adopted to monitor a specific variable or event. A WSN can integrate Wi-Fi or Bluetooth to extend its operation and coverage, according to the monitored event. Different WSN applications are based on the IEEE

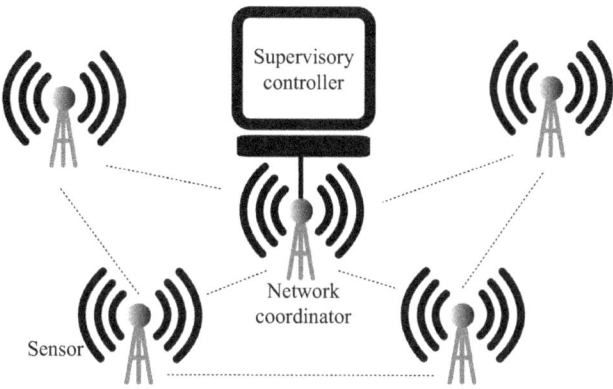

Figure 14.1   Example of a typical WSN

802.15.4 network layers. IEEE 802.15.4 is a low rate wireless personal area network (LR-WPAN) standard for radio communication, which typically operates in a reduced geographical area with low data rates (around 250 kbps), which leads to reduced energy consumption (and can increase the lifetime of the network due to the reduced energy consumption of the sensor batteries) [33].

Different variables (as humidity, temperature, pressure, wind, vibration, sound, among others) can be monitored by a WSN, as described in technical literature [33,34]. A WSN node integrates, besides wireless communication, sensing, data collection and processing, with an attached power supply on a single chip. This can be defined as an IED [9,30,35].

A wireless sensor network is composed by sensor nodes (deployed in the region to be monitored) communicating through a wireless connection with other nodes and with the network coordinator, also known as master. Data gathered by the nodes can be sent and processed by the network coordinator. The master node is usually integrated with a control station in order to perform predefined management and control tasks in the networks. These tasks are coordinated by different algorithms according to the network's structure. Moreover, sensor nodes can be fixed or mobile. Figure 14.1 illustrates a WSN with sensor nodes sending its measurements to the master node [33].

Due to its technical characteristics, the IEEE 802.15.4 is the basis for different wireless communication standards, which specify the upper layers of the OSI model according to its requirements. There are several standards based on the IEEE 802.15.4: ZigBee, ISA100.11a, WirelessHART, MiWi, 6LoWPAN, among others [36–38].

ZigBee is the most popular standard-based wireless technology designed to address low-cost, low-power wireless sensor, and control networks requirements. ZigBee is a specification for a suite of high-level communication protocols using small, low-power digital radios based on the IEEE 802.15.4 standard for wireless personal area networks (WPANs) [39,40].

The ZigBee protocol enables communication using multiple network topologies, including star, tree, and mesh. The technology defined by the ZigBee specification is intended to be simpler and less expensive than other WPANs, such as Bluetooth. ZigBee also operates in the ISM bands.

IEEE 802.15.4.g standard is under development by the Smart Utility Networks Task Group. The goal is to define a physical layer (PHY) for applications in the AMI infrastructures in the bands of 700 MHz to 1 GHz and in the ISM band of 2.45 GHz [41].

14.2.2   *Public cellular communication networks*

Electrical utilities can benefit from existing cellular networks such as 3G and 4G (and 5G in the next years) by installing new wireless communication networks. This alternative is cost attractive. The provisions of wide-area coverage and accommodation of hundreds of thousands of users for future SGs are dominant characteristics of a cellular network. It is expected that for the transmission of information from homes back to operations centers in SG, cellular network will play a leading role [27].

The advantages of cellular communication networks includes, among others, the cost savings by electric utilities, that would be required for installing cables or wireless systems; and mobility features made available by the cellular communication, which are especially suitable in SG for providing connectivity to moving vehicles. As restrictions from cellular communication networks, it can be mentioned the limited coverage (regarding to the fixed base stations) and RF interference responsible for poor performance of cellular communication technology. Additionally, security issues must be considered as consumer's data information can be forwarded in the cellular bands and could be accessed for unauthorized person.

Considering new perspectives for 5G communication, cognitive radio is one of the evaluated techniques that can improve the spectrum efficiency of cellular communication. The concept of cognitive radio (in which devices can monitor the available spectrum to detect potential unused band to enable opportunistic transmissions) is extended to different areas of electrical engineering [29].

New approaches investigate cognitive radio to reduce or even to eliminate interference inside the power lines [29,42]. Different work groups are focusing on standards to work with TV white spaces for SG applications [41]. The wireless interconnection of the AMI is fundamental for power operators and SG users [43], and the concept of cognitive radio was proposed as an hypothesis for establishing this link. Based on the white spaces available, the communication between the SMs and the other components of the SG can be more reliable [44]. New topologies and test beds based on spectrum sensing and cognitive radio are under evaluation [45,46].

14.3   Hybrid network architecture practical application

Traditionally, supervisory control and data acquisition (SCADA) systems have been used to monitor and control the electricity grid, which are based on communication technology with limited bandwidth, mainly wired communication systems, based on copper or/and OF medium, with high installation and maintenance cost. Different approach has been given in recent researches in SG, where several smart transmission networks, smart control center and smart solutions [13] should be controlled, demanding complex communication with flexible, cost-efficient, and also very reliable data in a wide-area coverage.

To a massive data volume be processed, the SG infrastructure should support different communication technologies – wired- and wireless-based solutions. For instance, the communication between IED or others to the utility grid can be carried out through: the already available infrastructure for communication, or, through a combination of existed and new communication. In this context, communication technologies widespread in industrial plants can be adopted in the EPS. For example, wireless networks as well as sensor networks are being adapted for the new intelligent and interconnected system.

For some applications, wireless communications have some advantages over wired technologies, such as low-cost infrastructure, and easier deployment in difficult or unreachable areas. However, the nature of the communication path may cause the signal to attenuate disproportionately. On the other hand, wired solutions do not face severe interference problems and usually are not battery powered.

The requirements for SG deployment can be met by applying a set of these communication technologies, providing a balanced tradeoff between investments and benefits. Giustina *et al.* [15] propose a hybrid infrastructure based on PLC, Wi-Fi, and OF to enable services for grid management. The choice for the particular technology to connect any node in the power grid is related to the application and its requirements [47]. In this view, hardly a single communication meets all the requirements of an SG. Thus, HNA are highlighted, as each subsystem can use a different technology. An HNA is a network architecture based on wired [e.g., Ethernet, Electronics Industries Association (EIA)-RS232, universal serial bus, CAN or PLC] and wireless [e.g., Wi-Fi, RF, cellular networks, WSN and wireless mesh networks (WMN)] communication technologies [9,15].

The HNA should provide reliable data in order to enhance all the future SG benefits. The future of electrical energy systems concepts makes of SGs an essential component. Such concepts are characterized as:

- intelligence on the network;
- interconnectivity/intercommunication among the various elements of the network;
- bidirectionality of power flow and data;
- distributed, renewable energy production, e.g., solar and wind; and
- entirely new components, e.g., electric vehicles, and operation concepts (virtual power plants).

The HNA architecture includes two different approaches: (i) the possibility of redundant communication paths (i.e., the same information is sent over two or more communication media), so increasing reliability; or (ii) cover different paths with different communication, e.g., adding a few wires to a wireless sensor network, one can reduce the average power consumption per sensor node, while providing a nonuniform power consumption across all sensor nodes, resulting in a better network life [3,15].

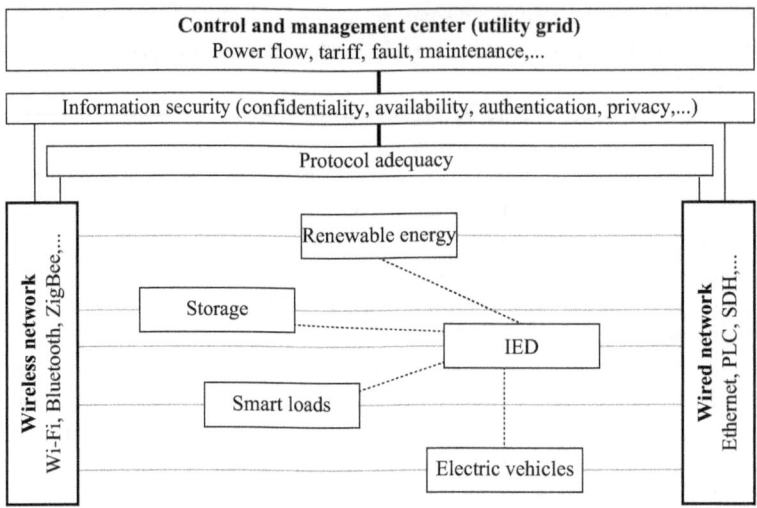

*Figure 14.2 An integrated communication technology framework for an SG
network with wired and wireless communication in redundancy*

First, the HNA is employed some/several times, in the same equipment/systems,
in order to provide a redundant structure. For a limited impact on the HNA deployment cost, a flexible communication infrastructure is required so that it is able to
support new services, while playing an increasingly relevant role on future deployment scenarios [3,48]. Also, in this approach, the supervisory control should handle
with different time responses of the communications.

Figure 14.2 illustrates the first approach, where communication perspective lies
in the core of the customer integration. First of all, a communication infrastructure
between home devices and the IED should be set up, so that IED could get information
from the devices and take any required actions to adjust local consumption considering the customer preferences. Second, a communication link between IED and the
utility grid should be established, allowing customers and the utility grid being bidi-
rectionally notified regarding the current electricity prices, customer behavior, and
power outages. Therefore, AMI communication can be represented by a hierarchical
multilayer architecture, in which redundancy happens in all the different systems, as
demonstrated by both boxes, wireless and wired communications.

In the second approach, the communication environment can be classified into
different areas. In Figure 14.3, three distinct communication networks are illustrated: home area network (HAN) for defining the interconnections among devices;
neighborhood area network (NAN) when referring to the interconnections among
costumers; and wide area network (WAN) for describing the interconnections among
utilities. In all such networks, a different communication technology can be applied,
including Ethernet, OF, Wi-Fi, PLC, satellites, and so forth [13].

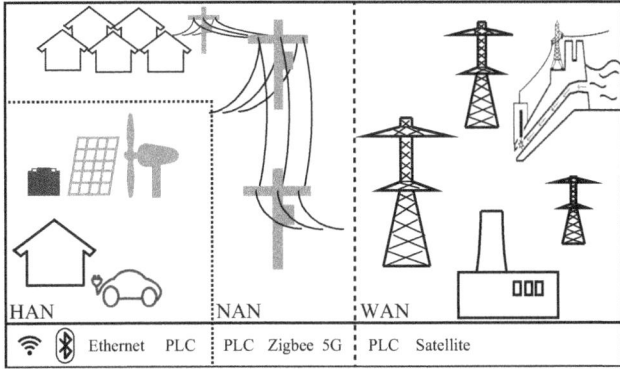

Figure 14.3   *Distinct communication areas: HAN, NAN, and WAN (each one with its particular communication requirements and some protocols alternatives)*

14.4   Practical results

The complete system was presented in R&D report [49]. The block diagram in Figure 14.4 illustrates a microgrid connected to the feeder. The feeder side sees the microgrid as one point of common coupling (PCC). The power utility management is based on an SC. The SC is responsible to control the active or reactive power flow.

The considered microgrid structure consists of:

1. IED based on a microprocessor, which was presented in [6], operating as communication modules, and also performing the power-flow management functions (active and reactive power) and bidirectional metering;
2. A switch to switch on/off the microgrid from the feeder;
3. Renewable energy sources: solar panel and/or wind generator;
4. Converters to perform the interconnection to the grid;
5. Alternatively, an energy storage and its converter;
6. Few controllable units.

The generation power unit could be provided by any renewable source, as solar or wind. The energy storage can be used only as emergency support or as voltage leveling [50]. Each unit communicates and is controlled based on an IED. In addition, the microgrid can operate independently or connected to the feeder. The SC can be achieved using an IED.

The IED should have a data acquisition system that allows monitoring of the electric power-grid variables (satisfying measurements requirements). The IED should process the data and then define the electric system characteristics. Finally, a communication system should allow reliable data exchange between the involved units and the SC.

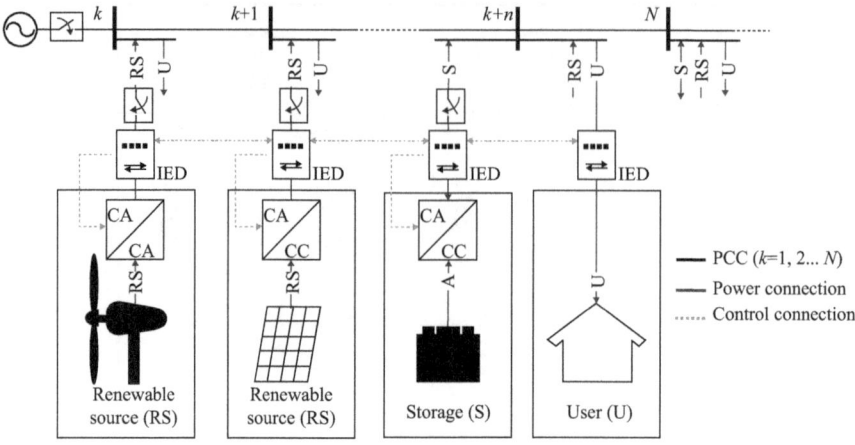

Figure 14.4   Microgrid system

The communication was based on HNA. In HNA, most of the critical data, such as voltage and current, were transmitted redundantly (wireless and wired), thus it is more difficult to occur packet losses. Nevertheless, due to the adoption of HNA, a more robust communication is provided. The communication technologies involved in the project were PLC, CAN, Ethernet based on OF, cellular and RF (free protocol).

The experimental results were implemented in a power energy utility, and the communication scheme is illustrated in Figure 14.5. The IEDs were monitoring some consumers/minigeneration, and the information was sent to a data concentration based on PLC. The PLC used as the power electric system was already available. An underground substation was also monitored in order to prevent inundation. The transformer is sealed to protect the equipment from water. As the transformer is sealed, there was no possibility to choose the communication, as the only point of access was the electrical wire connected to it. So, the PLC was used as well to monitor the transformer variables, such as pressure, temperature, and water. The PLC was connected to a bridge where the communication changes, because the transformer electrical wire is short. Thus, from the bridge, a twisted-pair cable based on CAN was used. The underground data output was performed via wireless solution, that sent the data to the data concentration. Finally, the data concentration sent all the received data to the power energy utility, where the SC is installed. The SC can receive two different data, status or monitoring. The status was developed for small signal, so it can be sent via wireless and wired communication. The monitoring was based on big packets, so Ethernet was used.

Considering that Ethernet and cellular are technologies sufficiently tested by the operators and CAN is widely tested by the automobile industry, the results with the free protocol RF (wireless) and PLC are presented. Some tests and improvements have been done; those results and improvements, for each technology, are explained in this section.

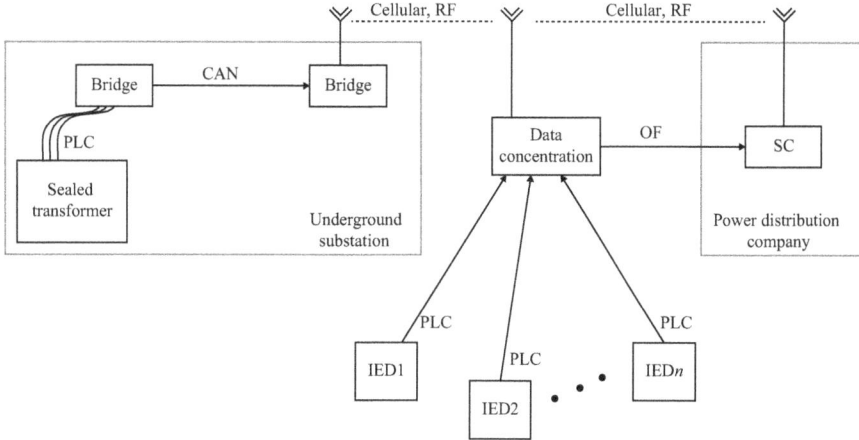

Figure 14.5 Hybrid communication applied to smart grid monitoring for a power energy utility

14.4.1 Wireless results

Usually, the wireless sensor network presents two significant problems: (i) energy consumption (battery lifetime) and (ii) interference (data loss). The problem (i) has already been studied by [9]. In this present work, problem (ii) is investigated, in which an efficient sequence of tests for analyzing the interference issues in the wireless network was performed.

The wireless tests were conducted considering one IED as master and others as slaves. The IED consists of a single node, composed by a controller and a wireless transceiver nRF-24L01+.[1] The nRF-24L01+ is a transceiver, based on Nordic Semiconductor.[2] This wireless module has an integrated antenna, and 126 channels operating in a frequency range of 2.4 until 2.525 GHz ISM band. For each channel, a bandwidth of less than 1 MHz (@250 kbps or @1 Mbps) is assigned; or a smaller than 2 MHz bandwidth (@2 Mbps).

The IED can operate like a master or a slave. When the IED operates as slave, it acquires data and sends to the master IED when requested. When the IED operates as master, it performs the supervisory controller functions by collecting data from the slave IED and sending it to the control center. The IED master received at least 10 times the slave IED packages, so to simplify the results, the average received data are presented in percentages.

Initially, for carrying out such tests, it was necessary to establish some metrics, [51], namely:

- configuration of the antennas (position/direction) of the tested transceiver modules;
- distance between the units;

---

[1] nRF-24L01 is a TM from Nordic Semiconductor.
[2] Nordic Semiconductor is a TM.

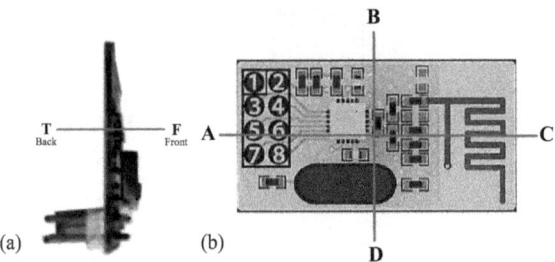

*Figure 14.6   Transceiver view to identify the orientation of the communication:
(a) side and (b) frontal*

- existence of obstacles between them;
- type of environment in which they operate, open or closed environments;
- possibility of interference in the communication by other elements using the same frequency of 2.4 GHz (as with wireless networks and other transceivers) for cordless phone or microwave apparatus;
- possibility of interference caused by climatic conditions (particularly wind);
- combination of all these parameters between these and the tested units.

The transceivers views are illustrated in Figure 14.6 so as to identify the orientation in the wireless communication.

Two distances: short distance, under 15 m, and long distance, above 15 m, were considered and described below. Furthermore, obstacles were considered (brick, glass, or metal walls) separating the receiver and transmitter units; ventilated and locations under climate change; places with possible interfering signals, like mobile phone network signals, other communications via RF or wireless communication, microwave, switching equipment.

Table 14.1 shows the test results. The table is split by tests and specifies the scenario considered for the transceiver communication. When different data is compared, the best result is emphasized.

From the short distance tests, even in low interference, if two slaves IED are sending data (which can occurs if some slaves are incorrectly configured with the same address as other slave IED), the reception of the transmitted packages is reduced to 99%. This reduction can be addressed due to packages conflict, but it was too small for consideration. However, when wireless routes are near to the slave IED and in the same communication channel, higher influence is observed, reaching 96% of receiving data.

Still for short distance, the antennas orientation was analyzed. Position TF–FT has been presented as the most efficient. This can be concluded by the results in the first line of Table 14.1, in which AC–AC is 99.10%, CA–CA is 99.85%, and TF–FT is 100%. The best position is explained by the radio pattern, which is provided by the antenna design. Thus, the same orientation has been performed to long and medium distances.

Finally, considering short distances, regardless of the direction of the transceivers, it was noticed a reduction in transmitted/received (T/R) package ratio only with high

Table 14.1   Wireless results under different test scenarios

Test	Position	Distance	Barriers	Interference	Received packets
Short distance					
1	AC-CA,	1 m–15 m	No	No	99.10%, 99.85%, 100%
	CA-AC,			Low	99.10%, 99.90%, 99.95%
	TF-FT			High	96.50%, 95.00%, 96.5%
2	TF-FT		Brickwall (1, 2, 3)	Low	100.0%, 83.0%, 4.98%
			Glass		100.0%
			Aluminum		99.98%
			Steel		96.6%
Medium/long distance					
3	TF-FT	50 m	No	Low	100.00%
		100 m			99.30%
		200 m			83.53%
		300 m			39.97%

interference (just over 3%), as presented in Table 14.1. Therefore, the effect of packet loss will depend on the nature and intensity of the interference.

For switching equipment with low power, the influence in the communication was lower than the other interference sources, and the receiving rate was almost 100%. So, even in short distance, wireless networks with high signal strength cause attenuation in the efficiency of communication.

The barrier tests were conducted based on obstacle influences, as detailed in Table 14.1. Obstacles with low thick, independently from the used material, have low influence, although under medium and high thick, the efficiency was reduced to 83% and 5%. Note that, when more brick walls were considered, the distance has also increased.

Concerning to transceiver coverage, for indoor communication, it was possible to reach 15 m, which present a better result than expected, due to the transceiver features (10 m indoor). Therefore, for outdoor tests, it was clear that external factors, such as wind and low amplitude signals (which are difficult to block), influence the communication. In spite of these adversities, the outdoor tests could prove that the transceiver can reach almost and up to 75% of efficiency for distances up to 250 m. Likewise indoor tests, the transceiver operates for longer distances than recommended.

14.4.2   Power line communication tests results

The PLC was applied in the project, and the tests were performed in the low-voltage electrical power system, so narrowband was used. The results were obtained based on two commercial PLC modems: ECHELON/PL-3120[3] and PRIME[4]-ATMEL/ATPL210.[5]

---

[3] ECHELON/PL-3120 is a TM from Echelon Corporation.
[4] PRIME is a TM from PRIME Alliance AISBL.
[5] ATMEL/ATPL210 is a TM from Atmel Corporation.

The ECHELON/PL-3120 modem performs auto interconnection (self-installation), as well as communication protocols (point-to-point or multipoint), with an option to enable or not the CENELEC protocol. The ECHELON/PL-3120 supports CENELEC A-band and C-band operation with dual carrier frequency mode. However, the multipoint protocol is proprietary and written in ROM on the device.

The PRIME-ATMEL/ATPL210 technology also operates with CENELEC A-band. Companies and institutions developed PRIME communication standard for remote measurement. The objective was to obtain an open and non-proprietary public telecommunications solution, whose deployment in the EPS would allow not only support to the smart metering functionality but also to operate as a communication integrator for the SG infrastructure. Thus, it can establish a complete set of standards at international level that will allow interoperability between equipment and systems of different manufacturers.

Both PLC modems can operate as master or slave; thus, both have TX amplifier circuit and RX receiver circuit. The TX amplifier circuit is responsible for coupling the PLC frequency modulation to the EPS. The RX circuit operates for conditioning the received signal to the analog-to-digital converter (ADC) input. However, the data signal amplitude depends on the grid impedance, which varies depending on the point of installation of the PLC modem. So, the gains of the TX and RX circuits should be configurable. The selected PLC modems provide an automatic gain for those circuits, which should change according to the grid impedance and still satisfies the regulations of the maximum allowed amplitude of the signal.

Based on those two modems, the tests using PLC communication were performed under two scenarios:

1. underground cables and
2. overhead lines.

In the first scenario, the communication was extreme for operational status. Given the underground grid characteristics, a point-to-point communication was suitable. ECHELON/PL-3120 was used in this case, as there was no need to build a network with multiple points. The results were obtained without any modifications and using free license protocol.

Considering that the goal of the project was to implement a network of IEDs, the second scenario, in the overhead grid, was prioritized for presenting the results. The second scenario consists of a group of IEDs (one master and several slaves). Even though the ECHELON/PL-3120 modem has presented good results in point-to-point communication, its multiple points solution is proprietary (not free); so for this scenario, PRIME-ATMEL/ATPL210 was used. The proposed system was tested over 90 days, and the data was collected in intervals of 10 s. In this period, more than 2 GB of information was stored in a database.

In Figure 14.7, the results of the received packets measured during one week are shown. As it can be verified, there were many hours without any communication established, and there was higher data traffic during the night period. It was clear that the communication improved when the line loading and the number of connections/disconnections of equipment were lower.

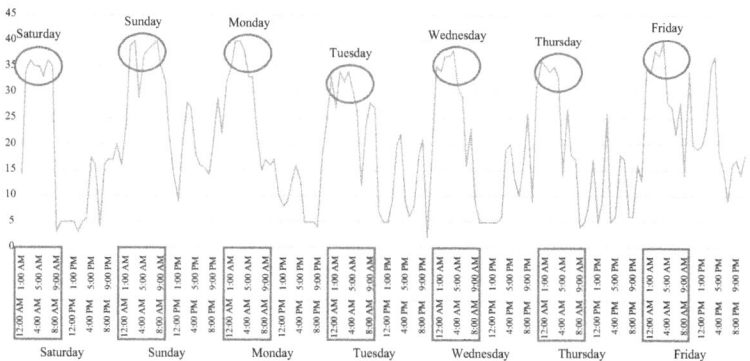

Figure 14.7   Average data transmission per hour in one week

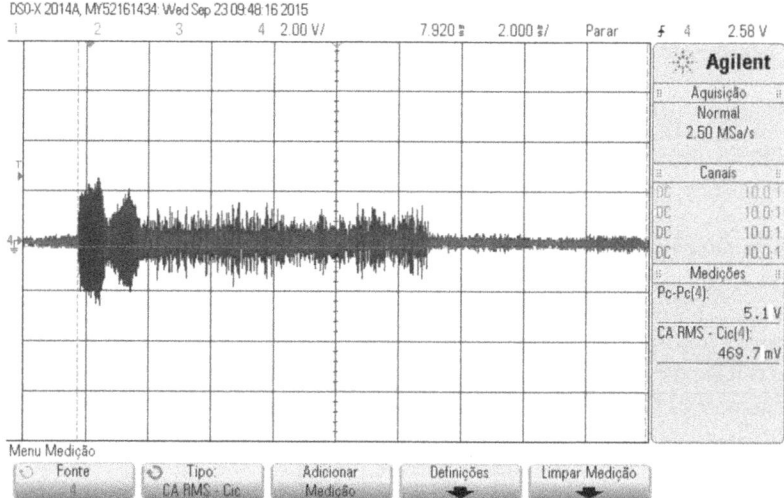

*Figure 14.8   PLC signal on the primary of the coupling transformer (hardware and
firmware in the manufacturer's standard)*

In the first scenario, regardless of day time, the communication was good. So,
in order to identify the issue in the second scenario, the signal in the primary of
the transformer was measured, and its results are illustrated in Figure 14.8. In those
practical results, it has been clear that the transmitted signals were much smaller than
the maximum level defined by CENELEC (peak voltage 134 dB μV–5 V) [23]. As
mentioned before, PRIME-ATMEL/ATPL210 has automatic coupling (TX circuit),
high impedance and low impedance channel; however, for the installed grid, it was
observed that always the high-impedance channel was selected.

In Figure 14.9, a result forcing the low impedance is depicted. Forcing the coupling circuit to operate in low impedance has amplified the signal level, but it has still
operated constrained to the maximum levels defined by CENELEC [23]. From the

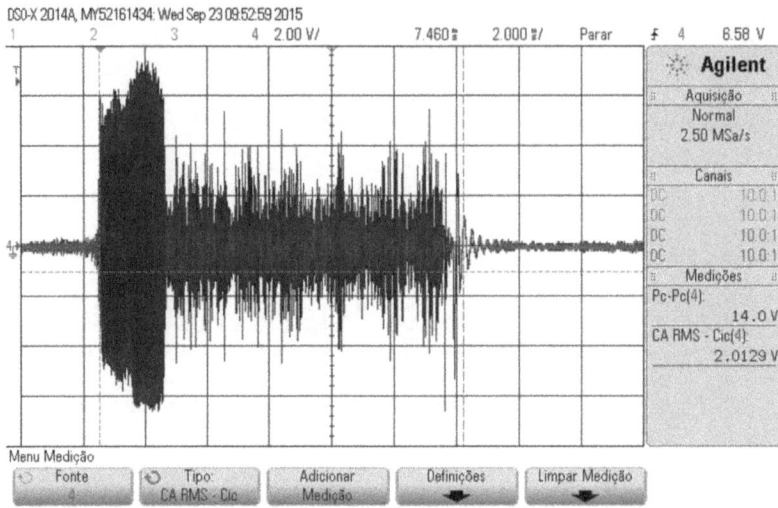

Figure 14.9   PLC signal on the primary of the coupling transformer (standard hardware, transmission forced by the low impedance branch)

manufacturer's configuration with automatic coupling, the signal level generated at the primary of the coupling transformer reached a peak-to-peak voltage of 5.1 V (peak voltage 2.51 V) and an RMS voltage of 469.7 mV (see Figure 14.8). When forcing the circuit to operate at low impedance, an increase in the level of the data signal was observed, the peak-to-peak voltage achieved 14 V (peak voltage 7 V), and an RMS voltage of 2.0129 V (see Figure 14.9). When comparing the values obtained in the modified circuit to the manufacture's circuit, an increase of 2.7× in the peak-to-peak voltage and 5× in the RMS voltage has been noted.

To ensure the correct operation, a modification in the high impedance circuit was performed, where the components of the circuit were replaced by the same components of the low impedance channel. As both circuits are equal, the automatic coupling does affect the signal amplitude.

The same verification was performed in the automatic conditioning selection (RX circuit). The received signal in the input of the ADC was measured and analyzed. It was also noted that the received signal was lower than the ADC maximum input, and the conditioning gain was configured to attenuate 12×. Considering the maximum transmitted signal levels regulated by CENELEC (5 V) and considering the ADC maximum input (±0.7 V), a modification in the circuit has been performed to attenuate only 5.8× instead of 12×. Note that, if the level of the signal is 5 V, the ADC voltage would result in 0.86 V; however, hardly the modems operate at the maximum voltage levels, and even if the transmitter operates at 5 V, the received signal amplitude would not be lower than this maximum, as the signal is attenuated by the line impedance. With this modification, it was possible to improve the reception sensitivity of the RX circuit of the PLC modem.

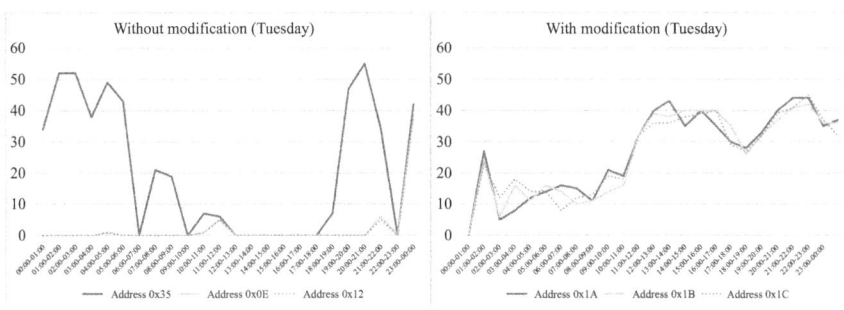

Figure 14.10   IEDs performance on Tuesday

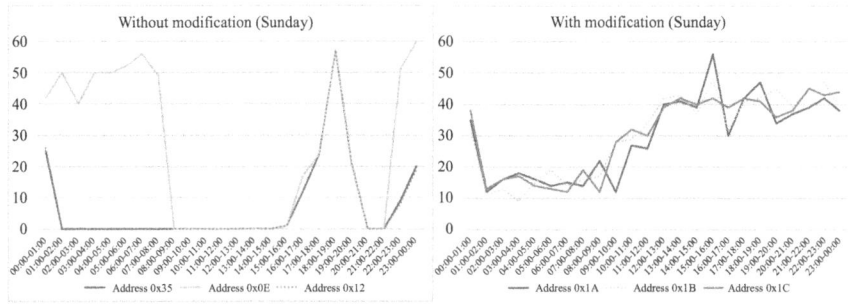

Figure 14.11   IEDs performance on Sunday

From the verification of the TX and RX signals, it was possible to identify some operational issues of the PRIME-ATMEL/ATPL210 that could be modified in order to optimize its performance:

• Wrong automatic selection of the coupling circuit (TX circuit), resulting in transmitting lower voltage signal data levels than necessary to establish a good communication.
• High attenuation of the received signal due to the automatic conditioning (RX circuit), thus identifying the data properly is a hard task.

To verify the improvement in the PLC communication under the modified PRIME-ATMEL/ATPL210 modems, new tests were performed, in the same time interval, with the modified and unmodified circuits. Figures 14.10 and 14.11 illustrate the received packets in one day, before and after the modifications. After the change, the IED had more effective communication. This is clear by the increase in the number of the daily received packages and the absence of intervals without receiving any packages.

Analyzing the data in Figures 14.10 and 14.11, it was verified that there was an improvement, for the modified circuit, in all evaluated metrics (average, minimum and maximum packets received per hour). The total number of received packets rose from 2,734 to 6,427, which represents an average increase of more than 200%. Likewise, there was a reduction in the average, minimum and maximum intervals without receiving any packages. The results were obtained in several days; to prove that there is no difference between week and weekend days, one result on a Tuesday and the other on a Sunday were illustrated.

In addition to having a better performance in all the metrics evaluated, the set of modified IEDs achieved a more stable transmission throughout the day, with no periods without any data packets. It is important to emphasize that the gains obtained in the amplitude of the signal levels have took into account the limits imposed by [23].

14.5   Conclusions and perspectives

The EPS has presented few changes since the "War of the Currents", even under an exponential increase of the load demand. However, in the last years, many efforts have been made to improve the electrical efficiency, to intensify the usage of renewable sources and to reduce CO_2 emissions. So, smart grid (SG) concept was formulated, in order to modernize the EPS in a controlled and safe way.

To achieve all those efforts, different issues have been pointed out. The increase of renewable sources results in intermittency, thus a load demand control is necessary. The control should manage not only the power flow but also the energy prices, dealing with the stakeholders. Also, the renewable source is usually installed in the residences, so the consumers become also a generator, and the usual unidirectional power flow becomes bidirectional. To deal with this scenario, robust and fast EPS controls are necessary.

The main objective of SG is to improve the EPS to achieve more efficiency, reliability, and safety, through an automatic control and modern communication. This is a new perception that integrates electricity, instrumentation, and communication. Besides, when SG is implemented, a bidirectional power flow is possible, so a consumer unit can also be a generator unit, modifying the strict unidirectional power flow. Furthermore, with the renewable source, ancillary services can be implemented in the EPS.

In the coming years, the study of the SGs will open new perspectives. For this, we can structure SGs into three major systems [3–11,15,35,47,52,53]:

- Intelligent infrastructure system – comprises energy, information, and communication:
  - generation of electricity, distribution, and consumption;
  - measurement of information, monitoring, and management; and
  - advanced communication technologies.
- Intelligent management system – provides advanced management and control services.
- Intelligent protection system – provides advanced analysis of grid reliability, fault protection, security, and privacy protection services.

A high interconnection of those areas is the key to consolidate the SG as an effective way to deal with all those efforts. The communication area is a vital role of the future power systems in order to achieve and integrate SG capabilities. The selection of the best communication is nontrivial, as each and every technology has its own limitations. In this chapter, some communication alternatives were presented and the available literature was analyzed.

The authors believe that hybrid communication should be implemented to ensure redundancy and reliability. For wired communication, PLC is a good solution when an average number of data packets and an average priority are required, because an existed infrastructure can be used (electrical wiring). However, as observed in the PLC results, a good tuning study should be performed to ensure a better data transfer. In cases of high priority and large amount of data are required, OF is the most reliable choice, but the costs are high. For wireless communication, the medium is the same regardless the protocol and the frequencies available, so to a better data communication, the environment should be analyzed considering barriers, and a cognitive solution can be used to select the best channel in the scenario.

References

[1]   Gungor VC, Lu B, Hancke GP. Opportunities and Challenges of Wireless Sensor Networks in Smart Grid. IEEE Transactions on Industrial Electronics. 2010;57(10):3557–3564.

[2]   Gungor VC, Sahin D, Kocak T, *et al.* Smart Grid Technologies: Communication Technologies and Standards. IEEE Transactions on Industrial Informatics. 2011;7(4):529–539.

[3]   Salvadori F, Gehrke CS, de Oliveira AC, *et al.* Smart Grid Infrastructure Using a Hybrid Network Architecture. IEEE Transactions on Smart Grid. 2013;4(3):1630–1639.

[4]   Palensky P, Dietrich D. Demand Side Management: Demand Response, Intelligent Energy Systems, and Smart Loads. IEEE Transactions on Industrial Informatics. 2011;7(3):381–388.

[5]   IEA, editor. Technology Road map How2Guide for Smart Grids in Distribution Networks. 1st ed. 9 rue de la Fédération 75739 Paris Cedex 15, France: IEA – International Energy Agency; 2015.

[6]   Salvadori F, Gehrke CS, Hartmann LV, *et al.* Efficient low cost and easy testing methodology for analysis of wireless communication applied for IEDs. In: 2016 IEEE International Instrumentation and Measurement Technology Conference Proceedings; 2016. p. 1–6.

[7]   Erol-Kantarci M, Mouftah HT. Wireless Sensor Networks for Cost-Efficient Residential Energy Management in the Smart Grid. IEEE Transactions on Smart Grid. 2011;2(2):314–325.

[8]   Amin SM. Smart Grid: Overview, Issues and Opportunities. Advances and Challenges in Sensing, Modeling, Simulation, Optimization and Control. European Journal of Control. 2011;17(5–6):547–567.

[9]   Salvadori F, De Campos M, Sausen PS, *et al.* Monitoring in Industrial Systems Using Wireless Sensor Network With Dynamic Power Management. IEEE Transactions on Instrumentation and Measurement. 2009;58(9):3104–3111.

[10]   Yang Q, Barria JA, Green TC. Communication Infrastructures for Distributed Control of Power Distribution Networks. IEEE Transactions on Industrial Informatics. 2011;7(2):316–327.

[11]   Malinowski A, Yu H. Comparison of Embedded System Design for Industrial Applications. IEEE Transactions on Industrial Informatics. 2011;7(2): 244–254.

[12]   Garau M, Celli G, Ghiani E, *et al.* Evaluation of Smart Grid Communication Technologies with a Co-Simulation Platform. IEEE Wireless Communications. 2017;24(2):42–49.

[13]   Shuaib K, Khalil I, Abdel-Hafez M. Communications in Smart Grid: A Review with Performance, Reliability and Security Consideration. Journal of Networks. 2013;8(6):1229–1240.

[14]   Mercurio A, Di Giorgio A, Cioci P. Open-Source Implementation of Monitoring and Controlling Services for EMS/SCADA Systems by Means of Web Services – IEC 61850 and IEC 61970 Standards. IEEE Transactions on Power Delivery. 2009;24(3):1148–1153.

[15]   Giustina DD, Rinaldi S. Hybrid Communication Network for the Smart Grid: Validation of a Field Test Experience. IEEE Transactions on Power Delivery. 2015;30(6):2492–2500.

[16]   Hrasnica H, Haidine A, Lehnert R. Broadband Powerline Communications Networks. Chichester: John Wiley and Sons; 2004.

[17]   Homeplug. Available from: http://www.homeplug.org [Accessed 10 Dec 2017].

[18]   Berger LT, Schwager A, Escudero-Garzas JJ. Power Line Communications for Smart Grid Applications. Journal of Electrical and Computer Engineering. 2013;2013:1–16.

[19]   Fang X, Misra S, Xue G, *et al.* Smart Grid – The New and Improved Power Grid: A Survey. IEEE Communications Surveys and Tutorials. 2012;14(4):944–980.

[20]   Hawker A. Security and Control in Information Systems: A Guide for Business and Accounting. Information Systems Textbooks. London: Routledge; 2000.

[21]   Shrestha D, Mestre X, Payaró M. Maximum-likelihood channel estimation in presence of impulsive noise for PLC systems. In: 2016 IEEE Global Conference on Signal and Information Processing (GlobalSIP); 2016. p. 20–24.

[22]   Sayed M, Tsiftsis TA, Al-Dhahir N. On the Diversity of Hybrid Narrowband-PLC/Wireless Communications for Smart Grids. IEEE Transactions on Wireless Communications. 2017;16(7):4344–4360.

[23]   CENELEC-EN50065-1. Signalling on low-voltage electrical installations in the frequency range 3 kHz to 148.5 kHz. Rue de la science, 23 B-1040 Brussels, Belgium: European Committee for Electrotechnical Standardization; 2001.

[24]   Hecht J. Great Leaps of Light. IEEE Spectrum. 2016;53(2):28–53.

[25] Agrawal GP. Optical Communication: Its History and Recent Progress. In: Al-Amri M., El-Gomati M., Zubairy M. (eds) Optics in Our Time. Cham: Springer; 2016.

[26] Levesque M, Maier M. Probabilistic Availability Quantification of PON and WiMAX Based FiWi Access Networks for Future Smart Grid Applications. IEEE Transactions on Communications. 2014;62(6):1958–1969.

[27] Khan F, ur Rehman A, Arif M, *et al.* A Survey of Communication Technologies for Smart Grid Connectivity. In: Computing, Electronic and Electrical Engineering (ICE Cube), 2016 International Conference on. IEEE; 2016. pp. 256–261.

[28] Gungor VC, Lambert FC. A Survey on Communication Networks for Electric System Automation. Computer Networks. 2006;50(7):877–897.

[29] Alencar MS, Carvalho FBS, Lopes WTA. Spectrum Sensing Techniques and Applications. New York, NY: Momentum Press; 2017.

[30] Gungor VC, Hancke GP. Industrial Wireless Sensor Networks: Challenges, Design Principles, and Technical Approaches. IEEE Transactions on Industrial Electronics. 2009;56(10):4258–4265.

[31] Wi-Fi Alliance. Available from: http://www.wi-fi.org [Accessed Dec 2017].

[32] Bluetooth Technology. Available from: http://www.bluetooth.com [Accessed Dec 2017].

[33] Carvalho FBS, Leal BG, Santos Filho JV, *et al.* Environmental Applications of Wireless Sensor Networks. Journal of Communication and Information Technology. 2012;2:14–19.

[34] Nayak A, Stojmenovic I. Wireless Sensor and Actuator Networks. New York, NY: John Wiley and Sons; 2010.

[35] Salvadori F, Gehrke CS, Hartmann LV, *et al.* Design of an Intelligent Electronic Device Based on TivaC Platform for Smart Grid Applications. In: 2016 IEEE International Instrumentation and Measurement Technology Conference Proceedings; 2016. p. 1–6.

[36] Ko J, Terzis A, Dawson-Haggerty S, *et al.* Connecting Low-Power and Lossy Networks to the Internet. IEEE Communications Magazine. 2011;49(4): 96–101.

[37] Yan Y, Qian Y, Sharif H, *et al.* A Survey on Smart Grid Communication Infrastructures: Motivations, Requirements and Challenges. IEEE Communications Surveys & Tutorials. 2013:5–20.

[38] Alam M, Kim J, Li YC, *et al.* Implementation of Wireless Industrial Networks for Industrial Smart Grids. In: 2014 International Conference on Advances in Energy Conversion Technologies (ICAECT). 2014; p. 83–87.

[39] Garcia-Hernandez J. Recent Progress in the Implementation of AMI Projects: Standards and Communications Technologies. In: 2015 International Conference on Mechatronics, Electronics and Automotive Engineering (ICMEAE); 2015. p. 251–256.

[40] IEEE. IEEE Draft Standard for Low-Rate Wireless Personal Area Networks (WPANs). IEEE P802154-REVc/D00, April 2015 (Revision of IEEE Std 802154-2011). 2015 Apr; p. 1–684.

[41]   Wang J, Ghosh M, Challapali K. Emerging Cognitive Radio Applications: A Survey. IEEE Communications Magazine. 2011;49(3):74–81.

[42]   Lu Y, Liu W, Li J, *et al.* Measurement and Cognitive Detection Method of Broadcast Radio Stations in Distribution Networks. In: 18th IEEE International Symposium on Power Line Communications and its Applications (ISPLC). 2014; p. 75–80.

[43]   Carvalho NB, Cidronali A, Gomez-Garcia R. White Space Communication Technologies. 1st Edition, Cambridge: Cambridge University Press; 2015.

[44]   Sum C, Villardi GP, Rahman MA, *et al.* Cognitive Communication in TV White Spaces: An Overview of Regulations, Standards, and Technology. IEEE Communications Magazine. 2013;51(7):138–145.

[45]   Qiu RC, Hu Z, Chen Z, *et al.* Cognitive Radio Network for the Smart Grid: Experimental System Architecture, Control Algorithms, Security, and Microgrid Testbed. IEEE Transactions on Smart Grid. 2011;2(4):724–740.

[46]   Bu S, Yu FR. Green Cognitive Mobile Networks With Small Cells for Multimedia Communications in the Smart Grid Environment. IEEE Transactions on Vehicular Technology. 2014;63(5):2115–2126.

[47]   2030-2011 IS. IEEE Guide for Smart Grid Interoperability of Energy Technology and Information Technology Operation with the Electric Power System (EPS), End-Use Applications, and Loads. 2011 Sep; p. 1–126.

[48]   Sharma G, Mazumdar RR. A Case for Hybrid Sensor Networks. IEEE/ACM Transactions on Networking. 2008;16(5):1121–1132.

[49]   Salvadori F, Gehrke CS, de Carvalho FBS. Micro Distributed Generation for Smart Grid: Study for Tools Development to the Power Flow Management (in Portuguese). João Pessoa – Brazil: Federal University of Paraíba – Electrical Engineering Department – Smart Grid Group; 2017. MCTI/CNPq/CT-ENERG #33/2013. National Council for Scientific and Technological Development (CNPq).

[50]   Gehrke CS, Lima AMN, Oliveira AC. Smart Control For Active Power Generation, Voltage Level And Harmonic Content Based On Photovoltaic Generators. In: Energy Conversion Congress and Exposition (ECCE), 2015 IEEE. vol. 1; 2015. p. 1–6.

[51]   Meng J. A Distributed Power Generation Communication System. In: Electrical and Computer Engineering, 2003. IEEE CCECE 2003. Canadian Conference on. vol. 1; 2003. p. 483–486.

[52]   Boccardi F, Heath RW, Lozano A, *et al.* Five Disruptive Technology Directions for 5G. IEEE Communications Magazine. 2014;52(2):74–80.

[53]   Collier SE. The Emerging Enernet: Convergence of the Smart Grid with the Internet of Things. IEEE Industry Applications Magazine. 2017;23(2): 12–16.

Chapter 15

Direct torque control for DFIG based wind turbines employing power line communication technology in smart grid environments

Samuel C. Pereira[1], Andre L. L. F. Murari[2],
Carlos E. Capovilla[2], Jose A. T. Altuna[2],
Rogerio V. Jacomini[3], Edmarcio A. Belati[2],
Alfeu J. Sguarezi Filho[2], and Ivan R. S. Casella[2]

This chapter proposes a control technique for a wind doubly fed induction generator (DFIG), based on direct torque control (DTC) with power references sent remotely via power line communication (PLC) technology. DTC achieves high dynamic performance, allowing independent control of DFIG electromagnetic torque and rotor flux magnitude. In this way, active and reactive power can be controlled by the voltage applied to the rotor independently. In order to operate in a smart grid (SG) environment, the proposed system employs PLC technology for transmitting the power references from the control center (CC) to the wind generator through power cables.

The complete control system (controller and PLC), implemented in an experimental test bench, is presented in this chapter with results that validated the control strategy and the proposed system as a whole.

15.1 Introduction

The sector of electricity and heat is, by far, the largest source of global CO_2, accounting for 42% of the total emission. In 2015, 81% of world's total primary energy supply came from fossil fuel combustion [1]. Due to the concern about environmental impacts, mainly the greenhouse effect, the use of renewable sources for generating electricity has increased in the last years. Currently, renewable generation is the fastest growing source of global electric power, being wind and hydropower the

[1]Department of Automation and Process Control, Federal Institute of São Paulo (IFSP Suzano), Brazil
[2]Center for Engineering, Modeling and Applied Social Sciences (CECS), Federal University of ABC (UFABC), Brazil
[3]Department of Electrical and Electronics, Federal Institute of São Paulo (IFSP Hortolândia), Brazil

largest contributors, together accounting for two-third of expected total increment from 2012 to 2040 [2].

As wind turbine generation systems (WTGS) are intended to be connected to the grid, and grid operators have to provide electricity within international standards, WTGS have to follow grid codes. These grid codes are related to voltage, reactive power, frequency control and fault ride-through capability [3]. Regarding this situation, a proper control of induction generators employed in WTGS is mandatory to ensure the power quality and system security over the dynamic behavior of grid conditions.

DFIG has been used to produce electricity in wind turbines for over two decades [4] and, since 2002, it represents the highest market share in WTGS [5]. This success is due to its characteristic of maintaining the amplitude and frequency of its output voltage at a constant value, even with rotor speed variations of $\pm30\%$ of its synchronous speed, and its ability to allow independent active and reactive power control [6].

DFIG power control is usually performed employing flux or stator voltage oriented vector control strategy [7]. Some research about proportional integral (PI) controllers and stator flux orientation with satisfactory dynamic performance were presented in [8,9].

DTC was presented in [10,11] as an alternative to field oriented control technique for induction machines. This strategy works with variable switching frequency [12, 13], providing a high-control performance. A strategy for reduction of DFIG torque oscillation, generated by the variable switching frequency, was proposed in [14]. In [15], PI controllers were employed to estimate a correction signal added to the angle of the stator flux and the magnitude of the rotor voltage.

In addition, an integral variable structure control applied to DTC and sliding mode control (SMC) of DFIG was presented with satisfactory results in [16,17]. The results have demonstrated that it is possible to reduce the torque and power oscillation of induction generators under unbalanced grid voltage conditions.

As wind is an intermittent energy source and grid conditions vary with power consumption characteristics of the consumers connected to the grid, the set-point parameters of the DFIG controllers, in response to these changes, need to be frequently updated to maintain the energy delivered by the generators under proper conditions. Then, the communication system between the CC and the controller at the wind generator has an important role in the performance of modern WTGS. The integration of CC (operator), wind generator and communication system, as shown in Figure 15.1, results in an SG concept that allows automatic and remote control of wind farms, according to the needs and dynamic characteristics of the electrical grid. As a satisfactory solution, wireless technologies were employed to transmit power references from the CC to wind turbines in [18–22].

In the other side, the use of PLC, a technology that allows transmitting data via power cables, reducing considerably time and costs for infrastructure installation, can be an interesting alternative, inside wind farms, for transmitting the reference information to wind turbines in order to implement an SG environment. In [23], a long range (10 km) PLC system with quaternary phase-shifting keying (QPSK)

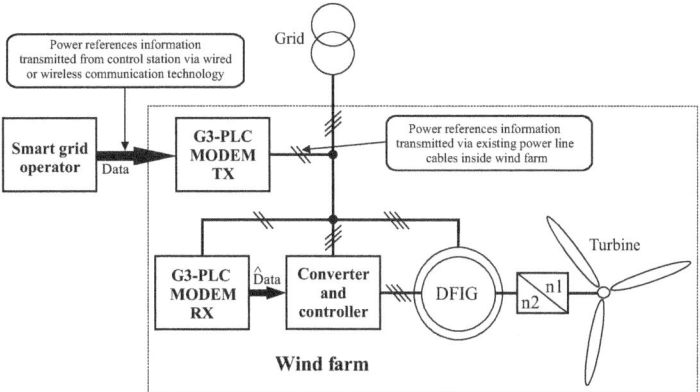

Figure 15.1   Proposed wind DFIG control system using PLC technology

modulation, in renewable energy systems, was proposed and simulated. In [24,25], practical communication tests were performed between an inverter and an electric motor using broadband PLC (BB-PLC). The results demonstrated that PLC has real potential for these kinds of applications.

Due to the high operating frequency and the higher attenuation through cable inductance and capacitance, that function like a second-order filter, BB-PLC signal has a short range (below 200 m) [26] which can be a limiting factor for applications in larger wind farms. In [27], communication tests via power line (field trial test) were performed with G3-PLC MODEMs, a narrowband PLC (NB-PLC) standard. Over medium voltage (MV) lines, a 6.4 km communication link was established, whereas in low voltage (LV) lines, that presents several devices connected (higher noise rate) and a lot of branches (with consequent impedance mismatches), a 200-m link was established (in [28], the longest transmission distance over LV line was 1.7 km). These tests also proved that G3-PLC signal is able to cross voltage transformers so that a communication link between a MODEM connected to the MV side and other to the LV was established, without the need of a coupling circuit at the MV/LV voltage transformer. In [26], G3-PLC was tested in a simulated industrial environment with satisfactory results (electric motors and inverter did not affect the communication between the MODEMs).

These results demonstrated that G3-PLC could be a solution for the last mile communication inside wind farms, i.e., the operator of the CC sends the control information to the wind farm, via any wired or wireless communication technology, and G3-PLC devices create a neighborhood area network inside the wind farm in order to deliver the control information from the communication link coming from the CC (operator) to the wind turbine controllers, as shown in Figure 15.1.

In this context, this chapter presents the analysis of a DTC, based on SMC technique, for wind DFIG, whose power references are transmitted through a G3-PLC communication system. The proposed control system offers interesting features that

make it suitable for applications in SG. Besides this introduction, this chapter is organized as follows: Section 15.2 shows the DFIG mathematical model and the DTC
principles, whereas Section 15.3 shows details of the proposed DFIG control system.
In Section 15.4, a brief introduction of PLC technology and characteristics of G3-PLC
is presented, the PLC standard employed in the proposed system. Lastly, the experimental results and validation of the proposed system are analyzed in Section 15.5,
and final conclusions are presented in Section 15.6.

15.2   DFIG mathematical model and DTC principles

The direct axis aligned with the stator flux vector allows $\psi_{sd} = |\vec{\psi}_{sdq}|$, $v_{sq} = |\vec{v}_{sdq}|$
and $\psi_{sq} = v_{sd} = 0$, so that, the mathematical model in the synchronous coordinate
system (dq) of DFIG is given by [29]

$$\vec{v}_{sdq} = R_s \vec{i}_{sdq} + \frac{d\psi_{sd}}{dt} + j\omega_s \psi_{sd} \tag{15.1}$$

$$\vec{v}_{rdq} = R_r \vec{i}_{rdq} + \frac{d\vec{\psi}_{rdq}}{dt} + j(\omega_s - n_{PP}\omega_{mec})\vec{\psi}_{rdq} \tag{15.2}$$

$$\vec{\psi}_{sdq} = \psi_{sd} = L_s \vec{i}_{sdq} + L_m \vec{i}_{rdq} \tag{15.3}$$

$$\vec{\psi}_{rdq} = L_m \vec{i}_{sdq} + L_r \vec{i}_{rdq} \tag{15.4}$$

where $\vec{v}, \vec{i}$ and $\vec{\psi}$ are, respectively, voltage, current and flux space vectors (s, r and
dq indicates, respectively, stator, rotor and rotating $d - q$ reference frame), n_{PP} is the
number of pole pairs, ω_s is the synchronous speed of stator, ω_{mec} is the mechanical
speed of rotor, R_s and R_r are the stator and rotor resistances, L_s and L_r are the stator
and rotor inductances and L_m is the mutual inductance between rotor and stator.

Combining (15.3) and (15.4), the stator and rotor currents can be calculated as a
function of the magnetic fluxes in accordance to

$$\vec{i}_{sdq} = \frac{\psi_{sd}}{\sigma L_s} - \frac{L_m}{\sigma L_s L_r} \vec{\psi}_{rdq} \tag{15.5}$$

$$\vec{i}_{rdq} = \frac{\vec{\psi}_{rdq}}{\sigma L_r} - \frac{L_m}{\sigma L_s L_r} \psi_{sd} \tag{15.6}$$

where the total dispersion coefficient is $\sigma = 1 - (L_m^2/L_s L_r)$.

In this way, the electromagnetic torque (T_e), the active (P_s) and reactive (Q_s)
power can be represented by

$$T_e = -\frac{3}{2} \frac{n_{PP}L_m}{\sigma L_s L_r} \Im(\vec{\psi}_{sdq}^* \cdot \vec{\psi}_{rdq}) \tag{15.7}$$

$$P_s = \frac{3}{2}\Re(\vec{v}_{sdq} \cdot \vec{i}_{sdq}^*) \tag{15.8}$$

$$Q_s = \frac{3}{2}\Im(\vec{v}_{sdq} \cdot \vec{i}_{sdq}^*) \tag{15.9}$$

where the symbol * indicates the complex conjugate of the vector and $\Re$ and $\Im$, respectively, real and imaginary operators.

Considering a balanced power grid, the voltage drop on stator resistance can be ignored in (15.1). Then, stator voltage is given by

$$\vec{v}_{sdq} = j\omega_s \psi_{sd} \tag{15.10}$$

Besides, it is possible to notice that the stator flux ψ_{sdq} is a constant determined by the magnitude of the supply voltage and the synchronous frequency ω_s. For DTC, it is possible to employ (15.7) to obtain P_s and Q_s control as a function of rotor flux. Then, T_e and P_s can be described as in [30]:

$$T_e = -n_{PP} k_\sigma \psi_{sd} \psi_{rq} \tag{15.11}$$

$$P_s = \omega_{mec} T_e \tag{15.12}$$

where $k_\sigma = 1.5 L_m / \sigma L_s L_r$.

Finally, Q_s, as a function of the fluxes, can be obtained by

$$Q_s = \frac{3}{2} \frac{L_m}{\sigma L_s L_r} v_{sq} \left(\frac{L_r}{L_m} \psi_{sd} - \psi_{rd} \right) \tag{15.13}$$

As a consequence, according to (15.11), (15.12) and (15.13), by means of T_e and rotor fluxes control, it is possible to control P_s and Q_s.

15.3 SMC technique

The SMC is a variable structure control that can be used as an alternative to the classical control theory for implementing a discontinuous control system [31]. The controllers of this work are based on DTC for three-phase variable reluctance motors presented in [32–34]. The switching surface of the SMC is defined by the error between the reference (set-point) and the current value of the controlled variable. The errors of T_e and ψ_{rd} are estimated according to

$$e_{Te} = T_{e_{ref}} - T_e \tag{15.14}$$

and

$$e_{\psi_{rd}} = \psi_{rd_{ref}} - \psi_{rd} \tag{15.15}$$

where $T_{e_{ref}}$ is the electromagnetic torque reference and $\psi_{rd_{ref}}$ is the rotor flux reference.

According to [31], the set S of switching surfaces is defined as

$$S = \begin{bmatrix} s_1 \\ s_2 \end{bmatrix} = \begin{bmatrix} e_{te} + k_{dd} \dfrac{de_{te}}{dt} \\ e_\psi + k_{dq} \dfrac{de_\psi}{dt} \end{bmatrix} \tag{15.16}$$

where k_{dd} and k_{dq} are constants defined according to the desired system response.

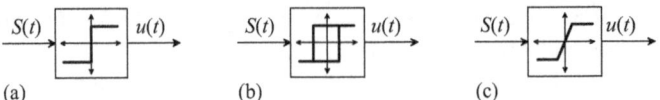

Figure 15.2   Eval functions used in SMC (a) signal (b) hysteresis (c) linear with saturation

The proposed DTC calculates the voltages to be applied to the rotor of DFIG with its stator directly connected to the electrical grid. Through the error, calculated between $T_{e_{ref}}$ and T_e, it is obtained the quadrature axis component of the rotor voltage vector. For calculating the rotor flux reference, it is used the SMC technique. The direct axis component of the rotor voltage vector is achieved by the error calculation between the reference $\psi_{rd_{ref}}$ and the estimated value ψ_{rd} of the direct component of the rotor flux vector. The errors processing is performed by a nonlinear controller based on SMC technique. Its behavior is described as follows:

$$v_{rd} = \left(k_{pd} + \frac{k_{id}}{s} \right) eval(s_1) \tag{15.17}$$

$$v_{rq} = \left(k_{pq} + \frac{k_{iq}}{s} \right) eval(s_2) \tag{15.18}$$

where k_{pd}, k_{pq}, k_{id} and k_{iq} are PI controller gains.

The error between the Q_s and its reference will be used for designing the sliding surface, as described in (15.16), being necessary only to replace the T_e error by Q_s error. Then, the PI controller processes the output of *eval* function to generate the reference flux. The *eval* function can be a signal function, hysteresis or linear with saturation, as shown in Figure 15.2 [35,36].

The signal function is simple and discontinuous, being easily implemented by means of a relay, that instantly calculates the S trajectory. However, SMC and the signal function, when working together, make the system to have a variable switching frequency and, consequently, increase the switching losses.

Meanwhile, the hysteresis function, by means of its hysteresis band, decreases the switching frequency, keeping, with controlled oscillation, the S trajectory operating in the neighborhood $\pm\Delta$ of the switching surface. The vibration will be as function of Δ.

Finally, the linear function with saturation has as main advantage, the reduction of the chattering phenomenon and, consequently, the steady-state error, when the trajectory state approaches the switching surface. However, properties like parametric invariance are reduced due to the mitigation of the curve near the switching surface [36].

Thanks to its advantages, linear function with saturation, illustrated in Figure 15.2(c), will represent the *eval* function in this work, being responsible for determining the system reaction as a function of its position in the state space.

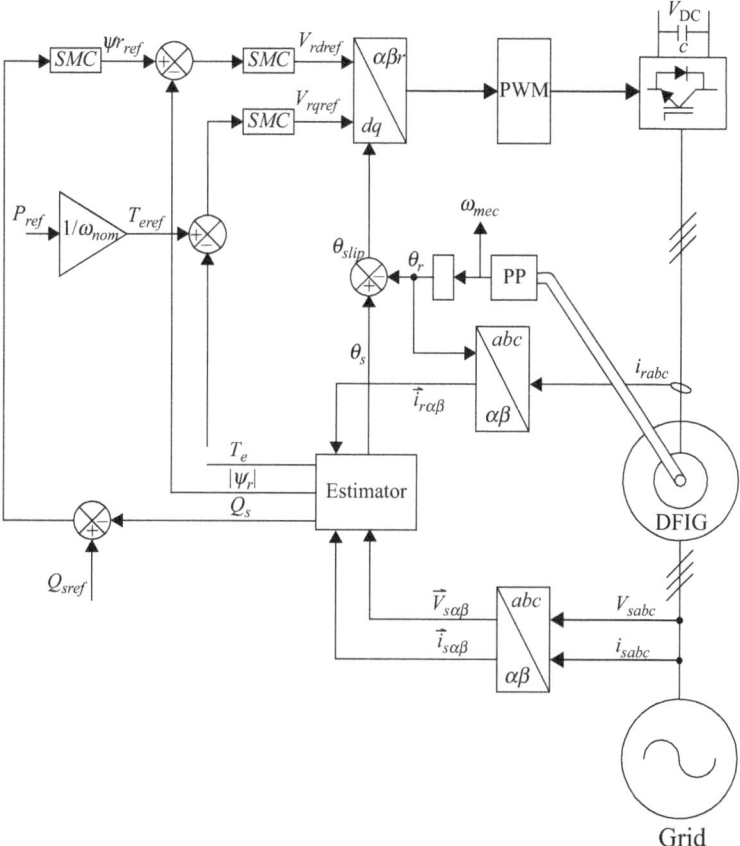

Figure 15.3   Block diagram of the Proposed DTC for DFIG based on SMC technique

This used *eval* function can be obtained by

$$eval(S) = \begin{cases} S \ ke & \text{if } \lim \min < S \ ke < \lim \max, \\ \lim \max & \text{if } S \ ke > \lim \max, \\ \lim \min & \text{if } S \ ke < \lim \min. \end{cases} \quad (15.19)$$

where k_e is the gain of *eval* function.

The proposed DTC for a DFIG based on SMC is shown in Figure 15.3. Analyzing carefully Figure 15.4, it is possible to note that the SMC blocks presented in Figure 15.3 correspond to the equations (15.14) to (15.19) and have the same basic structure for both torque and flux control, changing only the signals applied to the input and the resulting output.

Finally, the voltages applied to the rotor from (15.17) and (15.18) are transformed in a rotor stationary reference frame with the slip angle $(\delta_s - \delta_r)$. An important detail is that the DTC and SMC, when employed in machines with converters, present variable

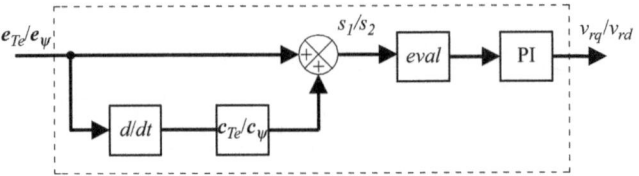

Figure 15.4   Detailed block diagram of the controller for torque (T_e) and flux (ψ) based on SMC technique

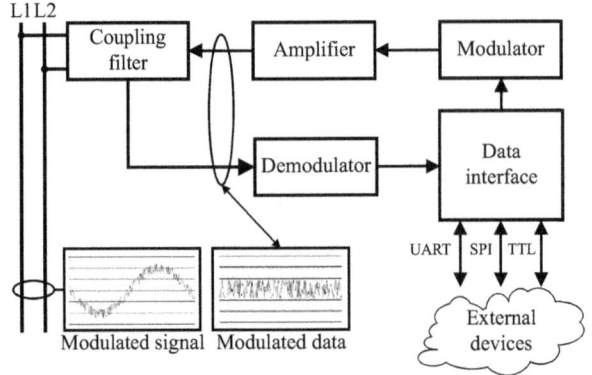

Figure 15.5   Simplified block diagram of a PLC MODEM

switching frequency operation. The variable switching frequency increases the losses inside the converter. To avoid this effect, the DTC was implemented by controllers based on SMC and PI techniques operating with fixed switching frequency by means of sinusoidal pulse width modulation (PWM) with third harmonic injection [36,37].

15.4   PLC principles

PLC is a technology that uses the power cables as a communication channel, taking advantage of an already existing infrastructure for transmitting data. The block diagram of a PLC MODEM is composed of a data interface, modulator/demodulator and a coupling interface for sending and receiving the PLC signal to the power cables, as shown in Figure 15.5. NB-PLC systems operate in the frequency range between 3 and 500 kHz, whereas BB-PLC systems between 1.8 and 250 MHz [38].

Power lines were not designed for data transmission, and several devices connected to it are sources of different types of noise so that it is not a proper communication channel at all. According to [39], it is expected, over a typical PLC channel below 500 kHz (NB-PLC), a signal-to-noise ratio (SNR) from −5 to 10 dB. Negative SNR means that the noise power is higher than the PLC signal power. Moreover, LV power lines have a lot of branches with several types of loads connected and also open ends, which results in impedance mismatches and consequent PLC signal

Table 15.1 Specifications of IEEE 1901–2010 and IEEE 1901.2 standards

Parameter	IEEE 1901–2010	IEEE 1901.2 (G3-PLC)
Frequency range	1.8–50 MHz	3–500 kHz
Data rate (maximum)	500 Mbps	500 kbps
Security	AES-128 encryption	AES-128 encryption
Range (maximum)	200 m	6 km (MV line) and 1.7 km (LV line)
Modulation	OFDM	OFDM
ROBO mode	Yes	Yes
Subcarrier modulation	BPSK, QPSK and QAM (8/16/64/256/1,024)	DBPSK, DQPSK and D8PSK

Source: References [27,28,41,42].

reflection in several points [40]. At the receiver, these reflections cause fading and distortion of PLC signal by means of a phenomenon known as multipath propagation. A typical 200 m link over LV networks can present attenuation of about 30 dB [27].

For overcoming these issues, several standards were developed for enabling PLC as a reliable communication technology. According to [26], before 2010, there were approximately 16 standards for BB-PLC and 18 for NB-PLC, being the standardization barrier for PLC technology. However, in 2010 and 2013, IEEE published, respectively, worldwide standards for BB-PLC (IEEE 1901–2010) and NB-PLC (IEEE 1901.2). It is important to mention that IEEE 1901.2 physical layer is based on the G3-PLC standard [26]. Table 15.1 presents the physical layer specifications of both IEEE standards.

In Table 15.1, it is possible to notice that IEEE 1901.2 (G3-PLC) has enough data rate for transmitting DFIG control parameters with a longer range in comparison to IEEE 1901.1, which can be helpful for applications in large wind farms. That is the main reason G3-PLC was chosen for the experimental tests.

Maxim Inc. initially developed G3-PLC in conjunction with Electricite Reseau Distribution France and aroused great interest for some applications in SG such as smart metering. It operates in different frequency bands reserved for NB-PLC applications such as the Committee for Electrotechnical Standardization (CENELEC) band (A: 3–95 kHz, B: 95–125 kHz, C: 125–140 kHz and D: 140–148.5 kHz), the Association of Radio Industries and Businesses (ARIB) band (10–450 kHz) and the Federal Communications Commission (FCC) band (10–490 kHz). In the FCC band, G3-PLC usually operates between 145 and 478 kHz, presenting the highest data rate [26].

G3-PLC employs Orthogonal Frequency Division Multiplexing (OFDM) associated with three different differential phase modulation schemes: differential binary phase shift keying (DBPSK), differential quadrature phase shift keying (DQPSK) and differential 8-phase shift keying (D8PSK). In all modulations, data are protected by an outer convolutional (CONV) code with code rate 1/2 and an appropriate inner shortened Reed Solomon (RS) code. After that, the coded information is suitably interleaved before being modulated. G3-PLC offers a robust operation (ROBO) mode, available only for the DBPSK scheme, that can further improve system robustness

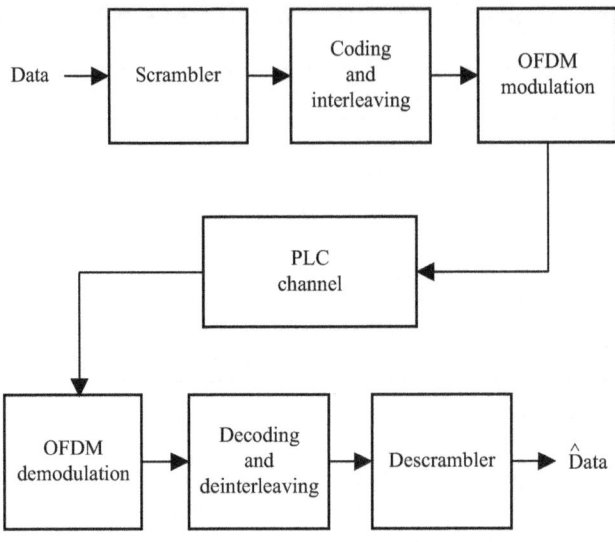

Figure 15.6   Block diagram of a G3-PLC MODEM

even in conditions with negative SNR by repeating the information inside a packet four times [27]. To illustrate, Figure 15.6 presents the block diagram of a G3-PLC system.

In [26], tests with G3-PLC MODEM operating in FCC band and ROBO mode have shown that G3-PLC standard, working with electric motors and inverters in the same environment, can offer an average of 34 kbps with a low packet error rate (0.4% maximum). IEEE 802.15.4, a popular low data rate wireless system, have an acceptable packet error rate from 1% to 5% [43]. Moreover, wireless signal has difficulty penetrating infrastructure such as basements, metal shielded cases, walls and metal and concrete structures, situation in which PLC signal has no limitations.

In order to implement an SG scenario, the power references were sent to the controllers, as data, via G3-PLC MODEM based on Max2992 (G3-PLC Transceiver) and Max2991 (Analog Front-End), both from Maxim Inc. As the MODEM uses the RS232 standard in its input data interface, it developed an analog to RS232 converter, based on the Atmel SAM3X8E ARM Cortex-M3 CPU, for sending and receiving the references as analog signals. The detailed block diagram of the MODEMs with their interfaces is presented in Figure 15.7. The G3-PLC MODEMs were configured to operate in the G3-PLC FCC band and ROBO mode, being separated from each other by a distance of 100 m, through a conventional 1.5 mm² cooper power cable.

15.5   Enabling SG concept with G3-PLC

The proposed controller, based on SMC technique, was implemented with a DFIG whose nominal parameters are presented in Table 15.2. The DFIG is coupled to a

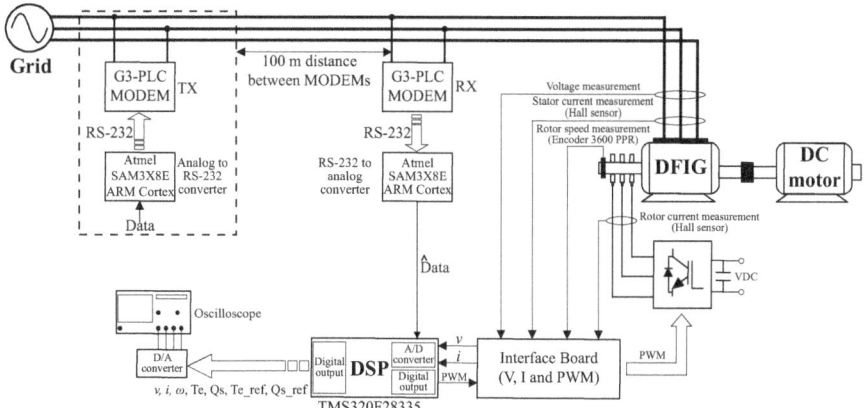

Figure 15.7 Experimental bench and G3-PLC MODEM connections

Table 15.2 DFIG nominal parameters

Parameter	Symbol	Value
Stator active power	P_N	3.5 kW
Phase voltage	V_N	220 V
Stator frequency	f	60 Hz
Number of poles pairs	n_{PP}	2
Stator resistance	R_s	1 Ω
Rotor resistance	R_r	3.13 Ω
Magnetizing inductance	L_m	191.7 mH
Stator inductance	L_s	201 mH
Rotor inductance	L_r	201 mH

direct current (DC) motor that activates the generator for testing the system (wind emulator). The controller uses a digital signal processor TMS320F28335 from Texas Instruments, electronic boards for signal conditioning and an encoder with 3,600 pulses per revolution.

The references were applied to the analog input of the G3-PLC TX MODEM (connected to a LV power line at 100 m distance from the controller) and sent to it via the analog output of the G3-PLC RX MODEM. This experimental test bench, including the G3-PLC MODEMs connections, is shown in Figure 15.7.

To validate the proposed system, two tests were performed with DFIG rotor speed at 1,675 rpm. The constant speed adopted due to the possible speed wind variations is supposed negligible during the time range of the tests. The generated references are $Q_{s_{ref}}$ and $P_{s_{ref}}$. $Q_{s_{ref}}$ is sent directly to the TX MODEM input, whereas $P_{s_{ref}}$ is first converted to Torque [$T_{e_{ref}}$ from (15.12)] to be sent to the TX MODEM input. Then, both information are sent to the controller through the PLC system. The conversion of $P_{s_{ref}}$ into $T_{e_{ref}}$ was performed to facilitate the experimental analysis of the torque response in the implemented DTC.

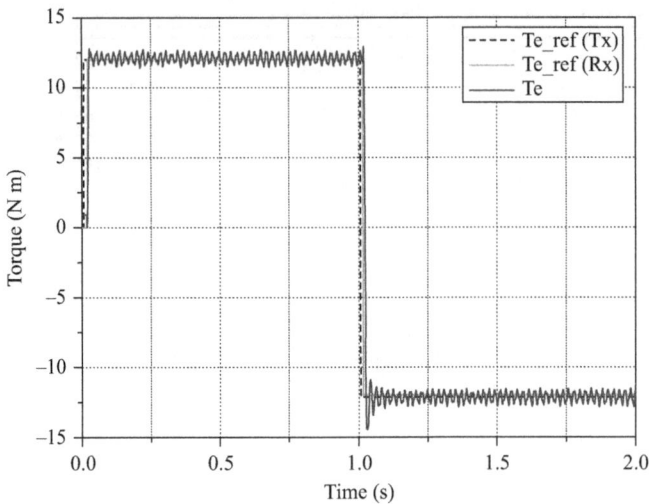

Figure 15.8　Transmitted and received $T_{e_{ref}}$ and the control system response at $Q_{s_{ref}} = 0$ kVAr

In the first test, it was applied to the controller, as a negative step function, a $P_{s_{ref}}$ from 2.1 to -2.1 kW ($T_{e_{ref}}$ from 12 to -12 N m) with $Q_{s_{ref}}$ constant at 0 kVAr. This torque range (positive and negative) was chosen to demonstrate that the torque control performance is satisfactory at four quadrants, ensuring the full operability of the system.

Figures 15.8 and 15.9 show the control references applied to the analog input of TX MODEM, the ones received from the analog output of RX MODEM and the control system response T_e and Q_s, respectively. According to these figures, it is possible to notice that the references received from PLC MODEM are a faithful reproduction of the original signal, with a negligible noise level and without any distortions or spikes. However, there is a delay between TX and RX references (approximately 15 ms), which is reflected in the controller response for $T_{e_{ref}}$ (Figure 15.8). This short delay does not affect the system dynamics and does not represent a concern to the grid operator, so that, the controller response is satisfactory, presenting a faithful reproduction of the received references.

In addition, Figure 15.9 presents a transient state in the Q_s response when $T_{e_{ref}}$ abruptly changes, even with $Q_{s_{ref}}$ constant (it occurs due to coupling between the model variables). Then, the independent control is only obtained at steady state (the transient response lasted about 30 ms). The behavior of rotor currents, stator voltage and current of one of the phases is presented in Figure 15.10. Analyzing the responses, it is noticed that the proposed control reaches the objectives with currents and voltage in accordance with the expected waveforms guaranteeing the operability of the proposal in this SG context.

In the second test, $P_{s_{ref}}$ is kept constant at 1.05 kW ($T_{e_{ref}} = -6$ N m) with $Q_{s_{ref}}$ varying from 1 to -1 kVAr (as in the first test, the received references are from

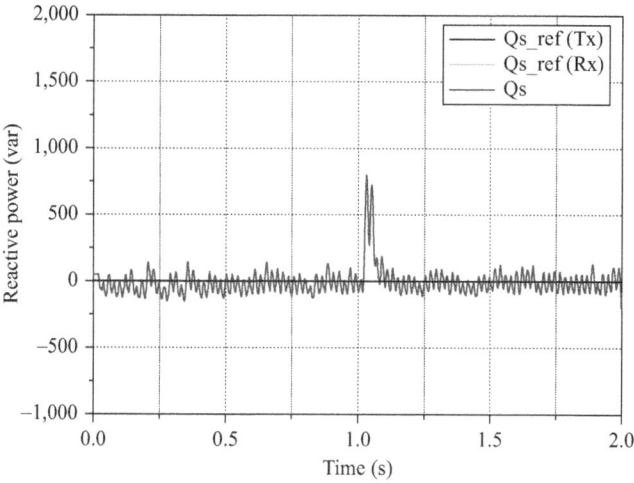

Figure 15.9 Transmitted and received $Q_{s_{ref}}$ and the control system response at $Q_{s_{ref}} = 0\ kVAr$

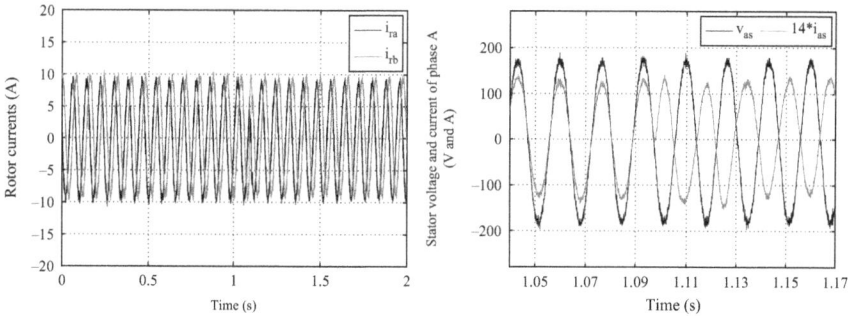

Figure 15.10 Rotor currents, stator voltage and current of phase A at $Q_{s_{ref}} = 0\ kVAr$

PLC MODEM). The control references, sent and received from PLC MODEMs, and the control system response are presented in Figures 15.11 and 15.12. As in the first test, the received references reproduce the transmitted ones with a small delay, as observed in Figure 15.11, which demonstrates that PLC system operates adequately and similarly for both conditions. The controller response is satisfactory with all references been followed. As expected, in this case, there is also a transient on the T_e controller response when $Q_{s_{ref}}$ changes, as observed in Figure 15.12. Again, the controller response reaches the steady state in a satisfactory way (few milliseconds).

The behavior of rotor currents and stator voltage and current of one of the phases is presented in Figure 15.13. As in the first test, all these responses are satisfactory, which ensures the system performance.

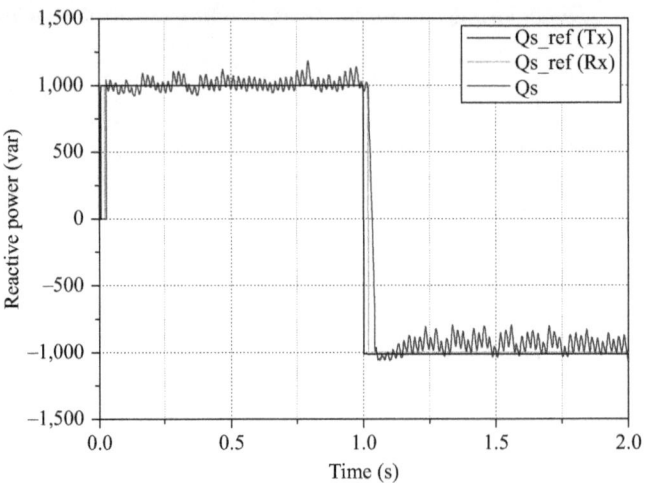

Figure 15.11 *Transmitted and received $Q_{s_{ref}}$ and the control system response at $T_{e_{ref}} = -6\,N\,m$*

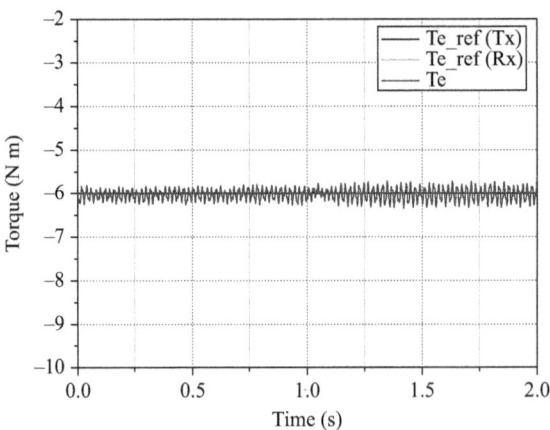

Figure 15.12 *Transmitted and $T_{s_{ref}}$ and the control system response at $T_{e_{ref}} = -6\,N\,m$*

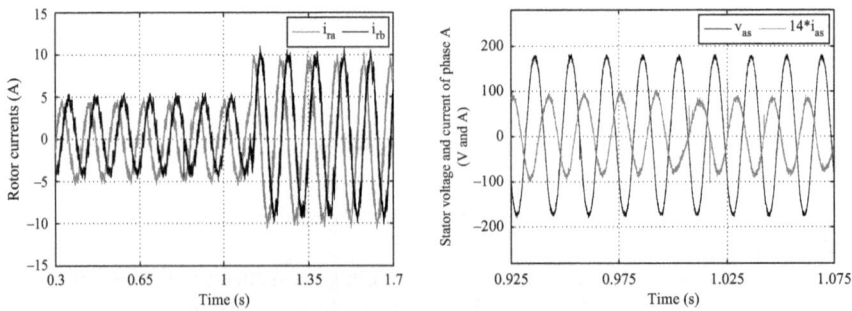

Figure 15.13 *Rotor currents, stator voltage and current of phase A at $T_{e_{ref}} = -6\,N\,m$*

15.6   Conclusion remarks

In this chapter, a DFIG control system was proposed, whose power references are transmitted from the CC to the wind generator through power lines by employing G3-PLC technology. As the power control is based on DTC, it is possible to control electromagnetic torque, rotor flux and powers of the DFIG by means of the voltage applied to the rotor terminals. This system has the potential to be used in wind farms by offering an efficient wind generator control associated with a cost-effective last mile communication link.

The presented system was also implemented in practice, and the performed experiments confirmed its viability, demonstrating additionally that NB-PLC technologies such as G3-PLC are a promising solution for this kind of application.

References

[1]   IEA. CO_2 Emissions From Fuel Combustion: Highlights. International Energy Agency; 2017.

[2]   EIA. International Energy Outlook 2016. U.S. Department of Energy; 2016. DOE/EIA-0484.

[3]   Colak I, Fulli G, Bayhan S, *et al.* Critical aspects of wind energy systems in smart grid applications. Renewable and Sustainable Energy Reviews. 2015;52(1):155–171.

[4]   Weng Y, Hsu Y. Reactive power control strategy for a wind farm with DFIG. Renewable Energy. 2016;94:383–390.

[5]   Saad NH, Sattar AA, Mansour AEM. Low voltage ride through of doubly-fed induction generator connected to the grid using sliding mode control strategy. Renewable Energy. 2015;80:583–594.

[6]   El-Naggar A, Erlich I. Analysis of fault current contribution of doubly-fed induction generator wind turbines during unbalanced grid faults. Renewable Energy. 2016;91:137–146.

[7]   Hopfensperger B, Atkinson DJ, Lakin RA. Stator-flux-oriented control of a doubly-fed induction machine with and without position encoder. IEE Proceedings – Electric Power Applications. 2000;147(4):241–250.

[8]   Cardenas R, Pena R, Alepuz S, *et al.* Overview of control systems for the operation of DFIGs in wind energy applications. IEEE Transactions on Industrial Electronics. 2013;60(7):2776–2798.

[9]   Moreira AB, Barros TAS, Teixeira VSC, *et al.* Power control for wind power generation and current harmonic filtering with doubly fed induction generator. Renewable Energy. 2017;107:181–193.

[10]   Takahashi I, Noguchi T. A new quick-response and high-efficiency control strategy of an induction motor. IEEE Transactions on Industry Applications. 1986;IA-22(5):820–827.

[11]   Depenbrock M. Direct self-control (DSC) of inverter-fed induction machine. IEEE Transactions on Power Electronics. 1988;3(4):420–429.

[12] Abad G, Lopez J, Rodriguez MA, *et al.* Doubly Fed Induction Machine: Modeling and Control for Wind Energy Generation Application. 1st ed. Piscataway, NJ: Wiley-IEEE Press; 2011.

[13] Zarean N, Kazemi H. A new DTC control method of doubly fed induction generator for wind turbine. In: 2012 Second Iranian Conference on Renewable Energy and Distributed Generation (ICREDG); 2012. p. 69–74.

[14] Zhang Y, Li Z, Wang T, *et al.* Evaluation of a class of improved DTC method applied in DFIG for wind energy applications. In: 2011 International Conference on Electrical Machines and Systems (ICEMS); 2011. p. 1–6.

[15] Pimple BB, Vekhande VY, Fernandes BG. A new direct torque control of doubly fed induction generator for wind power generation. In: 2010 India International Conference on Power Electronics (IICPE); 2011. p. 1–5.

[16] Chen SZ, Cheung NC, Wong KC, *et al.* Integral variable structure direct torque control of doubly fed induction generator. IET Renewable Power Generation. 2011;5(1):18–25.

[17] Chen SZ, Cheung NC, Chung K, *et al.* Integral sliding-mode direct torque control of doubly-fed induction generators under unbalanced grid voltage. IEEE Transactions on Energy Conversion. 2010;25(2):356–368.

[18] Capovilla CE, Casella IRS, Sguarezi AF, *et al.* A wind energy generator for smart grid applications using wireless-coded neuro-fuzzy power control. Computers and Mathematics with Applications. 2014;68(1):2112–2123.

[19] Costa FF, Sguarezi AJ, Capovilla CE, *et al.* Morphological filter applied in a wireless deadbeat control scheme within the context of smart grids. Electric Power Systems Research. 2014;107(1):175–182.

[20] Capovilla CE, Casella IRS, Sguarezi AJ, *et al.* Performance of a direct power control system using coded wireless OFDM power reference transmissions for switched reluctance aerogenerators in a smart grid scenario. IEEE Transactions on Industrial Electronics. 2015;62(1):52–61.

[21] Almeida LAL, Sguarezi AJ, Capovilla CE, *et al.* An impulsive noise filter applied in wireless control of wind turbines. Renewable Energy. 2016;86:347–353.

[22] Cardoso JG, Casella IRS, Sguarezi AJ, *et al.* SCIG wind turbine wireless controlled using morphological filtering for power quality enhancement. Renewable Energy. 2016;92:303–311.

[23] Kabalci E, Kabalci Y, Develi I. Modelling and analysis of a power line communication system with QPSK modem for renewable smart grids. International Journal of Electrical Power and Energy Systems. 2012;34(1):19–28.

[24] Ahola J, Kosonen A, Toukonen J, *et al.* A new approach to data transmission between an electric motor and an inverter. In: International Symposium on Power Electronics, Electrical Drives, Automation and Motion; 2006 May 23–26; Taormina, Italy. IEEE; 2006. p. 29–33.

[25] Ahola J, Kosonen A, Toukonen J, *et al.* Ethernet-based broadband power line communication between motor and inverter. IET Electric Power Applications. 2008;2:316–324.

[26]   Pereira SC, Caporali AS, Casella IRS. Power line communication technology in industrial networks. In: 2010 IEEE International Symposium on Power Line Communications and Its Applications (ISPLC); 2015 March 29–April 1; Austin (TX), USA. IEEE; 2015. p. 226–221.

[27]   Razazian K, Umari M, Kamalizad A, *et al.* G3-PLC specification for power-line communication: overview, system simulation and field trial results. In: 2010 IEEE International Symposium on Power Line Communications and Its Applications (ISPLC); 2010 March 28–31; Rio de Janeiro, Brazil. IEEE; 2010. p. 313–318.

[28]   Koch M. G3-PLC 500 kHz for smart metering. In: European Utility Week 2014 (EUW 2014); 2014 November 4–6; Amsterdam, Holland; 2014.

[29]   Filho AJS, de Oliveira Filho M, Filho ER. A predictive power control for wind energy. IEEE Transactions on Sustainable Energy. 2011;2(1):97–105.

[30]   Sen PC. Principles of Electric Machines and Power Electronics. 1st ed. Piscataway, NJ: John Wiley and Sons; 1997.

[31]   Utkin V, Guldner J, Shi J. Sliding Mode Control in Electromechanical Systems. 1st ed. Boca Raton, FL: CRC Press; 1999.

[32]   Lascu C, Boldea I, Blaabjerg F. Direct torque control of sensorless induction motor drives: a sliding-mode approach. IEEE Transactions on Industry Applications. 2004;40(2):582–590.

[33]   Barros TAS, Sguarezi AJ, Ruppert E. Direct power control for switched reluctance generator using a sliding mode controller for wind energy systems. In: Seminario Anual de Automacao, Eletronica Industrial e Instrumentacao (SAAEI 12); Guimaraes, Portugal; 2012.

[34]   Jacomini RV, Trindade FS, Sguarezi AJ, *et al.* Decoupled power control for doubly-fed induction generator using sliding-mode control. Przeglad Elektrotecniczny (Electrical Review). 2013;89:100–105.

[35]   Lazarini MV, Ruppert E. Induction motor control didactic set-up using sensorless and sliding mode DTC strategy. Eletronica de Potencia. 2008;13(4): 291–299.

[36]   Show SK, Parthiban P. Analysis of current controlled voltage source Inverter with Space Vector PWM and Sinusoidal PWM techniques. In: International Conference on Innovations in Information, Embedded and Communication Systems; 2015 March 19–20; Coimbatore, India. IEEE; 2015.

[37]   Gonzalo A, Jesus L, Miguel R, *et al.* Doubly Fed Induction Machine: Modeling and Control for Wind Energy Generation. Piscataway, NJ: John Wiley and Sons; 2011.

[38]   Galli S, Scaglione A, Wang Z. For the grid and through the grid: the role of power line communications in the smart grid. Proceedings of the IEEE. 2011;99(6):998–1027.

[39]   Lin J, Nassar M, Evans BL. Impulsive noise mitigation in powerline communications using sparse Bayesian learning. IEEE Journal on Selected Areas in Communications. 2013;31(7):1172–1183.

[40]   Gotz M, Rapp M, Dostert K. Power line channel characteristics and their effect on communication system design. IEEE Communications Magazine. 2004;42(4):78–86.

[41]   IEEE. IEEE Standard for Broadband over Power Line Networks: Medium Access Control and Physical Layer Specifications. Institute of Electrical and Electronics Engineers; 2010. 978-0-7381-6472-4.

[42]   IEEE. 1901.2-2013 – IEEE Standard for Low-Frequency (Less Than 500 kHz) Narrowband Power Line Communications for Smart Grid Applications. Institute of Electrical and Electronics Engineers; 2013. 978-0-7381-8793-8.

[43]   Lanzisera S, Mehta AM, Pister KSJ. Reducing average power in wireless sensor networks through data rate adaptation. In: IEEE International Conference on Communications; 2009 June 14–18; Dresden, Germany. IEEE; 2009.

Chapter 16

MIMO systems design for narrowband power line communication in smart distribution grids

Theofilos A. Papadopoulos[1], Andreas I. Chrysochos[2], and Grigoris K. Papagiannis[3]

In this chapter, data transmission in smart distribution grids (SDGs) is analyzed by means of multiple-input multiple-output (MIMO) narrowband (NB) power-line communication (PLC) systems, by applying orthogonal frequency-division multiplexing (OFDM) encoding. Regarding NB-PLC physical layer (PHY) modeling, multiconductor transmission line (MTL) theory is used for distribution lines channel characterization, whereas a measurement-based black-box method is adopted to characterize the medium-voltage (MV) to low-voltage (LV) path through the distribution transformers (DTs). OFDM bit-loading and transmit energy optimization processing techniques are applied across the OFDM subchannels of the spatial beams to improve the achievable data rates at the network nodes. Different scenarios are conducted to systematically evaluate the achievable data rates, revealing the possibility for an extensive and reliable application of NB-PLC communications in SDGs.

16.1 Introduction

Although the history of PLC technology dates back to the 1920s with the advent of the smart grid (SG) concept and the emerging communication needs, PLC is expected to play a significant and renewed role in building the SG vision [1]. Since 1920, several PLC designs have been developed, which can be generally classified according to the operating frequency band to the following categories [2–4]:

- Ultra NB-PLC, providing very low data rates (LDRs) (∼100 bps) in the range of 30–3,000 Hz, is the first PLC solution deployed by utilities for power system applications, e.g., speech, energy management and system protection signals.
- NB-PLC referring to signal transmission over the frequency range of 3–500 kHz [5–7]. NB-PLC includes the CENELEC band in Europe (3–148.5 kHz), the

---

[1] Department of Electrical and Computer Engineering, Democritus University of Thrace, Greece
[2] R&D Department, Cablel® Hellenic Cables S.A., Greece
[3] School of Electrical and Computer Engineering, Aristotle University of Thessaloniki, Greece

Federal Communications Commission (FCC) band in US (10–490 kHz), the Association of Radio Industries and Businesses (ARIB) band in Japan (10–450 kHz) and the Chinese band (3–500 kHz). NB-PLC technologies can be subdivided into:

- LDR-PLC, also known as Distribution Line Carrier or Power Line Carrier, employing single-line carrier technologies with data rates of several kbps;
- High data rate (HDR)-PLC, employing multicarrier technologies to support SG applications, offering data rates between tens of kbps to 500 kbps. For the standardization of HDR NB-PLC, the ITU-T G.hnem [8–10] and the IEEE 1901.2 [11] specifications have been proposed.

- Broadband (BB)-PLC or BB over power lines is a PLC technology mainly focusing on Internet access and in-home multimedia applications with frequencies up to 30 MHz (or 100 MHz in some cases). Currently, BB-PLC can provide to customers data rates up to 200 Mbps.
- Quasibroadband (QB)-PLC first deployed in China. QB-PLC combines the advantages of both NB-PLC and BB-PLC in the frequency range of 1 to 10 MHz, offering data rates up to 2 Mbps [12].

Among the available communication technologies, PLC is probably considered as the most cost-effective, requiring low-installation costs (utilizing the existing grid infrastructure as a communication channel) and almost no maintenance. Regarding SG applications, NB-PLC is one of the most favorable solutions, since it can ensure robust performance, meeting also the bandwidth, latency and security requirements of SG applications [1,2]. Moreover, compared to other PLC designs, NB-PLC can cover the long distances of distribution networks, especially in rural areas, using few repeaters. Other advantages of NB-PLC against other communication solutions is the utilization of the already installed advanced meter reading infrastructure, ease of upgrade to host HDR NB-PLC applications, efficient operation and independence from third parties [13–15].

In the frame of the SDG concept, NB-PLC can utilize both the MV and LV sides of the distribution grid as communication media to create an interactive environment between utilities, operators and customers, while offering new applications, which mainly concern [1,16]:

- Monitoring and detection: including measurements on equipment and network components, on-line diagnostics [17], islanding prevention [18], detection of high-impedance faults [19,20], advanced metering infrastructure, demand side management, smart metering, etc. [21].
- Operational services: energy saving solutions, remote control, emergency signaling, security systems, messaging, substation automation, vehicle-to-grid communications, etc.
- Network management: minimizing harmful events, network maintenance, ancillary services, enhanced protection schemes [22], etc.
- Communications: HDR solutions using multicarrier modulation schemes.

However, despite the significant potentialities, there are several obstacles that hinder the wide spread application of PLC in SDGs. The power grid was not originally designed for data transmission, thus may be a rather hostile environment for high-frequency (HF) signal transmission. Common problems are signal attenuation, the frequency-dependent nature of the channel and difficulty to establish an easy and safe communication path between the MV and LV sides of the grid [23,24]. Regarding the latter, DTs are considered as blocking elements of HF signals, minimizing the possibility for signal crossing and limiting the applicability of PLC in the whole extend of SDGs [25,26]. However, recent studies revealed that NB-PLC signals can be transmitted efficiently through the windings of power transformers, by utilizing specific spectrum windows in the range of the natural frequency of the transformer [27–30]. Under this concept, the most known NB-PLC specifications have exploited this solution to broaden the applicability of NB-PLC technology in SDGs [27,28,30].

Originally, single-carrier modulation schemes [frequency shift keying (FSK), binary phase shift keying (BPSK), quadrature phase shift keying (QPSK) and phase shift keying (PSK)] were used for LDR NB-PLC applications. However, due to technological advances in communications and the increasing demand for higher data rates, OFDM digital encoding was introduced as an attractive solution to efficiently utilize the frequency dependent NB-PLC channel bandwidth. Moreover, in most PLC system designs, the NB-PLC channel is considered as single-input–single-output (SISO) [31]. However, in multiphase wired configurations, MIMO designs can be established, consisting of the formed channels between the coupled pairs of wires. The advantage of MIMO systems is that they can achieve higher data rates and increased communication reliability, without any additional cost, regarding transmitter power and bandwidth. Therefore, by combining MIMO transmission and OFDM multiplexing, significantly, HDRs can be achieved considering NB-PLC designs [30,32–34]. Recently, MIMO-OFDM schemes have been integrated in PLC designs focusing mainly on in-home [31] and BB-PLC applications [35].

This chapter describes the design of MIMO OFDM NB-PLC systems in power distribution networks and systematically evaluates the achievable data rates of NB-PLC transmission. First, the PHY modeling of the MIMO NB-PLC system is presented. In particular, the distribution network, MV/LV transformer and total distribution channel models are described by means of transfer functions. Thereafter, the distribution network communication channel modeling and processing techniques are analyzed. Simulations of MV NB-PLC channels cascaded with MV/LV transformers are carried out to evaluate the achievable data rates. Finally, the conclusions of this chapter are presented.

16.2   PHY characteristics

The PHY characteristics of the complete MIMO NB-PLC system, shown in Figure 16.1, are investigated. The MIMO NB-PLC system under consideration consists of a 20-kV MV overhead distribution line and an MV/LV DT connected to the receiving line end. According to Kirchhoff's law, two transmit ports are considered; since the third depends on the remaining two inputs, while at the receiving side, three ports of signals are processed, resulting into a two-input/three-output MIMO PLC

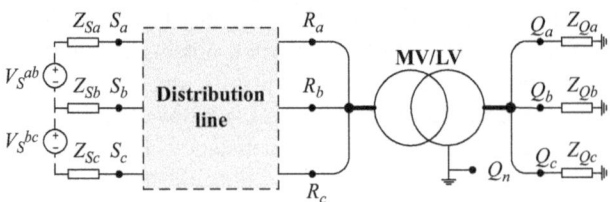

Figure 16.1   Distribution network configuration [30]

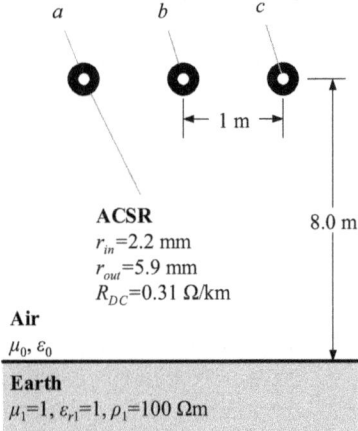

Figure 16.2   Overhead distribution line

design [33]. The source impedance of the MV transmitter unit and the loading condition at the LV side are represented by $\mathbf{Z}_S$ and $\mathbf{Z}_Q$, respectively. A description of the PHY modeling procedure of the system components in the NB-PLC range is given in the next subsections.

16.2.1   MV network modeling

MV overhead networks are mainly characterized by the line length, source impedance and termination load and thus can be assumed as time-invariant channels in the NB-PLC frequency range [5,28,33,36]. Considering the above assumptions, the modeling of the distribution network can be based on the MTL approach.

The geometric and electromagnetic properties of the examined MV overhead distribution line configuration are illustrated in Figure 16.2 [37]. The PLC signal voltages at the sending (S) and receiving (R) ends of the line at any frequency f are related as [30,33]

$$\begin{bmatrix} V_R^{ab} \\ V_R^{bc} \\ V_R^{ca} \end{bmatrix} = T \begin{bmatrix} V_S^{ab} \\ V_S^{bc} \end{bmatrix} \tag{16.1}$$

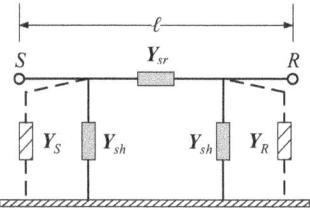

Figure 16.3   Distribution line two-port admittance representation

With respect to (16.1), voltages at S and R are phase-to-phase, i.e., between phases a, b and c, whereas T refers to the 3×2 distribution network transfer matrix, defined as

$$T = \begin{bmatrix} T_{11} & T_{12} \\ T_{21} & T_{22} \\ T_{31} & T_{32} \end{bmatrix} \tag{16.2}$$

where T_{ij} is the channel frequency response (CFR) from the jth to the ith node. Matrix T is determined in the frequency domain, using the two-port admittance representation of Figure 16.3. The nodal admittance matrix Y_n connecting nodes S and R is defined in (16.3). Termination conditions at the line ends are included by means of admittances Y_S and Y_R [38]:

$$Y_n = \begin{bmatrix} Y_{sr} + Y_{sh} + Y_S & -Y_{sr} \\ -Y_{sr} & Y_{sr} + Y_{sh} + Y_S \end{bmatrix} = \begin{bmatrix} A + Y_S & B \\ B & A + Y_R \end{bmatrix} \tag{16.3}$$

where

$$A = Z'T_v \, diag(\gamma) diag^{-1}(tanh(\gamma \ell)) T_i^T \tag{16.4}$$

$$B = -Z'T_v diag(\gamma)) diag^{-1}(sinh(\gamma \ell)) T_i^T \tag{16.5}$$

With reference to (16.4) and (16.5), Z' is the per-unit-length series impedance matrix, $diag(\gamma)$ is the diagonal modal propagation constant matrix, T_i is the current eigenvector matrix and ℓ is the line length [39,40]. Given Y_n, the voltages V_S and V_R as well as the injected currents I_S and I_R at ends S and R are related by

$$\begin{bmatrix} I_S \\ I_R \end{bmatrix} = Y_n \begin{bmatrix} V_S \\ V_R \end{bmatrix} \tag{16.6}$$

It should be noted that, the above formulation can extend to more complex network topologies of several nodes by means of nodal analysis [39].

The columns of T are determined by means of the superposition theorem, as analyzed in Figure 16.4. Sinusoidal signals in the NB range are applied sequentially at S, considering configurations (a) and (b), respectively, while recording the signal at R. Therefore, using configurations (a) and (b), the first and second columns of T can be determined [41].

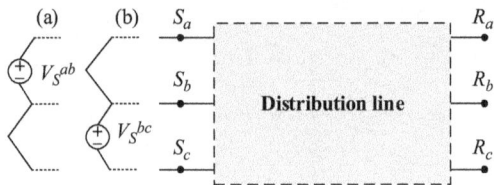

*Figure 16.4   Test configuration for the determination of the columns of **T**. Signal injection between phases (a) a − b and (b) b − c*

16.2.2   Transformer modeling

Next, the MV–LV distribution transformer channel is analyzed, using the measurement-based black-box model of [29]. Since the black-box model is used for the investigation of NB-PLC signal transmission, the following assumptions can be made [29]:

- Saturation and nonlinear effects are ignored.
- The frequency range is from 1 to 400 kHz.

The adopted black-box model formulation is described in (16.7) and is based on determining the signal transfer function by measurements. Briefly, HF signals are injected on one side of the transformer, while recording the corresponding responses at the other end.

$$\begin{bmatrix} V_Q^a \\ V_Q^b \\ V_Q^c \end{bmatrix} = F \begin{bmatrix} V_R^a \\ V_R^b \\ V_R^c \end{bmatrix} \tag{16.7}$$

where V_Q are the recorded voltages at the LV terminals as shown in Figure 16.1. The 3×3 transformer transfer matrix F is defined as

$$F = \begin{bmatrix} F_{11} & F_{12} & F_{13} \\ F_{21} & F_{22} & F_{23} \\ F_{31} & F_{32} & F_{33} \end{bmatrix} \tag{16.8}$$

Elements F_{ij} are determined using the experimental configuration of Figure 16.5. Self-elements of F refer to the signal transfer between the corresponding phases of the MV and the LV terminals, while the mutual elements express the transfer of the HF signals between different phases. Each column of F is calculated in a similar manner as the columns of T, described in the previous section.

Regarding the experimental setup, a signal generator is connected between the corresponding MV terminal of the transformer and ground, while the remaining MV terminals are grounded. Cyclic symmetrical behavior of the transformer voltages can be also assumed to simplify the procedure [29]. For all measurements, a logarithmic frequency scale is considered, while at resonance zones, the frequency interval is set to 1 kHz to improve accuracy. Each measurement is repeated 512 times and the recorded waveforms are averaged to achieve noise cancellation and increase the signal-to-noise ratio (SNR) [29].

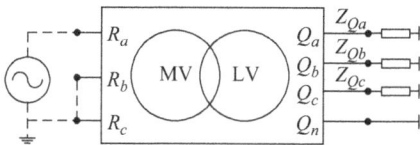

Figure 16.5 Transformer transfer function calculation experimental setup [30]

16.2.3 MV/LV NB-PLC channel model

The PHY characteristics of the complete MIMO NB-PLC system are calculated by means of the total channel transfer function matrix H, represented by the cascaded topology of the MV distribution line and the MV/LV transformer. The phase-to-phase voltages at the LV side of the transformer are expressed by means of (16.9), assuming that the input impedance on the transformer MV side is sufficiently high [30].

$$
\begin{bmatrix} V_Q^{ab} \\ V_Q^{bc} \\ V_Q^{ca} \end{bmatrix} = H \begin{bmatrix} V_S^{ab} \\ V_S^{bc} \end{bmatrix} \tag{16.9}
$$

where the 3×2 matrix H of the complete MV–LV PLC channel is derived by multiplying the transfer matrices T and F of the MV line and the transformer as

$$
H = KFK^{-1}T \tag{16.10}
$$

In (16.10), the transformation rotation matrix K defined in (16.11), is used to convert the phase-to-ground voltages of the transformer model to phase-to-phase voltages.

$$
K = \begin{bmatrix} 1 & -1 & 0 \\ 0 & 1 & -1 \\ -1 & 0 & 1 \end{bmatrix} \tag{16.11}
$$

16.2.4 Noise modeling

High noise levels may result in significant problems in communication systems in power grids. Therefore, accurate noise modeling is very important, focusing on the distinct characteristics and attributes of the examined part of the power grid [20]. In MV networks in the NB frequency range, the following general classes of noise are considered [42]:

- Colored background noise, containing all low-power spectrum density (PSD) noise sources not included in any other class.
- NB noise, caused by broadcasting radio sources.
- Impulsive noise:
    - Periodic impulsive noise asynchronous to the mains frequency, mainly caused by switched power supplies.
    - Periodic impulsive noise synchronous to the mains frequency, due to the switching of rectifier diodes of power supplies.
    - Asynchronous impulsive noise caused by switching transients.

Colored background, NB and periodic impulsive noise asynchronous to the mains frequency can be assumed as stationary, thus can be used in time-invariant modeling simulations [20,42]. On the other hand, the last two types of impulsive noise present significant fluctuations over short periods of time. Although the corresponding noise PSD might be significant, they can be only modeled using statistical models, while for capacity and data rate computations can be neglected for ease of simplicity in the simulations [33]. Based on the above, the noise models NM1 and NM2 with PSDs defined in (16.12) and (16.13) are considered, respectively. NM1 and NM2 correspond to best- and worst-case scenarios [30]. The PSDs are calculated by jointly fitting the adjusted power spectral densities of colored background, NB and periodic impulsive noise asynchronous to the mains measured in [43,44].

$$PSD_{NM_1} = 52.95e^{-0.0032f} - 115.7 \text{ dBm/Hz} \tag{16.12}$$

$$PSD_{NM_2} = 35.85e^{-0.00023f} - 136.96 \text{ dBm/Hz} \tag{16.13}$$

Note that in the above equations, f is expressed in kHz.

16.3   Communication channel model

In this section, the communication channel model of the MIMO-OFDM is described. The block diagram of the MIMO-OFDM system design is illustrated in Figure 16.6. The formulation of the channel response matrix is based on the following assumptions [30,33]:

- It can be considered as quasiconstant for a given topology, since it is mainly determined by the practically time-invariant distribution line and transformer characteristics.
- It is available at both channel ends. At the transmitter through a feedback channel from the receiver. At the receiver using a training sequence [45]. In this way, spatial channel diagonalization, optimized transmit power allocation and optimized bit loading techniques can be applied.

For the application of OFDM, the length of the cyclic prefix (CP) requires careful design to eliminate intersymbol-interference (ISI). Therefore, the CP length is at least equal to the channel impulse response (CIR) duration, in order to decouple the

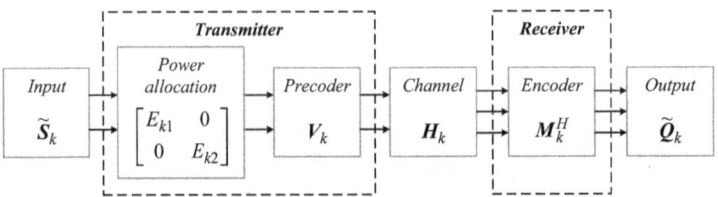

Figure 16.6   MIMO-OFDM NB-PLC system design [30]

MIMO CFR into N orthogonal frequency subchannels. The signal model at the kth subchannel is given by

$$Q_k = H_k S_k + n_k \tag{16.14}$$

where $k = 1 \ldots N$, Q_k is the received signal vector, S_k is the transmitted signal vector and n_k is the noise vector.

The equivalent MIMO CFR matrix H_k is the total channel matrix of (16.10) for each OFDM subchannel k. In the following subsections, the adopted signal-processing techniques and achievable data rate (ADR) calculations are described.

16.3.1 Spatial channel diagonalization

Singular value decomposition (SVD) is used to factorize H_k. By applying SVD, the channel throughput is maximized, considering that the transmitter precoding vectors are set equal to the right singular vectors, while the receiver vectors are set equal to the left singular vectors [33]. In this way, matrix diagonalization inside the determinant can be ensured; thus, the two spatial beams of the MIMO channel are decoupled by the precoding and equalization filters. Considering the above, the spatial matrix H_k is factorized:

$$H_k = M_k \Sigma_k V_k^H \tag{16.15}$$

where Σ_k is a 3×2 diagonal matrix with nonnegative real numbers and M_k as well as V_k are unitary matrices with dimensions 3×3 and 2×2, respectively. The MIMO channel throughput can be maximized by spatial decoupling of the two transmitted information streams using (16.15) [46]. Hence, the resulting filtered signal vector $\tilde{Q}_k$ is given by

$$\tilde{Q}_k \triangleq M_k^H Q_k = M_k^H H_k S_k + M_k^H n_k \tag{16.16}$$

and substituting transfer function H_k into (16.16)

$$\tilde{Q}_k = \Sigma_k \tilde{S}_k + n_k' \tag{16.17}$$

where $\tilde{S}_k$ is the information vector. With respect to (16.17), vectors S_k and n_k' are defined in (16.18) and (16.19), respectively.

$$S_k = V_k \tilde{S}_k \tag{16.18}$$

$$n_k' = M_k^H n_k \tag{16.19}$$

Hence, (16.17) is rewritten in (16.20), revealing that by applying SVD, each MIMO subchannel is decomposed to two spatially decoupled scalar channels, while the third output consists only of noise.

$$\tilde{Q}_k = \begin{bmatrix} \tilde{Q}_{k1} \\ \tilde{Q}_{k2} \\ \tilde{Q}_{k3} \end{bmatrix} = \begin{bmatrix} \lambda_{k1} & 0 \\ 0 & \lambda_{k2} \\ 0 & 0 \end{bmatrix} \begin{bmatrix} \tilde{S}_{k1} \\ \tilde{S}_{k2} \end{bmatrix} + \begin{bmatrix} n_{k1}' \\ n_{k2}' \\ n_{k3}' \end{bmatrix} \tag{16.20}$$

Within (16.20), λ_{k1} and λ_{k2} are the singular values.

16.3.2 Bit-loading optimization

OFDM bit-loading can be optimized to utilize efficiently the NB-PLC channel bandwidth compared to flat bit distributions across the transmission band, e.g., BPSK or QPSK as follows [30,33]:

- Neglect subchannels which cannot support at least 1 bit at the selected error rate.
- According to the SNR of the active subchannels, vary the signal constellation size across them.

16.3.3 Transmit energy optimization

To maximize the ADR of each SISO subchannel, transmit energy allocation optimization across frequency and space can be investigated. In NB-PLC channels, the transmit energy optimization across subchannels leads to negligible data rate gains and high complexity, thus energy optimization over only space can be considered, assuming equal energy allocation across all active frequency subchannels [33]. This is a near-optimal solution alternative to the optimal, though presenting less complexity and negligible data rate loss. Therefore, the energy E of each OFDM subchannel can be calculated by means of the water-filling algorithm as [47]

$$E = \frac{E_{tot}}{\overline{N}} = E_1 + E_2 \tag{16.21}$$

Within (16.21), E_{tot} is the total energy budget, $\overline{N}$ is the number of active subchannels, and E_1, E_2 is the energy of the first and second spatial beams, defined in (16.22) and (16.23), respectively.

$$E_1 = \frac{E}{2} + \frac{\sigma_{nk}^2 \Gamma}{2} \left(\frac{1}{\lambda_{k2}^2} - \frac{1}{\lambda_{k1}^2} \right) \tag{16.22}$$

$$E_2 = \frac{E}{2} - \frac{\sigma_{nk}^2 \Gamma}{2} \left(\frac{1}{\lambda_{k2}^2} - \frac{1}{\lambda_{k1}^2} \right) \tag{16.23}$$

where σ_{nk}^2 is the noise variance of the kth OFDM subchannel. Since the required bit number to achieve a certain probability of target error P_e is less than the theoretical capacity C, assuming error free transmission, the calculated bit loading is the capacity of the channel with SNR reduced by the SNR gap Γ defined in (16.24) [48].

$$\Gamma = \frac{\gamma_m}{3\gamma_c} \left(Q^{-1} \left(\frac{P_e}{4} \right) \right)^2 \tag{16.24}$$

With reference to (16.24), γ_c refers to the coding gain, γ_m refers to the desired system margin, P_e can be set to 10^{-7} for reliable data transmission, whereas function $Q(x)$ is defined as

$$Q(x) = \frac{1}{\sqrt{2\pi}} \int_x^\infty e^{(-u^2/2)} du \tag{16.25}$$

16.3.4 Achievable data rate calculation

According to the above, the ADR of the MIMO OFDM NB-PLC system is calculated by means of the gap approximation using two orthogonal spatial beams [48]. The bit-loading of the kth OFDM subchannel is given by

$$b_k = b_{k1} + b_{k2} = \frac{1}{2} \log_2 \left(\left(1 + \frac{\lambda_{k1}^2 E_1}{\sigma_{nk}^2 \Gamma} \right) \left(1 + \frac{\lambda_{k2}^2 E_2}{\sigma_{nk}^2 \Gamma} \right) \right) \qquad (16.26)$$

where b_{k1} and b_{k2} denote the bit-loading of the first and second spatial beam of the kth subchannel, respectively. Therefore, the ADR of the complete MIMO-OFDM NB-PLC channel is calculated by

$$ADR = \sum_{k=1}^{\overline{N}} b_k \Delta f \qquad (16.27)$$

where Δf is the OFDM sub-channel width.

16.4 NB-PLC channel PHY characteristics

The PHY characteristics of the MIMO NB-PLC system of Figure 16.1 are evaluated. NB-PLC signals of variable frequency are injected to the MV network and transmitted through the MV/LV transformer to the LV side, considering different operating conditions.

16.4.1 MV distribution line

In this section, the analysis is focused on the MV network, assuming the DT disconnected. In Figure 16.7, the calculated modal attenuation constant (α) and modal

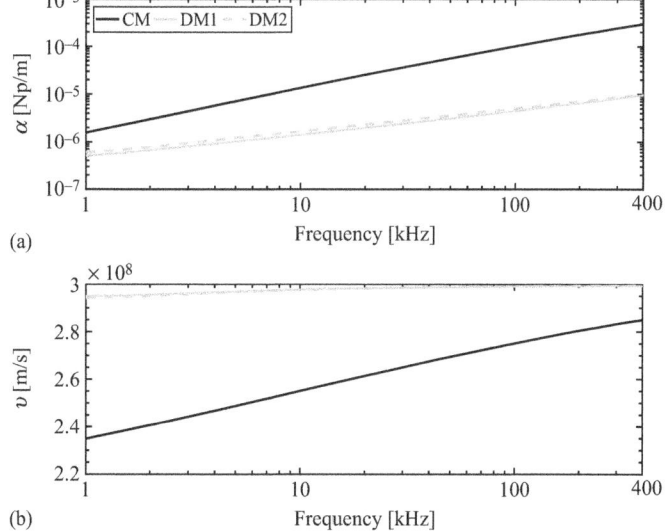

Figure 16.7   MV line modal (a) attenuation constant and (b) velocity

Table 16.1   Propagation characteristics of the MV line

Propagation characteristics	CM	DM1	DM2
Modal attenuation [Np/m]	1.56–$296 \cdot 10^{-6}$	5.07–$99.6 \cdot 10^{-7}$	6.00–$99.5 \cdot 10^{-7}$
Modal velocity [m/s]	2.35–$2.85 \cdot 10^8$	2.95–$2.99 \cdot 10^8$	2.94–$2.99 \cdot 10^8$

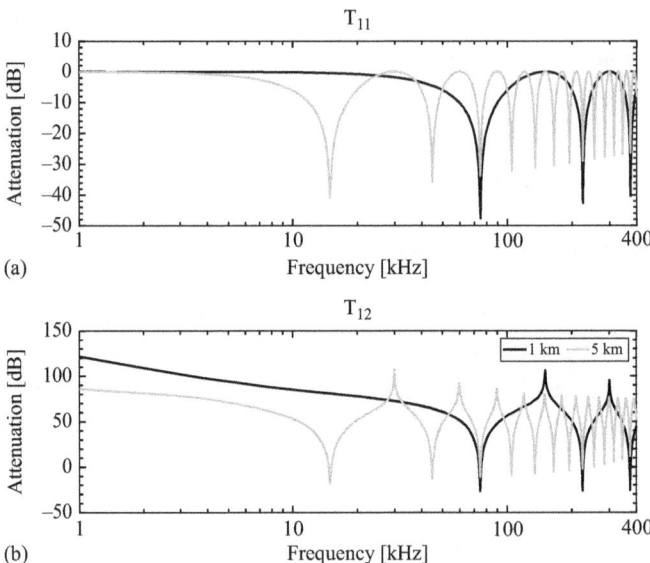

Figure 16.8   (a) Direct- and (b) cross-channel element of MV line for different line lengths

velocity (υ) of the MV line are illustrated in the NB frequency range of 1–400 kHz, while in Table 16.1, their lower and upper values are summarized. The three-phase MV overhead line is characterized by a common or ground mode (CM) and two differential or aerial modes (DM), i.e., DM1 and DM2 [40]. For wire-to-ground coupling, all modes are involved in the wave propagation, while for wire-to-wire propagation only the two DMs are excited [1]. As shown in Figure 16.7, the CM presents higher attenuation and lower velocity in the entire NB-PLC spectrum, while the two DMs almost overlap presenting a very similar behavior.

The frequency dependent behavior of the MV line is further investigated in Figure 16.8, where the effect of the line length on the attenuation of channel elements T_{11} and T_{12} of (16.2) is illustrated. The line length along with the impedances at both ends determine the signal quarter wavelength ($\lambda/4$) [49]. The resulting channel spectral peaks and notches can influence significantly the frequency selectivity of the active OFDM subchannels, and thus the MIMO NB-PLC design. The spectral peaks and notches occur at frequencies equal to the odd and even multiples of $\lambda/4$,

Table 16.2   Characteristics of the distribution transformers

DT	MV/LV ratio [kV/kV]	Rated power [kW]	u_k [%]	Vector group connection	Phase shift
1	20/0.4	50	4.17	Yzn	11
2	15/0.4	16	4.6	Dyn	1

respectively. Since signal injection occurs between the network phases, as shown in Figure 16.4, the channel spectral peaks and notches are mainly determined by the DM propagation and the corresponding characteristics [1].

It can be observed that, the increase of the line length results in an increasing number of channel peaks, while their amplitude decreases, due to the increasing attenuation introduced between the line ends. Therefore, the severe channel frequency selectivity makes the performance of the one-tap per-subchannel equalization superior and lower in complexity in comparison to the time-domain single-carrier equalization. In addition, the direct-channel elements T_{11} and T_{22} as well as the cross-channel elements T_{12} and T_{21} are highly correlated, due to the vertical symmetry of the examined distribution line. Moreover, since both excitation schemes of Figure 16.4 contribute equally to the voltage output V_R^{ca} of (16.1), T_{31} and T_{32} are almost identical to both T_{11} and T_{22}. This is justified by the strong coupling between the corresponding phase conductors.

16.4.2   MV/LV transformer

The individual performance of two DTs is investigated in the NB frequency range. Both transformers are three-phase, five-leg core-type oil-filled, with grounded neutral at the LV side, while their properties are given in Table 16.2.

The influence of the two DTs on the transmission of the PLC signal is analyzed in Figure 16.9, where the measured attenuation of F_{11} and F_{12} of (16.8) is illustrated. In this case, the LV terminals are open-circuited, and the MV line is disconnected [29]. Resonance frequencies for DT1 and DT2 are detected at 195 and 165 kHz, respectively. The natural frequency presents high sensitivity to the distinct characteristics of the DT, i.e., core geometry, material properties and installation topology; thus, it can be only determined for a given transformer installation [29,50,51]. In addition, it can be observed that spectrum windows suitable for NB-PLC signals crossing through the transformer are observed within the 150–400 kHz range, providing an average attenuation of 24 dB for both DT1 and DT2.

16.4.3   Complete distribution network

Finally, the transfer matrix $\boldsymbol{H}$ of (16.10) is calculated for the test cases (TCs) of Table 16.3, regarding different values of $\boldsymbol{Z}_S$ and $\boldsymbol{Z}_Q$. The MV line length varies from 0.25 to 5 km, while the two DTs are considered. In Figure 16.10, the attenuation of

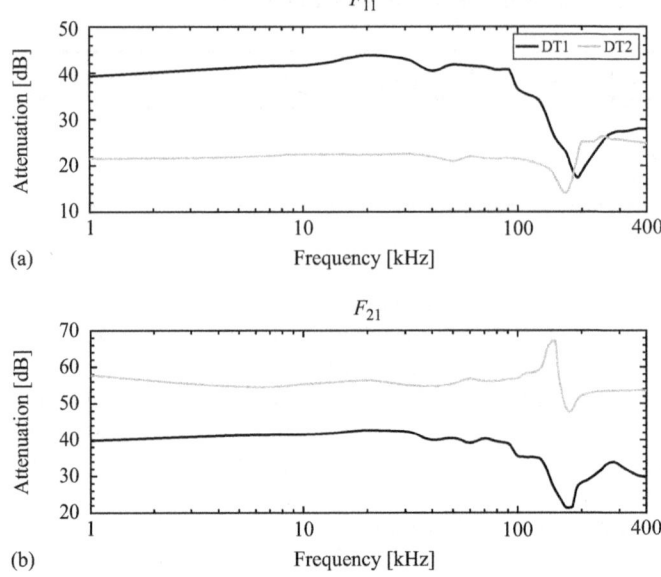

(a)

(b)

Figure 16.9　(a) F_{11} and (b) F_{21} elements of the examined transformers

Table 16.3　Summary of test cases

TC	Z_S [Ω]	Line length [km]	Z_Q [Ω]
1	0		Open-circuit (OC)
2	0		1,300
3	0	0.25, 0.5, 0,75, 1.00	700
4	0	1.25, 2.5, 3.75, 5.00	500
5	0		300
6	0		10

the direct-channel element H_{11} is depicted for TC1, assuming line lengths of 1- and 5-km. The MV/LV channel depends on both the signal frequency and line length, while the DTs influence the spectrum peaks and notches, resulting in significant signal degradation. However, possible communication zones in the NB-PLC band of the MV–LV channel are still observed.

16.5　Data rate results

The MIMO-OFDM ADR is evaluated assuming the OFDM subchannel width Δf equal to 1.5625 kHz [11]. The transmit PSD is −55 dBm/Hz, whereas γ_c and γ_m are both set equal to 6 dB [52]. Assuming a bandwidth of 400 kHz, the OFDM block size adopted by the IEEE 1901.2 standard is 512. Thus, information symbols are assigned

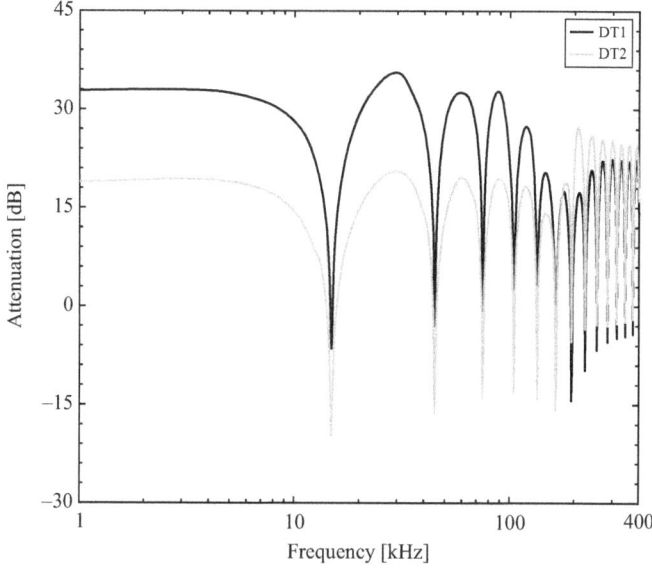

Figure 16.10   Attenuation of H_{11} for a 5-km line length

to $N = 256$ subchannels and the negative half of the OFDM symbol is the conjugate of the first half, since PLC is a baseband transmission scheme [30,33].

Although the choice of the CP length to be longer than the CIR duration completely cancels the ISI, this choice does not necessarily maximize system capacity. The CP length optimization represents a fundamental design trade-off, where a short CP results in a capacity loss due to ISI, while a long CP results in capacity loss due to CP overhead [44]. One of the proposed CP design approaches is to choose the CP length that captures at least 95% of the CIR power and ensure that the residual interference power caused by ISI can be neglected, compared with the noise power level. In the examined cases, the MV line lengths of Table 16.3 are considered, while the equivalent CIR includes the responses of the MV line and the transformer. A long CP length of 52 samples is used for line lengths longer than 1-km, which is optional [11]. For shorter line lengths, the standard CP length in IEEE 1901.2 of 30 samples is used to reduce the data rate loss due to the CP overhead.

In Figure 16.11, the ADR against the MV line is shown for noise models NM1 and NM2, and different line lengths, assuming the DT disconnected at the line end. Results reveal that the simple spatial power optimization rule of (16.22) and (16.23) can achieve HDR gains. Thus, the transmit power can be distributed equally over the set of the active subchannels, which is defined as the set of the OFDM subchannels with high SNR levels, capable to support at least 1 bit at the assumed SNR gap Γ.

Figure 16.12 shows the ADR at the LV side of DT1 and DT2 for TC1. The more complex behavior of the data rate results, compared to the corresponding of Figure 16.11, can be justified considering the MV line resonance frequencies at odd multiples of $\lambda/4$ in the NB-PLC range, summarized in Table 16.4 for all examined

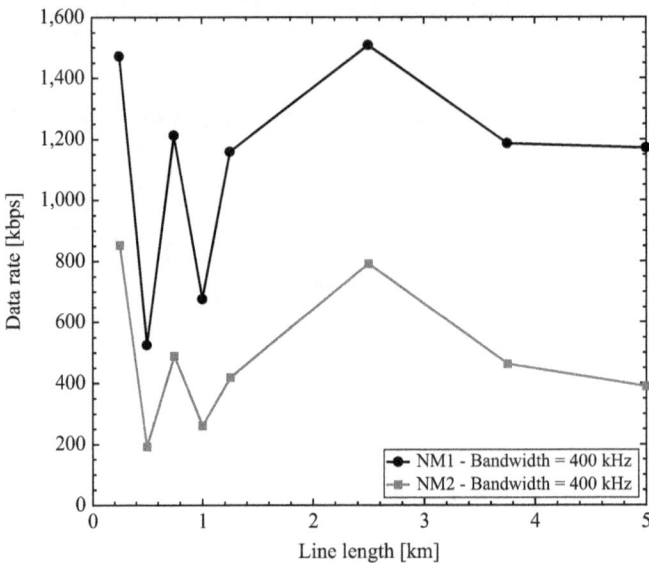

Figure 16.11 *ADR for the MV line assuming both noise models and different line lengths*

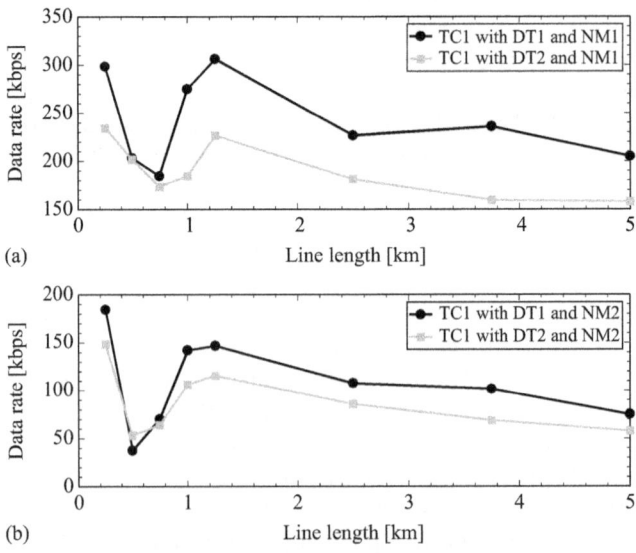

(a)

(b)

Figure 16.12 *ADR for the distribution network assuming both noise models, DTs, and different line lengths [30]*

Table 16.4   Resonance frequencies of MV line

Length [km]	Frequencies of resonance peaks [kHz]				
	1st	2nd	3rd	4th	5th
0.25	299.3	–	–	–	–
0.50	149.6	–	–	–	–
0.75	99.70	299.3	–	–	–
1.00	74.70	224.4	374.2	–	–
1.25	59.80	179.5	299.3	–	–
2.50	29.90	89.70	149.9	209.5	269.4
3.75	19.90	59.80	99.70	139.6	179.5
5.00	14.90	44.80	74.70	104.7	134.6

line lengths [49]. The attenuation of the channel transfer function reaches its minimum when the natural frequency of the transformer matches one of the resonance frequencies of the MV line [50]. The lower the order of the line resonance frequency, the lower is the calculated channel attenuation. For the 1.25-km line length, the second resonance frequency value is close to the transformer natural frequency of 180 kHz, and thus HDRs are achieved. For line lengths lower than 1.25-km, few line resonance frequencies are included in the spectrum window of 150–400 kHz, thus lower data rates are calculated compared to the 1.25-km line length. For lines longer than 1.25-km, the magnitude of the resonance peaks decreases with line length and peak order, while the influence of noise also becomes more significant. As an example, the fifth resonance peak for the case of 3.75-km is observed at 179.5 kHz which almost coincides with the natural frequency of both DTs. Although the ADR is high, it is clearly lower than that of the 1.25-km line case.

The adoption of transmit power optimization over space is further quantified in Figure 16.13 for TC2 and TC4, assuming noise model NM1. It is shown that the spatial power optimization rule of (16.22) and (16.23) can result into satisfactory data rate gains in the order of 15%–150%. Smaller data rate gains occur in scenarios where the two eigenvalues are close to each other, setting the flat power distribution near optimal.

In Figure 16.14, the data rate gains of the 2×3 MIMO-OFDM over the SISO-OFDM transmission systems are evaluated for TC1. A substantial data rate increase is observed due to the spatial diagonalization and multiplexing. In this case, optimized bit loading is considered for both techniques over the 400 kHz bandwidth. It can be concluded that additional gain in the ADR of the MIMO-OFDM design can probably compensate the potential costs of the second coupler needed in relation to the SISO system. Therefore, the MIMO design can be considered very attractive, based also on the expected learning curve of the coupler technology and the proliferation of PLC-based SG communications [53].

In Figure 16.15, the ADR at the LV side of DT1 is calculated against the effective impedance of all examined TCs, considering varying MV line lengths. The effective impedance value depends on several parameters, such as frequency, line length and

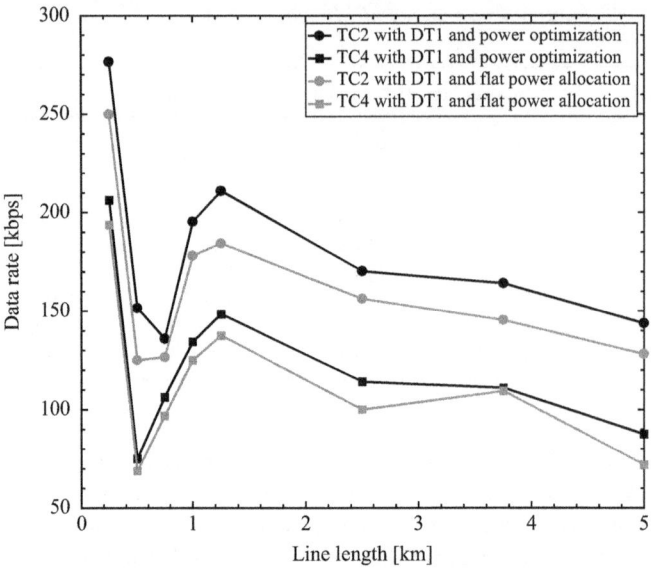

Figure 16.13   Effect of spatial power allocation optimization on ADRs [30]

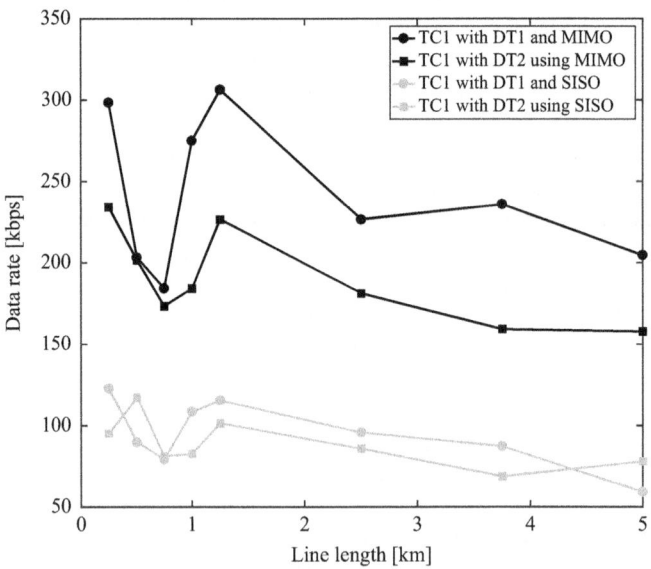

Figure 16.14   Comparison of SISO-OFDM and MIMO-OFDM ADR

load impedance, taking values from few Ω to several kΩ. The ADR increases with the effective impedance. Specifically, data rates higher than 100 kbps are obtained for values of the effective impedance higher than the distribution line characteristic impedance, i.e., 700 Ω for most examined line lengths.

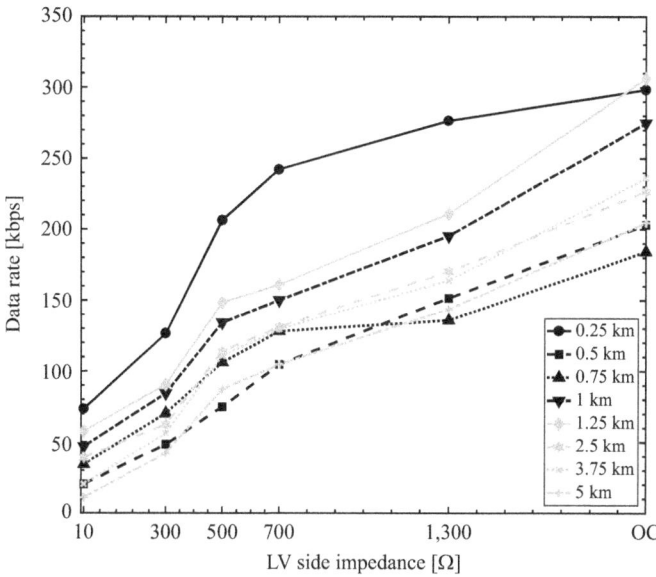

Figure 16.15   Effect of the LV side effective impedance on ADR [30]

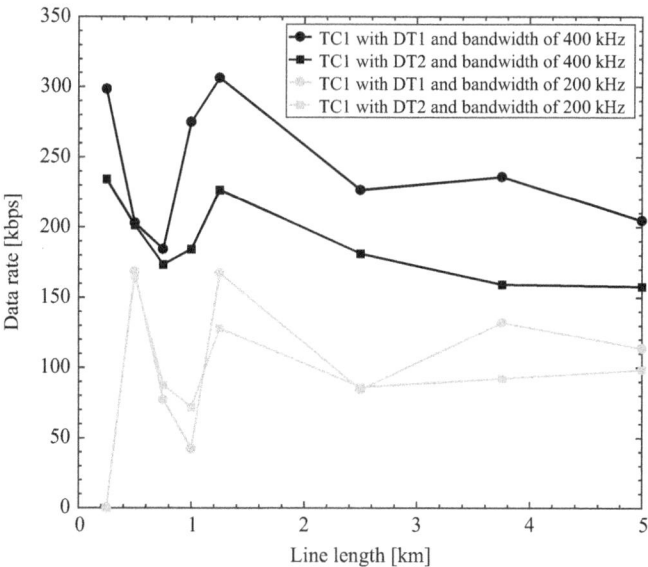

Figure 16.16   MIMO-OFDM ADR for different bandwidths

In all previous cases, results are based on the assumption of a transmission bandwidth of 400 kHz. Figure 16.16 shows the ADRs considering 200 and 400 kHz bandwidth, using adaptive bit loading and optimized spatial power allocation for TC1. Although the bandwidth is only doubled, the data rate percentage gain for a 1-km line length for both DTs is more than 100%. The reason is that the bit loading profile for the 400 kHz bandwidth case exhibits three transmission windows instead of the single transmission window for the 200 kHz bandwidth case. For the 0.25 km line length, the resonance frequency of the line is 299.3 kHz as shown in Table 16.4. Considering a 200 kHz bandwidth, this frequency is not included in the spectrum window, and thus significant data rate degradation is observed.

16.6 Conclusions

In this chapter, MIMO-OFDM NB-PLC data transmission over an MV–LV communication channel in SDGs has been investigated. According to the results of the conducted TCs, the main conclusions can be summarized as:

- The strong frequency-dependent behavior of the MV–LV channel favors the use of MIMO-OFDM transmission.
- Reliable MV–LV communication channel paths can be achieved especially in cases where the MV line resonance frequencies match the natural frequencies of DTs.
- MIMO-OFDM design leads to significantly higher data rates compared to the conventional SISO-OFDM systems.
- Spatial power optimization based on water-filling improves the data rate gains at the LV side.
- NB-PLC channel characteristics in the CENELEC band cannot take advantage of the MV–LV signal crossing through DTs, since transformer resonance frequencies are usually higher than the available frequency range. However, under different frequency band specifications or in systems operating under the IEEE 1901.2-2015, MV–LV MIMO-OFDM NB-PLC designs are feasible, resulting in data rates suitable for the reliable operation of SG applications.

Acknowledgments

The work of A. I. Chrysochos was conducted in the framework of the act "Support of Post-Doc Researchers" under the Operational Program "Human Resources Development, Education and Lifelong Learning 2014–2020", which is implemented by the State Scholarships Foundation and cofinanced by the European Social Fund and the Hellenic Republic.

The authors would like to acknowledge Prof. Naofal Al-Dhahir and Dr. Ahmed ElSamadouny from the University of Texas at Dallas for their valuable cooperation on the PLC communication channel modeling.

References

[1] Papadopoulos TA, Kaloudas CG, Chrysochos AI, *et al.* Application of Narrowband Power Line Communication in Medium Voltage Smart Distribution Grids. IEEE Transactions on Power Delivery. 2013;28(2):981–988.

[2] Galli S, Scaglione A, Wang Z. For the Grid and Through the Grid: The Role of Power Line Communications in the Smart Grid. Proceedings of the IEEE. 2011;99(6):998–1027.

[3] Lampe L, Tonello AM, Swart TG, editors. Power Line Communications: Principles, Standards and Applications from Multimedia to Smart Grid. New York (USA): John Wiley Sons, Ltd; 2016.

[4] Sharma K, Saini LM. Power-line Communications for Smart Grid: Progress, Challenges, Opportunities and Status. Renewable and Sustainable Energy Reviews. 2017;67:704–751.

[5] Nassar M, Lin J, Mortazavi Y, *et al.* Local Utility Power Line Communications in the 3–500 kHz Band: Channel Impairments, Noise, and Standards. IEEE Signal Processing Magazine. 2012;29(5):116–127.

[6] Lin J, Nassar M, Evans BL. Impulsive Noise Mitigation in Powerline Communications Using Sparse Bayesian Learning. IEEE Journal on Selected Areas in Communications. 2013;31(7):1172–1183.

[7] Berger LT, Schwager A, Schneider D, editors. MIMO Power Line Communications: Narrow and Broadband Standards, EMC, and Advanced Processing. New York (USA): CRC Press; 2017.

[8] Narrowband Orthogonal Frequency Division Multiplexing Power Line Communication Transceivers – Power Spectral Density Specification [Online]. ITU-T, Recommendation G.9903; 2013. Available from: http://www.itu.int/rec/T-REC-G.9901-201211-I/en.

[9] Narrowband Orthogonal Frequency Division Multiplexing Power Line Communication Transceivers for ITU-T G.hnem Networks [Online]. ITU-T, Recommendation G.9903; 2013. Available from: http://www.itu.int/rec/T-REC-G.9902.

[10] Narrowband Orthogonal Frequency Division Multiplexing Power Line Communication Transceivers for G3-PLC Networks [Online]. ITU-T, Recommendation G.9903; 2013. Available from: http://www.itu.int/rec/T-REC-G.9903.

[11] IEEE Standard for Low-Frequency (Less Than 500 kHz) Narrowband Power Line Communications for Smart Grid Applications – Amendment 1. IEEE Std 19012a-2015 (Amendment to IEEE Std 19012-2013). 2015 Oct. p. 1–28. Available from: https://ieeexplore.ieee.org/document/7286946/

[12] Jianming L, Bingzhen Z, Liang G, *et al.* Current Situations and Future Developments of PLC Technology in China. In: 2012 IEEE International Symposium on Power Line Communications and Its Applications; 2012. p. 60–65.

[13] A Systems View of the Modern Grid: Integrated Communications. Washington DC (USA): U.S. Dept. Energy; 2007. v2.0 Integrated Communications. Conducted by the National Energy Technology Laboratory. Available from: https://www.smartgrid.gov/files/a_systems_view_of_the_modern_grid.pdf

[14] Wouters PAAF, van der Wielen PCJM, Veen J, *et al.* Effect of Cable Load Impedance on Coupling Schemes for MV Power Line Communication. IEEE Transactions on Power Delivery. 2005;20(2):638–645.

[15] Gungor VC, Lambert FC. A Survey on Communication Networks for Electric System Automation. Computer Networks. 2006;50:877–897.

[16] Cataliotti A, Daidone A, Tinè G. Power Line Communication in Medium Voltage Systems: Characterization of MV Cables. IEEE Transactions on Instrumentation and Measurement. 2008;23(4):1896–1902.

[17] Korki M, Hosseinzadeh N, Vu HL, *et al.* A Channel Model for Power Line Communication in the Smart Grid. In: 2011 IEEE/PES Power Systems Conference and Exposition; 2011. p. 1–7.

[18] Benato R, Caldon R. Application of PLC for the Control and the Protection of Future Distribution Networks. In: 2007 IEEE International Symposium on Power Line Communications and Its Applications; 2007. p. 499–504.

[19] Zamora JJ, Zamora I, Mazon AJ, *et al.* Optimal Frequency Value to Detect Low Current Faults Superposing Voltage Tones. IEEE Transactions on Power Delivery. 2008;23(4):1773–1779.

[20] Milioudis AN, Andreou GT, Labridis DP. Detection and Location of High Impedance Faults in Multiconductor Overhead Distribution Lines Using Power Line Communication Devices. IEEE Transactions on Smart Grid. 2015;6(2):894–902.

[21] Mak S, Radford D. Communication System Requirements for Implementation of a Large Scale Demand Side Management and Distribution Automation. IEEE Transactions on Power Delivery. 1996;11(2):683–689.

[22] Milioudis AN, Andreou GT, Labridis DP. Enhanced Protection Scheme for Smart Grids Using Power Line Communications Techniques—Part I: Detection of High Impedance Fault Occurrence. IEEE Transactions on Smart Grid. 2012;3(4):1621–1630.

[23] Papadopoulos TA, Papagiannis GK, Dokopoulos PS. Low-Voltage Distribution Line Performance Evaluation for PLC Signal Transmission. IEEE Transactions on Power Delivery. 2008;23(4):1903–1910.

[24] Lefort R, Vauzelle R, Courtecuisse V, *et al.* Influence of the MV/LV Transformer Impedance on the Propagation of the PLC Signal in the Power Grid. IEEE Transactions on Power Delivery. 2017;32(3):1339–1349.

[25] Tran-Anh T, Auriol P, Tran-Quoc T. High Frequency Power Transformer Modeling for Power Line Communication Applications. In: 2006 IEEE PES Power Systems Conference and Exposition; 2006. p. 1069–1074.

[26] Kikkert CJ. MV to LV transformer PLC bypass coupling networks for a low cost Smart Grid rollout. In: 2011 IEEE PES Innovative Smart Grid Technologies; 2011. p. 1–6.

[27] Black WC. Data Transmission Through Distribution Transformers without Bypass Components. In: 2010 IEEE International Symposium on Power Line Communications and Its Applications; 2010. p. 13–17.

[28] Cataliotti A, Cosentino V, Di Cara D, *et al.* Measurement Issues for the Characterization of Medium Voltage Grids Communications. IEEE Transactions on Instrumentation and Measurement. 2013;62(8):2185–2196.

[29] Papadopoulos TA, Chrysochos AI, Nousdilis AI, *et al.* Simplified Measurement-based Black-box Modeling of Distribution Transformers Using Transfer Functions. Electric Power Systems Research. 2015;121:77–88.

[30]   Papadopoulos TA, Chrysochos AI, ElSamadouny A, *et al.* MIMO-OFDM Narrowband-PLC in Distribution Systems: Impact of Power Transformers on Achievable Data Rates. Electric Power Systems Research. 2017;151:251–265.

[31]   Hashmat R, Pagani P, Zeddam A, *et al.* MIMO Communications for Inhome PLC Networks: Measurements and Results up to 100 MHz. In: 2010 IEEE International Symposium on Power Line Communications and Its Applications; 2010. p. 120–124.

[32]   Hsieh SC, Ku TT, Chen CS, *et al.* Broadcasting Control of Intelligent Air Conditioners Using Power-Line-Carrier Technology. IEEE Transactions on Industry Applications. 2015;51(2):1890–1896.

[33]   Chrysochos AI, Papadopoulos TA, ElSamadouny A, *et al.* Optimized MIMO-OFDM Design for Narrowband-PLC Applications in Medium-Voltage Smart Distribution Grids. Electric Power Systems Research. 2016;140:253–262.

[34]   ElSamadouny A, El Shafie A, Abdallah M, Al-Dhahir N. Secure Sum-Rate-Optimal MIMO Multicasting Over Medium-Voltage NB-PLC Networks. IEEE Transactions on Smart Grid. 2018;9(4):2954–2963.

[35]   Lazaropoulos AG. Broadband Over Power Lines Systems Convergence: Multiple-input Multiple-output Communications Analysis of Overhead and Underground Low-voltage and Medium-voltage BPL Networks. ISRN Power Engineering. 2013;2013:30.

[36]   Anatory J, Theethayi N, Kissaka M, *et al.* The Effects of Load Impedance, Line Length, and Branches in the BPLC—Transmission-Lines Analysis for Medium-Voltage Channel. IEEE Transactions on Power Delivery. 2007;22(4):2156–2162.

[37]   Papadopoulos TA, Batalas BD, Radis A, *et al.* Medium Voltage Network PLC Modeling and Signal Propagation Analysis. In: 2007 IEEE International Symposium on Power Line Communications and Its Applications; 2007. p. 284–289.

[38]   Papadopoulos TA, Kaloudas CG, Papagiannis GK. A Multipath Channel Model for PLC Systems Based on Nodal Method and Modal Analysis. In: 2007 IEEE International Symposium on Power Line Communications and Its Applications; 2007. p. 278–283.

[39]   Chrysochos AI, Papadopoulos TA, Papagiannis GK. Enhancing the Frequency-domain Calculation of Transients in Multiconductor Power Transmission Lines. Electric Power Systems Research. 2015;122:56–64.

[40]   Wedepohl LM, Mohamed SET. Multiconductor Transmission Lines. Theory of Natural Modes and Fourier Integral Applied to Transient Analysis. Proceedings of the Institution of Electrical Engineers. 1969;116(9):1553–1563.

[41]   Schwager A, Schneider D, Bäschlin W, *et al.* MIMO PLC: Theory, Measurements and System Setup. In: 2011 IEEE International Symposium on Power Line Communications and Its Applications; 2011. p. 48–53.

[42]   Zimmerman M, Dostert K. Analysis and Modeling of Impulsive Noise in Broadband Powerline Communications. IEEE Transactions on Electromagnetic Compatibility. 2002;44(1):249–258.

[43] Tao Z, Xiaoxian Y, Baohui Z, *et al.* Research of Noise Characteristics for 10-kV Medium-Voltage Power Lines. IEEE Transactions on Power Delivery. 2007;22(1):142–150.

[44] Tao Z, Xiaoxian Y, Baohui Z, *et al.* Statistical Analysis and Modeling of Noise on 10-kV Medium-Voltage Power Lines. IEEE Transactions on Power Delivery. 2007;22(3):1433–1439.

[45] Minn H, Al-Dhahir N, Li Y. Optimal Training Signals for MIMO OFDM Channel Estimation in the Presence of Frequency Offset and Phase Noise. IEEE Transactions on Communications. 2006;54(10):1754–1759.

[46] Ginis G, Cioffi JM. Vectored Transmission for Digital Subscriber Line Systems. IEEE Journal on Selected Areas in Communications. 2002;20(5):1085–1104.

[47] A Multicarrier Primer [Homepage on the Internet]. USA: Stanford University; 1991 [cited 2017 Aug 12]. Available from: http://www.stanford.edu/group/cioffi/documents/multicarrier.pdf.

[48] Oksman V, Zhang J. G.HNEM: The New ITU-T Standard on Narrowband PLC Technology. IEEE Communication Magazine. 2011;49(12):36–44.

[49] Chrysochos AI, Papadopoulos TA, Papagiannis GK. Rigorous Calculation Method for Resonance Frequencies in Transmission Line Responses. IET Generation, Transmission Distribution. 2015;9(8):767–778.

[50] Gustavsen B. Study of Transformer Resonant Overvoltages Caused by Cable-Transformer High-Frequency Interaction. IEEE Transactions on Power Delivery. 2010;25(2):770–779.

[51] Cataliotti A, Cosentino V, Di Cara D, *et al.* Oil-Filled MV/LV Power-Transformer Behavior in Narrow-Band Power-Line Communication Systems. IEEE Transactions on Instrumentation and Measurement. 2012;61(10):2642–2652.

[52] Tonello AM, D'Alessandro S, Lampe L. Cyclic Prefix Design and Allocation in Bit-Loaded OFDM over Power Line Communication Channels. IEEE Transactions on Communications. 2010;58(11):3265–3276.

[53] Finamore WA, Ribeiro MV, Lampe L. Advancing Power Line Communication: Cognitive, Cooperative, and MIMO Communication. In: Brazilian Telecommunications Symposium; 2012.

Index